Sustainable Fuel
Technologies Han

Sustainable Fuel Technologies Handbook

Edited by

Suman Dutta

Department of Chemical Engineering, Indian Institute of Technology (ISM), Dhanbad, India

Chaudhery Mustansar Hussain

Department of Chemistry & Environmental Sciences, New Jersey Institute of Technology (NJIT), Newark, New Jersey, United States

ELSEVIER

ACADEMIC PRESS

An imprint of Elsevier

Academic Press is an imprint of Elsevier
125 London Wall, London EC2Y 5AS, United Kingdom
525 B Street, Suite 1650, San Diego, CA 92101, United States
50 Hampshire Street, 5th Floor, Cambridge, MA 02139, United States
The Boulevard, Langford Lane, Kidlington, Oxford OX5 1GB, United Kingdom

British Library Cataloguing-in-Publication Data

A catalogue record for this book is available from the British Library

Library of Congress Cataloging-in-Publication Data

A catalog record for this book is available from the Library of Congress

ISBN: 978-0-12-822989-7

For Information on all Academic Press publications
visit our website at https://www.elsevier.com/books-and-journals

Publisher: Brian Romer
Acquisitions Editor: Lisa Reading
Editorial Project Manager: Hannah Makonnen
Production Project Manager: Kamesh Ramajogi
Cover Designer: Mark Rogers

Typeset by MPS Limited, Chennai, India

Contents

List of contributors **xi**
Preface **xv**

Section I Modern Perspective of Sustainability 1

1 Overview of sustainable fuel and energy technologies **3**
Shireen Quereshi, Prashant Ram Jadhao, Ashish Pandey, Ejaz Ahmad and K.K. Pant
1.1 Introduction 3
1.2 Sustainable technologies for energy production 4
1.3 Sustainable technologies for energy, fuel and chemicals production 12
1.4 Summary 18
Acknowledgment 19
References 19

Section II Biomass Energy Technology and Biorefinery 27

2 The second- and third-generation biofuel technologies: comparative perspectives **29**
Preetha Ganguly, Rwiddhi Sarkhel and Papita Das
Graphical abstract 29
2.1 Introduction 29
2.2 Biorefineries 31
2.3 First-generation biorefineries 32
2.4 Second-generation biorefineries 33
2.5 Third-generation biorefineries 41
2.6 SWOT analysis 44
2.7 Conclusions 45
Acknowledgment 46
References 46

3 Biomass, biorefinery, and biofuels **51**
Swarnalatha Mailaram, Pankaj Kumar, Alekhya Kunamalla, Palkesh Saklecha and Sunil K. Maity
3.1 Introduction 51

3.2 Traditional biofuels 54
3.3 Hydrocarbon biofuels 61
3.4 Fuel additives 72
3.5 Conclusion 79
Abbreviations 80
References 80

Section III Hydro Power Technology **89**

4 Hydropower technology **91**
Nitai Pal and Faizan A. Khan
4.1 Introduction 91
4.2 Hydropower technologies 92
4.3 DAMS 101
4.4 Reservoirs 104
4.5 Small Scale hydropower plant 105
4.6 Micro-hydropower plant 106
4.7 Benefits and shortcomings of micro hydropower plants 106
4.8 Small hydro 108
4.9 Turbines 109
4.10 Conclusion 118
References 118

Section IV Wind Power Technology **121**

5 Wind power technology **123**
Subhadeep Bhattacharjee
5.1 Historical background 123
5.2 Energy and power from the wind 125
5.3 Wind resource feasibility 136
5.4 Wind energy conversion devices 142
5.5 Wind electrical generators 153
5.6 Speed control strategies for wind turbines 161
5.7 Environmental impact and public perception 164
5.8 Offshore wind energy 166
5.9 Wind energy applications 167
5.10 Wind turbine economics 167
5.11 Conclusion 168
References 168

Section V Solar Power Technology 171

6 Enabling solar photovoltaics penetration in highly dependent African fossil fuel markets 173
Ogechi Vivian Nwadiaru, Tonny Kukeera and David Ebanehita
6.1 Introduction 173
6.2 Energy transition in Africa 176
6.3 Solar photovoltaics deployment in oil producing African countries 186
6.4 Institutional and economic barriers to solar deployment 187
6.5 Outlook and recommendations 191
6.6 Conclusion 193
References 194

Section VI Geothermal Power Technology 199

7 An up-to-date perspective of geothermal power technology 201
Abel F. Hernández Ochoa, Alfonso Aragón Aguilar and José M. Franco Nava
7.1 Introduction 201
7.2 Preliminary survey stage 202
7.3 Exploration stage 204
7.4 Prefeasibility stage 217
7.5 Exploratory drilling 218
7.6 Well drilling 219
7.7 Project planning 223
7.8 Field development 223
7.9 Reservoir monitoring at start-up and operation 226
7.10 Energy conversion 228
7.11 Conclusion 234
Acknowledgments 235
References 235

Section VII Marine Power Technology 239

8 Marine power technology—wave energy 241
Abdus Samad and R. Suchithra
8.1 Introduction 241
8.2 Wave energy 242
8.3 Types of wave energy converters 243
8.4 Power take-off systems for wave energy converters 250
8.5 Future of wave energy 262
8.6 Summary 262
References 262

Section VIII Hydrogen Fuel 269

9 Renewable hydrogen production by water electrolysis 271
Ernesto Amores, Mónica Sánchez, Nuria Rojas and Margarita Sánchez-Molina
9.1 Introduction 271
9.2 Principles of water electrolysis 272
9.3 Alkaline water electrolysis 280
9.4 Proton exchange membrane water electrolysis 289
9.5 Solid oxide water electrolysis 295
9.6 Electrolysis system technology 300
9.7 Integration with renewable energies 306
References 308

10 Microbial production of hydrogen 315
Ioanna Ntaikou
10.1 Molecular basis of microbial hydrogen production 315
10.2 Hydrogen-producing microorganisms 317
10.3 Processes of microbial hydrogen production 321
10.4 Bioreactors 329
10.5 Storage and purification 331
10.6 Conclusion 331
Abbreviations 332
References 332

11 Hydrogen energy 339
Zuo-Yu Sun
11.1 Introduction 339
11.2 Production technologies for hydrogen gas 339
11.3 Combustion properties of hydrogen 346
11.4 Problems relating to using hydrogen: Explosion potential and nox emissions 359
11.5 Conclusion 362
References 363

Section IX Emerging Technologies (Nanotechnology) 367

12 Multifaceted usage of miniaturized energy technologies for sustainable energy harvesting 369
Siddharth Thakur, Ashok Kumar Dasmahapatra and Dipankar Bandyopadhyay
12.1 Introduction 369
12.2 Types of miniaturized energy technologies 371
12.3 Internet of things 398
12.4 Summary 400
References 400

Section X Design, Scale-up, and Commercialization of Sustainable Fuel Technologies 407

13 Communicating energy consumption: developing a low-cost, real-time, SMS feedback tool for off-grid household electricity consumption 409
Tonny Kukeera, Noara Kebir, Setu Pelz and Sebastian Groh
13.1 Introduction 409
13.2 Consumption information access in off-grid households 412
13.3 Solar power system 414
13.4 General process setup 418
13.5 Conclusion 426
Acknowledgment 427
References 427

14 Recent developments in hydrogen fuel cells: Strengths and weaknesses 431
Natarajan Rajalakshmi, Rengarajan Balaji and Sundararajan Ramakrishnan
14.1 Fuel cells based on hydrogen 431
14.2 Strategic energy plan 452
14.3 Technology barriers 453
14.4 Conclusion 453
Acknowledgments 454
References 454

15 Energy efficiency in buildings 457
Janmejoy Gupta and Manjari Chakraborty
15.1 Background: building in the wake of an energy crisis 457
15.2 Energy use in buildings: energy modeling and energy audits 458
15.3 Passive solar techniques 462
15.4 Active solar techniques 466
15.5 Energy-efficient landscaping 468
15.6 Concept of green buildings 476
15.7 Conclusion 480
References 480

Section XI Economic & Lifecycle assessment of Sustainable Fuel Technologies 481

16 Toward sustainable long-term energy planning for cities: an economic and environmental assessment of sustainable fuel technologies in the city of Donostia-San Sebastián 483
Eneko Arrizabalaga, Diego García-Gusano and Patxi Hernandez
16.1 Introduction 483
16.2 Conventional approaches in city energy planning 485

16.3 A new conceptual framework for long-term energy scenarios for cities 486
16.4 Conclusion 507
References 508

Section XII Green & Sustainable Future in Energy Sector (Zero Waste & Zero Emissions) 511

17 Oxy-fuel power cycles promoting the transition to green and sustainable future in the energy sector 513
N.D. Rogalev, Yu. K. Petrenya, V.O. Kindra and A.N. Rogalev
17.1 Introduction 513
17.2 Reducing CO_2 emissions in the power production industry 515
17.3 Carbon capture technologies for thermal power plants 516
17.4 Oxy-fuel combustion cycles with carbon dioxide recirculation: energy efficiency and emission levels 520
17.5 Conclusion 536
Acknowledgment 536
References 536

18 Integrated photocatalytic hydrogen production and pollutants or wastes treatment: prospects and challenges 541
Rohini Singh and Suman Dutta
18.1 Introduction 541
18.2 Utilization of pollutants or waste as sacrificial agents 542
18.3 Pros and cons of integrated systems 545
18.4 Reactors for an integrated system 545
18.5 Prospects and challenges 545
18.6 Conclusion 547
Acknowledgments 548
References 548

Conclusion 551
Concluding remarks by the editors 553
Index 557

List of contributors

Ejaz Ahmad Department of Chemical Engineering, Indian Institute of Technology (ISM) Dhanbad, India

Ernesto Amores Centro Nacional del Hidrógeno (CNH2), Puertollano, Ciudad Real (Spain)

Alfonso Aragón Aguilar National Institute of Electricity and Clean Energies, Cuernavaca, Mexico

Eneko Arrizabalaga TECNALIA, Basque Research and Technology Alliance (BRTA), Astondo bidea Edificio 700, Derio, Spain

Rengarajan Balaji Centre for Fuel Cell Technology, International Advanced Research Centre for Powder Metallurgy and New Materials, IIT Madras Research Park, Chennai, India

Dipankar Bandyopadhyay Department of Chemical Engineering, IIT Guwahati, Guwahati, India

Subhadeep Bhattacharjee Department of Electrical Engineering, National Institute of Technology (NIT), Agartala, India

Manjari Chakraborty Birla Institute of Technology, Mesra, Ranchi, India

Papita Das Department of Chemical Engineering, Jadavpur University, Kolkata, India; School of Advanced Studies in Industrial Pollution Control Engineering, Jadavpur University, Kolkata, India

Ashok Kumar Dasmahapatra Department of Chemical Engineering, IIT Guwahati, Guwahati, India

Suman Dutta Department of Chemical Engineering, Indian Institute of Technology (ISM), Dhanbad, India

David Ebanehita Department of Energy Management, ESCP Europe Business School Paris, Paris, France

José M. Franco Nava National Institute of Electricity and Clean Energies, Cuernavaca, Mexico

Preetha Ganguly Department of Chemical Engineering, Jadavpur University, Kolkata, India

Diego García-Gusano TECNALIA, Basque Research and Technology Alliance (BRTA), Astondo bidea Edificio 700, Derio, Spain

Sebastian Groh School of Business and Economics, North South University, Dhaka, Bangladesh

Janmejoy Gupta Department of Architecture, School of Planning and Architecture (SPA), Vijayawada, India

Patxi Hernandez TECNALIA, Basque Research and Technology Alliance (BRTA), Astondo bidea Edificio 700, Derio, Spain

Abel F. Hernández Ochoa National Institute of Electricity and Clean Energies, Cuernavaca, Mexico

Prashant Ram Jadhao Department of Chemical Engineering, Indian Institute of Technology Delhi, India

Noara Kebir MicroEnergy International GmbH, Berlin, Germany

Faizan A. Khan Department of Electrical Engineering, Indian Institute of Technology (Indian School of Mines), Dhanbad, India

V.O. Kindra Department of Innovative Technologies of High-Tech Industries, National Research University "Moscow Power Engineering Institute", Moscow, Russia

Tonny Kukeera Pan African University Institute of Water and Energy Sciences (Including Climate Change), Tlemcen, Algeria; MicroEnergy International GmbH, Berlin, Germany; School of Business and Economics, North South University, Dhaka, Bangladesh; Microenergy Systems Research Group, Technical University of Berlin, Berlin, Germany; Working Group of Economic and Infrastructure Policy, Technical University of Berlin, Berlin, Germany

Pankaj Kumar Department of Chemical Engineering, Indian Institute of Technology Hyderabad, Kandi, India; Department of Chemical Engineering, Birla Institute of Technology and Science, Pilani, Hyderabad Campus, Hyderabad, India

Alekhya Kunamalla Department of Chemical Engineering, Indian Institute of Technology Hyderabad, Kandi, India

Swarnalatha Mailaram Department of Chemical Engineering, Indian Institute of Technology Hyderabad, Kandi, India

Sunil K. Maity Department of Chemical Engineering, Indian Institute of Technology Hyderabad, Kandi, India

Ioanna Ntaikou Foundation for Research and Technology, Institute of Chemical Engineering Sciences, Patras, Greece

Ogechi Vivian Nwadiaru Microenergy Systems Research Group, Technical University of Berlin, Berlin, Germany; Working Group of Economic and Infrastructure Policy, Technical University of Berlin, Berlin, Germany

Nitai Pal Department of Electrical Engineering, Indian Institute of Technology (Indian School of Mines), Dhanbad, India

Ashish Pandey Department of Chemical Engineering, Indian Institute of Technology Delhi, India

K.K. Pant Department of Chemical Engineering, Indian Institute of Technology Delhi, India

Setu Pelz MicroEnergy International GmbH, Berlin, Germany; Department of Energy and Environmental Management, Europa-Universität Flensburg, Berlin, Germany

Yu. K. Petrenya PJSC "Power Machines", St. Petersburg, Russia

Shireen Quereshi Department of Chemical Engineering, Indian Institute of Technology Delhi, India

Natarajan Rajalakshmi Centre for Fuel Cell Technology, International Advanced Research Centre for Powder Metallurgy and New Materials, IIT Madras Research Park, Chennai, India

Sundararajan Ramakrishnan Centre for Fuel Cell Technology, International Advanced Research Centre for Powder Metallurgy and New Materials, IIT Madras Research Park, Chennai, India

A.N. Rogalev Department of Innovative Technologies of High-Tech Industries, National Research University "Moscow Power Engineering Institute", Moscow, Russia

N.D. Rogalev Department of Thermal Power Plants, National Research University "Moscow Power Engineering Institute", Moscow, Russia

Nuria Rojas Centro Nacional del Hidrógeno (CNH2), Puertollano, Ciudad Real (Spain)

Palkesh Saklecha Department of Chemical Engineering, Indian Institute of Technology Hyderabad, Kandi, India

Abdus Samad Wave Energy and Fluids Engineering Laboratory, Department of Ocean Engineering, Indian Institute of Technology Madras, Chennai, India

Mónica Sánchez Centro Nacional del Hidrógeno (CNH2), Puertollano, Ciudad Real (Spain)

Margarita Sánchez-Molina Centro Nacional del Hidrógeno (CNH2), Puertollano, Ciudad Real (Spain)

Rwiddhi Sarkhel Department of Chemical Engineering, Jadavpur University, Kolkata, India

Rohini Singh Department of Chemical Engineering, Sitarambhai Naranji Patel Institute of Technology & Research Centre, Bardoli, India

R. Suchithra Wave Energy and Fluids Engineering Laboratory, Department of Ocean Engineering, Indian Institute of Technology Madras, Chennai, India

Zuo-Yu Sun Hydrogen Energy and Space Propulsion Laboratory, School of Mechanical, Electronic and Control Engineering, Beijing Jiaotong University, Beijing, P.R. China

Siddharth Thakur Department of Chemical Engineering, IIT Guwahati, Guwahati, India

Preface

It is said that fuel is essential for the development of any nation. Utilization of fuel is inevitable for any society. The burning of conventional and nonrenewable fuel causes environmental pollution. Therefore it is desirable to minimize the usage of fossil fuels. The rapid depletion of fossil fuels also causes a price hike. Sustainable fuel and energy technologies help us to use energy resources efficiently as well as to develop systems with less energy consumption. This handbook discusses the various sources of renewable energy, their storage, and efficient utilization. The book is designed in such a way that it covers all the renewable energy resources such as hydropower, wind power, geothermal energy, and marine power technology. The recyclable and clean fuels such as biomass and hydrogen is also included. Harvesting and utilization of solar energy by various means, namely photovoltaic cell and solar hydrogen, is discussed in detail. Numerous research works show that the required energy in the near future will be produced mostly by hydrogen, biomass and biomass-derived fuel, and solar energy. Hence, the emphasis is on these fuel and energy resources. Development of energy-efficient building and awareness among citizens, thereby saving energy/fuel consumption. Readers will get a clear idea about green and sustainable energy by using the oxy-fuel combustion cycle, multifaceted nano- or microscale technologies, and fuel cells. We hope students, professors, and industry personnel will enjoy this book as a valuable resource.

The content of this book was finalized after many discussions among us. Reviewers' comments also helped to modify the content of this book. We are thankful to all reviewers for their useful suggestions and recommendations. The chapters are written by expert researchers, scientists, and professors. We are grateful to all the authors and coauthors for their valuable contribution. We acknowledge the support of Lisa Reading (Senior Acquisition Editor – Energy), Ali Afzal-Khan (Editorial Project Manager), Kamesh Ramajogi, and all staff members of Elsevier who were involved in this project.

Editors
Suman Dutta
Chaudhery Mustansar Hussain

Section I

Modern Perspective of Sustainability

Overview of sustainable fuel and energy technologies

1

Shireen Quereshi[1], Prashant Ram Jadhao[1], Ashish Pandey[1], Ejaz Ahmad[2] and K.K. Pant[1]
[1]Department of Chemical Engineering, Indian Institute of Technology Delhi, India,
[2]Department of Chemical Engineering, Indian Institute of Technology (ISM) Dhanbad, India

1.1 Introduction

Fuels and energy are essential needs of human civilization after food, water, and air. Indeed, the evolution of human civilization cannot be imagined without fuels and energy that caused a drastic change when humans first learned to ignite fire, thereby generating energy from wood. Interestingly, until 1700 timber or wood remained the primary source of energy and then the usage of coal exceeded it by the late 1780s [1]. In contrast, several rapid developments took place over the past three centuries, for example, steam engines, coal-fired power plants, improvement in methods of oil extraction and crude refining technologies that drastically changed the wood and coal-based fuel consumption pattern. As a result, it does not take more than two hundred years to cater to 80% of the energy needs of human civilization from fossil-derived crude oil compared to the replacement of wood by coal which took 100s of 1000s of years. Coal is still a significant part of the cycle that is used to fulfill energy demands. However, the application of coal was primarily limited to power plants to produce electricity until recent developments such as the development of coal-to-chemicals technologies [2]. However, these emerging technologies for coal to chemical productions are yet to be tested at commercial scale; thus, it is improbable that coal can serve as a replacement for fossil-based fuel soon. Moreover, recent improvement in natural gas, shale gas, and coal bed methane gas extraction technologies have further narrowed the road for coal utilization [3]. It should be noted that a wide variety of fuels are derived from fossil-based crude oil, including liquified petroleum gas, gasoline, diesel, kerosene, and other transportation fuels. In fact, the majority of chemicals produced in the world are directly or indirectly made up of fossil-derived crude oil. Moreover, processing crude oil for use in fuel applications does not necessarily require a chemical reaction and mostly are unit operations that involve mostly physical separation processes. Thus the chances of crude oil-loss and by-products formation during downstream processing are small. These are some of the reasons why fossil-derived crude oil has remained the most used fuel and energy source since the previous century.

Sustainable Fuel Technologies Handbook. DOI: https://doi.org/10.1016/B978-0-12-822989-7.00001-9

It is also expected that the world's fossil-fuel reserves may not be able to cater to the energy demands of the exponentially growing population for much longer [4]. Moreover, there are numerous challenges that human civilization is facing because of the excessive utilization of fossil-derived crude oil. The exponential increase in greenhouse gases, primarily CO_2 emission, is one such major problem that has resulted in severe environmental damage, thereby causing a threat to not only human civilization but the entire ecosystem [5,6]. It is to be noted that the consumption of fossil fuel does not only cause the emission of greenhouse gases but also SO_x, NO_x, CO, and unburnt hydrocarbons that are significant contributors to environmental degradation.

Thus, there have been numerous efforts by the scientific community and various governments agencies to put a curb on the emission of harmful gases by implementing several measures such as implementation of Bharat Stage VI norms in India and Euro VI in Europe. Although such restrictions may be able to reduce pollutants and greenhouse gases emissions partially, complete elimination of the problem remains a distant task. Another important aspect of fossil fuel is its availability in different demographic regions. It is well-known that fossil-fuel reserves are not uniformly distributed and engender sociopolitico conflict, economic imbalance, and dependency of some nations on others [7]. Also, there is a massive gap in the demand and supply chain worldwide, which eventually affects the overall growth and lifestyle of human society. Therefore it is of utmost importance to reduce dependency on fossil-derived fuel and include sustainability factors in fuels and energy production technologies. This can be achieved via energy transition from fossil-fuel sources to other sources that can establish a low-carbon economy [8]. However, energy transition has been a difficult task so far and not many milestones have been achieved. Historically low prices of crude oil during past few years have further eliminated any possibilities of the reduction in its usage and indeed, have caused a drastic increase in CO_2 emissions further [9]. However, alternate energy sources such as solar, wind, hydropower, nuclear, and biorenewable energy have emerged as potential contenders to overcome existing challenges due to use of fossil fuels.

1.2 Sustainable technologies for energy production

Unlike fossil fuels, not all energy sources can produce chemicals, fuel, and energy via direct processing methods. For example, solar, wind, hydropower, and nuclear sources of energy can fulfill our energy demands but cannot produce chemicals or any other fuel. Although solar energy is used by plants to produce food via the photosynthesis process, any human-made process for this purpose is yet to be seen on a massive scale. Thus solar power is considered only as a sustainable source of energy and not any other fuel. Similarly wind energy is also a clean source to cater to our energy demands but it cannot produce any chemical or other fuel except electrical energy. Hydropower technologies are used to produce energy and they cannot be used to generate any other fuel. In contrast, nuclear fuels may undergo nuclear fission to produce another nuclear fuel in isotopic form but because of the risk associated with processing nuclear fuel and recent disasters at some of the nuclear plants in Japan, it is not considered as a sustainable fuel source for discussion in this chapter.

The term sustainable energy refers to forms of energy which sources are expected to last unceasingly and contribute to all species sustainability [10]. It is also worth noting that most energy sources are indirectly derived from solar energy. For example, wind energy originates from solar energy due to the movement of the air. It is well-known that the movement of air is caused due to uneven heating of the Earth by the sun combined with Earth's rotation and pressure gradient. Similarly, hydropower energy from river water is due to water evaporation by the sun followed by its condensation in the form of rain. Thus only solar, wind, hydropower, and biomass energy are considered for discussion in this chapter. Accordingly, technologies used to harness energy from these sources have been considered as sustainable technologies for fuel and energy production. In other words, the technologies which are not employed at the cost of our future generation's resources may be called sustainable technologies for fuel and energy production.

1.2.1 Solar energy capturing and usage

Solar energy is undoubtedly one of the cleanest and most freely available sources of energy in nature. In general, it is uniformly distributed worldwide; thus, it has a high potential to reduce the dependency on crude oil, end geo-conflict, and improve the lives of the unprivileged population. Solar energy can reduce global warming as it does not emit CO_2 and other greenhouse gases, thereby resulting in a clean and pollution-free environment. Therefore most nations have formulated policies to promote the use of solar energy [11]. It is estimated that 1×10^{18} kWh solar energy is available to the human civilization, which much higher than the energy that could be obtained from either fossil fuels or nuclear energy [4]. The typical application of solar energy involves greenhouses, drying, heating and cooling, solar-assisted irrigation, wastewater treatment, space applications, rooftop solar systems, and photovoltaic (PV) applications [12].

Rooftop installations of solar energy harnessing systems are widespread and can be seen on top of most government buildings. Further efforts are directed toward promoting the installation of rooftop solar systems to fulfill household energy demands. In general, a series of PV panels are installed on the roof of the building. The PV panels usually consist of solar cells and modules regulated by a mechanism that is capable of handling both electrical and mechanical connections. Solar cells are generally fabricated from semiconducting materials such as silicon that can absorb light and induce electricity. The absorption of light causes the jumping of electrons from a low-energy state to high-energy state due to the energy gained from sunlight. Eventually, the movement of electrons from lower to higher state causes defects and free electrons in the semiconducting device, which in turn generates electricity [13]. The produced electricity is distributed using grids. However, fluctuation in solar energy and light intensity due to several reasons causes a decrease in the efficiency of such PV devices. Therefore solar flux tracking devices are installed to harness maximum solar energy. A solar flux tracking device maintains the position of PV panels at the desired angle mostly in the direction of the sun with tolerability to 10 degrees deviation and perpendicular to solar rays [14]. Another way to improve solar energy harnessing efficiency is by improving the properties and cost of semiconducting materials. For example, cost reduction can be done by using mono- and multicrystalline materials, development of new silicon materials other than silicon wafers, development of thin-film materials such as amorphous silicon, copper indium diselenide, cadmium telluride, and so forth [15]. Other potential materials for harnessing solar energy include silicon carbide, gallium nitride, and diamonds. The PV devices based on semiconducting materials can be divided into five major categories: (1) crystalline silicon, (2) thin-film, (3) concentrating PV, (4) organic PV, and (5) third-generation PV [16]. Notably, solar energy recovery efficiency can also be improved by installing PV/T (thermal) collectors where T collectors harness the heat energy from radiation not captured by the PV cells. It also recovers the waste energy from the PV cells [17–19].

Like rooftop installations for harnessing solar energy, another critical technology is solar heating, which is generally the most convenient way to convert solar energy to heat and storing it for further applications. This is achieved by using an air-heating system based on phase-change material and has a higher storage capacity compared to regular heat storage devices. Such types of systems can be used for solar drying of crops, high moisture-containing energy materials, and agro residue. Another interesting application is solar drying of wet or pond fly ash. This wild have significant application for power plants that do not run at the same capacity throughout the year. For example, cement industries currently use fly ash, which can run out with seasonal variations in the plant operational capacity. In that case, dried pond fly ash can be used to fulfill cement industry demands. However, drying pond fly ash leads to a higher cost of production. Thus there is considerable scope for the implementation of solar drying technologies in cement industries to reduce the cost of pond fly ash drying.

Small scale solar cooling and heating systems have recently gained momentum and attracted the general population. Solar heaters are commonly used in urban areas for heating the air and water in winter to avoid cold. In contrast, solar cooling systems are mostly employed in rural areas either due to the unavailability of electricity or frequent power cuts. As many agricultural products are perishable and stored at a low temperature to avoid damage, solar energy derived cooling systems can be very beneficial for the rural population. In general, there are two types of solar energy cooling devices: (1) sorption refrigerators and (2) solar thermal-mechanical refrigerators [12]. Similarly to solar energy derived cooling and heating systems, solar-energy assisted irrigation is another critical area of consideration. Here solar energy harnessed in the form of electricity using PV cells is transferred to a motor that, in turn, runs a pump. The capacity of the pump, need for water, and electricity generation can be controlled using a microprocessor-based device or automation for desired parameter set points. Solar energy is also used in greenhouses for the controlled production of plants and food crops in a closed structure. In greenhouses, light and energy are harnessed from solar energy to maintain a constant temperature during day and night, especially in colder regions where sunlight availability is less. Other prominent applications of solar energy include applications in aviation sector such as solar energy–driven airplanes and spacecraft, seawater treatment, desalination, wastewater treatment, and direct vehicular applications in the near future. However, several challenges such as the high cost of PV panels, solar flux tracking, use of toxic chemicals, and discarded PV cell recycling are yet to addressed and thereby limit its usage at a larger scale and in day-to-day activities.

1.2.2 Capturing wind energy and its challenges

Capturing wind energy is another important sustainable technology that has enormous potential to reduce the current environmental load and meet energy demands without emitting anything harmful into the atmosphere. Also, despite the high abundance of solar energy, wind energy harvesting is higher compared to solar energy; thus, a larger share of sustainable energy comes from wind and not from the sun [4]. A roadmap for 2050 shows that the wind energy share in total renewable energy would be 24% compared to solar energy using PV cells at 15% [20]. It is also to be noted that a decrease in fossil fuel combustion–based energy production has been witnessed over the past two decades. In contrast, a steady growth in the sustainable production of wind energy has been observed [21]. Moreover, nearly 23 billion tons of CO_2 emission can be saved by using wind energy if the current growth rate of wind energy usage continues [22]. Furthermore, the current rate of wind energy capturing technologies also has the potential to create millions of jobs.

The overall growth of both sectors depends on social acceptability, especially for the community wind energy concept [23]. There are numerous factors that negatively affect the harnessing of wind energy. Most of these barriers are location-specific, which inhibits the development of a universal solution to promote harnessing wind energy. For example, the high initial cost in India, UAE, Lithuania, Turkey, and China, lack of information or distorted information in Belarus, South Africa, Kyrgyzstan, and Bangladesh, the inefficient transmission system in Germany, Russia, and Spain, lack of support instruments in Taiwan, Nigeria, Norway, Lebanon, Estonia, Chile, Canada, and Argentina, local population opposition in France, unfavorable regulations in Denmark, New Zealand, and the USA, powerful lobbying against wind energy in Japan, the Czech Republic, and Poland, lack of accurate wind data in Pakistan and the United Kingdom are some of the major hurdles that hamper the wider employability of wind energy in these countries [22].

There are technical issues with using wind energy as well, which led to some of the above problems. For example, noise caused by wind turbines is a problem and thus attracts opposition from the population living nearby such installations [24]. This can be overcome by installing wind farms far away from residential areas. It is also speculated that the excessive deployment of wind energy turbines may affect local flora and fauna, especially the concern with birds dying when flying into a turbine. One possible solution for saving birds can be installing nets so that birds cannot approach the wind turbines. There are also speculated changes in local climate due to the installation of windmills, although this is yet to be studied in detail. For example, the excessive harnessing of wind energy may cause a change in weather, such as drought, causing high temperatures at night and low temperatures during the day. However, these hypotheses are yet to be studied in detail to ascertain whether they are valid for all wind turbines installations. Overall, considering the current environmental threat due to excessive exploitation of fossil fuels, the application of wind energy cannot be avoided for saving the whole world from further environmental damage [24].

In general, a wind energy harnessing unit works on the principle of electromagnetic induction. It consists of aerodynamic turbine blades that create mechanical energy from the movement of airwaves followed by mechanical energy conversion to electrical energy using a permanent magnet. Since the turbine has a crucial role to play in harnessing wind energy, the turbine system alone accounts for 71% of the total capital cost for land installations [25]. Interestingly, 25% of the total capital cost for a unit is used for the turbine blades. Thus any improvement in turbine blade materials and design will have a direct impact on the cost of wind energy unit installations. Proposed design changes include load reduction by a slender design of blades, deflections, and prevention of natural frequencies. Similarly, the material of construction of the turbine that costs approximately 30% of the total turbine blade cost need to be considered where exists a higher possibility of improvement. There are several other methods for the sustainable production of wind energy, such as the piezoelectric cantilever model reported recently [26]. Currently, the conventional electromagnetic model for electricity production from wind energy remains the most employed technology on land and offshore.

1.2.3 Hydropower energy capturing and challenges

Although solar energy and wind energy are abundant in nature, the hydropower energy share in terms of total production and consumption in world renewable energy is much higher than these two. Hydropower is the largest perpetual of renewable energies [27]. 2014 world energy consumption data suggests that hydropower had an 8.79% share compared to 3.17% share of renewable energies from solar, wind, and biomass [1]. A further study suggests that 76% electricity of all renewable sources comes from hydropower, which is approximately 16.4% of total electricity production from all sources worldwide [28]. Moreover, hydropower production and consumption are growing linearly [4]. Canada is one of the leading players in implementing the use of hydropower energy, which accounts for 63% of the country's total power generation. Some possible reasons for large-scale production of hydropower compared to solar and wind energy can be attributed to its availability in all geographic areas, efficient and matured conversion technologies, longer plant life and lower operational costs, water cost does not very much, and flexibility to produce as per demand. Like solar and wind energies, hydropower is a carbon-neutral and sustainable source of energy. Moreover, hydropower conversion system fabrication requires negligible toxic materials compared to the fabrication of solar energy devices. The major components of hydropower production systems are made of iron, and are thus easier to recycle and reuse. However, it should be noted that hydropower production necessarily requires interference in the natural flow of water which may have its own implications in future.

A series of factors such as quality of water, impact on aquatic lives, protection of rare and endangered species, protection of culture, fish lives, watershed protection, recreational facilities, community acceptance, local population displacement, and the effect on agricultural land, flora, and fauna need consideration to define its sustainability [29,30]. Although hydropower is produced in more than 163 countries, the significant share comes from five or six countries including Canada, China, Brazil, the United States, India, Japan, Russia, Norway and Turkey etc. This is an indicator that despite higher efficiency and broader acceptability, hydropower production technologies are yet to reach their boom [31]. For example, some of developed countries have exploited up to 70% of their hydropower sources [32]. In contrast, less than 20% of hydropower energy potential is exploited in developing countries [33]. It is also reported that Asia has the highest untapped hydropower energy, yet Asia also has the highest number of people without access to electricity (more than 1400 million). Moreover, hydropower production technologies deployment in Europe is also slow [34]. Thus it essential to develop and employ hydropower production technologies at a massive scale to ensure quality of life and electricity access to most of the population worldwide. Undoubtedly, the full utilization of untapped hydropower will enhance the social and economic condition of people. Hydropower production technologies would be helpful in meeting the carbon-credit requirements of several nations as well as emission-reduction goals [35]. In addition, water stored in large reservoirs may be used for irrigation, water supply, and flood control. Overall, available hydropower utilization at its fullest capacity will solve environmental and economic issues.

Hydropower is generally produced by converting the potential energy of water into electricity. A typical hydropower plant consists of large dams constructed to store water, a hydro turbine, and a generator. At one side of the dam, hydro turbines and generators are installed while the other side of the dam holds water, thereby forming an artificial lake. The water falls on the hydro turbine which converts the potential energy of water to kinetic energy, which, in turn, operates a generator that produces electricity, thus potential energy is converted into electrical energy [36]. These processes can achieve up to 90% efficiency concerning water potential energy ($\rho g \Delta z$) [37]. The loss in energy can be attributed to water wheels that are not efficient and several challenges such as turbidity, frictional losses, and partial filling of buckets. Therefore the thermodynamic limit is not attained in many hydropower plants despite the whole process being reversible. Moreover, water buckets are usually capable of handling solid suspension containing dirty water coming from the dam. It is also not possible to construct large dams everywhere due to societal issues such as resettlement, land problem and the nonavailability of water.

The recent trend in sustainable hydropower production is directed toward the construction and installation of small hydropower units. The implementation of smaller hydropower units will result in decentralized electricity production, thereby reducing transmission losses. In general, hydropower plants are classified into four main categories: (1) large (> 1000 kW), (2) small (500–1000 kW), (3) mini (100–500 kW) and micro (<100 kW) [38]. Unlike large hydropower units that necessarily requires big lakes or dams, small hydropower units can be installed in rivers, canals, and other following water systems used for agricultural purposes. The implementation of small hydropower projects is highly recommended for rural areas or places that do not have water-storage systems, which is one of the major reasons for the shift from large-scale hydropower production to small-scale hydropower production [39]. A typical small hydropower unit consists of a canal or leat, penstock, weir, and forebay (also known as a settling tank). A weir is primarily used to control and divert water from the main river to the intake of the leat or canal of the small hydropower system [40]. This water then settles in the forebay to remove particulate matters. The outlet of the forebay is equipped with a metallic mesh filter so that large solid particles will not enter the turbine. The penstock is generally a pressure pipe that is used to move the water from the forebay to a small hydropower turbine. A spillway is also provided to divert water back to the river when the turbine is shut down or not working. Small hydropower units are further categorized into (1) low head (<5 m) and (2) high head (approximately 15–20 m) units. Hydropower turbines are categorized into (1) impulse turbines (subcategorized as Pelton, Turgo, and Crossflow) or (2) reaction turbines, further subcategorized to propeller, Francis, Kinetic/free flow, and Archimedes screw [40].

The impulse turbine usually generates electricity from the kinetic energy of water, whereas the reaction turbine uses pressure and movement of water to produce electricity. The Pelton turbine contains a series of buckets, while a cross-flow turbine contains a drum-like rotor. The Turgo turbine is similar to Pelton but uses a differently shaped bucket. In the reaction turbines list, the propeller can be identified by its axial-flow arrangement having three to six blades. The Francis turbine is the most widely used turbine in industry. It can be a radial, axial, or mixed-radial runner mounted on a spiral casing. The kinetic turbine, as the name suggests, produces electricity from kinetic energy and not from the potential energy of the water. The Archimedes screw turbine is custom designed as per the requirement of process and bespoke in installation. There has been much development in decentralized small-scale hydropower units. However, seeing population density such technologies can serve, it would be appropriate to say that the small hydropower units would be the future of sustainable energy production.

1.2.4 Other clean and sustainable energy sources and production technologies

Geothermal, marine, and tidal energies are another vital class of sustainable energies. It is estimated that geothermal energy consumption would surpass energy consumption from large hydropower plants by 2024 [41]. Like solar energy, geothermal energy also has a wide range of applications that includes electricity production, space heating, and cooling, spas related to heath care, fish farming, greenhouse heating, hot baths, heat pumps, and in industries. However, 42% of the direct use of geothermal energy is for swimming, bathing, and spas [42]. Since geothermal energy is dependent on the heat content inside the Earth, its implementation is a function of geographic locations. As of 2001, 58 countries were using geothermal energy in some form, and 21 countries were producing electricity from it [42]. Iceland is the major exploiter of geothermal energy which catered 50% of its energy demand. Geothermal energy is also a carbon-free energy source; thus, the fullest exploitation of geothermal energy may result in saving 1000 million tons of CO_2 emissions [43]. Geothermal energy is generally extracted from the heat content inside the Earth using sophisticated power extraction units based on the Rankine cycle. The Earth's internal heat is produced from nuclear reactions inside the Earth's crust. To harness this energy, holes are drilled deep into the Earth and water is then injected into these holes. When the water comes into contact with higher temperatures, it absorbs the energy and the heated water is converted to steam which eventually produces electricity via conversion of thermal energy to kinetic energy, and then from kinetic energy to electricity [44]. The first commercial geothermal energy extraction unit dates back to 1913 [42].

There are several other methods for commercial power production from geothermal energy sources, such as the Rankine cycle, and Organic Rankine cycle (ORC). However, such methods were not effective in harnessing the energy from low-temperature sources. Thus researchers attempted to enhance the efficiency of ORC by making suitable changes in the configuration and using additional equipment. For example, the addition of a vapor absorption chiller (VAC) enhances the efficiency of ORC, or an absorption heat transformer (AHT) can be coupled with ORC to enhance the efficiency of the process by 40%–70% [45]. Besides VAC and AHT, several other improved designs have been proposed and implemented over the past three decades, including flash cycle with ORC, development of binary cycle process, and development of combined cycle processes. All geothermal harnessing units can be broadly classified into three broad categories based on their capacity to handle different ranges of geothermal energy source temperatures: (1) ORC used for low enthalpy up to 160°C, (2) binary two-phase configuration used for medium enthalpy up to 160°C to 190°C, and (3) combined geothermal cycle unit used for high enthalpy above 190°C [45]. There are certain technical, economic, societal, and policy-related challenges associated with geothermal energy extraction. One such major challenge is the low-conversion efficiency scaling of equipment, unclear understanding of regulations, higher cost, and stringent land regulations compared to oil and natural gas, mitigation of clay particles, long-distance power transmissions, and maintenance issues [46]. The heat content inside the Earth varies from place to place, thus the cost of production may vary drastically as some places require deeper drilling than others to reach the desired temperature [47].

Like solar, wind, and geothermal energy, marine energy is another important and sustainable source of energy [48]. Noteworthy, marine energy is also known as ocean energy that includes tidal energy, wave energy, thermal energy, ocean-current energy, and salinity-gradient energy. An indirect source of wave energy can be attributed to wind and solar energy as waves are produced due to the combined effects of these energy sources. Thus the extraction of wave energy can also be counted in the exploitation of wind and solar energy. Wave energy possesses kinetic and potential energy; thus, it is capable of producing electricity from both kinetic energy as well as potential energy. A wide range of wave energy extraction systems are used and can be broadly categorized into three units: (1) offshore system, (2) onshore system, and (3) shoreline system, which is widely used [49]. Since wave energy possesses kinetic and potential energy, several mechanisms have been developed to harness both forms of energy. One method is similar to the windmill flow energy extraction technique where the kinetic energy of the water is used to rotate a turbine connected with a generator that converts it into electricity. The other method for electricity production involves the conversion of wave energy to high-pressure fluid. An electromagnetic system without mechanical connections can also be used to produce electricity from wave energy.

Unlike wave energy which is primarily due to solar and wind energies, tidal energy is due to gravitational forces related to the rotation of the Earth with respect sun, as well as the moon's rotation around the Earth. The combined effect of Earth, sun, and moon position along with gravitational forces causes the water in the ocean to rise and fall, creating tidal currents. Since tidal currents move forward, equipment similar to wind turbines is used to produce electricity from this motion [48]. Thermal energy is another important source of marine energy because most of the Earth's surface has water; therefore, maximum solar energy is absorbed by the oceans. Ocean energy converters, similar to geothermal extraction units, are used to harness this energy [49,50]. It is to be noted that solar-energy absorption by water also leads to the generation of ocean currents in which energy can be harnessed using turbines such as those used in small hydropower production units. Salinity-gradient energy is caused due to the osmotic pressure difference between freshwater and saline water. It is to be noted that although marine energy has great potential, there are several ecological and environmental concerns such as disturbances to benthic habitats, flow alteration, sediments and nutrients transport, artificial reefs, marine protected areas, biofouling, electromagnetic fields, collision risk, underwater noise, and societal impact that need consideration before marine energy extraction [51–53].

1.3 Sustainable technologies for energy, fuel and chemicals production

Although solar, wind, hydropower, geothermal, and marine sources can produce energy, they cannot be used for direct production of fuel or chemicals. Thus the complete replacement of fossil fuels with these energy resources may not be feasible. Biorenewable energy resources have shown promising potential to produce energy, fuel, and chemicals. Biomass can be classified as nonlignocellulosic or lignocellulosic. The nonlignocellulosic biomass represents inedible oil, algal biomass, and other sources that have negligible lignin [54]. In contrast, biomass resources such as wood, timber, sugarcane bagasse, rice husk, and wheat straw have a significant amount of lignin and are classified as lignocellulosic biomass. Major compositional constituents of lignocellulosic biomass are cellulose (composed of C_6 sugar polymers), hemicellulose (composed of C_6 and C_5 sugar polymers), lignin (composed of aromatic compound polymers,) and ash content [55]. The majority of biomass constituents are oxygenated hydrocarbons that have the potential to serve as fuel and chemicals. However, due to variations in composition and complex molecular structure, a wide range of sustainable technologies such as direct combustion, gasification, pyrolysis, hydrolysis, fermentation, enzymatic processes, and catalytic routes are employed to produce fuel and chemicals from the biomass. There could be several ways of classifying these biomass conversion processes as a plethora of technologies are available. However, for the ease of industrial applicability and based on operating conditions, we have broadly classified all biomass conversion technologies into two major categories (which are further divided into several subcategories): (1) high-temperature processes and (2) low-temperature processes.

1.3.1 High-temperature biomass conversion technologies

Any biomass conversion process that operates above 300°C is considered a high-temperature process in this discussion. Some of the major high-temperature processes are combustion, gasification, and pyrolysis that are primarily meant for energy and fuel production. In all such processes, the solar energy stored in the form of carbon–hydrogen bonds in the biomass is released by breaking of bonds when subjected to severe high temperature. For example, biomass molecules get completely broken down when subjected to direct combustion (at 1000°C to 2000°C) in the presence of air, and burning the separated hydrogen causes a massive production of energy that is used for household cooking and in industrial applications [56–58]. However, direct combustion of biomass also causes the emission of pollutants, greenhouse gases, as well as particulate matters which necessitated the development of other models for harnessing biomass energy. The combustion of biomass can produce only energy and cannot produce chemicals. Therefore gasification emerged as a potential substitute for the biomass combustion process and can be used to produce energy as well as chemicals. Syngas is the primary product obtained from biomass which contains heat that can be used for electricity production. In contrast, it can also be converted to liquid fuel via Fischer Tropsch synthesis reaction [59]. However, the production of chemicals and liquid fuels from biomass gasification products is a challenging task. Thus, pyrolysis emerged as another major contender for liquid fuel production from biorenewable resources. Pyrolysis processes require a lower temperature range (up to 500°C) compared to the gasification process which usually requires a temperature range of 800°C–1300°C [60,61]. Each of the high-temperature processes has its pros and cons which are discussed next.

1.3.1.1 Biomass combustion and challenges

Biomass combustion has been used 100s of 1000s years. It is a one-pot, simple process that does not require specific equipment or instruments and can be used for everyday household applications. However, a very low moisture content in the biomass is necessary for its combustion. In contrast to the general claim of being one-pot process, biomass combustion reaction involves several steps that take place during the combustion. Volatile materials such as tars and gases are initially produced, followed by the combustion of volatiles and char, respectively. Biomass decomposition during combustion starts in the 160°C to 250°C temperature range to remove moisture and light volatiles, followed by thermal decomposition of hemicellulose and cellulose in the temperature range of 300°C to 500°C. In contrast, lignin decomposition takes place at a wide range of temperatures starting from 150°C up to 500°C. However, the overall process temperature goes beyond 1000°C due to the heat released by the breaking of biomass compound bonds.

Combustion can be classified into two broad categories: (1) flaming combustion and (2) smoldering combustion. Flaming combustion primarily decomposes volatile compounds by mixing them with air to produce heat energy and light [62]. CO, CO_2, and H_2O are some of the major products produced due to flaming combustion. Smoldering combustion burns residual biochar slowly and does not produce any flames and is therefore sometimes called glowing combustion. One possible reason for smoldering combustion can be attributed to oxygen attack on biochar surface and moves inside. However, due to mass transfer limitations in oxygen diffusion inside the biochar and cooling down of process over time, this combustion always remains incomplete unless it is done in a controlled manner. However, there are several challenges associated with direct combustion technology such as dioxins formation, particulate emission, greenhouse gases emission, NOx and SOx emissions, metal–metal interaction due to high-temperature process, and ash handling, which has encourage the scientific community to search for alternative processes [62–72].

1.3.1.2 Biomass gasification

Biomass gasification is also a high-temperature process and is usually carried out in a controlled environment. In general, either oxygen, air, CO_2, or steam is used for incomplete combustion of biomass above 800°C to yield carbon monoxide (CO), hydrogen (H_2), and a few other gases such as methane in trace amounts [73,74]. Gasification is one of the most employed methods for hydrogen production from biomass [75]. However, produced hydrogen gas needs extensive cleaning using scrubbers and purification by either pressure swing adsorption or membrane-based separation technologies. Since gasification processes produce a massive amount of heat energy, it is also used worldwide for electricity production from biorenewable resources. Indeed, gasification of biomass is considered one of the most promising options for decentralized electricity production especially in countries like India. Biomass integrated gasification combined cycle has high energy efficiency [76].

There are several gasifier designs available that can be used for the production of electricity as well as hydrogen gas from the biomass, including downdraft fixed-bed gasifier, updraft fixed-bed gasifier, bubbling fluidized bed gasifier, circulating fluidized bed gasifier, entrained flow gasifier, dual fluidized bed gasifier, two-stage fixed-bed gasifier, stratified downdraft fixed-bed gasifier, stratified twin bed downdraft/updraft gasifier, floating fluidized bed gasifier, and plasma gasifier [61,77]. However, it should be noted that most of the gasifier designs have not been successful at commercial scale, and hence, small-scale gasifiers are usually recommended for hydrogen or electricity production from biomass. Two-stage fixed-bed gasifiers, plasma gasifiers, and stratified downdraft gasifiers are the most efficient for biomass gasification and energy recovery.

Biomass gasification end products, CO and H_2, together are called syngas, which can also serve as potential feedstock to produce a wide range of chemicals and liquid fuels such as methanol, ethanol, dimethyl ether, and synthetic natural gas [78]. For example, our group demonstrated the production of higher hydrocarbons from biomass-derived CO_2 rich syngas [59]. Several other research groups have reported the production of synthetic paraffinic kerosene and synthetic kerosene with aromatics from syngas using Fischer Tropsch reaction in the presence of a catalyst [79]. Synthetic paraffinic kerosene can be blended in jet fuels by up to 50%. Synthetic kerosene with aromatics can be used as a fuel source in energy generation where fuel oil is used. In this regard, several Fe and Co based catalytic materials have been synthesized and tested by different research groups and companies. The activity of a synthesized catalyst can be tuned using a wide range of metal-oxide supports and promoters for the selective production.

However, like any other sustainable process, biomass gasification technology also faces several challenges starting from the feedstock preparation which involves drying, shape and size optimization, type of feedstock, reaction conditions such as residence time optimization, gasifying agents selection, air–fuel ratio and equivalent air, reaction temperature, and pressure optimization, as well as postreaction challenges such as tar removal, gas cleaning, and syngas conversion [80–84]. Large-scale gasifier design and ash handling are other significant challenges that limit the application of biomass gasification technology in many cases. Therefore, the research community recommends the integration of gasification technology with pyrolysis, where the latter will produce gas fuels and char that will become feedstock for gasifiers to yield less tar [74]. Another method for reforming tar could be using metals present in biomass or char as a catalyst or externally supplied catalytic materials [85]. However, despite several publications and the availability of gasifiers, biomass gasification technologies are still considered immature;[86] thus, a whole new effort has to be directed toward designing improved gasifiers to enhance the sustainability factor in biomass gasification technologies.

1.3.1.3 Biomass pyrolysis

Gasification can be termed controlled and partial combustion, while pyrolysis can also be referred to as gasification in the presence of an inert environment. However, pyrolysis cannot be limited to gasification in an inert environment as there are several other reactor designs including fixed-bed pyrolizers, auger pyrolizers, twin-screw pyrolizers, microwave, and others [87]. In pyrolysis, biomass is fed into the reactor and oxygen is removed, mostly by using nitrogen as an inert gas. Thereafter, heating of the reactor is started which causes devolatilization of lighter volatiles in the gaseous form at 150°C–250°C temperature, followed by decomposition of hemicellulose, cellulose, and lignin at a temperature range of 225°C–326°C, 300°C–425°C, and 100°C–800°C, respectively [88]. The end product obtained from pyrolysis are gaseous fuel, liquid fuel, and solid fuel (in the form of biochar). Pyrolysis process operating conditions can be optimized to yield the desired product. However, reactions involved in biomass pyrolysis are very complex and vary from process to process and feedstock to feedstock [89].

Therefore, pyrolysis process based on end-product requirement, in general, is classified into three broad categories, (1) slow, (2) intermediate and (3) fast/flash pyrolysis, based on the need for desired product yield, which affects the operating conditions as well [90,91]. Slow pyrolysis is usually employed to produce solid fuel in the form of briquettes or char via slow heating of the biomass for a prolonged period at a temperature range below 400°C. The most common examples of slow pyrolysis are VMR ovens, biomass heated in a kiln or pit, Reichert process, Lambiotte process, tubular kilns, screw pyrolizers, and agitated drum kilns [92]. In contrast, fast pyrolysis is usually carried out above 500°C for a short-residence time, usually a few minutes. The fast pyrolysis process is used to produce liquid fuel, whereas slow pyrolysis is used for solid fuel production. Most of the reactors used for fast pyrolysis are a fluidized bed, stirred or moving beds, or vacuum pyrolizers. When the biomass residence time is very short, usually in a few seconds, then fast pyrolysis is also termed as flash pyrolysis. Similarly, intermediate pyrolysis is particularly used to define the pyrolysis in a specific type of screw reactor which produces liquid and solid products. Pyrolysis liquid fuel has a high oxygen content and therefore requires oxygen removal using a suitable catalyst via a hydrodeoxygenation reaction [93]. However, it is also observed that by putting a reformer in the intermediate screw pyrolizer, the char produced in the process itself acts as in-situ catalyst for oxygen removal [60]. This type of process is also referred to as thermo-catalytic reforming.

Besides operating temperature and time, the bio-oil and other products yield from the pyrolysis process also depends on the type of feedstock used, state of feedstock, size of feedstock, feed rate, and feed composition. The composition of biomass directly affects the composition of bio-oil [87]. The pretreatment of biomass feedstock also changes the end-product yield. For example, physical pretreatment methods such as sizing of the feedstock helps to improve bio-oil yield whereas the thermal pretreatment process helps to remove moisture, to produce less oxygen content in the resulting bio-oil [94]. Several chemical pretreatment methods such as acid and alkali pretreatment [95], ammonia fiber expansion, ionic liquid, and deep eutectic solvent methods have been developed. Chemical pretreatment removes lignin and minerals from the biomass; thus, a lower temperature is required to obtain metal-free bio-oil. The future recommendations for pyrolysis processes are directed toward in-situ catalytic upgradation of the produced bio-oil.

1.3.2 Low-temperature biomass conversion processes

It is evident that the high-temperature processes for biomass conversion are mainly meant for energy and fuel production. Although syngas can be used to produce some essential chemicals, the conversion process is challenging. Also, high-temperature processes specifically require sophisticated equipment, high cost due to unique material of construction, careful handling, and several other cumbersome measures. Therefore low-temperature biomass conversion attracted the attention of the scientific community. The significant advantage of low-temperature processes is that it can be used to produce chemicals besides producing fuels. Moreover, low operating temperature equipment does not necessarily require high-grade steel, thus lowering the plant setup cost. These fuels and chemicals can be produced via biological or thermo-catalytic routes. Both routes are used to produce an array of chemicals; however, some chemicals can be produced via either route. Low-temperature processes are used to produce multipurpose platform chemicals that can further produce an array of new commodity chemicals, pharmaceuticals, medicines, polymers, fragrances, food preservatives, and so forth, as well as fuel. All these biological and thermo-catalytic routes are further divided into several subcategories; however, this chapter is limited to the broad application of these routes as each of molecules produced through these routes follows a different subroute. For a detailed review of product-specific processes, readers may refer to the US Department of Energy report for chemical production from biomass published in 2004 and 2006 [96,97].

1.3.2.1 Biological routes for fuel and chemicals production

Biological routes are used to produce some essential building-block chemicals such as lactic acid, aspartic acid, glutamic acid, succinic acid, itaconic acid, 3-hydroxy propionic acid, ethanol, and bio hydrocarbons [96]. However, out of all these biologically produced chemicals, only ethanol has the potential to serve as fuel. Ethanol in fuel application is primarily used for blending to enhance the octane rating. In some cases, up to 85% ethanol can be blended in gasoline as its research octane number (109) and motor octane number (90) are comparable with the average octane of regular gasoline (91)([98]. For example, Brazil has allowed 100% anhydrous ethanol for its vehicular applications without blending with gasoline. Several other chemicals such as ethylene and acetic acid can be produced from ethanol. Ethanol is also used in human consumption grade applications and was initially not considered in the US DOE top 12 building-block chemicals due to its consideration under super commodity chemicals until it was added to the DOE list in 2010 [99]. Ethanol is produced via a fermentation process which is the most widely employed process at commercial scale to produce chemicals from biomass using biological routes. The use of ethanol in vehicular application to a limit of 85% blending may result in conflict with human consumption if the production of ethanol is not increased drastically [100]. This could be one of the reasons that many nations have not allowed more than 15%–20% blending. Besides ethanol, butyl butyrate is another outstanding fuel-grade biologically derived product that can be used as aviation fuel. There are several reasons why butyl butyrate can be used as an aviation fuel, such as comparable flash and fire points, comparable cloud and pour point, as well as the freezing point [101]. The cloud point and pour point are the most essential criteria for aviation fuels [102]. Butyl butyrate can be produced from biorenewable resources using collagen-immobilized lipase or via the fermentation route [103]. Lipases are usually referred to as biocatalysts as they have the capability to operate at lower temperatures compared to thermo-catalytic routes. However, the presence of biogenic impurities in such processes may be a problem for further upgrades of these products [104].

1.3.2.2 Catalytic routes for fuels and chemicals production

Most of the chemicals identified by the US DOE cannot be used directly as a fuel. Thus further conversion of these chemicals into fuel-grade products is required [105]. However, some processes can directly produce fuels from biomass. Some of the important biomass-derived fuels are alkyl levulinates (AL) such as methyl levulinate (ML), ethyl levulinate (EL), propyl levulinate (PL), butyl levulinate (BL), butyl butyrate (BB), and ring oxygenates such as 5-ethoxymethylfurfural (EMF), 2,5-dimethylfuran (DMF), 2-methyl furan (MF), γ-valerolactone (GVL), and long-chain hydrocarbons [106,107]. Detailed insight into the reaction mechanism of these products' formation suggests that all possible reactions that may occur during their production are depolymerization, isomerization, dehydration, etherification, Diels Alder reaction, and hydrogenation reaction. Thus all the catalytic materials used in such processes can be divided into three major categories: Brønsted acid catalysts required for depolymerization, dehydration, etherification, and Diels Alder reaction, Lewis acid catalyst for isomerization reaction, and precious or transition metal catalysts for hydrogenation reaction. However, it should be noted that the Lewis acids due to the positive charge on metals may create a Brønsted acidic environment. Therefore the proposed rule-of-thumb may not be accurate in some specific cases; thus, care should be taken while choosing the catalyst. Some of the major catalysts used for AL and EMF production are mineral acids, metal salts, ionic liquids, sulfated metal oxides, sulfonated catalysts, heteropolyacids, zeolites, resins, and dichalcogenides [108–115]. In contrast, DMF, MF, GVL, and long-chain hydrocarbon production requires metal catalysts which are mostly Pt, Pd, Ru, Rh, Ir, and Ni supported over carbon or metal oxides are employed [106]. Some metal oxides such as ZrO_2, MgO, and γ-Al_2O_3 were also found to be useful in GVL production under different reaction conditions.

1.4 Summary

In this chapter the most widely used sustainable technologies (i.e., solar energy, wind energy, hydropower, geothermal energy, marine energy, and bioenergy) have been discussed. It is observed that all solar energy, wind energy, hydropower, geothermal energy, and marine energy are clean sources of carbon-free energies that can help to reduce greenhouse gas and other pollutant emissions into the environment. However, these technologies face severe societal, ecological, and technical challenges which limits the implementation of such technologies. Also, these sources can only produce energy and cannot produce fuel for later use. In contrast, biorenewable resources such as biomass can produce energy, fuel, and chemicals. However, these emit greenhouse gases, particulate matter, and other toxic compounds that are yet to be adequately addressed. Although CO_2 emission may not be important in the case of biorenewable energy as biomass-derived energy forms a closed carbon chain in the environment, the emission of other pollutants are of great concern. Overall, all sustainable fuel and energy production technologies have advantages and disadvantages. However, considering the enormous environmental benefits, some of the disadvantages can be ignored until an appropriate solution is found.

Acknowledgment

SQ thanks the Science and Engineering Research Board, Government of India, for the National Post-Doctoral Fellowship (PDF/2019/002631). EA thanks IIT ISM Dhanbad for Faculty Research Grant (FRS) for financial support for the project relevant to this topic. KKP would like to thank the Petrotech Society of India for Chair Professorship.

References

[1] C. Zou, Q. Zhao, G. Zhang, B. Xiong, Energy revolution: from a fossil energy era to a new energy era, Nat. Gas. Ind. B 3 (1) (2016) 1–11.

[2] H.H. Schobert, C. Song, Chemicals and materials from coal in the 21st century, Fuel 81 (1) (2002) 15–32.

[3] S.H. Mohr, J. Wang, G. Ellem, J. Ward, D. Giurco, Projection of world fossil fuels by country, Fuel 141 (2015) 120–135.

[4] N. Abas, A. Kalair, N. Khan, Review of fossil fuels and future energy technologies, Futures 69 (2015) 31–49.

[5] M. Höök, X. Tang, Depletion of fossil fuels and anthropogenic climate change-a review, Energy Policy 52 (2013) 797–809.

[6] D. Mboowa, S. Quereshi, C. Bhattacharjee, K. Tonny, S. Dutta, Qualitative determination of energy potential and methane generation from municipal solid waste (MSW) in Dhanbad (India), Energy 123 (2017) 386–391.

[7] M.J. Burke, J.C. Stephens, Political power and renewable energy futures: a critical review, Energy Res. Soc. Sci. 35 (2018) 78–93. no. October 2017.

[8] R. Fouquet, P.J.G. Pearson, Past and prospective energy transitions: insights from history, Energy Policy 50 (2012) 1–7.

[9] C. Day, G. Day, Climate change, fossil fuel prices and depletion: the rationale for a falling export tax, Econ. Model. 63 (2017) 153–160. no. August 2016.

[10] H. Lund, Chapter 1 Introduction, Renewable Energy Systems: The Choice and Modeling of 100% Renewable Solutions 2010, 1–12, Elsevier. Available from: https://doi.org/10.1016/B978-0-12-3750280.00001-7.
[11] K.H. Solangi, M.R. Islam, R. Saidur, N.A. Rahim, H. Fayaz, A review on global solar energy policy, Renew. Sustain. Energy Rev. 15 (4) (2011) 2149–2163.
[12] N. Kannan, D. Vakeesan, Solar energy for future world: a review, Renew. Sustain. Energy Rev. 62 (2016) 1092–1105.
[13] M.A. Green, Photovoltaic principles, Phys. E 14 (2002) 11–17.
[14] P. Roth, A. Georgiev, H. Boudinov, Cheap two axis sun following device, Energy Convers. Manage 46 (2005) 1179–1192.
[15] T.M. Razykov, C.S. Ferekides, D. Morel, E. Stefanakos, H.S. Ullal, Solar photovoltaic electricity: current status and future prospects, Sol. Energy 85 (8) (2011) 1580–1608.
[16] F. Obeidat, A comprehensive review of future photovoltaic systems, Sol. Energy 163 (2018) 545–551. no. July 2017.
[17] A.M. Eltamaly, H.M.H. Farh, M.F. Othman, A novel evaluation index for the photovoltaic maximum power point tracker techniques, Sol. Energy 174 (2018) 940–956. no. September.
[18] A. Gholami, I. Khazaee, S. Eslami, M. Zandi, E. Akrami, Experimental investigation of dust deposition effects on photo-voltaic output performance, Sol. Energy 159 (2018) 346–352. no. October 2017.
[19] M. Mussard, M. Amara, Performance of solar photovoltaic modules under arid climatic conditions: a review, Sol. Energy 174 (2018) 409–421. no. September.
[20] D. Gielen, F. Boshell, D. Saygin, M.D. Bazilian, N. Wagner, R. Gorini, The role of renewable energy in the global energy transformation, Energy Strategy Rev. 24 (2019) 38–50. no. June 2018.
[21] J.L. Peters, T. Remmers, A.J. Wheeler, J. Murphy, V. Cummins, A systematic review and meta-analysis of GIS use to reveal trends in offshore wind energy research and offer insights on best practices, Renew. Sustain. Energy Rev. 128 (2020) 109916. no. August 2019.
[22] J. Ramsi, F. Diógenes, M. Valentim, J. Claro, J.C. Rodrigues, Barriers to onshore wind energy implementation: a systematic review, Energy Res. Soc. Sci. 60 (2020) 101337. no. October 2019.
[23] J. Baxter, C. Walker, G. Ellis, P. Devine-wright, M. Adams, R. Smith, Energy research & social science scale, history and justice in community wind energy : an empirical review, Energy Res. Soc. Sci. 68 (2020) 101532. no. July 2019.
[24] M. Shahzad, A. Jafer, M. Bilal, M. Sohail, N. Ali, M.N. Iqbal, Environmental impact and pollution-related challenges of renewable wind energy paradigm – a review, Sci. Total. Environ. 683 (2019) 436–444.
[25] C. Mone et al., Cost of wind energy review 2015 cost of wind energy review, National Renewable Energy Laboratory (NREL/TP-6A20-66861, 2017.
[26] Y. Gong, Z. Yang, X. Shan, Y. Sun, T. Xie, Capturing flow energy from ocean and wind. Energies (2019) 12, 2184. Available from: https://doi.org/10.3390/en12112184.
[27] F. Manzano-agugliaro, A. Alcayde, F.G. Montoya, A. Zapata-sierra, C. Gil, Scientific production of renewable energies worldwide: an overview, Renew. Sustain. Energy Rev. 18 (2013) 134–143.
[28] M.E. Bildirici, S.M. Gökmeno, Environmental pollution, hydropower energy consumption and economic growth: evidence from G7 countries, Renew. Sustain. Energy Rev. 75 (2017) 68–85. no. October 2016.

[29] J. Liu, J. Zuo, Z. Sun, G. Zillante, X. Chen, Sustainability in hydropower development – a case study, Renew. Sustain. Energy Rev. 19 (2013) 230–237.
[30] A. Botelho, P. Ferreira, F. Lima, L.M.C. Pinto, S. Sousa, Assessment of the environmental impacts associated with hydropower, Renew. Sustain. Energy Rev. 70 (2017) 896–904. no. May 2015.
[31] Ö. Can, J.Ø. Nielsen, Is the hydropower boom actually taking place ? A case study of a South East European country, Bosnia and Herzegovina, Renew. Sustain. Energy Rev. 110 (2019) 278–289. no. December 2018.
[32] A. Shaktawat, S. Vadhera, no. 0123456789 Risk Management of Hydropower Projects for Sustainable Development: A Review, Springer, Netherlands, 2020.
[33] R. Sipahutar, S. Masreah, M. Sodik, Renewable energy and hydropower utilization tendency worldwide, Renew. Sustain. Energy Rev. 17 (2013) 213–215.
[34] B. Wagner, C. Hauer, H. Habersack, Current hydropower developments in Europe, Curr. Opin. Environ. Sustain. 37 (2020) 41–49. no. i.
[35] H. Nautiyal, V. Goel, Sustainability assessment of hydropower projects, J. Clean. Prod. 265 (2020) 121661.
[36] H. Production et al., Hydrogen production by electrical energy, Sustainable Hydrogen Production. 2016, 99–161, Elsevier. Available from: https://doi.org/10.1016/B978-0-12-801563-6.00003-0.
[37] T.U. States, et al., Hydroelectricity hydroelectricity, Future Energy (2014) 215–235.
[38] D. Kilama, Review of small hydropower technology, Renew. Sustain. Energy Rev. 26 (2013) 515–520.
[39] J. Xu, T. Ni, Integrated technological paradigm-based soft paths towards sustainable development of small hydropower, Renew. Sustain. Energy Rev. 70 (2017) 623–634. no. July 2016.
[40] O. Paish, Small hydro power : technology and current status, Renew. Sustain. Energy Rev. 6 (2002) 537–556.
[41] N.L. Panwar, S.C. Kaushik, S. Kothari, Role of renewable energy sources in environmental protection: a review, Renew. Sustain. Energy Rev. 15 (3) (2011) 1513–1524.
[42] I.B. Fridleifsson, Geothermal energy for the benefit of the people, Renew. Sustain. Energy Rev. 5 (3) (2001) 299–312.
[43] R. Bertani, Geothermal power generation in the world 2010–2014 update report, Geothermics 60 (2016) 31–43.
[44] H. Kalusekara, V. Seynulabdeen, A review of geothermal energy for future power generation, in: Proceedings of the 2019 5th International Conference on Advances in Electrical Engineering (ICAEE), 26–28 September, Dhaka, Bangladesh, 2019, pp. 26–28.
[45] M. Soltani, et al., A comprehensive review of geothermal energy evolution and development, Int. J. Green. Energy 16 (13) (2019) 971–1009.
[46] K. Wang, B. Yuan, G. Ji, X. Wu, A comprehensive review of geothermal energy extraction and utilization in oilfields, J. Pet. Sci. Eng. 168 (2018) 465–477. no. December 2017.
[47] E. Barbier, Nature and technology of geothermal energy: a review, Renew. Sustain. Energy Rev. 1 (1–2) (1997) 1–69.
[48] T. Wilberforce, Z. El Hassan, A. Durrant, J. Thompson, B. Soudan, A.G. Olabi, Overview of ocean power technology, Energy 175 (2019) 165–181.
[49] M. Melikoglu, Current status and future of ocean energy sources: a global review, Ocean. Eng. 148 (2018) 563–573. no. December 2017.

[50] F. Zabihian, A.S. Fung, Review of marine renewable energies: case study of Iran, Renew. Sustain. Energy Rev. 15 (5) (2011) 2461–2474.
[51] P.A.J. Bonar, I.G. Bryden, A.G.L. Borthwick, Social and ecological impacts of marine energy development, Renew. Sustain. Energy Rev. 47 (2015) 486–495.
[52] M. Winskel, A. McLeod, R. Wallace, R. Williams, Energy policy and institutional context: marine energy innovation systems, Sci. Public. Policy 33 (5) (2006) 365–376.
[53] A. Uihlein, D. Magagna, Wave and tidal current energy – a review of the current state of research beyond technology, Renew. Sustain. Energy Rev. 58 (2016) 1070–1081.
[54] S. Quereshi, E. Ahmad, K.K. Pant, and S. Dutta, Recent advances in production of biofuel and commodity chemicals from algal biomass, In: S.K. Gupta, A. Malik, and F. Bux, (Eds), Algal Biofuels: Recent Advances and Future Prospects, 2017, Springer, 394–416. Available from: https://doi.org/10.1007/978-3-319-51010-1.
[55] S. Quereshi, S. Dutta, Synthesis of biomass derived platform chemicals & fuel additives, in: Proceedeings of 5th Asia Oceana Conference on Green and Sustainable Chemistry, RSC, 2015, p. 134.
[56] L. Wang, J.E. Hustad, Ø. Skreiberg, G. Skjevrak, M. Grønli, A critical review on additives to reduce ash related operation problems in biomass combustion applications, Energy Procedia 20 (1876) (2012) 20–29.
[57] J.M. Jones, M. Pourkashanian, A. Williams, D. Hainsworth, A comprehensive biomass combustion model, Renew. Energy 19 (1–2) (2000) 229–234.
[58] J. Li, M.C. Paul, P.L. Younger, I. Watson, M. Hossain, S. Welch, Characterization of biomass combustion at high temperatures based on an upgraded single particle model, Appl. Energy 156 (2015) 749–755.
[59] E. Sonal, S. Ahmad, Upadhyayula, K.K. Pant, Biomass-derived CO2 rich syngas conversion to higher hydrocarbon via Fischer-Tropsch process over Fe–Co bimetallic catalyst, Int. J. Hydrogen Energy 44 (51) (2019) 27741–27748.
[60] E. Ahmad, N. Jäger, A. Apfelbacher, R. Daschner, A. Hornung, K.K. Pant, Integrated thermo-catalytic reforming of residual sugarcane bagasse in a laboratory scale reactor, Fuel Process. Technol. 171 (2018) 277–286.
[61] R. Thomson, P. Kwong, E. Ahmad, K.D.P. Nigam, Clean syngas from small commercial biomass gasifiers; a review of gasifier development, recent advances and performance evaluation, Int. J. Hydrogen Energy (2020). no. xxxx.
[62] M. Zhang, A. Buekens, X. Li, Dioxins from biomass combustion: an overview, Waste Biomass Valoriz. 8 (1) (2017) 1–20.
[63] J.L. Míguez, J.C. Morán, E. Granada, J. Porteiro, Review of technology in small-scale biomass combustion systems in the European market, Renew. Sustain. Energy Rev. 16 (6) (2012) 3867–3875.
[64] A.A.A. Abuelnuor, et al., Characteristics of biomass in flameless combustion: a review, Renew. Sustain. Energy Rev. 33 (2014) 363–370.
[65] N.C. Cruz, F.C. Silva, L.A.C. Tarelho, S.M. Rodrigues, Critical review of key variables affecting potential recycling applications of ash produced at large-scale biomass combustion plants, Resour. Conserv. Recycl. 150 (2019) 104427. no. August.
[66] A. Akhtar, V. Krepl, T. Ivanova, A combined overview of combustion, pyrolysis, and gasification of biomass, Energy Fuels 32 (7) (2018) 7294–7318.
[67] E.D. Vicente, C.A. Alves, An overview of particulate emissions from residential biomass combustion, Atmos. Res. 199 (2018) 159–185. no. August 2017.
[68] M. Mladenović, M. Paprika, A. Marinković, Denitrification techniques for biomass combustion, Renew. Sustain. Energy Rev. 82 (2018) 3350–3364. no. October 2016.

[69] S.V. Vassilev, C.G. Vassileva, Y.C. Song, W.Y. Li, J. Feng, Ash contents and ash-forming elements of biomass and their significance for solid biofuel combustion, Fuel 208 (2017) 377–409.
[70] M.M. Parascanu, F. Sandoval-Salas, G. Soreanu, J.L. Valverde, L. Sanchez-Silva, Valorization of Mexican biomasses through pyrolysis, combustion and gasification processes, Renew. Sustain. Energy Rev. 71 (2017) 509–522. no. April 2016.
[71] L.J.R. Nunes, J.C.O. Matias, J.P.S. Catalão, Biomass combustion systems: a review on the physical and chemical properties of the ashes, Renew. Sustain. Energy Rev. 53 (2016) 235–242.
[72] R. García, C. Pizarro, A. Álvarez, A.G. Lavín, J.L. Bueno, Study of biomass combustion wastes, Fuel 148 (2015) 152–159.
[73] E. Ahmad, A. Vani, K.K. Pant, An overview of fossil fuel and biomass-based integrated energy systems: co-firing, co-combustion, co-pyrolysis, co-liquefaction and co-gasification, in: Sonil Nanda, P.K. Sarangi, D.-V.N. Vo (Eds.), Fuel Processing and Energy Utilization, first ed., CRC Press, New York, 2019.
[74] V.S. Sikarwar, et al., An overview of advances in biomass gasification, Energy Environ. Sci. 9 (10) (2016) 2939–2977.
[75] M.A. Salam, K. Ahmed, N. Akter, T. Hossain, A review of hydrogen production via biomass gasification and its prospect in Bangladesh, Int. J. Hydrog. Energy 43 (32) (2018) 14944–14973.
[76] B. Buragohain, P. Mahanta, V.S. Moholkar, Biomass gasification for decentralized power generation: the Indian perspective, Renew. Sustain. Energy Rev. 14 (2010) 73–92.
[77] S.K. Sansaniwal, K. Pal, M.A. Rosen, S.K. Tyagi, Recent advances in the development of biomass gasification technology: a comprehensive review, Renew. Sustain. Energy Rev. 72 (2017) 363–384. no. December 2015.
[78] A. Molino, V. Larocca, S. Chianese, D. Musmarra, A. Review, Biofuels production by biomass gasification: a review, Energies 11 (811) (2018) 1–31.
[79] M. Shahabuddin, T. Alam, B.B. Krishna, T. Bhaskar, G. Perkins, A review on the production of renewable aviation fuels from the gasification of biomass and residual wastes, Bioresour. Technol. 312 (2020) 123596. no. May.
[80] M.P. Morales, P. Mun, J.A. Ruiz, M.C. Jua, Biomass gasification for electricity generation: review of current technology barriers, Renew. Sustain. Energy Rev. 18 (2013) 174–183.
[81] S.K. Sansaniwal, M.A. Rosen, S.K. Tyagi, Global challenges in the sustainable development of biomass gasification: an overview, Renew. Sustain. Energy Rev. 80 (2017) 23–43. no. May.
[82] E.R. Widjaya, G. Chen, L. Bowtell, C. Hills, Gasification of non-woody biomass: a literature review, Renew. Sustain. Energy Rev. 89 (2018) 184–193. no. February.
[83] A.S. Mednikov, A review of technologies for multistage wood biomass gasification, Therm. Eng. 65 (8) (2018) 531–546.
[84] I.L. Motta, N.T. Miranda, R.M. Filho, M. Regina, W. Maciel, Biomass gasification in fluidized beds: a review of biomass moisture content and operating pressure effects Ingrid, Renew. Sustain. Energy Rev. 94 (2018) 998–1023. no. June.
[85] D. Xu, et al., Mini-review on char catalysts for tar reforming during biomass gasi fi cation: the importance of char structure, Energy Fuels 34 (2020) 1219–1229.
[86] A.F. Kirkels, G.P.J. Verbong, Biomass gasification: still promising ? A 30-year global overview, Renew. Sustain. Energy Rev., 15, 2011, pp. 471–481.

[87] R.E. Guedes, A.S. Luna, A.R. Torres, Operating parameters for bio-oil production in biomass pyrolysis: a review, J. Anal. Appl. Pyrolysis 129 (2018) 134–149. no. July 2017.
[88] J.Y. Yeo, B. Lai, F. Chin, J.K. Tan, Y.S. Loh, Comparative studies on the pyrolysis of cellulose, hemicellulose, and lignin based on combined kinetics, J. Energy Inst. 92 (1) (2019) 27–37.
[89] F. Collard, J. Blin, A review on pyrolysis of biomass constituents: mechanisms and composition of the products obtained from the conversion of cellulose, hemicelluloses and lignin, Renew. Sustain. Energy Rev. 38 (2014) 594–608.
[90] S. Zhou, B. Pecha, M. Van Kuppevelt, A.G. Mcdonald, M. Garcia-perez, Slow and fast pyrolysis of Douglas-fir lignin: importance of liquid-intermediate formation on the distribution of products, Biomass Bioenergy 66 (2014) 398–409.
[91] S. Zhou, B. Pecha, M. Van Kuppevelt, A.G. Mcdonald, M. Garcia-perez, Slow and fast pyrolysis of Douglas-fir lignin: importance of liquid-intermediate formation on the distribution of products, Biomass Bioenergy (2014) 1–12.
[92] P.A. Brownsort, Biomass pyrolysis processes: review of scope, control and variability, UK Biochar Research Centre Report, 2009.
[93] L. Dai, et al., Integrated process of lignocellulosic biomass torrefaction and pyrolysis for upgrading bio-oil production: a state-of-the-art review, Renew. Sustain. Energy Rev. 107 (2019) 20–36. no. February.
[94] T. Kan, V. Strezov, T.J. Evans, Lignocellulosic biomass pyrolysis: a review of product properties and effects of pyrolysis parameters, Renew. Sustain. Energy Rev. 57 (2016) 1126–1140.
[95] S. Wang, G. Dai, H. Yang, Z. Luo, Lignocellulosic biomass pyrolysis mechanism: a state-of-the-art review, Prog. Energy Combust. Sci. 62 (2017) 33–86.
[96] J. Holladay, J. White, A. Aden, J. Bozell, Top value added chemicals from biomass volume I — results of screening for potential candidates from sugars and synthesis gas top value added chemicals from biomass volume I: results of screening for potential candidates (2004).
[97] J.E. Holladay, J.F. White, J.J. Bozell, D. Johnson, Top value-added chemicals from biomass volume II — results of screening for potential candidates from biorefinery lignin (2007), Pacific Northwest National Laboratory (PNNL) and the National Renewable Energy Laboratory (NREL).
[98] J. Baeyens, Q. Kang, L. Appels, R. Dewil, Y. Lv, T. Tan, Challenges and opportunities in improving the production of bio-ethanol, Prog. Energy Combust. Sci. 47 (2015) 60–88.
[99] J.J. Bozell, G.R. Petersen, Technology development for the production of biobased products from biorefinery carbohydrates—the US Department of Energy's 'Top 10' revisited, Green. Chem. 12 (4) (2010) 539–554.
[100] A. Singh, D. Pant, N.E. Korres, A. Nizami, S. Prasad, J.D. Murphy, Key issues in life cycle assessment of ethanol production from lignocellulosic biomass : challenges and perspectives, Bioresour. Technol. 101 (13) (2010) 5003–5012.
[101] R.W. Jenkins, M. Munro, S. Nash, C.J. Chuck, Potential renewable oxygenated biofuels for the aviation and road transport sectors, Fuel 103 (2013) 593–599.
[102] G.R. Wilson, T. Edwards, E. Corporan, R.L. Freerks, Certification of alternative aviation fuels and blend components (2013) Energy & Fuels, ACS. Available from: https://doi.org/10.1021/ef301888b.

[103] S. Dewei, C. Min, C. Haiming, Collagen-immobilized lipases show good activity and reusability for butyl butyrate synthesis, Appl. Biochem. Biotechnol. 180 (5) (2016) 826–840.
[104] M.I. Alam, S. Gupta, E. Ahmad, and M.A. Haider, Integrated bio- and chemocatalytic processing for biorenewable chemicals and fuels. Sustainable Catalytic Processes. 2015, Elsevier. Available from: https://doi.org/10.1016/B978-0-444-59567-6.00006-6.
[105] M.I. Alam, et al., Development of 6-amyl-α-pyrone as a potential biomass-derived platform molecule, Green. Chem. 18 (24) (2016) 6399–6696.
[106] A. Bohre, S. Dutta, B. Saha, M.M. Abu-Omar, Upgrading furfurals to drop-in biofuels: an overview, ACS Sustain. Chem. Eng. 3 (7) (2015) 1263–1277.
[107] S. Quereshi, S. Dutta, T.K. Naiya, Catalytic conversion of lignocellulosic biomass into fuels and value-added chemicals, in: S. Nanda, P.K. Sarangi, D.-V.N. Vo (Eds.), Fuel Processing and Energy Utilization, CRC Press, 2019, pp. 21–40.
[108] E. Ahmad, M.I. Alam, K.K. Pant, M.A. Haider, Catalytic and mechanistic insights into the production of ethyl levulinate from biorenewable feedstocks, Green. Chem. 18 (18) (2016) 4804–4823.
[109] D. Gupta, E. Ahmad, K.K. Pant, B. Saha, Efficient utilization of potash alum as a green catalyst for production of furfural, 5-hydroxymethylfurfural and levulinic acid from mono-sugars, RSC Adv. 7 (67) (2017).
[110] S. Quereshi, E. Ahmad, K.K. Pant, S. Dutta, Insights into the metal salt catalyzed ethyl levulinate synthesis from biorenewable feedstocks, Catal. Today 291 (2017) 187–194.
[111] S. Quereshi, E. Ahmad, K.K. Pant, S. Dutta, Insights into microwave-assisted synthesis of 5-ethoxymethylfurfural and ethyl levulinate using tungsten disulfide as catalyst, ACS Sustain. Chem. Eng. (2019).
[112] E. Ahmad, M.I. Alam, K.K. Pant, M.A. Haider, Insights into the synthesis of ethyl levulinate under microwave and nonmicrowave heating conditions, Ind. Eng. Chem. Res. 58 (35) (2019) 16055–16064.
[113] S. Quereshi, E. Ahmad, K.K. Pant, S. Dutta, Synthesis and characterization of zirconia supported silicotungstic acid for ethyl levulinate production, Ind. Eng. Chem. Res. 58 (35) (2019) 16045–16054.
[114] E. Ahmad, T.S. Khan, M.I. Alam, K.K. Pant, M. Ali Haider, Understanding reaction kinetics, deprotonation and solvation of brønsted acidic protons in heteropolyacid catalyzed synthesis of biorenewable alkyl levulinates, Chem. Eng. J. 400 (2020) 125916. no. January.
[115] S. Quereshi, K.K. Pant, Unfolding the role of molybdenum disulfide as a catalyst to produce platform chemicals from biorenewable resources, Biomass Conversion and Biorefinery (2020). Available from: https://doi.org/10.1007/s13399-020-00888-7. Springer.

Section II

Biomass Energy Technology and Biorefinery

The second- and third-generation biofuel technologies: comparative perspectives

2

Preetha Ganguly[1], *Rwiddhi Sarkhel*[1] *and Papita Das*[1,2]
[1]Department of Chemical Engineering, Jadavpur University, Kolkata, India, [2]School of Advanced Studies in Industrial Pollution Control Engineering, Jadavpur University, Kolkata, India

Graphical abstract

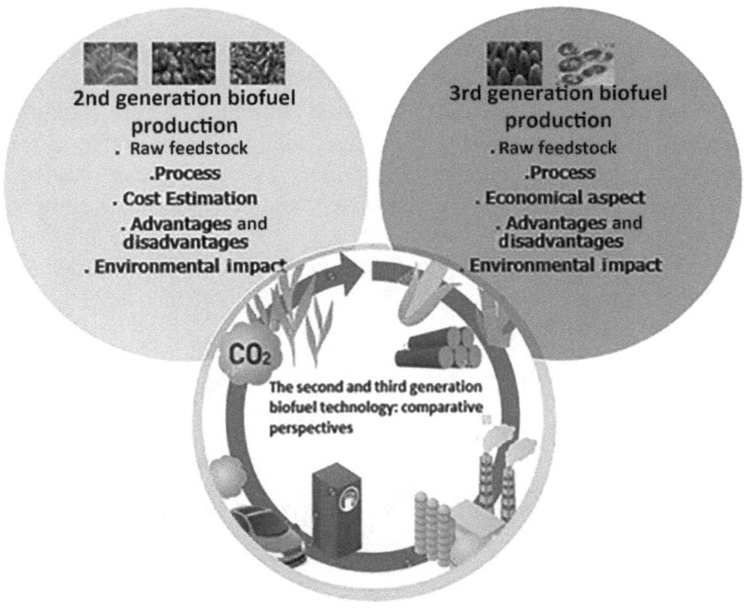

2.1 Introduction

Depletion of present crude oil resources is a key concern from an economical and energy point of view. According to a report submitted by renewable energy policy network [1], in 2011 it is estimated that approximately 78% of energy used

Sustainable Fuel Technologies Handbook. DOI: https://doi.org/10.1016/B978-0-12-822989-7.00002-0

worldwide came from fossil fuel, 3% is derived from nuclear energy, and the rest from renewable resources such as solar, hydrothermal, wind, geothermal, and biomass. Currently, approximately 85 million barrels of fossil fuel oil is refined every year for energy production. At the present rate the need for crude oil will be 116 million barrels annually by 2030. Increasing energy demand may result in the depletion of crude oil reserves around the globe and therefore alternative energy resources are required [2]. The energy obtained from different resources is represented in Fig. 2.1.

The US Department of Energy (DOE) and the US Department of Agriculture (USDA) has made it mandatory that about 15%–20% of liquid transportation fuel, 25% of materials and chemicals, and 5% of energy and heat energy should come from raw biomass feedstock by 2022 [3].

Biochemical and bioenergy generated from nonedible biomass like lignocellulosic feedstock provide various benefits socially and economically [4], including: (1) being sustainable and recyclable; (2) no food versus fuel competition; (3) indirectly help in reducing carbon dioxide emissions into the atmosphere [1] due to the decrease in air pollution from burning feedstock in fields; (6) facilitating the increase of local economic development; (7) creating new opportunities for scientists, researchers, process engineers, fermentation specialists, and engineers; and (8) bringing fuel security for the countries dependent on dwindling crude oil resources.

Very little biofuel are also produced from the noncellulosic raw material by transformation and transesterification of used animal fats and cooking oils [5]. Other types of biofuels are mainly from sugar that can be available in the markets as biofene, bisabolene, and butanol. Biorefinery technology can only be useful if lignocellulosic biomass can be used completely and convert all the components into value-added products [6].

It has been estimated that about 15 billion gallons of this biofuel is extracted from starch-based raw material, while the rest of biofuel that is 21 billion gallons

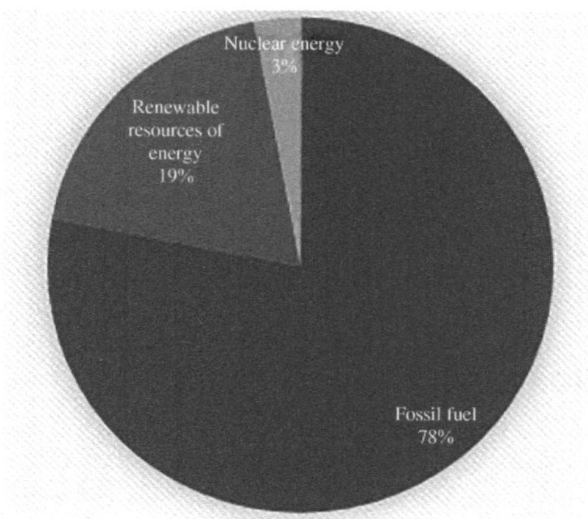

Figure 2.1 Current energy scenario.

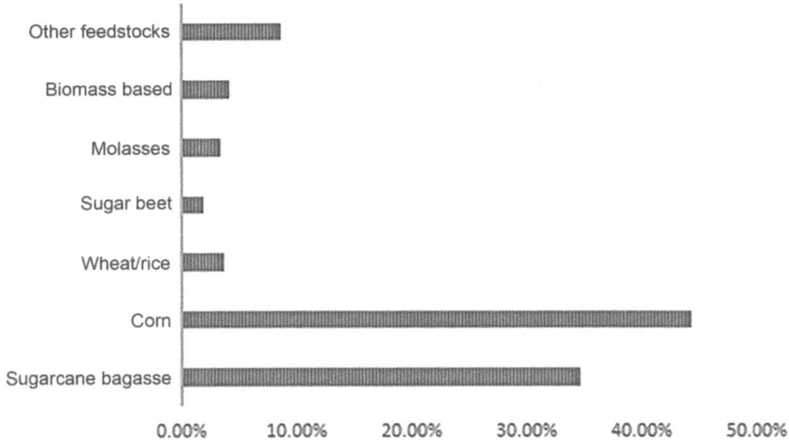

Figure 2.2 Different feedstocks for biofuel production.

(mostly ethanol) are generated by utilizing lignocellulosic biomass [3]. Bioethanol can be produced from various feedstocks as shown in Fig. 2.2.

2.2 Biorefineries

The manufacturing technologies that can convert bio-based feedstocks to various value-added products that include energy, food, materials, chemicals, and feed and are known as biorefinery technologies [7,8]. Petroleum processing is analogous to the biomass refining. In the petroleum refineries, the fossil extracted oil is refined to various by-products such as diesel, kerosene, and petrol and also to various chemical predecessors like butanol (for producing various materials). The energy is obtained by directly burning the feedstock or by thermochemical conversion of feedstock. Presently biofuel generation is done by utilizing sugar-rich agricultural feedstock or by hydrolyzing starch into simple sugars mainly derived from edible food grains. The sugar extracted is further fermented into bioethanol using bacterial or yeast strains [1].

An example of the biorefinery concept is the production of bioethanol that can be used as an additive to gasoline. Bioethanol production depends on the substrate used, for example, lignocellulosic biomass, starch-based biomass, or algae. Lignocellulosic ethanol has gained immense importance because of its dependence on inexpensive and renewable feedstock. It has also been estimated that about billion tons of raw biomass are available for biofuel generation [9]. Microalgae, on the other hand, can accumulate approximately 60% of oil per dry weight of the microalgae under nonsuitable or stressful conditions [10]. Therefore algae have been attractive sources for the potential generators of biodiesel or bioethanol and also

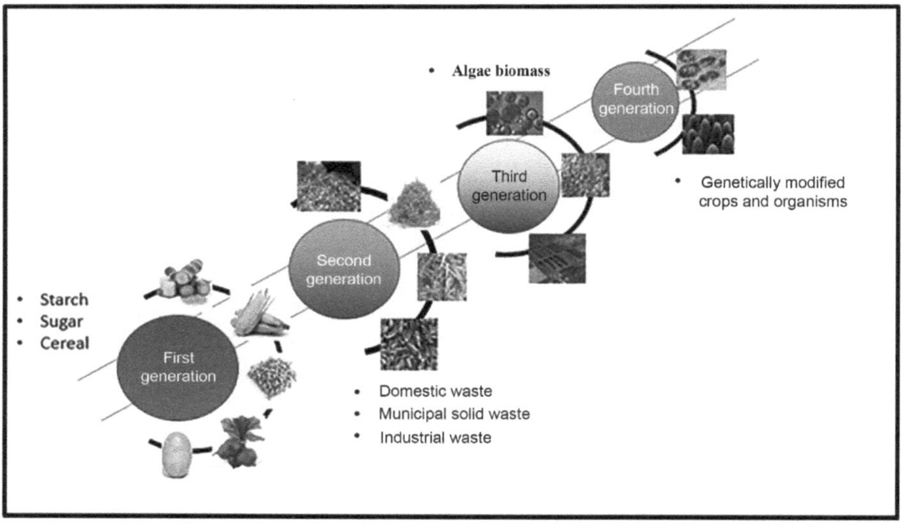

Figure 2.3 Different biorefineries.

other types of biofuels [11–13]. The different generations for the production of biofuel are depicted in Fig. 2.3.

2.3 First-generation biorefineries

The raw material used in the first-generation biorefineries includes sugarcane, barley, rye, wheat, corn, cassava, soybean, sweet sorghum, sugar beet, or sweet sorghum. In first-generation biorefineries the sugar from the sweet sorghum, sugarcane, and sugar beet were extracted utilizing water [14]. These sugars were then subsequently processed utilizing either chemical or biological catalyst such as propionic acid, lactic acid, poly (3-hydroxyalkanoate acid) (PHA/PHB), or fuels (ethanol, butanol methanol) [15]. The process of bioethanol production has been represented in Fig. 2.4. However, the processing of feedstock such as wheat, soybean, rye, or corn needs additional refining after which biofuel chemicals are generated.

Among all the biofuels, bioethanol is the major biofuel around the globe and is increasing at a rate of about 20%/year only in the United States alone [16]. The other leading ethanol-producing countries include Brazil, France, Canada, Japan, and China. The feedstocks utilized in these countries are mainly wheat, sweet sorghum, and cassava juices. In addition to biofuel, the other by-products extracted include food or animal feed [17,18]. Some of the other important bio-based materials generated in 1st generation biorefinery are dyes, polymers, adhesives, detergents, cardboard, paper, sorbents, cleaning compounds, and paint pigments [19–21]. Despite the high production of biofuel along with the other value-added products, the first-generation biorefineries became unpopular because of their

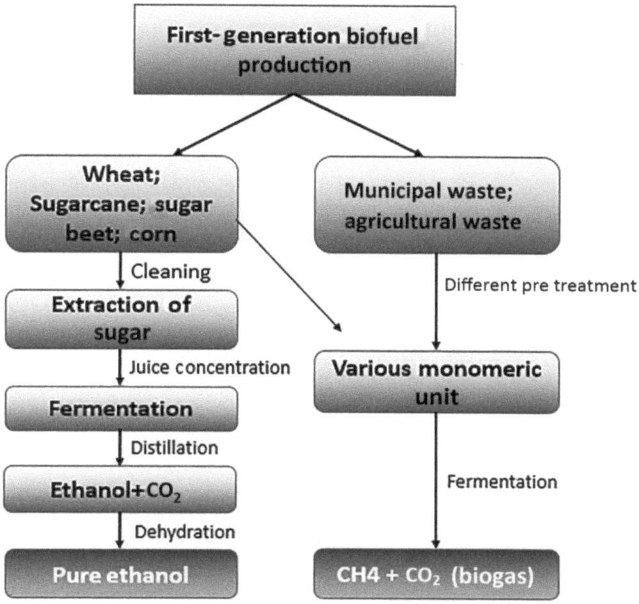

Figure 2.4 Process of biofuel production in first-generation biorefinery.

dependence on the edible food reserves. The first-generation biorefineries are creating a food versus energy competition [22] and therefore an alternative was needed and led to the advancement to second-generation biorefineries.

2.4 Second-generation biorefineries

Generating biofuel from edible materials has major problems due to food scarcity. Hence developing second-generation technologies to generate biofuel from nonedible food materials is ongoing [23]. In the BTL conversion process, the first step is the subjection of feedstock to gasification or pyrolysis to produce synthetic gas or syngas (such as the mixture of hydrogen and carbon monoxide). These gasses can be reformed to biofuel either by biological conversion or by a catalytic process such as the Fischer–Tropsch reactions. The other biochemical route for biofuel generation includes the conversion of the sugar polymers present in the feedstock (cellulose or hemicellulose part of the plant cell) to monosaccharides [24]. The third and the final route is a hybrid pathway whereby utilizing biochemical process chemical intermediate is produced which are finally transformed into high value-added products using a thermochemical pathway. Some of the observable funding and investments are made by the European Union (EU) and US Union. These funding agencies are establishing various bioenergy centers to perform dedicated research for the generation of cost-effective biofuels [25].

A recent report estimated that the cost for bioethanol generation shown in the second-generation biorefineries is twice or thrice more expensive than that of fossil fuel on an energy equivalent basis [26]. Various challenges for transforming lignocellulosic feedstocks into biofuel and other value-added products utilizing biochemical platforms are needed [27–29] to minimize the cost of production. The challenges are mainly confined to the areas of: (1) raw material production; (2) feedstock organization; (3) development of energy and cost-effective technologies (biological or chemical hydrolysis and microbial fermentation); (4) value-added product generation; (5) biofuel distribution; (6) cost-effectiveness; (7) social acceptance; and (8) environmental impact.

2.4.1 Raw feedstock for bioethanol production

Second-generation biorefineries mainly include the lignocellulosic biomass as their feedstock, but can also rely on the use of industrial waste products such as crude glycerol and whey. This lignocellulosic biomass is relatively inexpensive and found in abundance [30]. Lignocellulosic feedstock is considered a sustainable and renewable resource of carbon and is mainly found in raw plant matter [31]. However, the production of total reducing sugar by lignocellulosic conversion is harder than the conversion of starch-based feedstock. These include dedicated crops that may grow in soil grasses such as *Miscanthus giganteus* [32] and *M. sinensis* or switchgrass [33] as well as herbaceous crops. Agricultural by-products, such as wheat straw [34], cereal straw stover [33], rice husk [35], corn cob [36] bagasse, and groundnut shell from the food industry have also shown as potential sources of lignocellulosic feedstock as represented in Fig. 2.5.

From other reports it has been viewed that many researchers are focusing on forest biomass and forest-based woody by-products—hardwood chips (oak) [37], softwood trimmings (pine) [38], sawdust [39], bark [40]—and also waste from gardens and parks such as branches [41], grasses [42], and leaves [36]. Industrial wastes can also be used as potential feedstock and may include spent grains from distilleries [43] and brewers' spent grains [44]—this includes municipal solid wastes containing cellulose as the main component [45]. The available lignocellulosic feedstocks wastes can be utilized for bioethanol generation without any additional land requirement or influence on edible food and fiber crop.

2.4.2 Process of bioethanol production

There are two principal routes for the generation of liquid bioenergy from lignocellulosic feedstock.

Thermochemical conversion

The steps involved are heating the feedstock in the presence of oxygen (in different concentrations) as shown in Fig. 2.6. The major advantage is the complete conversion of the organic components. In this technology pyrolysis/gasification generates

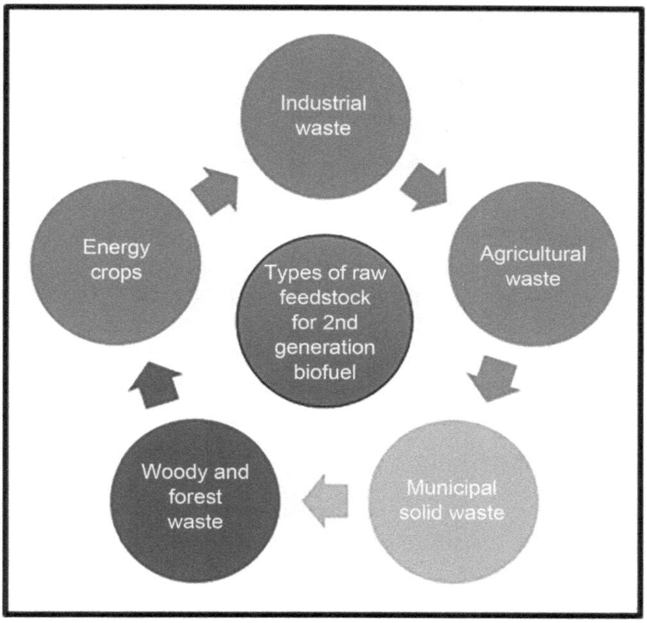

Figure 2.5 Types of raw feedstock for second-generation biofuel.

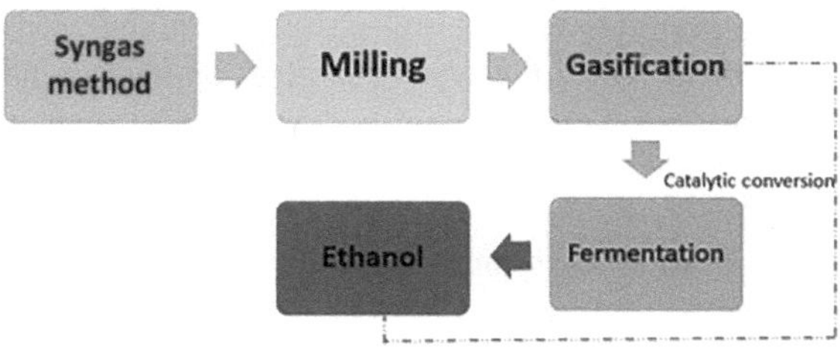

Figure 2.6 Thermochemical method of bioethanol generation.

a synthetic gas (CO + H2) by which long-chain carbon biofuel like ethanol, synthetic diesel, or aviation fuel can be regenerated based on Fischer–Tropsch conversion technology.

Biochemical conversion

In this technique microorganisms and enzymes are utilized for conversion of cellulose and hemicellulose to reducing sugar.

The biochemical conversion technique is the common method for biofuel generation because of the efficiency of feedstock conversion and high selectivity [23,30]. The basic steps involved in liquid biofuel generation is pretreatment of

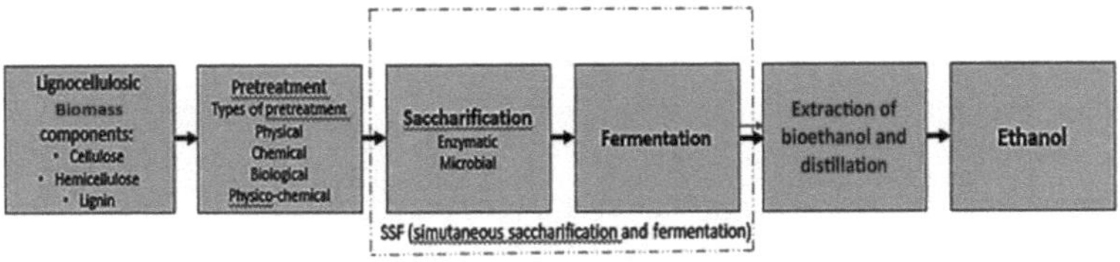

Figure 2.7 Biochemical method for bioethanol generation.

lignocellulosic feedstock, biological/enzymatic hydrolysis, fermentation of the reduced sugar by strains of specific microorganism extraction of bioethanol, and its downstream processing as shown in Fig. 2.7. The biochemical pathway for biofuel generation from lignocellulosic waste will be discussed next.

2.4.2.1 Pretreatment

The crystalline structure of lignin is the barrier for cellulose surface, and also makes the cellulose component of the plant cell recalcitrant to hydrolysis [46]. So the main aim of the pretreatment is to remove the lignin component without harming the main structure of the cellulose [47–49]. The pretreatment step in bioethanol production has become the most important step for the biofuel conversion process.

2.4.2.1.1 Types of pretreatment

The pretreatment of lignocellulosic raw feedstock is mainly divided into biological, chemical, and physical pretreatment. During physicochemical pretreatment, the last two methods are used often. Different types of microorganisms are used for the biological treatment to improve the enzymatic hydrolysis. The main advantage is that it does not generate toxic by-products, is cost-effective, environment-friendly, energy-efficient, and does not need the recycling of chemicals [50]. The release of lignolytic enzymes like laccase and peroxidase takes place while delignification by white rot fungi disrupts the lignin component of the cell wall.

Chemical pretreatment includes treatment with an alkali (base) (e.g., NaOH, KOH), treatment with ionic liquid (green solvent), diluted or concentrated acid (H_2SO_4, HCL), and with oxygen (wet oxidizer). Physical pretreatment reduces the size of the feedstock and is a cost-effective technology. The mechanical method reduces the cell wall and size of the particle by grinding and milling [51]. Physicochemical pretreatment includes liquid hot water (LHW), steam explosion (autohydrolysis), ammonia recycle percolation (ARP), ammonia fiber explosion (AFEX), or processing with supercritical fluid carbon dioxide (CO_2) [51]. The main advantages of steam treatment improve cellulose digestibility and solubilize the hemicellulose component. Steam explosion is a hydrothermal technique that utilizes chemical factors (steam and catalyst) and mechanical factors. The aim of the steam explosion is the dissociation of plant fibers through explosive decomposition under high temperature and pressure for 1–5 min. The most commonly used technique of processing agricultural residue and herbaceous feedstock is steam explosion, but this may contribute to the degradation of polysaccharides, lignin, and hemicellulose into inhibitory substances such as phenolic and levulinic acid [52].

The composition of lignocellulosic feedstock is the main determinant of pretreatment technology. For agricultural and hardwood waste with low lignin content and high pentose level, thermal degradation is not advisable because of the susceptibility of pentose denaturation. Straw and bagasse are processed mainly by steam explosion. Ammonia water and wet oxidation can be used for the pretreatment of sugarcane bagasse.

2.4.2.2 Inhibitors and their impact on microorganisms

The inhibitors may cause serious weakening of the fermentative microorganism, causing their death and a critical environment. The inhibitors can increase the period of the lag phase that may cause the low-growth rate of microbes(fermentation) and loss of cell density that may finally reduce ethanol production [53]. Inhibitors consist of various components, mainly phenolic compounds and weak organic acids. The inhibitors mainly interfere with the integrity and function of the cell membrane [54]. Accumulation of anions of weak acids may inhibit the growth of *Saccharomyces cerevisiae* cells. The microbes affected by the furan derivatives have shown low-specific growth rates, lesser volumetric ethanol productivity, and lower cell mass [55]. Detoxification is required for their removal and there are various techniques. The detoxification methods include ion exchange, molecular sieve, over-liming extraction with an organic solvent, and steam stripping [56]. Overcoming of the hydrolysates disrupts the inhibitor and causes the inhibitory substance to precipitate. The detoxification of these lignocellulosic inhibitors can be achieved by using activated carbon, lignite, or an abundance of lime.

2.4.2.2.1 Enzymatic hydrolysis

The main aim of enzymatic hydrolysis is the release of monomeric sugar units from polysaccharides present in the plant cell wall. The quality and quantity of the sugar yield after the hydrolysis step depends on the raw feedstock utilized and the pretreatment technique applied [57]. The main carbon source for the microorganism responsible for fermentation is a monosaccharide. After the pretreatment and enzymatic hydrolysis, the main sugars broken-down from the raw feedstock are glucose and xylose. The hydrolysis is of two types, namely biological or chemical. In chemical hydrolysis, the enzyme is extracted from the microorganism and utilized further. The main types of hydrolytic enzymes are cellulose that digests the cellulosic part of the biomass. The raw biomass is difficult to digest by enzymes, therefore a high amount of hydrolytic enzymes are required [58].

Trichoderma reesei can produce exo-cellobiohydrolase (CBH), β-glucosidase (BGL), endoglucanase (EG), and specifically cellulases [59]. The internal amorphous cellulose regions are attacked by EGs resulting in depolymerization of the cellulose structure. The further breakdown of the β-glucan molecules into cellobiose units is carried by the function of exoglucanase. The β-glucosidase attacks the cellobiose and breaks it into two glucose units [60]. The cost of enzyme cellulases is high, so reutilization of this enzyme may be considered reduce the cost [23]. An alternative to the above technique is the direct utilization of the microorganism. In this technology, the downstream processing cost will be lower as microorganism will be used in place of enzyme. The utilization of hemicellulases like exoxylanases, endoxylanases, xylosidases, and other enzymes such as arabinosidases and esterases can enhance the enzymatic hydrolysis of the raw lignocellulosic biomass. The complete process of hydrolysis becomes effective when the cellulases, hemicellulases, and other accessory enzymes interact with each other [61].

2.4.2.3 Fermentation technique of lignocellulosic hydrolysates

The main objective of the process is too effectively covert the pentose and hexose sugars to bioethanol utilizing microorganisms such as yeasts and bacteria. Susceptibility toward lignocellulosic hydrolysate is found in the ethanol-producing microbes depending on the different parameters such as temperature, pH, nutrient supply, aeration, and the microbe strain. The poor yield of lignocellulosic hydrolysate is determined in the microbial strains that can effectively ferment the xylose sugar. Fermentation may lead to stressful situations such as a shift toward anaerobic growth condition, limitation in oxygen and nutrients, alcohol accumulation, and a decrease in pH [62]. Yeast growth and viability can be inhibited by osmotic stress due to ethanol accumulation and sugar and ion in hydrolysate [63]. Osmotic shock may be due to the high level of sugar during the fermentation process [64,65]. The performance of alcohol fermentation can be negatively influenced by a severe condition such as ethanol stress and oxidative.

2.4.2.4 Enzymatic saccharification and fermentation

Enzymatic saccharification of hemicellulose and cellulose in the presence of a microbial strain can be executed separately (separate saccharification and fermentation) (SHF), or simultaneously (hydrolysis and fermentation) (SSF). The SSF method of ethanol production is very cost-effective. Significant improvement in the technology is made with regards to reduce the inoculum volume, increase the feedstock loading, and confirm the cofermentation of the pentose and hexose sugar during SSF fermentation [66,67]. Mesophilic yeast strains show a lower growth rate at high temperatures than thermophilic and thermotolerant yeasts. Therefore for the process of SSF fermentation ethanologenic and thermotolerant yeast strain is suggested. Utilizing genetically modified microbial strains this problem can be overcome as they can ferment at the higher temperatures. Some of the microbial strains found to be thermophilic include *Candida*, *Saccharomyces*, *Wickerhamomyces*, and *Pichia*.

The main advantage of SSF fermentation is that feedback inhibition can be prevented because the sugar molecule produced after the process of hydrolysis is immediately fermented by the microbial yeast strains. The other advantages of SSF include the microbial contamination is decreased as the sugar is fermented simultaneously to ethanol. Two separate vessels are utilized for the SHF process, which makes this process uneconomical. It may also cause feedback inhibition in the end product and thereby harm hydrolysis. Simple carbohydrates can cause end-product inhibition.

2.4.3 Cost estimation

2.4.3.1 Production costs

A common characteristic that separates first-generation biofuel production from the second is in turn of high capital cost per unit of generation of fuel. In the case of first-generation biofuel production, the capital cost is higher than second generation fuel. The low cost of the raw material will somewhat reduce the high cost of production. For second-generation biofuel production via the biochemical route, some necessary points included are advances in engineering processes and microorganisms, as well as low cost raw feedstock such as bagasse, switchgrass, and sawdust that are estimated to be the prime competitors for bioethanol generation in next 10–20 years. Cellulosic bioethanol generation with presently known technologies in various developing countries is advantageous because of the possibility of the lower cost of sustainable biomass production.

By utilizing thermochemical systems for second-generation biofuel production, it is estimated that with further demonstration and development efforts this method will be commercially utilized for ethanol production within a few years. The thermochemical route for bioethanol production is more economical than commercial oil prices. In developing countries where biomass harvesting cost is lower (due to climatic conditions) and where labor and construction costs are less, thermochemical bioethanol is more economically competitive.

2.4.3.1.1 Cost prediction

Observations were made that notable reduction is presumed in all the three major costs which include conversion, enzyme and feedstock cost for second-generation biofuel production. For the longer term production of bioethanol, it is needed to reach both the raw feedstock cost and large-scale production cost below corn-ethanol costs. The most challenging task is the large supply volume of the feedstock at a cost below US$30/ton. The NREL laboratory assumed that by the end of 2012, the raw feedstock and process costs will be reduced by 40%, whereas the enzymatic cost will decrease by 70%. From these estimations it can be concluded that lignocellulosic bioethanol production has become competitive with fossil fuel costs. Currently it is competitive for US$100/barrel but in 2031 it has been estimated to be competitive at a crude cost of US$75.

The processing or conversion cost reductions could be a result of different efficiency attempts. The largest of these reductions in cost can come from different aspects such as efficiencies in different pretreatment methods, integrated bioprocessing methods, and optimal utilization of cellulose enzyme cost. It has been expected that by reducing energy utilization, decreasing feedstock requirements, initial capital cost, and enzymatic cost it can be possible to reach the target-cost value. Based on these data and utilizing different reference sources, the cellulosic bioethanol cost can be estimated as shown in the following table.

2.4.4 Advantages and disadvantages

Advantages	Disadvantages
Use of nonfood feedstock	Not produced on commercial scale
Environment-friendly and produces less greenhouse gases	Current transport, harvesting, and storage systems are inadequate for processing biomass on a commercial basis
No competition for land	Extensive processing of the raw feedstock is needed to make ethanol
Does not produce any coproducts	Not suitable for biodiesel production
Can be grown on marginal lands	
High net energy yields up to 540%	
No, or less, need for fertilizers	

2.4.5 Environmental aspect

After the combustion of bioethanol the carbon dioxide released is reused by plant material because plants recycle the carbon dioxide to synthesis cellulosic components during photosynthesis.

The toxic gases released after the combustion of petroleum sources are much less for bioethanol [68]. Bioethanol derived from lignocellulosic biomass on combustion does not produce harmful gases therefore do not add to greenhouse gas (GHG). [69]. The main advantage of utilizing lignocellulose-derived bioethanol is the reduction of GHG pollution [7] as shown in Fig. 2.8. The complete combustion of bioethanol is possible because it constitutes 35% of the oxygen that reduces particulate outflow that, in turn, reduces the health hazard. Research conducted by Bang-Quan et al. in the year 2003 on bioethanol-blended diesel concluded that the addition of ethanol to diesel simultaneously reduces the high-heating value, octane number, kinematic viscosity, and aromatics fractions. These factors conduct the complete combustion of bioethanol with less emission of various GHG. Subramanian et al. [70] in their research study concluded that bioethanol blended to diesel significantly reduces 41% in particulate matter and approximately 5% in NO_x emission. One of the major disadvantages of utilizing ethanol as fuel is that acetaldehyde emissions are at a higher rate although it has a less adverse effect on human health than those of formaldehyde emitted from gasoline. Life-cycle assessment and the environmental impact of bioethanol generation is a theoretical methodology and framework for the evaluation of the production system on environmental factors. Lignocellulosic feedstock utilization to bioethanol offers various environmental benefits in terms of global warming impact and nonrenewable energy consumption. However, bioethanol production from lignocellulosic feedstock also tends to increase

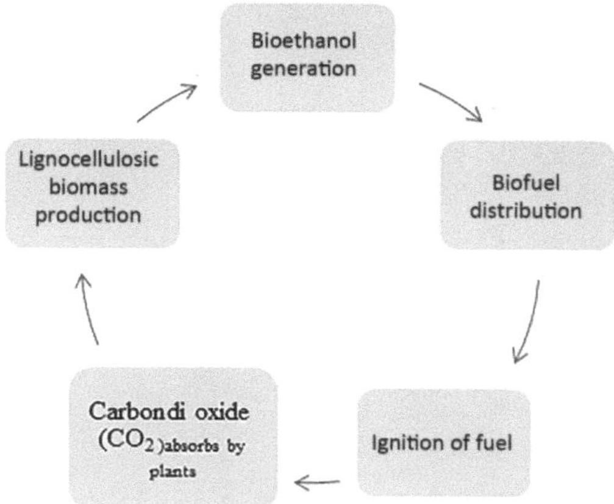

Figure 2.8 CO_2 cycle for bioethanol.

eutrophication and acidification because a large amount of phosphorus and nitrogen are emitted after farming crops.

2.5 Third-generation biorefineries

Third-generation bioethanol (TGB) or biofuels can be defined as the fuels that can be generated from algal biomass [71,72]. Algae comprises a diversified group of microorganisms including microalgae, macroalgae (seaweed), and cyanobacteria (formerly known as blue-green algae). TGB has been produced from the potential feedstock that can be directly converted into energy [73] Usually bioethanol production from algae depends on various factors such as technology and marine environment [74,75]. Biofuels produced from the algae depend on the lipid content of the microorganisms [76].

There are more than 72,000 species of algae [77]. Algae may grow in freshwater, wastewater, or brackish water. The major advantages of algae in biofuel production includes, they can grow rapidly under harvesting cycles (measured in days), and need less nutrients to grow [78]. Algae can grow naturally in different types of environments. The rate of growth can be varied by increasing the levels of different growth conditions, such as temperature, acidity, and enrichment of nutrient levels. Algae generating high-energy yield is considered as a raw material for third-generation biofuels [79]. Many products such as ethanol, butanol, biodiesel, jet fuel, syngas, bio-oil, a wide variety of chemical feedstocks, and natural fertilizers can be produced from algal biomass. The combustion of algal biomass can generate electricity.

2.5.1 Technique of algal oil extraction

Algae can be grown in various environments and conditions as depicted in Fig. 2.9.

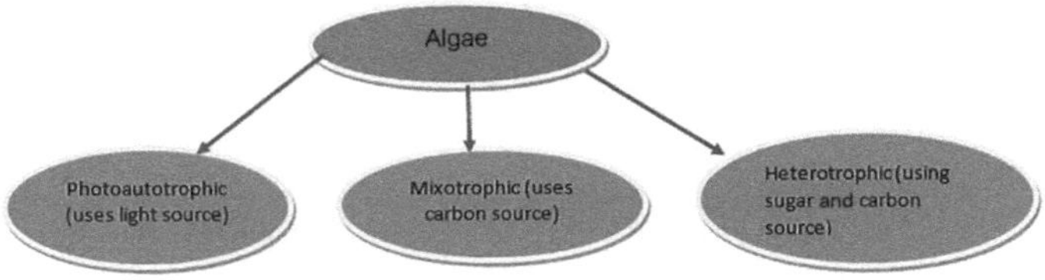

Figure 2.9 Categorization of algae under various growth conditions.

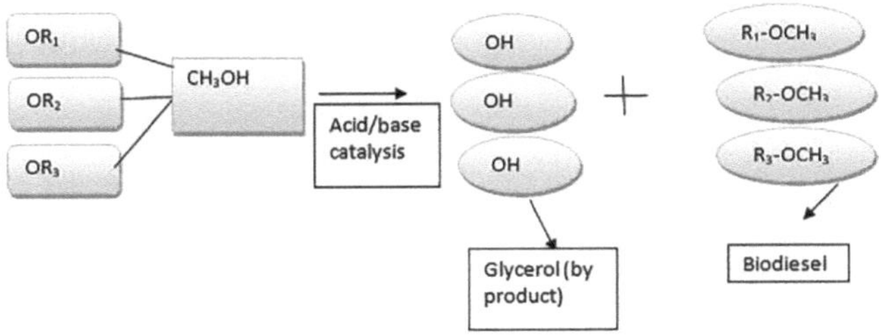

Figure 2.10 Structure of triacylglycerol and its conversion to biodiesel via acid- or base-catalyzed transmethylation reaction.

2.5.1.1 Extraction of oil from algal biomass

Cell walls comprising of lipids and fatty acids have been produced from the unicellular microalgae. Different methods of oil extraction such as mechanical and solvent extraction can be used to extract oil [80]. The structure of triacylglycerol and its conversion to biodiesel via acid- or base-catalyzed transmethylation reactions is shown in Fig. 2.10.

2.5.1.2 Conversion of algae to biofuels

Conversion of algae to biofuels or bioethanol requires different options, such as:

1. algal biomass conversion to energy products,
2. metabolites extraction, and
3. algal secretions are directly collected and then processed.

Third-generation biofuel initiatives focus on the extraction of metabolites, whereas fourth-generation biofuel initiatives focus on the algal secretions. The process of biofuel production from algae is represented in Fig. 2.11.

2.5.1.3 Steps in production of biofuels from algae

The processing steps to produce biofuels from algae after the extraction methods are applied for the conversion of algae to liquid fuels include harvesting, dewatering, extraction, and processing to energy products and coproducts [81]. The composition of the algal oil part present in the biomass is around 30% and the rest is an

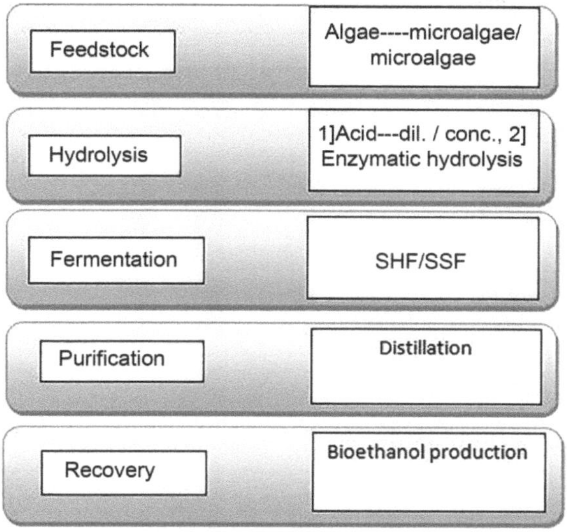

Figure 2.11 The general description of processes in bioethanol production from algae.

algal by-product. This by-product can be further utilized as feedstock nutrients for animals, fish, and water plants, as well as for pharmaceutical ingredients, cosmetics, toiletries, and fragrant products [82]. Various types of renewable biofuels could be obtained and further bio-oil and syngas production is possible [83]. There are three different ways of conversion technologies: biochemical, chemical, and thermochemical conversion to utilize algal biomass to energy sources as shown in Fig. 2.12.

2.5.2 Environmental impact

Biofuels produced from renewable organic material, has created keen interest for the researchers to alternate the crude fossil fuel. Biofuels have the ability in reducing some undesirable aspects of production and usage of fossil fuels. This includes GHG pollutant emissions and depletion of resources in limited amount. The constantly increasing demand for biofuels might also increase income of agricultural farms because of the production of crops like soybeans which has a high yield of oil production. However, biofuel production from first and second generation biomass may also give rise to several undesirable effects. The flaws in the generation of biofuel include a change in land-use patterns that increase GHG emissions, air and water pollution, and the rate of increase in the food costs [84]. Bioethanol is capable of substituting a petroleum-derived product for running engines, gasoline. But, the utmost challenge in the bioethanol industry is the need for establishing a process with an environment-friendly approach and an economically feasible process of production. Due to the effect on the food security and production process, first- and second-generation bioethanol are nonviable. Bioethanol from micro- and

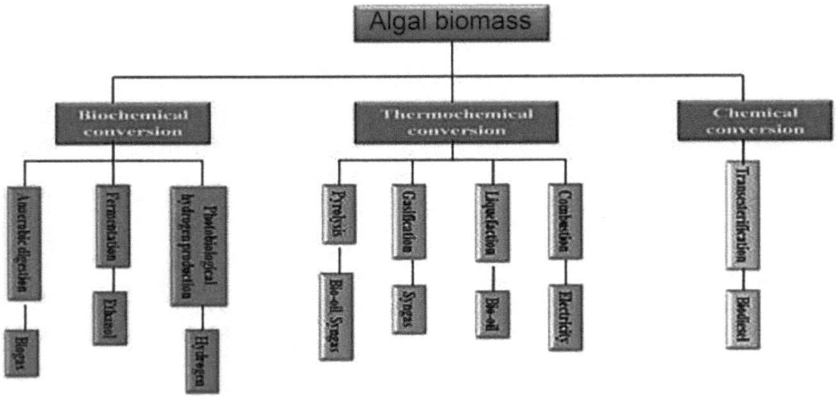

Figure 2.12 Algal biomass conversion process for biofuel production.

macroalgae commonly can be represented as TGB, emerge as an alternative for replacing the existing fossil fuels. These problems and concerns governed the TGB feedstock from a marine microorganism (i.e., algae). The combination of two types of algae (microalgae and macroalgae) serves as a sustainable feedstock for bioethanol gaining worldwide attention and an increase in environmental impact. The utilization of algae to produce bioethanol has evolved as research and has increased in recent years and is expected to become the major drives in the bioethanol industry.

2.6 SWOT analysis

	Second-generation bioethanol	Third-generation bioethanol
Strength	Feedstock costs are low	Algal biofuel possesses very low to approximately zero CO_2 emission
	The feedstock used for second-generation biofuel production is not part of the food chain, therefore there is no food versus crops competition	Algae are useful for carbon apprehend and sequestration
	Latest cultivation techniques, like mixed cropping, are very advantageous for biodiversity	Algae can produce different kinds of value-added products (such as oxygen), useful for various purposes by humans
	Perennial plants are used (instead of annual plants for starch and sugar) this prevents soil erosion and is useful for groundwater conservation	The adaptability of algae to different types of growing locations, media, and aquatic bodies (freshwater, brackish and marine environment)
	Combustion of bioethanol generates low CO emissions	Algae are ubiquitous in nature

	Ethanol can be blended with gasoline	The biofuel generated from algae have higher Octane Number (CN)
Weakness	No experiences for large-scale production of bioethanol from lignocellulosic waste	The commercial scale of production is still relatively expensive. Unresolved environmental and social issues are still a problem because of inadequate funding
	Bioethanol processing from cellulose is energy-intensive	Obtaining desired species attributes such as harvest and lipid is a critical step
Opportunities	Research carried out for optimizing the process of bioethanol production from cellulose	Can provide small scale business opportunities
	Due to the utilization of the entire plant, higher yields per hectare are expected	The mixing of biofuel with petroleum fuel will consequently reduce the sulfur content of the fuel and in turn, create better lubrication quality and lower aromatic properties
Threat	Limited land is available for cultivating	The capital and operational costs are high
	In major European countries, the use of bioethanol is not yet established therefore limited infrastructure for distribution of ethanol exists	The market acceptability is yet not certain

2.7 Conclusions

Ethanol production from sugarcane molasses or bagasse is leading production in the market thus reducing biofuel production in one of the ethanol-producing countries like Brazil. Production of ethanol from corn is also paradoxically similar to the increasing value of food on the world's market. In each case, the feedstock is getting very expensive, leading to an increase in keen interest toward the second-generation biofuels that are produced from solid, forest, agricultural, animal, or municipal wastes. Various techniques are being researched around the world to produce second-generation biofuels, but they rely only on two distinctive pathways: either thermochemical or biochemical pathway. The thermochemical pathway relates to the thermal treatment of biomass at a lower concentration of oxidizing agent (either carbon dioxide, water, air, or oxygen). Temperatures ranging from 300°C to 1000°C, produces a solid biofuel, biochar. Syngas and gasoline may be produced at high temperatures. The biochemical pathway is based on the use of microorganisms (mainly yeasts) for the conversion of biomass into biofuels. The process requires pretreatment for the conversion of cellulosic macromolecules into biomass. Although the biofuel production from second-generation processes has proved efficient, the biochemical approach completely utilizes the generated carbon content from the lignocellulosic biomass to produce a wide range of biofuels and other bio-based commodities. Third-generation biofuels are mostly related to algae and the production of algal oils and the extraction of ethanol from algal biomass. The future of biofuels may be a combination of the three generations to cope with the worldwide demand due to the reduction in the world's oil resources.

Acknowledgment

All the authors are thankful to Jadavpur University for the support at each level of the review. P.D. and R.S. are thankful to RUSA II for financial assistance for executing the project work, and P.G. is thankful to TEQUIP (TEQUIP phase III), Jadavpur University for financial assistance.

References

[1] Mohr, S. Raman, Lessons from first generation biofuels and implications for the sustainability appraisal of second-generation biofuels, Energy Policy 63 (2013) 114–122.
[2] R. Lee, J.M. Lavoie, From first- to third-generation biofuels: challenges of producing a commodity from a biomass of increasing complexity, Anim. Front. 3 (2) (2013) 6–11.
[3] R.D. Perlack, L.L. Wright, A.F. Turhollow, R.L. Graham, B.J. Stokes, D.C. Erbach, Biomass as Feedstock for a Bioenergy and Bioproducts Industry: The Technical Feasibility of a Billion-Ton Annual Supply, U.S. Department of Energy & U.S. Department of Agriculture, Oak Ridge, TN, 2005.
[4] H.C. Greenwell, M. Loyd-Evans, C. Wenner, Biofuels, science and society, Interface Focus. 3 (2012) 1–4.
[5] T.M. Mata, A.A. Martins, N.S. Caetano, Valorisation of waste frying oils and animal fats for biodiesel production, Adv. Biofuel. Bioprod., 4, 2013, pp. 671–693.
[6] Rabinovitch-Deere, J.W.K. Oliver, G.M. Rodriguez, S. Atsumi, Synthetic biology and metabolic engineering approaches to produce biofuels, Chem. Rev. 113 (7) (2013) 4611–4632.
[7] Demirbas, Biofuels securing the planet's future energy needs, Energy Convers. Manage. 50 (9) (2009) 2239–2249.
[8] S.N. Naik, V.V. Goud, P.K. Rout, A.K. Dalai, Production of first and second-generation biofuels: a comprehensive review, Renew. Sustain. Energy Rev. 14 (2) (2010).
[9] H. Li, M. Wu, L. Xu, J. Hou, T. Guo, X. Bao, et al., Evaluation of industrial Saccharomyces cerevisiae strains as the chassis cell for second-generation bioethanol production, Microb. Biotechnol. 8 (2) (2015) 266–274.
[10] Y. Chisti, Biodiesel from microalgae, Biotechnol. Adv. 25 (2007) 294–306.
[11] L.L. Beer, E.S. Boyd, J.W. Peters, M.C. Posewitz, Engineering algae for biohydrogen and biofuel production, Curr. Opin. Biotechnol. 20 (2009) 264–271.
[12] S.A. Scott, M.P. Davey, J.S. Dennis, Biodiesel from algae: challenges and prospects, Curr. Opin. Biotechnol. 21 (2010) 277–286.

[13] R.H. Wijffels, M.J. Barbosa, An outlook on microalgal biofuels, Science 329 (2010) 796–799.
[14] M.A. Martin, First generation biofuels compete, N. Biotechnol. 27 (5) (2010) 596–608.
[15] S.T. Yang, H.A. El-Enshasy, N. Thongchul (Eds.), *Bioprocessing Technologies in Integrated Biorefinery for Production of Biofuels, Biochemicals and Biopolymers from Biomass,* 2013, John Wiley & Sons, New York, p. 1.
[16] G.R. Timilsina, A. Shrestha, How much hope should we have for biofuels? Energy 36 (4) (2011) 2055–2069.
[17] R.J. Bothast, M.A. Schlicher, Biotechnological processes for conversion of corn into ethanol, Appl. Microbiol. Biotechnol. 67 (1) (2005) 19–25.
[18] K.A. Rosentrater, A review of corn masa processing residues: generation, properties, and potential utilization, J. Waste Manag. 26 (3) (2006) 284–292.
[19] F. Cherubini, S. Ulgiati, Crop residues as raw materials for biorefinery systems—a LCA case study, Appl. Energ. 87 (1) (2010) 47–57.
[20] B. Kamm, P.R. Gruber, M. Kamm (Eds.), *Biorefineries Industrial Processing and Products*, **2**, 2006, Wiley-VCH, New York, 97.
[21] B. Kamm, M. Kamm, P. Gruber, S. Kromus, Biorefinery systems—an overview. *In: Biorefineries-industrial Processes and Products: Status Quo and Future Directions* **1**, 2006, 1–40.
[22] V.B. Agbor, N. Cicek, R. Sparling, A. Berlin, D.B. Levin, Biomass pretreatment: fundamentals toward application, Biotechnol. Adv. 29 (2011) 675–685.
[23] R.E.H. Sims, W. Mabee, J.N. Saddler, M. Taylor, An overview of second-generation biofuel technologies, Bioresour. Technol. 101 (6) (2010) 1570–1580.
[24] V. Balan, S. Kumar, B. Bals, S.P.S. Chundawat, M. Jin, M. Dale, in: A. Monti (Ed.), Biochemical and Thermochemical Conversion of Switchgrass to Biofuels in Switchgrass: A Valuable Biomass Crop for Energy, 7, Springer, London, 2012, pp. 153–186.
[25] V. Balan, D. Chiaramonti, S. Kumar, Review of US and EU initiatives towards development, demonstration and commercialization of lignocellulosic biofuels, Biofuels Bioprod. Bioref. 7 (6) (2013) 732–759.
[26] M.A. Carriquiry, X. Du, G.R. Timilsina, Second generation biofuels: economics and policies, Energy Policy 39 (7) (2011) 4222–4234.
[27] S.K. Hoekman, Biofuels in the U.S.—challenges and opportunities, Renew. Energy 34 (1) (2009) 14–22.
[28] V. Menon, M. Rao, Trends in bioconversion of lignocellulose: biofuels, platform chemicals & biorefinery concept, Prog. Energy Combust. Sci. 38 (4) (2012) 522–550.
[29] L. Luo, E. Van der Voet, G. Huppes, Biorefining of lignocellulosic feedstock—technical, economic and environmental considerations, Bioresour. Technol. 101 (13) (2010) 5023–5032.
[30] S. Achinas, G.J.W. Euverink, Consolidated briefing of biochemical ethanol production from lignocellulosic biomass, Electron. J. Biotechnol. 23 (2016) 44–53.
[31] J.H. Kim, D.E. Block, D.A. Mills, Simultaneous consumption of pentose and hexose sugars: an optimal microbial phenotype for efficient fermentation of lignocellulosic biomass, Appl. Microbiol. Biotechnol. 88 (5) (2010).
[32] W.C. Lee, W.C. Kuan, Miscanthus as cellulosic biomass for bioethanol production, Biotechnol. J. 10 (6) (2015) 840–854.
[33] V.S. Chang, W.E. Kaar, B. Burr, M.T. Holtzapple, Simultaneous saccharification and fermentation of lime-treated biomass, Biotechnol. Lett. 23 (16) (2001) 1327–1333.

[34] C. Hongzhang, L. Liying, Unpolluted fractionation of wheat straw by steam explosion and ethanol extraction, Bioresour. Technol. 98 (3) (2007) 666–676.
[35] B.C. Saha, L.B. Iten, M.A. Cotta, Y.V. Wu, Dilute acid pretreatment, enzymatic saccharification, and fermentation of rice hulls to ethanol, Biotechnol. Prog. 21 (3) (2005).
[36] E.I. Melekwe, S.A. Lateef, G. Rowland, E. Ekpeyong, Bioethanol production potentials of corn cob, waste office paper and leaf of *Thaumatococcus daniellii*, Br. J. Appl. Sci. Technol. 17 (4) (2016).
[37] P. Perego, A. Converti, E. Palazzi, M. Del Borghi, G. Ferraiolo, Fermentation of hardwood hemicellulose hydrolysate by *Pachysolen tannophilus*, *Candida shehatae* and *Pichia stipitis*, J. Ind. Microbiol. 6 (3) (1990).
[38] M. Galbe, G. Zacchi, A review of the production of ethanol from softwood, Appl. Microbiol. Biotechnol. 59 (6) (2002) 618–628.
[39] R.B. Stoffel, P.V. Neves, F.E. Felissia, L.P. Ramos, L.M. Gassa, M.C. Area, Hemicellulose extraction from slash pine sawdust by steam explosion with sulfuric acid, Biomass Bioenergy 107 (2017) 93–101.
[40] B. Frankó, M. Galbe, O. Wallberg, Influence of bark on fuel ethanol production from steam-pretreated spruce, Biotechnol. Biofuels 8 (2015) 15.
[41] C. Sasaki, R. Okumura, C. Asada, Y. Nakamura, Steam explosion treatment for ethanol production from branches pruned from pear trees by simultaneous saccharification and fermentation, Biosci. Biotechnol. Biochem. 78 (1) (2014) 160–166.
[42] S.I. Njoku, J.A. Iversen, H. Uellendahl, B.K. Ahring, Production of ethanol from hemicellulose fraction of cocksfoot grass using Pichia stipitis, Sustain. Chem. Process. 1 (2013) 13.
[43] W. Kricka, T.C. James, J. Fitzpatrick, U. Bond, Engineering Saccharomyces pastorianus for the co-utilisation of xylose and cellulose from biomass, Microb. Cell Fact. 14 (2015) 61.
[44] S. Wilkison, K.A. Smart, S. James, D.J. Cook, Bioethanol production from brewers spent grains using a fungal consolidated bioprocessing (CBP) approach, BioEnergy Res. 10 (1) (2017) 146–157.
[45] J. Prasetyo, K. Naruse, T. Kato, C. Boonchird, S. Harashima, E.Y. Park, Bioconversion of paper sludge to biofuel by simultaneous saccharification and fermentation using a cellulase of paper sludge origin and thermotolerant *Saccharomyces cerevisiae* TJ14, Biotechnol. Biofuels 4 (2011) 35.
[46] N. Mosier, C. Wyman, B. Dale, R. Elander, Y.Y. Lee, M. Hotzapple, et al., Features of promising technologies for pretreatment of lignocellulosic biomass, Bioresour. Technol. 96 (2005) 73–686.
[47] M. FitzPatrick, P. Champagne, M.F. Cunningham, R.A. Whitney, A biorefinery processing perspective: treatment of lignocelluosic materials for the production of value—added products, Bioresour. Technol. 101 (2010) 8915–8922.
[48] C. Liu, F. Wang, A.R. Stiles, C. Guo, Ionic liquids for biofuel production: opportunities and challenges, Appl. Energy 92 (2012) 406–414.
[49] C.E. Wyman, B.E. Dale, R.T. Elander, M. Holtzapple, M.R. Ladisch, Y.Y. Lee, Comparative sugar recovery data from laboratory scale application of leading pretreatment technologies to corn stover, Biores. Technol. 96 (18) (2005) 2026–2032.
[50] P. Kumar, D.M. Barrett, M.J. Delwiche, P. Stroeve, Methods for pretreatment of lignocellulosic biomass for efficient hydrolysis and biofuel production, Ind. Eng. Chem. Res. 48 (8) (2009) 3713–3729.

[51] G. Brodeur, E. Yau, K. Badal, J. Collier, K.B. Ramachandran, S. Ramakrishnan, Chemical and physicochemical pretreatment of lignocellulosic biomass: a review, Enzyme Res. (2011).
[52] P. Alvira, M.J. Negro, I. Ballesteros, A. González, M. Ballesteros, Steam explosion for wheat straw pre-treatment for sugars production, Bioethanol 2 (1) (2016) 66–75.
[53] E. Albers, C. Larsson, A comparison of stress tolerance in YPD and industrial lignocellulose-based medium among industrial and laboratory yeast strains, J. Ind. Microbiol. Biotechnol. 36 (8) (2009) 1085–1091.
[54] E. Palmqvist, B. Hahn-Hägerdal, Fermentation of lignocellulosic hydrolysates. II: inhibitors and mechanisms of inhibition, Bioresour. Technol. 74 (1) (2007) 25–33.
[55] J.R. Weil, B. Dien, R. Bothast, R. Hendrickson, N.S. Mosier, M.R. Ladisch, Removal of fermentation inhibitors formed during pretreatment of biomass by polymeric adsorbents, Ind. Eng. Chem. Res. 41 (24) (2002) 6132–6138.
[56] L. Olsson, B. Hahn-Hägerdal, Fermentation of lignocellulosic hydrolysates for ethanol production, Enzyme Microb. Technol. 18 (5) (1996) 312–331.
[57] J. Choudhary, S. Singh, L. Nain, Bioprospecting thermotolerant ethanologenic yeasts for simultaneous saccharification and fermentation from diverse environments, J. Biosci. Bioeng. 123 (3) (2017) 342–346.
[58] R. Sindhu, P. Binod, A. Pandey, Biological pre-treatment of lignocellulosic biomass—an overview, Bioresour. Technol. 199 (2016) 76–82.
[59] E. Battaglia, S.F. Hansen, A. Leendertse, S. Madrid, H. Mulder, I. Nikolaev, et al., Regulation of pentose utilisation by AraR, but not XlnR, differs in *Aspergillus nidulans* and *Aspergillus niger*, Appl. Microbiol. Biotechnol. 91 (2) (2011) 387–397.
[60] T.M. Enari, P. Markkanen, Production of cellulolytic enzymes by fungi, Adv. Biochem. Eng. 5 (1977) 1–24.
[61] P. Alvira, M.J. Negro, M. Ballesteros, Effect of endoxylanase and α-l-arabinofuranosidase supplementation on the enzymatic hydrolysis of steam exploded wheat straw, Bioresour. Technol. 102 (6) (2011) 4552–4558.
[62] F. Cardona, P. Carrasco, J.E. Pérez-Ortín, M. del Olmo, A. Aranda, A novel approach for the improvement of stress resistance in wine yeasts, Int. J. Food Microbiol. 114 (1) (2007) 83–91.
[63] L.J. Jönsson, B. Alriksson, N.O. Nilvebrant, Bioconversion of lignocellulose: inhibitors and detoxification, Biotechnol. Biofuels 6 (2013) 16.
[64] D.E. Briggs, C.A. Boulton, P.A. Brookes, R. Stevens, Brewing: Science and Practice, Woodhead Publishing Limited, Cambridge, UK, 2004.
[65] J. Spencer, T.G. Phister, K.A. Smart, D. Greetham, Tolerance of pentose utilising yeast to hydrogen peroxide-induced oxidative stress, BMC Res. Notes 7 (2014) 151.
[66] A. Singh, S.I. Olsen, A critical review of biochemical conversion, sustainability and life cycle assessment of algal biofuels, Appl. Energy 88 (10) (2011) 3548–3555.
[67] S. McIntosh, Z. Zhang, J. Palmer, H.H. Wong, W.O.S. Doherty, T. Vancov, Pilot-scale cellulosic ethanol production using eucalyptus biomass pre-treated by dilute acid and steam explosion, Biofuel Bioprod. Bioref. 10 (4) (2016) 346–358.
[68] C.E. Wyman, N.D. Hinman, Ethanol: fundamentals of production from renewable feedstocks and use as transportation fuel, Appl. Biochem. Biotechnol. 24/25 (1990) 735–775.
[69] B. Foody, Ethanol from Biomass: The Factors Affecting Its Commercial Feasibility, Iogen Corporation, Ottawa, ON, 1988.
[70] K.A. Subramanian, S.K. Singal, M. Saxena, S. Singhal, Utilization of liquid biofuels in automotive diesel engines: an Indian perspective, Biomass Bioenergy 29 (2005) 65–72.

[71] L. Brennana, P. Owendea, Biofuels from microalgae—a review of technologies for production, processing, and extractions of biofuels and coproducts, Renew. Sustain. Energy Rev. 14 (2010) 557–577.
[72] C.Y. Chen, K.-L. Yeh, R. Aisyah, J. Lee, J.S. Chang, Cultivation, photobioreactor design, and harvesting of microalgae for biodiesel production: a critical review, Bioresour. Technol. 102 (2011) 71–81.
[73] K.W. Jun, H.S. Roh, K.S. Kim, J.S. Ryu, K.W. Lee, Catalytic investigation for Fischer–Tropsch synthesis from bio-mass derived syngas, Appl. Catal. A 259 (2004) 221–226.
[74] A. Firoz, D. Abhijit, R. Roesfiansjah, M. Saleh, H. Moria, B. Abdul, Biofuel from algae – is it a viable alternative, Procedia Eng. 49 (2012) 221–227.
[75] M.Y. Menetrez, An overview of algae biofuel production and potential environmental impact, Environ. Sci. Technol. 46 (2012) 7073–7085.
[76] Y. Liang, N. Sarkanyet, Y. Cui, Biomass and lipid productivities of *Chlorella vulgaris* under autotrophic, heterotrophic, and mixotrophic growth conditions, Biotechnology 31 (2009) 1043–1049.
[77] J.P. Maity, J. Bundschuh, C.Y. Chen, P. Bhattacharya, Microalgae for third generation biofuel production, mitigation of greenhouse gas emissions and wastewater treatment: present and future perspectives—a mini review, Energy (2014) 1–10.
[78] M. Guiry, How many species of algae are there? J. Phycol. 48 (5) (2012) 1057–1063.
[79] C. Joshi, A. Nookaraju, New avenues of bioenergy production from plants: green alternatives to petroleum, J. Pet. Environ. Biotechnol. 3 (7) (2012).
[80] A. Demirbas, Use of algae as biofuel sources, Energ. Convers. Manage. 51 (2010) 2738–2749.
[81] L. Lardon, A. Hélias, B. Sialve, J.P. Steyer, O. Bernard, Life-cycle assessment of biodiesel production from microalgae, Environ. Sci. Technol. 43 (17) (2009) 6475–6481.
[82] S. Kraan, Mass cultivation of carbohydrate rich microalgae, a possible solution for sustainable biofuel production, Mitig. Adapt. Strateg. Glob. Change 18 (2013) 27–46.
[83] L. Chaudhary, P. Pradhan, N. Soni, P. Singh, A. Tiwari, Algae as a feedstock for bioethanol production: new entrance in biofuel world, Int. J. Chem. Technol. 6 (2014) 1381–1389.
[84] G. Dragone, B. Fernandes, A.A. Vicente, J.A. Teixeira, Third generation biofuels from microalgae, *In: Current Research, Technology and Education Topics in Applied Microbiology and Microbial Biotechnology*, 2010, pp. 1315–1366, ed: A. Mendez vilas, (Madrid: Formatex).

Biomass, biorefinery, and biofuels

3

Swarnalatha Mailaram[1], *Pankaj Kumar*[1,2], *Alekhya Kunamalla*[1], *Palkesh Saklecha*[1] *and Sunil K. Maity*[1]
[1]Department of Chemical Engineering, Indian Institute of Technology Hyderabad, Kandi, India, [2]Department of Chemical Engineering, Birla Institute of Technology and Science, Pilani, Hyderabad Campus, Hyderabad, India

3.1 Introduction

Biomass is the only renewable carbon resource in the whole planet. It is well-accepted that fossil fuels, such as coal, natural gas, and petroleum, originate by decomposition of biomass over millions of years. These fossil fuels are the primary source of energy in our society. Currently, it supplies more than 80% of the energy in the world. On the other hand, transportation fuels are primarily produced from petroleum. Gasoline, diesel, and jet fuel are the leading transportation fuels. Transportation fuel is one of the vital energy sectors and plays an extremely important role in our daily life. Currently, this sector alone consumes about 28% of the energy in the world. Petroleum reserves are, however, depleting rapidly while the demands of petroleum products are growing owing to the rapid rise in population. Consequently, the price of petroleum products is increasing continuously. Moreover, the use of petroleum has adverse impacts on the Earth's environment. The availability of renewable transportation fuels is thus essential for the sustainability of our civilization. Biomass, being the origin of petroleum, has the potential to produce transportation fuels and are known as biofuels. At present, biomass provides more than 10% energy in the world. In India, renewables contribute about 27% of the energy with 22% share of biomass and waste alone [1]. Considering the enormous potential of biomass, India set a target of 20% mixing of biofuels with transportation fuels [2]. The European Renewable Energy Council projected to replace 50% of the energy in the world from renewables by 2040. In this context, this chapter covers various technologies for the production of biofuels from biomass.

3.1.1 Biomass

Any organic matter that is available on a renewable basis is considered as biomass. It includes agricultural crops and residues, forest wastes and residues, and municipal and industrial wastes. Dedicated biomass crops, such as jatropha, switchgrass, and so forth, are also grown for producing energy and are known as energy crops. Generally, the chemical structure of biomass is fairly complex and varies depending

Sustainable Fuel Technologies Handbook. DOI: https://doi.org/10.1016/B978-0-12-822989-7.00003-2

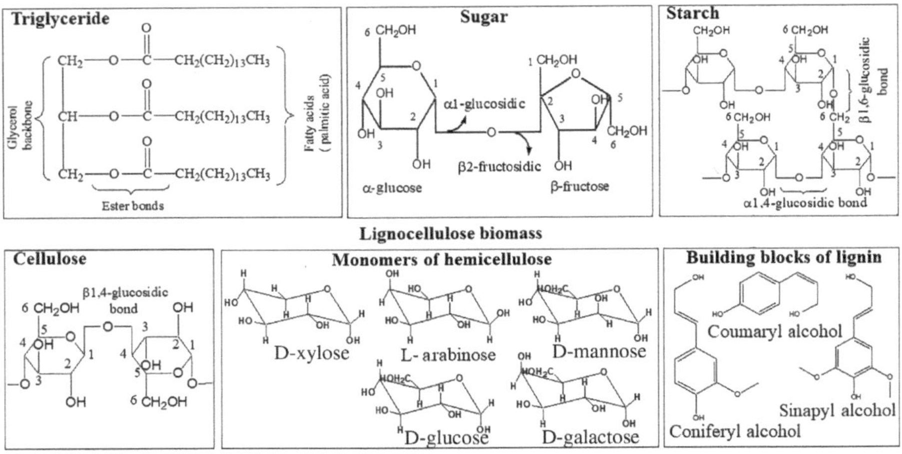

Figure 3.1 Chemical structures of biomass.

on the source and type of biomass. The biomass used in biorefinery is classified as triglyceride, sugar and starch, and lignocellulose based on their chemical structure, as shown in Fig. 3.1 [3,4].

3.1.1.1 Triglyceride

This biomass includes vegetable oil (both edible and inedible), waste cooking oil, microalgae oil, and animal fat. In this biomass, one glycerol molecule is bonded with three fatty acids by ester bonds. Some of the triglycerides, such as inedible oil, also contain a substantial quantity of free fatty acids. These fatty acids are linear with 8–24 carbon chain length. Generally, 16 and 18 carbon chain length dominates in most of the triglycerides. The fatty acids are either saturated or unsaturated. Usually microalgae oil and animal fat contain a substantial amount of unsaturated fatty acids.

3.1.1.2 Sugar and starch

Sugars are derived from sugarcane, sugar beet, and sweet sorghum, while starch is obtained from corn, wheat, barley, and so forth. Sugar is a disaccharide of α-glucose and β-fructose that are bonded by α1-glucosidic β2-fructosidic bond. Starch is a semicrystalline and high molecular-weight (10^6–10^8 g/mol) biopolymer of α-glucose that are linked by α1,4-glucosidic and β1,6-glucosidic bonds. Former bonds are present in amylose, while both types of bonds exist in amylopectin. Starch commonly comprises 75%–80% amylopectin and 20%–25% amylose. It is immiscible in cold water, but soluble in hot water.

3.1.1.3 Lignocellulose

This biomass includes forest wastes (bark, needle, leaves, etc.) and agriculture residues (straw, husk, etc.). It is composed of 40%–50% cellulose, 25%–35% hemicellulose, and 15%–25% lignin. Cellulose is a linear and crystalline biopolymer of β-glucose with a molecular weight of 10^6 g/mol or more. The monomers in cellulose are bonded by β1,4-glucosidic bonds. It is insoluble in water and organic solvent. Hemicellulose is composed of pentose (arabinose and xylose) and hexose (mannose, glucose, and galactose) sugars. It is a relatively low-molecular weight (about 150 monomers) amorphous biopolymer with short branching. The hydroxyl groups of sugars are also acetylated in this biopolymer. The monomers in hemicellulose are bound by β1,4 and β1,3 glycosidic bonds. It is highly soluble in water. Lignin is a highly cross-linked, three-dimensional biopolymer. It comprises three dissimilar phenylpropane building blocks: sinapyl alcohol, coniferyl alcohol, and coumaryl alcohol. It is soluble in the organic solvent but insoluble in water.

3.1.2 Biorefinery

Petroleum refinery mainly produces transportation fuels, fuel oil, and feedstock (naphtha) for petrochemical industries. Naphtha and natural gas are converted into a vast range of organic chemicals, polymers, and commodity products in petrochemical industries. A new concept has been developing for manufacturing biofuels, organic chemicals, polymers, and other products from biomass. This concept is analogous to petroleum refinery and petrochemical industries and uses biomass as the feedstock instead of nonrenewable petroleum. This manufacturing concept is known as biorefinery. Biorefinery also provides heat and electricity. Biorefinery is usually based on the particular chemical structure of biomass and is thus classified as triglyceride biorefinery, sugar and starch biorefinery, and lignocellulose biorefinery [3,4].

3.1.3 Biofuels

Biorefineries produce vast ranges of biofuels. Biodiesel, bio-ethanol, and biobutanol are the most popular biofuels at the moment (Fig. 3.2). In this chapter, these biofuels are referred to as traditional biofuels. Traditional biofuels contain structural oxygen and hence provide less fuel mileage than existing transportation fuels. They are also facing the issues of incompatibility with engines and petroleum refinery facilities and are thus mixed with existing transportation fuels to a certain extent for application in unmodified combustion engines. For example, biodiesel is mixed with diesel up to 20(v/v)%. Bio-ethanol is mixed with gasoline up to 15(v/v)%. A new manufacturing concept is developing to obtain hydrocarbon-analogous biofuels from biomass to overcome these problems. This manufacturing concept is called hydrocarbon biorefinery. The biofuels produced in this biorefinery are analogous to the current transportation fuels and are known as hydrocarbon biofuels, including green gasoline, green diesel, and green jet fuel. They offer fuel mileage similar to

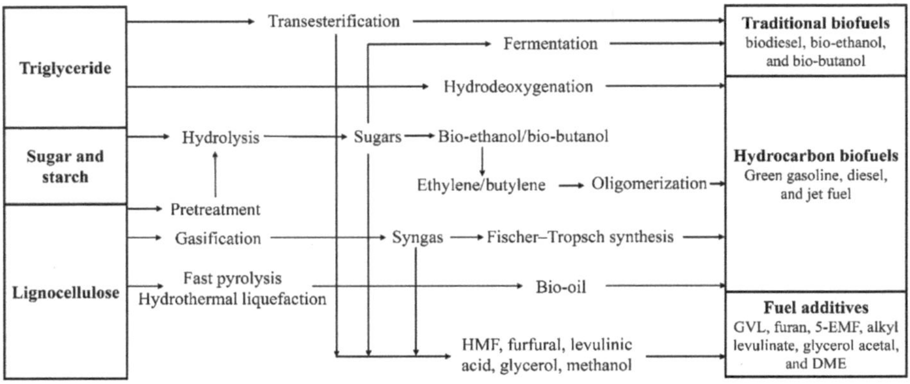

Figure 3.2 Various processes for manufacturing biofuels from biomass.

gasoline, jet fuel, and diesel and are well-suited with existing engines and refinery facilities. Hydrodeoxygenation (HDO) of triglycerides, fast pyrolysis followed by HDO of bio-oil, Fischer–Tropsch synthesis (FTS), oligomerization of olefins, and HDO of the condensed products from C–C bond-forming reactions are some of the routes for the production of hydrocarbon biofuels (Fig. 3.2). Additionally, biomass is converted to several building-block chemicals in the biorefinery to meet the demands of organic chemicals similar to petrochemical industries. These chemicals are known as platform chemicals. 5-Hydroxymethylfurfural (HMF), furfural, levulinic acid (LA), and glycerol are some of the important platform chemicals. These platform chemicals and methanol are converted to various compounds, such as γ-valerolactone (GVL), furanic compounds, 5-ethoxymethylfurfural (5-EMF), alkyl levulinates, glycerol acetal, and dimethyl ether (DME). The furanic compounds include 2-methylfuran (2-MF), 2-methyltetrahydrofuran (2-MTHF), 2,5-dimethylfuran (2,5-DMF), and 2,5-dimethyltetrahydrofuran (2,5-DMTHF). These compounds have suitable fuel properties and can be blended with gasoline or diesel to a certain extent. These biofuels are referred to as fuel additives.

3.2 Traditional biofuels

3.2.1 Biodiesel

Rudolf Diesel first used vegetable oil as engine fuel in 1900 [5]. It, however, triggered the problems of cold-weather starting, injector coking, gumming of injectors, lines and filters, engine knocking, among others, due to its high viscosity and low volatility. Transesterification of vegetable oil with an alcohol has emerged as a promising process for overcoming these problems. This process produces fatty acid alkyl esters with glycerol as a coproduct. These alkyl esters display diesel-like fuel properties, commonly known as biodiesel. Biodiesel further reduces the emission of CO, hydrocarbons, and particulate matters compared to diesel [6]. The methyl and ethyl esters of fatty acids are quite common as biodiesel. The ethyl ester has a lower cloud/pour point, higher oxidative stability, better lubricity, and higher cetane number compared to methyl ester. However, methanol is cheaper than ethanol and can be easily separated from the reaction mixture. Methanol is thus widely used in this process [7].

3.2.1.1 Catalysts

The transesterification reaction is carried out using an alkali, acid, or enzyme catalyst. The acid-catalyzed reaction is slow and requires high temperature and an excessive amount of alcohol (around 30:1 alcohol/oil mole ratio) to drive the equilibrium toward the product. The enzymatic reaction occurs at mild temperatures. However, the inhibition of enzyme activity by alcohol, a low yield of biodiesel, prolonged reaction, and the high cost of the enzyme makes enzymatic reaction unsuitable for large-scale applications. The alkali-catalyzed reaction is, however, relatively fast and occurs at mild reaction conditions: 303K−333K, atmospheric pressure, 6:1 alcohol/oil mole ratio. NaOH-catalyzed reaction is thus commonly employed in large-scale applications. The noncatalytic reaction under supercritical conditions is another approach for the simultaneous transesterification of triglycerides and esterification of free fatty acids. However, it involves the operation at high temperature and high pressure [8]. The microwave-assisted reaction is an emerging concept that reduces energy consumption and inhibits catalyst deactivation [9].

3.2.1.2 Process variables

Low alcohol/oil mole ratio results in a low conversion, while the high alcohol/oil mole ratio poses the difficulty in the separation of esters from glycerol [10]. Inedible feedstock, such as low-grade oil, waste cooking oil, fat, and microalgae oil, contain a large quantity of free fatty acid. These free fatty acids form soap by reacting with NaOH. The soap forms emulsion that poses difficulty in the separation of the esters from glycerol. Acid-catalyzed, enzymatic, and noncatalytic (supercritical) reactions are, however, less sensitive to free fatty acids. Generally, the feedstock with high free fatty acid content is pretreated with an acid catalyst to reduce free fatty acid content below 1%. The presence of water in feedstock leads to the hydrolysis of the ester to free fatty acid. Anhydrous alcohol is thus used in this reaction.

3.2.1.3 Process and economics

In the transesterification process, vegetable oil, alcohol, and catalyst are first sent to the reactor (operated at about 333K) [11]. Following the reaction, esters are separated from glycerol and excess methanol by a settling tank. The glycerol-rich phase, being heavier than the biodiesel-rich phase, is separated from the bottom of the settling tank. The biodiesel stream contains a small quantity of methanol and traces of the catalyst. Methanol is removed by a stripper and the resulting stream is subjected to neutralization, water washing, and drying to obtain purified biodiesel. Glycerol

(50%) stream contains excess methanol and catalyst. The catalyst is removed by neutralization. The methanol is separated by either vacuum flash or evaporation. The crude glycerol (85%) is further refined to 99.5%–99.7% purity by vacuum distillation. The methanol separated from biodiesel and glycerol may contain a small amount of water that is removed by distillation. For alkali and acid-catalyzed continuous process, the biodiesel break-even price from various feedstock is in the range of US$2.10–2.90 per gal [12]. The feedstock price and co-product revenue contribute 60%–75% and 10% of the biodiesel cost, respectively [13]. Low-grade oils are cheaper than vegetable oil but requires additional pretreatment with the added cost and process complexity. The microalgae oil is very expensive. Biodiesel produced from microalgae oil is thus costly compared to the diesel and biodiesel obtained from other triglycerides.

3.2.1.4 Biodiesel specification

The biodiesel may contain traces of alcohol and mono-, di-, and tri-glycerides. The presence of a high level of alcohol in biodiesel degrades rubber gaskets and seals. The presence of glycerides (measured by conradson carbon residue) in biodiesel escalates the pour/cloud point and causes fuel filter plugging [10]. The presence of fatty acids (measured by neutralization number) in biodiesel is also undesirable. In general, the cetane number, pour/cloud point, and viscosity of biodiesel are higher than diesel (Table 3.1). A standard biodiesel specification is thus required for assured engine performance and its successful induction in the market. Austria first defined the biodiesel standard followed by Germany, Italy, France, Czech Republic, and the United States. The biodiesel specification in Austria is as follows: density = 0.85–0.89 g/cm^3 at 288K, viscosity = 3.5–5.0 mm^2/s at 313K, flash point = 373K, cold filter plugging point = 273/268K, cetane number $\geq$ 49, neutralization number (mg KOH/g) $\leq$ 0.8, and conradson carbon residue = 0.05% [10].

3.2.2 Bio-ethanol

Bio-ethanol is the most important biofuel in the biorefinery. It is produced from either sugar or starch biomass. For example, Brazil and India produce ethanol from sugarcane, while corn is the feedstock for ethanol in the United States and EU. United States, Brazil, and the EU are the leading producer of ethanol [20]. United States and Brazil together contributed nearly 84% of the global bio-ethanol production in 2017 [21].

3.2.2.1 Fuel properties

Bio-ethanol has a broader flammability limit and higher octane number, flame speed, and heat of vaporization than gasoline (Table 3.1). These characteristics permit a higher compression ratio, leaner burn engine, and shorter burn time. Bio-ethanol contains 35% structural oxygen that facilitates clean-burning, reduction in emission, and improvement in combustion efficiency [22,23]. For E85 and higher

Table 3.1 Fuel properties of traditional biofuels [14–19].

Property	Biodiesel	Bio-ethanol	Bio-butanol	Diesel	Gasoline
Composition	$C_{12}-C_{24}$ fatty acid methyl esters	C_2H_5OH	C_4H_9OH	$C_{12}-C_{25}$	C_5-C_{12}
Molar mass, g/mol	–	46.07	74.12	198.4	95–120
Specific gravity at 293K	0.88 (288K)	0.789	0.809	0.84	0.739
Water, ppm	0.05% max	–	–	–	–
C, wt.%	77	52.2	65	–	86
H, wt.%	12	13.0	13.5	–	14
O, wt.%	11	34.8	21.5	0	0
S, ppm	<1	–	–	<10	–
Boiling point, K	455–610	350.3	390.7	513–633	308–413
Flash point, K	373–443	285.5	308	338–341	230
Cloud point, K	268–288	–	–	268	–
Pour point, K	258–289	–	–	–	–
Air/fuel ratio, wt/wt	12.5	9	11.1	–	14.8
Lowering heating value, MJ/kg	28.12	26.7	33.1	42–46	44–46
Calorific value, MJ/L	31–33	21.2	29.2	38.6	33.5
Density, kg/m^3	860–894	794	809	810–890	720–750
Reid vapor pressure, kPa	<0.6	17	2.3	–	45–90
Cetane number	50–60	5–8	~25	40–55	5–20
Research octane number	–	107	105–113	20–30	95
Motor octane number	–	89	80	20–30	90
Auto ignition temp, K	409	696	638	483–523	298.7
Heat of vaporization, MJ/kg	–	0.83	0.71	0.25–0.6	0.26–0.36
Kinematic viscosity, mm^2/s at 293K	3.5–5.2 (313K)	1.5	3.6	1.9–4.1	0.4–0.8

blend, the reduction in greenhouse gas emission is more than 40% [24]. However, bio-ethanol is highly hygroscopic, slightly corrosive, and toxic to the ecosystem [4]. It has low-flame luminosity and Reid vapor pressure and high-water solubility. It has a lower energy density than gasoline, owing to the existence of structural oxygen. Bio-ethanol is thus incompatible with the existing infrastructures, including the combustion engine. Hence, bio-ethanol is mixed with gasoline for use in unmodified combustion engines. Some countries, such as Brazil, however, developed the flexi-fuel engine that can run using pure ethanol.

3.2.2.2 Processes

For sucrose feedstock, biomass is crushed to extract sugar juice. The corn (65%–76% starch) is processed through dry milling in which the powder form of grains is heated with water at 358K. Starch is then liquefied using α-amylase that converts starch into short-chain dextrins. Saccharification (pH 4.5 and 338K) is carried out using the gluco-amylase enzyme. In contrast, the rigid lignin cell wall protects carbohydrates in lignocellulose biomass. The cellulosic biomass is thus processed through milling, followed by pretreatment. The pretreatment breaks the lignin barrier, reduces cellulose crystallinity, and enhances the accessibility of carbohydrates for hydrolysis. The pretreatment is, however, very expensive, and has an enormous influence on the overall yield of ethanol. For example, pretreatment improves the hydrolysis yield to 90% from merely 20% (without pretreatment) [25]. The polysaccharides are then hydrolyzed to fermentable sugar. Hydrolysis is accomplished using either enzyme or dilute acid. The enzymatic hydrolysis is, however, commonly used (pH 4.8 and 318K–323K) [26]. Cellulase and hemicellulase enzymes are used for hydrolysis of cellulose and hemicellulose, respectively.

The sugar, starch, and cellulose are composed of hexose sugars, while both hexose and pentose sugars exist in hemicellulose. The hexose sugars are traditionally fermented by Baker's yeast. *Saccharomyces cereviseae* is the most common organism (at 306K and pH 4–4.8). The maximum yield of ethanol is 0.48 g per g glucose [25]. *Pichia stipitis*, *P. segobiensis*, *Candida shehatae*, *Pachysolen tannophilus*, and *Hansenula polymorpha* are some of the organisms for fermentation of pentose sugars. These organisms, however, suffer from the drawback of the slow fermentation rate. The genetic modification of the microorganisms is thus done to ferment both hexose and pentose sugars. Metabolically engineered strain recombinant *Escherichia coli* (KO11), *Saccharomyces cerevisiae* 1400 (pLNH33), *Zymomonas mobilis* are some examples of these. The high sugar and ethanol concentration and inhibitory fermentation products are toxic to the organism. Less than 10% ethanol concentration (typically 4%–4.5%) is thus maintained in fermentation broth to reduce the stress on the enzyme.

The fermentation broth is sent to the beer stripper and rectification column to obtain 95% ethanol–water azeotrope mixture. The ethanol is further purified to fuel-grade ethanol by azeotropic distillation using benzene or ethylene glycol as entrainer followed by dehydration using the molecular sieve [27]. There are several process alternatives for bio-ethanol production: (1) simultaneous saccharification and fermentation that perform enzymatic hydrolysis and fermentation in the same reactor; (2) cofermentation, in which hexose and pentose sugars are fermented by single microorganism; (3) simultaneous saccharification and co-fermentation; and (4) consolidated bioprocessing, in which cellulose production, cellulose hydrolysis, and fermentation happens in a single step [28]. These process variations aim to reduce the investment cost and have a lower risk of inhibition and contamination [25].

3.2.2.3 Challenges and economics

The choice of feedstock is a crucial factor in this process and contributes 35%–40% of the bio-ethanol production cost [29]. Despite the low-cost of feedstock, bioethanol from lignocellulose biomass is facing several barriers in the way of commercialization. The pretreatment is the most prominent factor that contributes to about 20% of the bio-ethanol manufacturing cost [25]. The cost-effective pretreatment method and reduction in enzyme cost can reduce the cost of lignocellulosic ethanol significantly [29]. The separation of ethanol from fermentation broth is energy-intensive. Hybrid and integrated technologies, such as membrane separation and gas-stripping, are developed to reduce energy consumption. For the oil price of US$60 per bbl, the estimated production cost of lignocellulosic ethanol is US$0.62 L^{-1} of gasoline-equivalent (LGE), which is expensive than gasoline (US$0.42 per LGE) and corn ethanol (US$0.5 per LGE) [30]. For the increase in the oil price to US $120 per bbl, bio-ethanol derived from corn (US$0.7 per LGE) and lignocellulose biomass (US$0.73 per LGE) are, however, cheaper than gasoline (US$0.78 per LGE).

3.2.3 Bio-butanol

Butanol was known as a fermentation product even before 1900. The n-butanol, however, gained notable attention as a biofuel only in 2005, when a car was driven successfully with pure n-butanol. In 2007 BP and DuPont started the production of bio-butanol in Great Britain [14].

3.2.3.1 Fuel properties

Bio-butanol has 25% higher energy content compared to bio-ethanol. Thus it provides a higher fuel mileage than bio-ethanol. It is highly hydrophobic and can be blended to gasoline at any ratios or used in pure form without modifying the existing combustion engine [14,31]. These properties further allow the blending of bio-butanol with gasoline before storage and transportation [14]. It is less corrosive and has lower vapor pressure than bio-ethanol. This factor makes the bio-butanol safe in handling and transportation. Iso-butanol has a higher research octane number (105–113) and less toxic than n-butanol and has necessary fuel properties similar to n-butanol [4]. It is thus considered as a potential biofuel in the future.

3.2.3.2 Acetone-butanol-ethanol fermentation

Bio-butanol is obtained by acetone-butanol-ethanol (ABE) fermentation of sugars using *Clostridia acetobutylicum* bacteria. The sugars are derived from either sugar, starch, or lignocellulose biomass, as discussed in the bio-ethanol section. The clostridium strain produces ABE in 3:6:1 mass ratio by anaerobic fermentation. ABE fermentation occurs in two phases: acidogenesis and solventogenesis [14]. The acetic acid, butyric acid, CO_2, and H_2 are produced in the acidogenesis phase. It increases acid-stress if pH is not controlled. The conversion of these acids to ABE occurs in the solventogenesis phase. Strains, such as *C. acetobutylicum* ATCC 824, *C. beijerinckii* NCIMB 8052, produce 9–12 g/L butanol from 40–60 g/L sugar in a batch fermentation [15]. Nonclostridium strains, such as *E. coli, L. brevis, P. putida, L. buchneri, and S. cerevisiae*, produce butanol by either synthetic biology or reducing/eliminating acetone and ethanol [32]. Solventogenic clostridia cannot tolerate more than 20 g/L solvent [33,34]. More than 60 g/L sugar concentration is also toxic to the culture. The developing butanol-tolerant strain by genetic modification is required to improve butanol productivity. Further research is also needed for the simultaneous digestion of hexose and pentose sugars or reducing/ eliminating acetone and ethanol.

3.2.3.3 Fermentation engineering

ABE fermentation is carried out in three modes: batch, fed-batch, and continuous. The batch process is widely used in industries. The low yield and loss of substrate and residual cell culture are the major drawbacks in this process. In the fed-batch process, nutrients and substrates are continuously provided to the fermenter, and the product is recovered after the fermentation [15]. Fed-batch mode has a drawback of the low solvent yield with low-butanol concentrations and limitations of using diluted substrates along with the requirement of longer time for production of butanol. The continuous mode is, however, preferred because of the high butanol yield and high productivity at the dilution rate of 0.1 h^{-1} [15]. Among free cell, immobilized cell, and cell recycling and bleeding continuous processes, the immobilized cell process showed 15.8 g/L/h solvent productivity, which is almost 40 times the batch process [31]. Two-stage or multistage continuous fermentation is an effective technique to improve productivity as the growth of cells occurs in one fermenter and the solvent synthesis in the subsequent fermenters [15].

3.2.3.4 Separation of acetone-butanol-ethanol

The concentration of ABE in the fermentation broth is typically between 2% and 4%. The low concentration is due to the toxicity of solvents on the microbes. Besides, ABE forms azeotropic mixtures. Concentrating ABE from 2% to 40% by a beer stripper is thus energy-demanding and involves huge capital investment [34]. The in situ product recovery techniques are thus employed to enhance butanol concentration, reduce solvent toxicity on microorganisms, reduce energy consumption, and improve economics. These technologies include liquid–liquid extraction, pervaporation, perstraction, gas stripping, and adsorption. Membrane separation processes, such as pervaporation and perstraction, are very selective and efficient, but associated with the high cost and membrane fouling. Moreover, these methods cannot separate ABE into individual solvents [33]. For the separation of 0.5% butanol, hybrid pervaporation-distillation requires only 7.4 MJ/kg of butanol [33,35,36]. Whereas distillation alone requires 79.5 MJ/kg for the same separation, which is almost twice the energy content of butanol (36 MJ/kg). The adsorption is an energy-efficient technique and can be scaled-up to the industrial level. The difficulty in desorption and regeneration and reuse of adsorbents are, however, the bottlenecks in this technique. In gas stripping, concentrated sugar solution can be used in the fermenter that increases ABE productivity [37]. The adsorption and gas stripping techniques suffer from low butanol selectivity. The energy consumption for gas stripping, followed by the separation of ABE by distillation is around 14–31 MJ/kg [37]. Liquid–liquid extraction is highly selective and easy to operate. The toxicity of the extractant to the microorganism is, however, the major bottleneck in this technique. A novel dual extraction process was thus developed for reducing the toxicity to the microorganism [38]. This technique is the most energy-efficient and requires less than 5 MJ/kg of butanol [38].

3.2.3.5 Economics

The feedstock contributes more than 65% of the bio-butanol production cost [39]. The pretreatment and hydrolysis are not required for sugar biomass. The capital investment for this feedstock is thus 37% lower compared to cellulosic biomass [39]. For sugarcane and cellulosic feedstock, the estimated unitary production cost lies between US\$0.59–0.75 kg^{-1} of bio-butanol for the plant capacity of 0.01 million tons butanol per annum [39].

3.3 Hydrocarbon biofuels

3.3.1 HDO of vegetable oil

The triglycerides are made up of fatty acids ranging between C_8 and C_{24} (dominating C_{16} and C_{18}) [40]. The oxygen from this feedstock is thus removed to produce hydrocarbon biofuel and known as green diesel. The pyrolysis and catalytic cracking over the zeolite catalyst can remove oxygen from vegetable oil. However, these routes have several limitations, such as lower hydrocarbons yield and the massive loss of carbon as gas [41]. HDO is thus an attractive alternative for removing oxygen in the form of water and CO. It is similar to the hydroprocessing process of petroleum refinery [40,42]. This route produces a high yield of green diesel with minimal carbon loss. The benefits of this process are as follows: co-processing with petroleum fractions and the utilization of established hydrotreatment technology and its infrastructure. HDO is a catalytic process that occurs over supported metal catalysts in the H_2 atmosphere. The temperature and pressure in this process are generally lying in the range of 523K–773K and 25–150 bar, respectively [40,43]. The properties of green diesel are also similar to diesel and superior compared to biodiesel, as shown in Table 3.2.

3.3.1.1 Reaction mechanism

Generally, unsaturated vegetable oil first undergoes saturation in the hydrogen atmosphere over the metal catalysts (Fig. 3.3). The saturated vegetable oil then decomposes to corresponding fatty acids and propane. These fatty acids are reduced by hydrogen to corresponding fatty aldehyde over supported metal catalyst. The resultant fatty aldehyde then follows two different routes. In the first route, fatty aldehyde undergoes the decarbonylation reaction with the formation of hydrocarbon and one mole of CO. In the second route, fatty aldehyde undertakes the HDO route,

Table 3.2 Fuel properties of green diesel [44,45].

	Diesel	Biodiesel	Thermally cracked	Green diesel
Specific gravity	0.84	0.88	–	0.78
Viscosity at 313K, cP	1.9–4.1	30–40	10.2 cSt	3.81
Sulfur, ppm	<10	<1	–	<1
Heating value, MJ/kg	42–46	38	40.3	44
Cloud point, K	268	268–288	–	253–293
Cetane number	40–55	50–60	43	70–90
Flash point, K	338–341	373–443	–	411
Fire point, K	529	409	–	418
Pour point, K	–	258–289	280.35	282
Stability	Good	Marginal	–	Good
O, wt.%	Nil	11	–	0.53
H/C	2.0	–	–	1.97

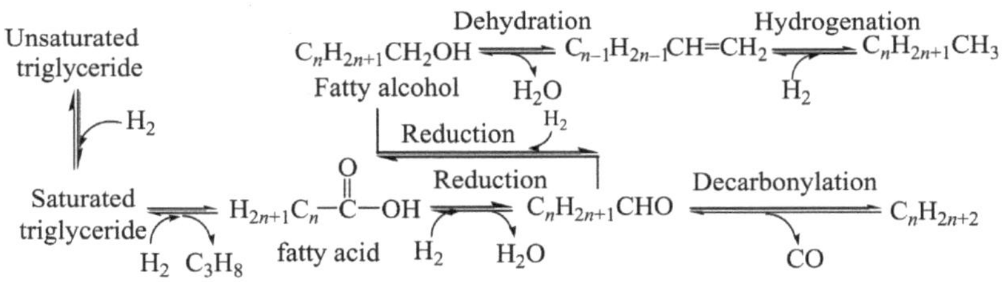

Figure 3.3 Reaction mechanism for HDO of vegetable oil.

that is, the reduction of fatty aldehyde followed by dehydration of fatty alcohol and hydrogenation [44,46]. The decarbonylation route is dominant over supported monometal (Ni) catalyst, resulting in the formation of alkane with the loss of one carbon [46]. The HDO route is, however, dominant over supported bimetal (NiMo and CoMo) catalysts, reducible oxide, and metal dispersed on acidic support [40,46].

3.3.1.2 Catalysts

Supported noble metals (Pt, Pd, Ru) showed excellent catalytic activity for HDO of vegetable oil [47]. However, noble metals are not economical for large-scale applications due to their high cost. Therefore, the HDO reaction was studied extensively over inexpensive metal catalysts, such as Ni, Co, NiMo, and CoMo [43]. The sulfided form of γ-Al_2O_3-supported CoMo and NiMo catalysts are quite active and selective to C_{17} and C_{18} alkanes [48]. However, their application has some limitations, such as sulfur contamination of the liquid product and rapid decline in activity due to sulfur leaching. The reduced, phosphide and nitride form of transition metal catalysts are thus employed for HDO reactions [43,44]. The reduced form of transition metals, such as Ni, Co, NiMo, and CoMo, showed high catalytic activity with long-run stability. For example, HDO of stearic acid was studied using reduced γ-Al_2O_3-supported Ni, Co, NiMo, and CoMo catalysts at 543K and 20 bar H_2 pressure [46]. These catalysts showed more than 90% conversion in 6 h reaction time [46]. The C_{17} and C_{18} alkanes were dominating products over monometal (Ni) and bimetal (NiMo and CoMo) catalysts, respectively, while C_{17} and C_{18} alkanes were comparable over Co catalyst.

3.3.1.3 Supports

The various supports, such as Al_2O_3, SiO_2, TiO_2, activated carbon, and so forth, have been reported for this reaction [40,43]. Highly acidic catalysts enhance catalytic activity as well as C—C bond cracking with a high selectivity to lower alkanes [40,43]. The moderately acidic catalyst decreases the C—C bond cracking and increases the yield of liquid hydrocarbons. For HDO of palm oil, moderately acidic Ni/SAPO-11 catalyst showed a low tendency of cracking and high isomerization activity (80%) with about 80% yield of liquid hydrocarbons [49]. The highly acidic support exhibits high selectivity toward HDO products, while mild and neutral supports favors decarbonylation products. The high reaction temperature also enhances cracking reactions [40]. High pressure, however, promotes the HDO reaction [48]. The mesoporous catalyst demonstrates the superior catalytic activity due to strong metal-support interaction. Yenumala et al., studied HDO of karanja oil using ordered mesoporous nickel-alumina composite catalysts [50]. The composite catalyst showed 47% conversion of karanja oil in 6 h reaction time compared to only 40% over mesoporous alumina and alumina supported nickel catalysts.

3.3.1.4 Economics and commercial processes

There are two possible ways for manufacturing green diesel from vegetable oil [51]. In the first way, vegetable oil directly undergoes HDO (direct HDO). In the second way (two-step HDO), vegetable oil undertakes hydrolysis with the formation of fatty acid followed by HDO of fatty acids. The estimated green diesel manufacturing cost from the karanja oil was reported as US\$0.798 and US\$0.84 kg^{-1} for two-step and direct HDO, respectively [51]. Nest Oil, NextBTL, Petrobras, and Honeywell UOP (Eni Ecofining) are the major players for manufacturing green diesel from triglycerides [52]. There are two operating plants of Nest oil in Finland and Singapore with an annual capacity of 0.17 and 0.8 million-ton green diesel, respectively. Another commercial unit for green diesel production is operational in Italy by UOP/Ecofining. A similar plant with an annual capacity of more than 130 million gallons is commissioned at the Diamond Green Diesel facility in Norco, Louisiana.

3.3.2 Fast pyrolysis of lignocellulosic biomass

Pyrolysis is an old technology for turning wood into charcoal. It is the thermal decomposition of carbonaceous materials in the absence of oxygen. In this process, three products are formed from biomass: solid called char, liquid called bio-oil, and

Table 3.3 Types of pyrolysis based on operating conditions and their major products [53].

	Temperature, K	Heating rate	Residence time	Major product	Yield, %
Carbonization	673	–	Days[a]	Solid (charcoal)	35
Torrefaction	563	–	5–60 min[a]	Solid (charcoal)	80
Gasification	1023–1173	Fast	Very high[b]	Gas	85
Intermediate	773	Fast	10–30 s[b]	Liquid	50
Fast pyrolysis	773–923	Fast	1–2 s[b]	Liquid	75

[a]Solid residence time.
[b]Vapor residence time.

Table 3.4 Properties of bio-oil and hydroprocessed bio-oil [53,54].

Physical properties	Bio-oil	Hydroprocessed bio-oil
Moisture content	25%	–
pH	2.5	–
Specific gravity	1.2	0.93
Elemental analysis		
C	56%	87.9%
H	6%	8.9%
O	38%	3%
N	0%–0.1%	0.4%
HHV	17 MJ/kg	41.4 MJ/kg
Viscosity	40–100 mPa s	–
Solid (char)	0.1%	–
Vacuum distillation residue	Up to 50%	–

noncondensable gas. The quantity of these products depends on reaction conditions, such as temperature, residence time, and heating rate. Accordingly, pyrolysis is categorized into different modes, as shown in Table 3.3 [53]. The bio-oil is the dominant product in fast pyrolysis. This process is quite simple and can convert all forms of lignocellulose biomass into bio-oil. Moreover, the transportation of bio-oil to the centralized biorefinery is much easier compared to the transportation of lightweight biomass, such as agricultural wastes and residues. It also has enormous potential to strengthen the rural economy. This process is thus gaining tremendous importance in the past 40 years. The bio-oil is a dark-brown color liquid. It is a mixture of a vast range of oxygenated compounds, such as phenolics, ketone, alcohol, aldehyde, furans, sugars, carboxylic acid, ester, other oxygenated organic compounds, alkene, and aromatics, with a significant amount of water. The composition of bio-oil depends on the types of feedstock and reaction conditions. The bio-oil has a typical heating value of 17 MJ/kg (Table 3.4).

3.3.2.1 Process conditions

Fast pyrolysis is operated at around 773K–923K with less than 2 s residence time. Typical yields in this process are about 75% bio-oil, 12% char, and 13% gas. The lower temperatures favor char formation, while higher temperatures promote the generation of gas. Fast pyrolysis is thus operated at a moderate temperature. Furthermore, a short vapor residence time is maintained to reduce secondary reactions, which is responsible for gas formation. The char acts as a catalyst for cracking of pyrolysis vapor. The rapid removal of char is thus critical. Fast quenching of pyrolysis vapor is also essential to avoid the cracking reaction at elevated temperatures. The small biomass particle of below 5 mm is desirable for faster heat transfer. The dry biomass with less than 10% moisture content is used to minimize the water content in bio-oil.

3.3.2.2 Reactors and commercial processes

The reactor system is the heart of this process and contributes about 10%–15% of the capital investment [53]. Therefore, research has been focused on developing different types of reactors for the fast pyrolysis of a vast range of biomass (Table 3.5) [55]. The bubbling and circulating fluidized-bed reactors are commonly used in this process owing to their well-understood technology, efficient heat transfer, and a high yield of bio-oil [56]. Rapid thermal processing of UOP and Ensyn joint venture developed a process using the circulating fluidized-bed reactor to convert lignocellulosic biomass to bio-oil. This technology has been used in North America for the past 25 years [56]. The application of bio-oil in this process is limited to fuel for heating and chemicals for the food industry. KiOR, another US-based company, also developed the fast pyrolysis process using the circulating fluidized-bed reactor. In this process, the HDO of bio-oil is further carried out to produce transportation fuels. Empyro BV constructed a bio-oil production plant using the rotating cone reactor in Netherlands with a design capacity of 0.024 million tons per year. This technology targets the cofiring of bio-oil for boiler and furnace.

3.3.2.3 Hydroprocessing of bio-oil

Bio-oil suffers from poor fuel qualities for use as transportation fuel, such as corrosiveness, low heating value, high viscosity, and thermal instability. The elimination of oxygen from bio-oil by hydroprocessing has thus been a subject of research for the past 25 years. Hydroprocessing is performed at high hydrogen pressure (about 200 bar) and 673K to produce gasoline-range fuel [53]. Sulfided NiMo and CoMo catalysts were employed initially in this process. These catalysts were, however, facing many fundamental issues, such as instability owing to high-water content in bio-oil and rapid catalyst deactivation due to sulfur leaching. The carbon is alternative support for this reaction [57]. Attention later turned into precious metals, such as ruthenium, palladium, platinum, among others. These metals are more active than the sulfided NiMo and CoMo catalysts at lower reaction temperatures.

Table 3.5 Characteristics of fast pyrolysis reactors [55].

Reactor	Bio-oil yield (wt.%)	Status	Complexity	Feed size	Carrier gas	Scale-up	Gas quality
BFB	75	Commercial	Medium	Very small	High	Easy	Low
CFB	75	Commercial	High	Very small	High	Easy	Low
Rotating cone	70	Commercial	High	Very small	Low	Medium	High
Entrained flow	60	Laboratory	Medium	Very small	High	Easy	Low
Ablative	75	Laboratory	Low	Large	Large	Difficult	High
Auger	60	Pilot-scale	Medium	Medium	Medium	Medium	High

BFB, Bubbling fluidized-bed; *CFB*, circulating fluidized-bed.

Moreover, these metals can be easily dispersed on nonalumina supports, such as carbon and titania, to overcome water instability issues [57]. The higher heating value of hydroprocessed bio-oil is about 41.4 MJ/kg, which is almost similar to diesel (Table 3.4).

3.3.3 Fischer–Tropsch synthesis

In 1923, Fischer and Tropsch observed the reaction between CO and H_2 at high-pressure over alkalized iron turnings [58]. This discovery led to modern FTS. This process is performed in a fixed-bed reactor using syngas with the 2:1 H_2/CO ratio at 10–60 bar and 473K–623K [59,60]. FTS integrated with biomass gasification is commonly known as biomass-to-liquid process [59]. It produces fuels in the range of fuel gas, LPG, gasoline, jet fuel, diesel, and paraffin wax. The product consists of olefins, paraffins, oxygenated compounds, and aromatics [61,62]. Linear α-olefins are the dominant product. The fuels obtained from FTS are free from sulfur and contain only a small quantity of aromatics and nitrogen compounds. Therefore, FT-fuels exhibit a lower level of emissions [59]. The contaminants present in syngas, such as particulates, tar, alkali compounds, HCl, H_2S, HCN, NH_3, and COS, can poison the FTS catalyst. Therefore, syngas is cleaned before FTS. FTS proceeds through the surface polymerization via chain growth, propagation, termination, and desorption after initiation of monomer CH_2 [59].

3.3.3.1 Catalysts

Fe, Co, Ni, and Ru metals display catalytic activity for FTS [61]. Ru is, however, expensive, and Ni produces a large quantity of CH_4. These factors make them unsuitable for large-scale applications. Co, being a hydrogenation catalyst, produces more hydrogenated products. Co is dispersed on Al_2O_3, SiO_2, and TiO_2 and promoted by Pt, Ru, and Re. Alkali enhances the catalytic activity of the fused iron catalyst [63]. Alkali also prevents CH_4 formation with the enhancement of carbon deposition and the formation of organic acid. Co is more active in FTS than Fe with higher turnover rates [59]. Hence, Co catalyst is preferred over Fe catalyst.

3.3.3.2 Process conditions

The H_2/CO ratio, known as the usage ratio, is the key factor in FTS. The low partial pressure of CO deters chain-growth and enhances the desorption of alkene [61]. On the other hand, the high partial pressure of H_2 favors the chain-termination to paraffins. The lighter and saturated hydrocarbons thus rise with the increase in the usage ratio. The usage ratio requirement for Co catalysts is about 2.06–2.16 [59]. For Fe catalyst, the water-gas shift reaction occurs, thereby changing the overall usage ratio. At low temperatures, the usage ratio requirement is around 1.7. At elevated temperatures, the reverse water-gas shift reaction occurs with the $H_2/(2CO + 3CO_2)$ ratio requirement of around 1.05. The branched, hydrogenated, and lower carbon number products rise with the increase in reaction temperature [61]. The ketones and aromatics also increase with increasing reaction temperature [62]. The Co catalyst shows higher selectivity to CH_4 at high reaction temperatures than Fe catalyst [61]. The FTS process is operated under two different modes: low-temperature FT (LTFT) and high-temperature FT (HTFT) [61,62]. LTFT is operated at 473K–523K and suitable for high-quality diesel. LTFT is carried out in a slurry reactor using Co catalyst. The yield of straight-run diesel is around 18%–20% with 45%–50% wax. Wax is further hydrocracked to obtain diesel (80%), naphtha (15%), and C_1-C_4 gas (5%). The combined diesel showed a high cetane number of around 73. HTFT is operated at 573K–623K over the Fe catalyst. HTFT is suitable for gasoline. HTFT produces around 40% gasoline, 20% propene and butenes, 15% C_1-C_2, and 5% aromatics. HTFT is operated in a fluidized-bed reactor in the presence of Fe catalyst at about 613K for maximum gasoline yield.

3.3.3.3 Commercial processes and economics

The first biomass-to-liquid process was developed in 1996 by CHOREN Industries in Germany, with a plant capacity of 0.015 million tons of liquid fuel per year [59]. The Indian Institute of Science in India designed biomass-to-liquid process (40–45 barrels per day) based on oxy-steam biomass gasifier followed by FTS in a packed-bed reactor over Co catalyst. The diesel cost was in the range of US\$0.6–0.8 L^{-1}, which was almost similar to the diesel price in India. The pretreatment, gasification, and gas-cleaning together account for nearly 75% of the capital investment [60]. The diesel can also be produced by methane autothermal reforming, followed by FTS. For the natural gas price of US\$0.5 per MMBtu, the production cost of diesel in this process was nearly US\$22–25 per barrel for the plant capacity 30,000 barrel per day [64]. The estimated breakeven price of liquid fuels was about US\$75, US \$36, and US\$60 per barrel for biomass, natural gas, and coal, respectively [65].

3.3.4 Oligomerization of olefins

Oligomerization is a well-known process in the petroleum refinery, in which olefins are converted to fuel-range hydrocarbons. Bio-olefins, such as ethylene and butylenes, are obtained from bio-ethanol and bio-butanol by the dehydration reaction, respectively. These bio-olefins can be oligomerized to obtain hydrocarbon biofuels.

3.3.4.1 Process parameters

Oligomerization reaction occurs at 423K–773K and 1–50 bar [66]. It is an exothermic reaction that releases 1.046–1.381 kJ/g of olefin [66]. Low temperature (below 573K) and high pressure (30–50 bar) favors diesel-range hydrocarbons, while high temperature and ambient pressure are suitable for gasoline-range hydrocarbons. Cracking and branching reactions occur above 490K and 573K, respectively. The oligomerization reaction occurs over an acidic catalyst. Solid phosphoric acid, zeolite and modified zeolite, sulfonic acid-functionalized material, transition-metal complex, metal oxide, and organometallic complex are some of the catalysts used in this reaction [66–68]. Zeolite (Mobil olefins to gasoline and distillate process) and solid phosphoric acid (catapoly process) are commonly used in industrial processes [66]. Solid phosphoric acid produces highly branched olefins due to skeletal isomerization [69]. The branching is, however, much less over HZSM-5 owing to pore geometry constraint [70]. The reaction proceeds through the carbenium ion and 1,2 insertion mechanism over acid and metal catalyst, respectively [71]. The propene and 1-hexene oligomerization was carried out over HZSM-5 at 538–543K, 48 bar, and 1 h^{-1} LHSV and achieved more than 95% conversion [70]. The motor-gasoline obtained from propene showed a higher degree of branching with a higher-octane number compared to 1-hexene. The reactivity of olefins decreases with chain length (ethylene > propylene > 1-butene > 1-hexene) and double bond position (1-butene > 2-butene) [67].

3.3.4.2 Zeolite catalyst

One-dimensional zeolite, in which pores are parallel to the crystal length, limits the diffusion and causes poisoning at the pore mouth [66]. Hierarchical zeolite, having both mesopore and micropore, improves the accessibility of active sites. Micropore causes catalyst deactivation, while mesopore enhances catalytic activity. Large pores favor branched and long-chain oligomers [72]. Submicrometer and nanosized one-dimensional zeolites further improve catalytic activity with reduced deactivation. High porosity supports the formation of aromatics. The oligomerization reaction generally takes place on Brönsted acid sites inside the pores [73]. External acid sites are responsible for the branching of the oligomers. Strong Lewis acid sites favor side reactions with branched products, aromatics, and coke. The Si/Al ratio in the zeolites represents acid strength. The higher is the ratio, the higher is acid strength and vice versa. In the oligomerization reaction, the Si/Al ratio is generally in the range of 45–150. The lower Si/Al ratio favors short-chain oligomers, predominantly dimers. The presence of excessive water in the feed deactivates the catalyst irreversibly by dealumination. The hydrothermal stability of the catalyst is thus very important. The alkali treatment of HZSM-5 improves hydrothermal stability [74].

3.3.4.3 Commercial processes

Mobil Oil (now ExxonMobil) developed a Mobil olefins to gasoline and distillate process for the conversion of olefins to gasoline and distillate-range fuels using ZSM-5 catalyst [75]. This process is flexible and maximizes either gasoline or diesel, with more than 95% selectivity to gasoline and distillate [76]. It is achieved by changing reactor temperature and recycle composition with a gasoline/distillate ratio in the range of 0.2 to more than 100 [75,76]. Lurgi also developed a process using a zeolite catalyst for the conversion of olefins to diesel. UOP further established a plant for oligomerization of olefins-rich gas using solid phosphoric acid as the catalyst at 27–35 bar, 448K–508K, and 2.5 L/kg-h. The yield of gasoline in this process is 90%–97% of olefin [77]. The commercial process has also been initiated for the conversion of bio-butanol to jet fuel and diesel [78,79].

3.3.5 C–C bond formation reactions

The C–C bond-forming reactions are one of the key areas in the production of hydrocarbon biofuels. This reaction produces the fuel precursors with a targeted molecular-weight from the low-molecular-weight biomass-derived compounds. These oxygenated fuel precursors are subsequently subjected to HDO reaction to obtain hydrocarbon biofuels. The aldol-condensation and hydroxyalkylation-alkylation (HAA) are commonly used C–C bond formation reactions.

3.3.5.1 Aldol-condensation

Aldol-condensation is a base-catalyzed reaction among two carbonyl group containing compounds, of which one of them must contain at least one αH. Furfural and HMF are the carbonyl-group containing platform chemicals. However, they do not contain αH. Furfural and HMF are thus reacted with other αH-bearing carbonyl compounds, such as cyclopentanone, acetone, angelica lactone, 2-pentanone and 2-heptanone, hydroxyacetone, LA, and so forth. For example the reaction between HMF and acetone produces the C_{15} fuel precursor (Fig. 3.4). The reaction is carried out at moderate temperatures (303K–393K) and atmospheric pressure. The reaction is performed in either homogeneous, heterogeneous, or biphasic system. The biphasic system consists of an aqueous phase (NaOH and reactants) and the organic phase (tetrahydrofuran) [80]. The product formed in the aqueous phase is instantly extracted to the organic phase, thereby preventing side reactions. The addition of salt further improves the partition coefficient owing to the salting-out effect. Heterogeneous catalysts, such as hydrotalcite and layered double hydroxides (Mg-Zr, Mg-Al, Ca-Zr), showed promising catalytic activity in this reaction [81,82]. These catalysts, however, displayed low structural stability and high sensitivity toward ambient CO_2. The nitrogen-substituted NaY showed enhanced catalytic activity compared to NaY [83]. It, however, displayed instability owing to nitrogen

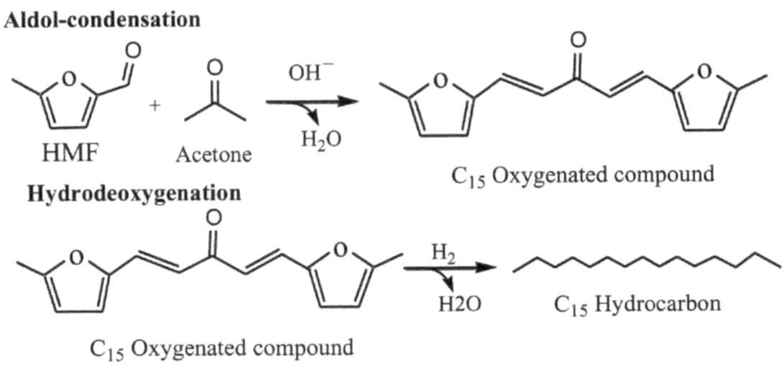

Figure 3.4 Aldol-condensation of HMF with acetone and HDO of the condensed product.

leaching from the framework. The MgO incorporated NaY, however, showed high catalytic activity and selectivity to the condensed product [84].

HDO of the fuel precursors is performed over the supported metal catalysts [85,86]. HZSM-5 has higher acid strength compared to HBeta, Al_2O_3, and SiO_2-Al_2O_3 [85]. Pd/HZSM-5 thus exhibited the highest catalytic activity with 90% carbon yield (C_9-C_{12} alkanes) at 533K and 60 bar H_2. HDO of aldol-product (cyclopentanone/cyclohexanone with furfural/HMF) over Pd/HBeta produced aviation-range fuel with good fuel properties (0.815−0.826 g/mL density and low viscosity) [86]. Barret et al., demonstrated simultaneous aldol-condensation of HMF/furfural with acetone and HDO of aldol-product over a bifunctional catalyst, Pd/MgO-ZrO_2 [87]. Optimal reaction conditions were 353K and 2:1 furfural:acetone mole ratio.

3.3.5.2 Hydroxyalkylation−alkylation (HAA) reaction

HAA reaction ensues between furan and αH-containing carbonyl compound in the presence of an acid catalyst under mild reaction conditions (313K−363K, 2−6 h) [88,89]. Furfural is the only αH-containing furanic platform molecule. The presence of the electronegative aldehyde group, however, deters furfural from self or cross HAA reaction. 2-MF, a hydrogenation product of furfural is, however, highly reactive for this reaction. HAA of 2-MF was thus extensively studied using various carbonyl compounds, such as furfural, HMF, acetone, butanal, hydroxyacetone, acetoin, cyclohexanone, mesityl oxide, angelica lactone, and ethyl levulinate [88−93]. For example, the HAA reaction between 2-MF and HMF produces the C_{16} fuel precursor (Fig. 3.5). These studies employed mineral acids, organic acids, and solid acid catalysts. Among the mineral and organic acids, triflic acid showed the highest catalytic activity [93]. The heteropolyacids displayed low selectivity to the HAA product owing to the over-alkylation reaction [91].

Nafion-212 displayed the highest catalytic activity (99.9% conversion) and the yield of HAA product (72.4%) [94]. Over Nafion-212, the product yield was increased from 2.1% to 41.8% for rising the reaction temperature from 273K to 323K [94]. The increase in product yield was, however, insignificant for the further increase in reaction temperature to 353K. The catalyst loading increases the 2-MF conversion and product yield due to the availability of more active sites. Sulfonated carbon was studied for HAA of furfural with 2-MF in a batch reactor [95]. More

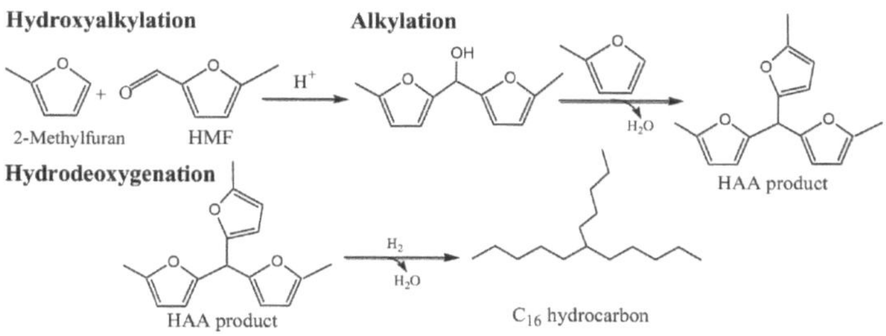

Figure 3.5 HAA of 2-MF with HMF and HDO of the HAA product.

than 90% 2-MF conversion was reported at optimum reaction conditions: 1:1 2-MF/furfural mole ratio, 323K, and 6 h reaction time. The HAA reaction over Zr/SBA-15 showed 93.9% product yield at 413K [96]. HDO of the HAA product was performed at 623K and 60 bar H_2 using activated carbon-supported noble metal catalysts [93]. 5 wt.% Pd/C (81.0%) and 5 wt.% Pt/C (81.6%) showed the higher carbon yield compared to 5 wt.% Ru/C (60.4%). These alkanes exhibited excellent fuel properties (0.7896 g/mL density and 173K pour point). Ni/HZSM-5 displayed high catalytic activity and excellent stability, with about 90% diesel-range alkanes (C_9-C_{14}) [92].

3.4 Fuel additives

The fuel additives are derived from methanol and platform chemicals, such as furfural, HMF, LA, and glycerol. The methanol is produced from biomass-derived syngas by FTS reaction. Quaker Oats first commercialized the furfural production process from oat hulls in 1921 [97]. This process was carried out in a batch reactor at 443K–458K using aqueous H_2SO_4 as the catalyst. The HMF is obtained by the acid-catalyzed dehydrocyclization reaction of cellulose. LA is produced by either rehydration of HMF or hydrolysis of furfuryl alcohol. About 10% glycerol is obtained as a by-product in the transesterification reaction. The properties of the fuel additives are compared with bio-ethanol, gasoline, and diesel, as shown in Table 3.6.

3.4.1 *γ-Valerolactone (GVL)*

GVL is produced by hydrogenation of LA over noble metal (Ru, Ir, Pd, and Pt) catalysts (Fig. 3.6) [103]. Among these noble metals, Ru showed the highest GVL yield (97%) at 423K and 54 bar H_2 [104]. The conversion of LA was almost 100% over 5% Ru/C, and the conversion remained stable for 10 days due to the higher dispersion of Ru nanoparticles on the support compared to Pt and Pd [103]. The transition metal catalysts, such as Cu-ZrO_2, and Cu-Al_2O_3, and hydrotalcite (CuAl, CuCr, and CuFe), were also investigated in this reaction [105,106]. More than 99.9% GVL selectivity was obtained over Cu-ZrO_2 catalysts [106]. Among hydrotalcite, CuCr showed the best catalytic performance: 91% GVL yield at 473K and 70 bar H_2 [105]. Among Al_2O_3, SiO_2, ZnO, ZrO_2, TiO_2, and MgO supported Ni catalysts, 30% Ni/SiO_2 showed the highest catalytic activity with 87% GVL yield and excellent time-on-stream stability at 523K and atmospheric pressure [107]. A process for hydrogenation of LA was also developed using in situ H_2 generated from formic acid. This process showed 96% GVL yield over the bifunctional Ru-P/SiO_2 catalyst [108].

3.4.2 *Furanic compounds*

Furanic compounds, such as 2-MF, 2-MTHF, 2,5-DMF, and 2,5-DMTHF, are derived by selective hydrogenation of furfural and HMF (Fig. 3.7). 2-MF and

Table 3.6 Properties of various fuel additives [98–102].

	Gasoline	Diesel	Bio-ethanol	EL	DME	GVL	2-MF	2-MTHF	2,5-DMF	2,5-DMTHF	5-EMF
Molecular formula	C_5–C_{12}	C_{12}–C_{25}	C_2H_5OH	$C_7H_{12}O_3$	C_2H_6O	$C_5H_8O_2$	C_5H_6O	$C_5H_{10}O$	C_6H_8O	$C_6H_{12}O$	$C_8H_{10}O_3$
Melting point, K	–	–	158.6	298	132	242	184	137	211	228	–
Boiling point, K	308–413	513–633	351.4	479.2	248.1	480	337	351	367	365	507
Flash point, K	230	338–341	285.5	363–367	232	369	251	261	272	300	368
Vapor pressure@ 293K, kPa	–	<1	5.95	0.033	510	0.066	13.9	13.6	7.07	6.64	0.007
Lower heating value, MJ/L	44–46 (MJ/kg)	42–46 (MJ/kg)	22.8	24.8	18.92	26.2	27.6	28.2	30.1	31	–
Density, g/mL	0.739	0.820	0.789	1.014	0.661	1.046	0.91	0.854	0.89	0.83	1.1
Research octane number	95	20–30	107	110	–	100	103	86	119	82	–
Motor octane number	90	20–30	89	102	–	100	86	73	88.1	–	–
Cetane number	5–20	40–55	–	< 10	55–60	< 10	–	–	9	17.2	–
Carbon, wt.%	86	–	52.2	58.3	52.1	60	73.2	69.7	75.0	72	62.3
Hydrogen, wt.%	14	–	13.0	8.3	13.0	8	7.3	11.7	8.3	12	6.5
Oxygen, wt.%	0	0	34.8	33.3	34.8	32	19.5	18.6	16.7	16	31.2
Enthalpy of vaporization, kJ/kg	260–360	250–600	838.3	306.7	460	442.4	357	364.4	332	348	–
Water miscibility, mg/mL	0	0	Miscible	45.6	Miscible	> = 1	3	150	1.47	Immiscible	8.3
Energy density, MJ/L	33.5	38.6	21.2	–	–	–	28.5	–	29.3	31.8	30.3

DME, Dimethyl ether; *EL*, ethyl levulinate, *GVL*, γ-valerolactone; *2-MF*, 2-methylfuran; *2-MTHF*, 2-methyltetrahydrofuran; *2,5-DMF*, 2,5-dimethylfuran; *2,5 DMTHF*, 2,5-dimethyltetrahydrofuran; *5-EMF*, 5-ethoxymethylfurfural.

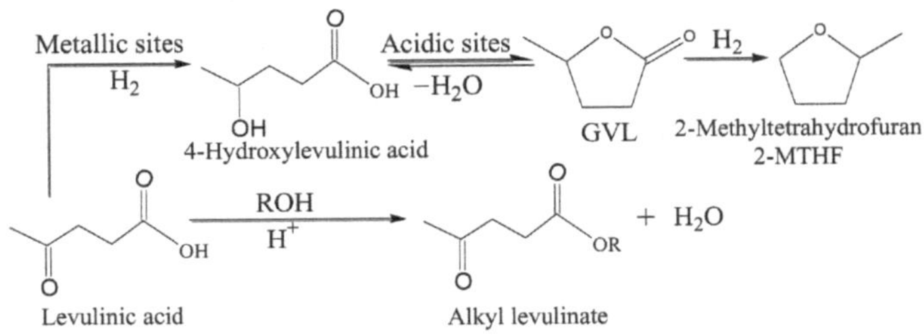

Figure 3.6 Reaction pathway for the production of GVL, 2-MTHF and alkyl levulinate from LA.

2-MTHF are obtained from furfural, whereas 2,5-DMF and 2,5-DMTHF are derived from HMF. These compounds are the potential candidates as the additive in gasoline due to their excellent fuel properties, such as high energy density, high octane number, and hydrophobic nature (Table 3.6) [109].

3.4.2.1 2-Methylfuran (2-MF)

The selective hydrogenation of furfural to 2-MF is carried out under vapor-phase in a fixed-bed reactor at 473K−573K, atmospheric pressure, and 2−29 hydrogen/furfural mole ratio. The reaction proceeds through the furfuryl alcohol intermediate. The reaction was studied using various metals (Cu, Fe, Zn, Pd, etc.) and bimetal catalysts (usually Cu-based catalysts) [110]. Among Raney-Cu, Cu/alumina, and $Cu_2Cr_2O_5$, the $Cu_2Cr_2O_5$ showed high catalytic activity and stability [98]. These catalysts showed about 85%−95% 2-MF yield. However, they deactivate slowly due to the deposition of coke. Cu-Zn-Al catalyst showed 99.9% conversion with 93% 2-MF yield at 498K and $0.7\ h^{-1}$ LHSV [98,111]. The silica-supported Ni-Fe catalyst displayed higher selectivity to 2-MF compared to Ni at 483K−523K and 1 bar hydrogen [112]. Zheng et al., studied this reaction using commercial (Cu/Zn/Al/Ca/Na = 59:33:6:1:1) and prepared (Cu/Cr/Ni/Zn/Fe = 43:45:8:3:1) catalysts at 473K−573K [113]. The commercial catalyst (87%) showed a higher 2-MF yield compared to the prepared catalyst. About 97% 2-MF yield was reported over CuZnCa/$\gamma-Al_2O_3$ at 508K, $0.32\ h^{-1}$ LHSV, and 10:1 hydrogen/furfural mole ratio [98].

3.4.2.2 2-Methyltetrahydrofuran (2-MTHF)

2-MTHF is highly hydrophobic. It has a higher heating value, higher density, low flammability, and lower toxicity compared to bio-ethanol. These characteristics allow the blending of 2-MTHF with gasoline up to about 70% [114]. The US Department of Energy further approved 2-MTHF as one of the P-Series fuels. 2-MTHF is produced by either selective hydrogenation of furfural (Fig. 3.7) or sequential hydrogenation of LA (Fig. 3.6). The selective hydrogenation of furfural is carried out under either vapor phase or liquid phase (20−80 bar hydrogen) using Ru, Re, Pd, Cu, and Ni catalyst. The direct conversion of furfural to 2-MTHF was carried out at 453K in the two-stage packed-bed reactor using two different

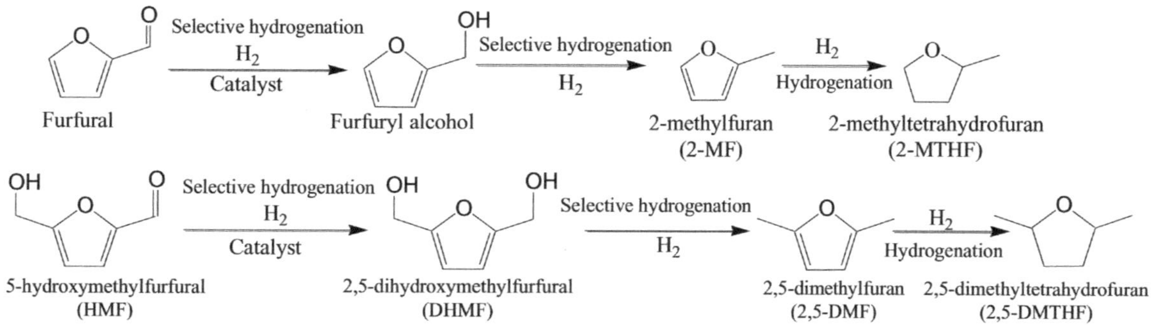

Figure 3.7 Reaction pathway for production of furanic compounds from furfural and HMF.

catalysts—the first reactor being Cu phyllosilicate and second being Pd/SiO_2 [109]. The Cu phyllosilicate favors C = O hydrogenolysis, while Pd/SiO_2 favors C = C hydrogenation reaction. The complete furfural conversion with 97.1% 2-MTHF yield was observed. In the second pathway, LA is hydrogenated to 4-hydroxypentoic acid, which upon dehydration, produces GVL. GVL is then hydrogenated to 1,4-pentanediol, which upon dehydration, yields 2-MTHF. The reaction is carried out over metal catalysts (Cu, Ni, and noble metals) [115]. The US Pacific Northwest National Laboratory patented one-step production of 2-MTHF from LA using PdRe/C catalyst [116]. The Cu/SiO_2 catalyst showed 64% selectivity to 2-MTHF at 538K and 10 bar H_2 [117]. The addition of Ni in Cu/SiO_2 promoted 2-MTHF selectivity (89%) and stability up to 320 h time-on-stream. The Cu/ZrO_2 catalyst showed high selectivity to 2-MTHF (93%) from GVL at 513K and 60 bar H_2 [118].

3.4.2.3 2,5-Dimethylfuran (2,5-DMF)

2,5-DMF has a higher boiling point, higher energy density, and lower water miscibility compared to 2-MF and 2-MTHF. It is thus an ideal candidate as a fuel additive. The selective hydrogenation of HMF is carried out in a batch reactor using either H-donor organic solvent or molecular hydrogen (up to 30 bar) at 423K–493K over supported metal (Ru, Pt, Pd, Ni, Cu, etc.) catalysts [119]. The reaction temperature requirement is, however, higher for Cu and Ni catalysts compared to Pd, Ru, and Pt [110,111]. The 2,5-DMF yield was in the following order: Pd/C > Ru/C > Cu/ZnO > Pt/C > Ni/C [110]. The two-step process was further developed for manufacturing 2,5-DMF from fructose: conversion of fructose to HMF in a biphasic reactor followed by vapor-phase hydrogenolysis of HMF over CuRu/C catalyst [111]. The minimum selling price of HMF and 2,5-DMF in this process was US\$1.33 and US\$2.02 L^{-1}, respectively, for the plant capacity of 300 tons of fructose per day [120].

3.4.2.4 2,5-Dimethyltetrahydrofuran (2,5-DMTHF)

2,5-DMTHF has a higher energy density than 2,5-DMF. 2,5-DMTHF was produced from fructose using a bifunctional catalyst ($RuCl_3 \cdot xH_2O$ and HI) [121]. HI converts the fructose to HMF by dehydration reaction, while hydrogenation of C = C and C = O bonds occurs over Rh [122]. The addition of benzene increased 2,5-DMTHF yield (75%) owing to the extraction of unstable reaction intermediates. The reaction was further studied over the Ni-Al_2O_3 catalyst at 453K and 12 bar H_2 [123]. The catalyst with the calcination temperature of 1123K showed the highest 2,5-DMTHF yield (97.4%). A combination of layered double hydroxides and nitrogen-doped, carbon-decorated, copper catalyst (NC-Cu/MgAlO) was employed for the hydrogenolysis of HMF [124]. At 423K and 30 min reaction time, 96.1% 2,5-DMF yield was observed. Prolonged reaction (nearly 2.8 h) resulted in 94.6% yield of 2,5-DMTHF.

3.4.3 5-Ethoxymethylfurfural (5-EMF)

5-Alkoxymethylfurfurals are the new generation fuel additive for diesel and jet fuel. For example, 5-EMF, having a high-boiling point and good-energy density, is suitable as a diesel fuel additive. This fuel blend also reduces the emission of particulate matters and SOx [111]. 5-EMF is produced from HMF by etherification reaction with ethanol. It can also be produced by one-pot dehydration of fructose, followed by the etherification reaction of HMF with ethanol. Ethyl levulinate and 1,1-diethoxyethane (acetal) are also coproduced in this reaction (Fig. 3.8). The reaction is operated at 448K–498K using either homogeneous or heterogeneous acid catalysts. The solid-acid catalysts include mesoporous materials, sulfonic acid resins, and organic-inorganic hybrid solid. At 413K and 2 h reaction time, 5-EMF yield was in the following order: Z-SBA-15 > MCM-41(Si/Al = 50) > SZ-SBA-15 > MCM-41 (Si/Al = 25) [111]. The strong Brönsted acidity (H_2SO_4 and Amberlyst15) favors ethyl levulinate, while weak acidity (pure SBA-15) promotes acetal formation. The strong Lewis acid (Zr-SBA-15) is, however, suitable for the formation of 5-EMF [111]. Branched and high molecular weight 5-alkoxymethylfurfurals, such as 5-isopropoxymethylfurfural, 5-t-butoxymethylfurfural, and 5-octyloxymethylfurfural, demonstrate even better cold-flow properties compared to 5-EMF [110].

3.4.4 Alkyl levulinates

Alkyl levulinates, such as methyl, ethyl, and butyl levulinate, are suitable as fuel additive [98,125]. They are generally produced from LA by esterification reaction with alcohol in the presence of an acid catalyst. A vast range of solid-acid catalysts, such as heteropolyacid, sulfonic acid functionalized SBA-15, sulfated zirconia, and resin was employed for this reaction. The mesoporous heteropolyacid silica showed 73% and 67% methyl and ethyl levulinate yield at 338K and 348K, respectively [126]. Among various heteropolyacids (phosphotungstic acid, silicotungstic acid, and phosphomolybdic acid), phosphotungstic acid exhibited the highest catalytic activity with 92% ethyl levulinate and butyl levulinate yield [127]. A series of sulfated metal oxides, such as ZrO_2, TiO_2, ZrO_2-TiO_2, and ZrO_2-Al_2O_3, were tested for one-step production of ethyl levulinate from glucose at 473K [128]. The higher temperatures (483K and above), however, produced undesirable side products. Alkyl levulinates are also produced from furfuryl alcohol by alcoholysis reaction in the presence of an acid catalyst (Fig. 3.9). This reaction was carried out in a batch

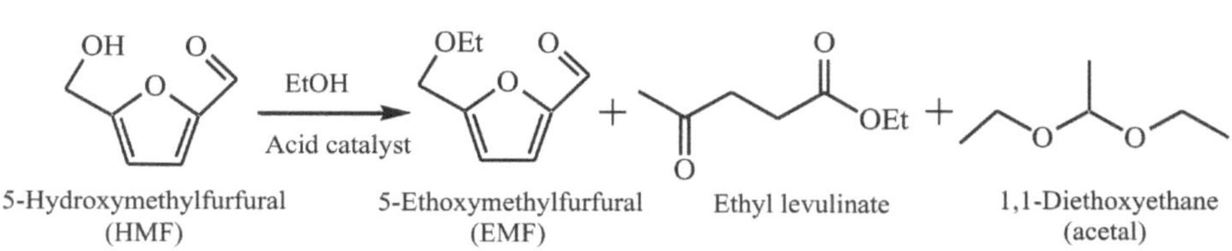

Figure 3.8 Etherification of HMF to 5-alkoxymethylfurfural.

reactor at 473K over the zeolite and sulfated metal oxide catalysts [129]. The ethyl levulinate yield was in the following order: $SO_4^{2-}/TiO_2 > SO_4^{2-}/SnO_2 > SO_4^{2-}/Al_2O_3 > SO_4^{2-}/ZrO_2 >$ HZSM-5 > HY > H-mordenite. Al_2O_3/SBA-15 showed almost 100% conversion of furfuryl alcohol with 94% selectivity to butyl levulinate at 383K [130].

3.4.5 Glycerol acetals

The glycerol acetals are produced by the acetalization of glycerol with aldehyde or ketone (Fig. 3.10). This reaction is carried out using either homogeneous (p-toluenesulfonic acid and H_2SO_4) or heterogeneous (zeolite, heteropolyacid, cationic resin, and mixed metal oxide) catalysts. Homogeneous catalysts are, however, highly corrosive [131]. SBA-15 supported molybdenum phosphate exhibited almost complete conversion of glycerol with 98% selectivity to solketal [132]. Catalyst leaching was, however, observed during the reaction. SBA-15 supported mixed metal oxide (Zr, Al, Ti, and Nb) was used as the catalyst for the acetalization of glycerol with acetone at room temperature [133]. The catalytic activity was decreased in the following order: Nb/SBA-15 > Zr/SBA-15 > Ti/SBA-15 > Al/SBA-15. The amberlyst-36 showed the best catalytic performance for glycerol acetalization with benzaldehyde and formaldehyde [134]. The hierarchical zeolites (BEA, MFI, and MOR) showed almost 100% selectivity to solketal at 343K and 1:1 glycerol/acetone mole ratio [135].

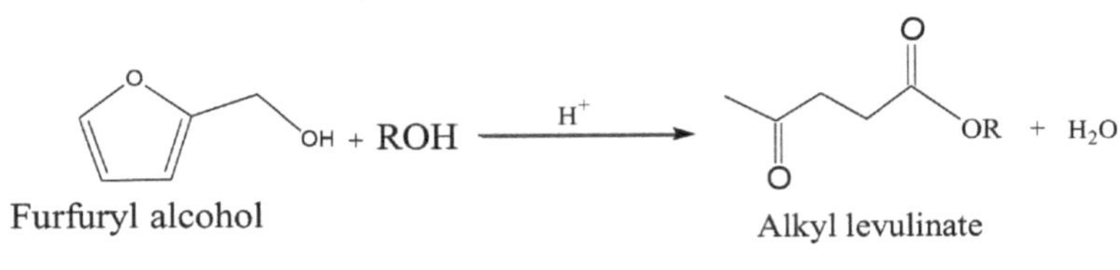

Figure 3.9 Alcoholysis of furfuryl alcohol to alkyl levulinate.

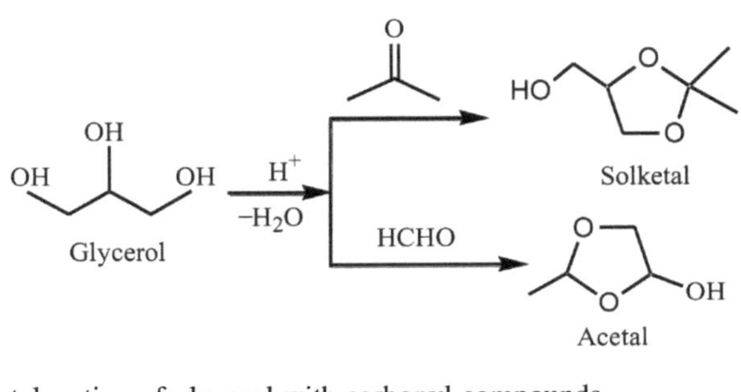

Figure 3.10 Acetalyzation of glycerol with carbonyl compounds.

3.4.6 Dimethyl ether (DME)

Methanol is converted to DME for application as a fuel additive. This reaction is carried out using a solid-acid catalyst, such as γ-Al_2O_3, sulfated zirconia, heteropolyacids, and zeolites [136]. Mobil pioneered this process in 1965 as a crucial step in the manufacturing of gasoline from methanol. The reaction was carried out in a fixed-bed reactor using silica-modified γ-Al_2O_3 as the catalyst at 573K and atmospheric pressure [137]. The 6% SiO_2/γ-Al_2O_3 showed no by-products and better catalytic performance compared to the untreated γ-Al_2O_3. The water adsorbed on Lewis acid sites of γ-Al_2O_3 decreases the catalytic activity. The zeolites exhibit high thermal stability, but high acid strength leads to the generation of undesirable by-products [138]. The selectivity to DME was reduced with the rise in the reaction temperature. For example, the selectivity to DME was reduced from 88.1% to 75.9% for increasing the reaction temperature from 643K to 673K over the K-HZSM-5 catalyst [139]. The heteropolyacids showed 89% methanol conversion and 100% DME yield at 423K with a stable catalyst performance up to 4 h time-on-stream [140].

3.5 Conclusion

This chapter presents an overview of various types of biomass and their chemistries, concepts of biorefinery, and different kinds of biofuels: traditional biofuels, hydrocarbon biofuels, and fuel additives. Biodiesel, bio-ethanol, and bio-butanol are traditional biofuels. Biodiesel and bio-ethanol are mixed with diesel and gasoline, respectively, while bio-butanol can be blended with gasoline in any ratio or used in pure form. Bio-butanol is currently facing several issues, such as low-solvent titer and expensive ABE separation. Bio-ethanol and bio-butanol from lignocellulosic biomass are also very expensive. The hydrocarbon biofuels are similar to gasoline, diesel, and jet fuel and compatible with current infrastructures. HDO of triglycerides produces high-quality green diesel at an affordable price. The HDO reaction dominates over NiMo and CoMo catalysts, while the decarbonylation reaction prevails over Ni catalyst. Both HDO and decarbonylation reactions are, however, significant over the Co catalyst. The mesoporous and composite catalyst enhances the catalytic performance. Fast pyrolysis is the potential process for the direct conversion of inexpensive and abundant lignocellulosic biomass to hydrocarbon biofuels. This process is carried out using either a bubbling fluidized-bed reactor or circulating fluidized-bed reactor at a moderate temperature with a short vapor residence time. This process is quite simple and economical and suitable for decentralized biorefinery. FTS produces all ranges of hydrocarbon biofuels and is carried out using promoted Fe and Co catalysts. HTFT and LTFT are suitable for gasoline and diesel, respectively. Brönsted acid sites are responsible for the oligomerization of olefins. Topology and acid strength of zeolites play an important role in the oligomerization reaction. The C−C bond-forming reaction produces high-molecular-weight oxygenated compounds that are subjected to HDO to obtain diesel or aviation fuel range alkanes. This process, however, involves multiple steps with the high cost of production. The fuel additives, such as GVL, furanic compounds, 5-EMF, alkyl levulinates, and glycerol acetal, are produced from platform chemicals, such as furfural, HMF, LA, and glycerol. The dimethyl ether is, however, produced from methanol. These processes also involve multiple steps with high production costs.

Abbreviations

ABE	acetone-butanol-ethanol
DME	dimethyl ether
2,5-DMF	2,5-dimethylfuran
2,5-DMTHF	2,5-dimethyltetrahydrofuran
5-EMF	5-ethoxymethylfurfural
FTS	Fischer–Tropsch synthesis
GVL	γ-valerolactone
HAA	hydroxyalkylation-alkylation
HDO	hydrodeoxygenation
HMF	5-hydroxymethylfurfural
HTFT	high-temperature Fischer–Tropsch synthesis
LA	levulinic acid
LGE	liter of gasoline-equivalent
LTFT	low-temperature Fischer–Tropsch synthesis
2-MF	2-methylfuran
2-MTHF	2-methyltetrahydrofuran

References

[1] <https://www.eia.gov/outlooks/aeo/pdf/0383(2012).pdf>.
[2] Government of India, National policy on biofuels, Ministry of New and Renewable Energy, 2008, pp. 1–18.
[3] S.K. Maity, Opportunities, recent trends and challenges of integrated biorefinery: part I, Renew. Sustain. Energy Rev. 43 (2015) 1427–1445.
[4] S.K. Maity, Opportunities, recent trends and challenges of integrated biorefinery: part II, Renew. Sustain. Energy Rev. 43 (2015) 1446–1466.
[5] A. Demirbas, Biodiesel from vegetable oils via transesterification in supercritical methanol, Energy Convers. Manag. 43 (2002) 2349–2356.
[6] Y. Zhang, M.A. Dubé, D.D. McLean, M. Kates, Biodiesel production from waste cooking oil: 1. Process design and technological assessment, Bioresour. Technol. 89 (2003) 1–16.
[7] O.S. Stamenković, A.V. Veličković, V.B. Veljković, The production of biodiesel from vegetable oils by ethanolysis: current state and perspectives, Fuel 90 (2011) 3141–3155.
[8] J.A. Melero, J. Iglesias, G. Morales, Heterogeneous acid catalysts for biodiesel production: current status and future challenges, Green. Chem. 11 (2009) 1285–1308.
[9] S. Nomanbhay, M.Y. Ong, A review of microwave-assisted reactions for biodiesel production, Bioengineering 4 (2017) 1–21.

[10] L.C. Meher, D. Vidya Sagar, S.N. Naik, Technical aspects of biodiesel production by transesterification—a review, Renew. Sustain. Energy Rev. 10 (2006) 248–268.
[11] J.V. Gerpen, Biodiesel processing and production, Fuel Process. Technol. 86 (2005) 1097–1107.
[12] Y. Zhang, M.A. Dubé, D.D. McLean, M. Kates, Biodiesel production from waste cooking oil: 2. Economic assessment and sensitivity analysis, Bioresour. Technol. 90 (2003) 229–240.
[13] L. Tao, A. Aden, he economics of current and future biofuels. In: D. Tomes, P. Lakshmanan, D. Songstad. (Eds), *Biofuels*. 2011, Springer, New York, NY, 37–69.
[14] P. Dürre, Biobutanol: an attractive biofuel, Biotechnol. J. 2 (2007) 1525–1534.
[15] A. Pugazhendhi, T. Mathimani, S. Varjani, E.R. Rene, G. Kumar, S.H. Kim, et al., Biobutanol as a promising liquid fuel for the future - recent updates and perspectives, Fuel (**253**, 2019) 637–646.
[16] S.-Y. No, Application of biobutanol in advanced CI engines – a review, Fuel 183 (2016) 641–658.
[17] A. Demirbas, Competitive liquid biofuels from biomass, Appl. Energy 88 (2011) 17–28.
[18] M.N.A.M. Yusoff, N.W.M. Zulkifli, B.M. Masum, H.H. Masjuki, Feasibility of bioethanol and biobutanol as transportation fuel in spark-ignition engine: a review, RSC Adv. 5 (2015) 100184–100211.
[19] J.C. Serrano-Ruiz, E. Ramos-Fernández, A. Sepúlveda-Escribano, From biodiesel and bioethanol to liquid hydrocarbon fuels: new hydrotreating and advanced microbial technologies, Energy Env. Sci. 5 (2012) 5638–5652.
[20] M. Balat, H. Balat, Recent trends in global production and utilization of bio-ethanol fuel, Appl. Energy 86 (2009) 2273–2282.
[21] Renewable Fuels Association, Ethanol strong: 2018 Ethanol industry outlook, <https://ethanolrfa.org/wp-content/uploads/2018/02/NECfinalOutlook.pdf>, 2018.
[22] M. Balat, H. Balat, C. Öz, Progress in bioethanol processing, Prog. Energy Combust. Sci. 34 (2008) 551–573.
[23] H.N. Abubackar, M.C. Veiga, C. Kennes, Biological conversion of carbon monoxide: rich syngas or waste gases to bioethanol, *Biofuels, Bioprod. Bioref.* **5**, 2011, 93–114.
[24] M. Morales, J. Quintero, R. Conejeros, G. Aroca, Life cycle assessment of lignocellulosic bioethanol: environmental impacts and energy balance, Renew. Sustain. Energy Rev. 42 (2015) 1349–1361.
[25] B.O. Abo, M. Gao, Y. Wang, C. Wu, H. Ma, Q. Wang, Lignocellulosic biomass for bioethanol: an overview on pretreatment, hydrolysis and fermentation processes, Rev. Env. Health 34 (2019) 57–68.
[26] M. Vohra, J. Manwar, R. Manmode, S. Padgilwar, S. Patil, Bioethanol production: feedstock and current technologies, J. Env. Chem. Eng. 2 (2014) 573–584.
[27] L.M. Vane, Separation technologies for the recovery and dehydration of alcohols from fermentation broths, Biofuels, Bioprod. Bioref. 2 (2008) 553–588.
[28] D. Carrillo-Nieves, M.J. Rostro Alanís, R. de la Cruz Quiroz, H.A. Ruiz, H.M.N. Iqbal, R. Parra-Saldívar, Current status and future trends of bioethanol production from agro-industrial wastes in Mexico, Renew. Sustain. Energy Rev. 102 (2019) 63–74.
[29] J.K. Saini, R. Saini, L. Tewari, Lignocellulosic agriculture wastes as biomass feedstocks for second-generation bioethanol production: concepts and recent developments, Biotech 5 (2015) 337–353.
[30] <https://www.oecd.org/berlin/44567743.pdf>.

[31] M. Kumar, K. Gayen, Biobutanol: The future biofuel, In: C. Baskar, S. Baskar, R. Dhillon, (Eds), *Biomass Conversion,* 2012, Springer, Berlin, Heidelberg.
[32] Y. Li, W. Tang, Y. Chen, J. Liu, Cf.F. Lee, Potential of acetone-butanol-ethanol (ABE) as a biofuel, Fuel 242 (2019) 673–686.
[33] K. Karimi, M. Tabatabaei, I.S. Horváth, R. Kumar, Recent trends in acetone, butanol, and ethanol (ABE) production, Biofuel Res. J. 8 (2015) 301–308.
[34] C. Xue, X.-Q. Zhao, C.-G. Liu, L.-J. Chen, F.-W. Bai, Prospective and development of butanol as an advanced biofuel, Biotechnol. Adv. 31 (2013) 1575–1584.
[35] S. Hu, P. Qin, T. Tan, Separation of butanol, acetone, and ethanol, in: H.N. Chang (Ed.), Emerging Areas in Bioengineering, first ed, Wiley-VCH Verlag GmbH & Co. KGaA, 2018, pp. 255–285.
[36] A. Friedl, Downstream process options for the ABE fermentation, FEMS Microbiol. Lett. 363 (2016) 329–336.
[37] N. Qureshi, H. Blaschek, Recovery of butanol from fermentation broth by gas stripping, Renew. Energy 22 (2001) 557–564.
[38] A. Kurkijärvi, J. Lehtonen, J. Linnekoski, Novel dual extraction process for acetone–butanol–ethanol fermentation, Sep. Purif. Technol. 124 (2014) 18–25.
[39] M. Kumar, Y. Goyal, A. Sarkar, K. Gayen, Comparative economic assessment of ABE fermentation based on cellulosic and non-cellulosic feedstocks, Appl. Energy 93 (2012) 193–204.
[40] P. Kumar, S.R. Yenumala, S.K. Maity, D. Shee, Kinetics of hydrodeoxygenation of stearic acid using supported nickel catalysts: effects of supports, Appl. Catal. A Gen. 471 (2014) 28–38.
[41] P. Kumar, M. Varkolu, S. Mailaram, A. Kunamalla, S.K. Maity, Biorefinery polyutilization systems: production of green transportation fuels from biomass, in: K.R. Khalilpour (Ed.), Polygeneration with Polystorage: For Chemical and Energy Hubs, Elsevier Inc., 2019, pp. 373–407.
[42] E. Furimsky, Catalytic hydrodeoxygenation, Appl. Catal. A Gen. 199 (2000) 147–190.
[43] M. Ameen, M.T. Azizan, S. Yusup, A. Ramli, M. Yasir, Catalytic hydrodeoxygenation of triglycerides: an approach to clean diesel fuel production, Renew. Sustain. Energy Rev. 80 (2017) 1072–1088.
[44] S.R. Yenumala, S.K. Maity, D. Shee, Hydrodeoxygenation of karanja oil over supported nickel catalysts: influence of support and nickel loading, Catal. Sci. Technol. 6 (2016) 3156–3165.
[45] T.N. Kalnes, T. Marker, D.R. Shonnard, K.P. Koers, Green diesel production by hydrorefining renewable feedstocks, Biofuels Technol. (2008) 7–11.
[46] P. Kumar, S.K. Maity, D. Shee, Hydrodeoxygenation of stearic acid using Mo modified Ni and Co/alumina catalysts: effect of calcination temperature, Chem. Eng. Commun. 207 (2020) 904–919.
[47] J.G. Immer, M.J. Kelly, H.H. Lamb, Catalytic reaction pathways in liquid-phase deoxygenation of C_{18} free fatty acids, Appl. Catal. A Gen. 375 (2010) 134–139.
[48] K. Ludek, D. Kubic, The comparison of Co, Ni, Mo, CoMo and NiMo sulfided catalysts in rapseed oil hydrodeoxygenation, React. Kinet. Mech. Catal. 122 (2017) 333–341.
[49] Q. Liu, H. Zuo, T. Wang, L. Ma, Q. Zhang, One-step hydrodeoxygenation of palm oil to isomerized hydrocarbon fuels over Ni supported on nano-sized SAPO-11 catalysts, Appl. Catal. A Gen. 468 (2013) 68–74.
[50] S.R. Yenumala, P. Kumar, S.K. Maity, D. Shee, Hydrodeoxygenation of karanja oil using ordered mesoporous nickel-alumina composite catalysts, Catal. Today 348 (2020) 45–55.

[51] S. Mailaram, S.K. Maity, Techno-economic evaluation of two alternative processes for production of green diesel from karanja oil: a pinch analysis approach, J. Renew. Sustain. Energy 11 (2019). 025906.

[52] S.B. Glisic, J. Pajnik, A.M. Orlović, Process and techno-economic analysis of green diesel production from waste vegetable oil and the comparison with ester type biodiesel production, Appl. Energy 170 (2016) 176–185.

[53] A.V. Bridgwater, Review of fast pyrolysis of biomass and product upgrading, Biomass Bioenergy 38 (2012) 68–94.

[54] S. Zhang, Y. Yan, T. Li, Z. Ren, Upgrading of liquid fuel from the pyrolysis of biomass, Bioresour. Technol. 96 (2005) 545–550.

[55] <http://task34.ieabioenergy.com/pyrolysis-reactors/>.

[56] A. Oasmaa, B. Van de Beld, P. Saari, D.C. Elliott, Y. Solantausta, Norms, standards, and legislation for fast pyrolysis bio-oils from lignocellulosic biomass, Energy Fuels 29 (2015) 2471–2484.

[57] D.C. Elliott, T.R. Hart, Catalytic hydroprocessing of chemical models for bio-oil, Energy Fuels 23 (2009) 631–637.

[58] C. Masters, The Fischer-Tropsch reaction, Adv. Organomet. Chem. 17 (1979) 61–103.

[59] S.S. Ail, S. Dasappa, Biomass to liquid transportation fuel via Fischer Tropsch synthesis - technology review and current scenario, Renew. Sustain. Energy Rev. 58 (2016) 267–286.

[60] M.J. Tijmensen, A.P. Faaij, C.N. Hamelinck, M.R. van Hardeveld, Exploration of the possibilities for production of Fischer Tropsch liquids and power via biomass gasification, Biomass Bioenergy 23 (2002) 129–152.

[61] M.E. Dry, The Fischer-Tropsch process: 1950-2000, Catal. Today 71 (2002) 227–241.

[62] M.E. Dry, Fischer-Tropsch reactions and the environment, Appl. Catal. A Gen. 189 (1999) 185–190.

[63] A.P. Steynberg, Introduction to Fischer-Tropsch technology, Stud. Surf. Sci. Catal. 152 (2004) 1–63.

[64] M.E. Dry, High quality diesel via the Fischer-Tropsch process - a review, J. Chem. Technol. Biotechnol. 77 (2002) 43–50.

[65] O.P.R. Van Vliet, A.P.C. Faaij, W.C. Turkenburg, Fischer–Tropsch diesel production in a well-to-wheel perspective: a carbon, energy flow and cost analysis, Energy Convers. Manag. 50 (2009) 855–876.

[66] O. Muraza, Maximizing diesel production through oligomerization: a landmark opportunity for zeolite research, Ind. Eng. Chem. Res. 54 (2015) 781–789.

[67] J. Skupinska, Oligomerization of α-olefins to higher oligomers, Chem. Rev. 91 (1991) 613–648.

[68] C.T. O'Connor, M. Kojima, Alkene oligomerization, Catal. Today 6 (1990) 329–349.

[69] A. de Klerk, D.O. Leckel, N.M. Prinsloo, Butene oligomerization by phosphoric acid catalysis: separating the effects of temperature and catalyst hydration on product selectivity, Ind. Eng. Chem. Res. 45 (2006) 6127–6136.

[70] A. de Klerk, Properties of synthetic fuels from H-ZSM-5 oligomerization of Fischer–Tropsch type feed materials, Energy Fuels 21 (2007) 3084–3089.

[71] C.P. Nicholas, Applications of light olefin oligomerization to the production of fuels and chemicals, Appl. Catal. A Gen. 543 (2017) 82–97.

[72] M. Sanati, C. Hörnell, S.G. Järäs, The oligomerization of alkenes by heterogeneous catalysts, Catalysis 14 (1999) 236–288.

[73] X. Zhang, J. Zhong, J. Wang, L. Zhang, J. Gao, A. Liu, Catalytic performance and characterization of Ni-doped HZSM-5 catalysts for selective trimerization of n-butene, Fuel Process. Technol. 90 (2009) 863–870.

[74] A.G. Gayubo, A. Alonso, B. Valle, A.T. Aguayo, J. Bilbao, Selective production of olefins from bioethanol on HZSM-5 zeolite catalysts treated with NaOH, Appl. Catal. B Env. 97 (2010) 299−306.
[75] S.A. Tabak, F.J. Krambeck, W.E. Garwood, Conversion of propylene and butylene over ZSM-5 catalyst, AIChE J. 32 (1986) 1526−1531.
[76] A.A. Avidan, Gasoline and distillate fuels from methanol, Stud. Surf. Sci. Catal. 36 (1988) 307−323.
[77] J.H. Gary, G.E. Handwerk, Petroleum Refining: Technology and Economics, fourth ed., Marcel Dekker, Inc, 2001, pp. 231−236.
[78] M.E. Wright, B.G. Harvey, R.L. Quintana, Highly efficient zirconium-catalyzed batch conversion of 1-butene: a new route to jet fuels, Energy Fuels 22 (2008) 3299−3302.
[79] https://www.prnewswire.com/news-releases/cobalt-and-the-naval-air-warfare-center-team-up-to-produce-a-renewable-jet-fuel-from-bio-n-butanol-143461676.html.
[80] R.M. West, Z.Y. Liu, M. Peter, C.A. Gärtner, J.A. Dumesic, Carbon-carbon bond formation for biomass-derived furfurals and ketones by aldol condensation in a biphasic system, J. Mol. Catal. A Chem. 296 (2008) 18−27.
[81] H. Liu, W. Xu, X. Liu, Aldol condensation of furfural and acetone on layered double hydroxides, Kinet. Catal. 51 (2010) 75−80.
[82] L. Faba, E. Díaz, S. Ordóñez, Aqueous-phase furfural-acetone aldol condensation over basic mixed oxides, Appl. Catal. B Env. 113−114 (2012) 201−211.
[83] W. Shen, G.A. Tompsett, K.D. Hammond, R. Xing, F. Dogan, C.P. Grey, et al., Liquid phase aldol condensation reactions with MgO-ZrO_2 and shape-selective nitrogen-substituted NaY, Appl. Catal. A Gen. 392 (2011) 57−68.
[84] X. Huang, Q. Zhang, T. Wang, Q. Liu, L. Ma, Q. Zhang, Production of jet fuel intermediates from furfural and acetone by aldol condensation over MgO/NaY, J. Fuel Chem. Technol. 40 (2012) 973−978.
[85] J. Yang, N. Li, S. Li, W. Wang, L. Li, A. Wang, et al., Synthesis of diesel and jet fuel range alkanes with furfural and ketones from lignocellulose under solvent free conditions, Green. Chem. 16 (2014) 4879−4884.
[86] Q. Deng, J. Xu, P. Han, L. Pan, L. Wang, X. Zhang, et al., Efficient synthesis of high-density aviation biofuel via solvent-free aldol condensation of cyclic ketones and furanic aldehydes, Fuel Process. Technol. 148 (2016) 361−366.
[87] C.J. Barrett, J.N. Chheda, G.W. Huber, J.A. Dumesic, Single-reactor process for sequential aldol-condensation and hydrogenation of biomass-derived compounds in water, Appl. Catal. B Env. 66 (2006) 111−118.
[88] A. Corma, O. de la Torre, M. Renz, N. Villandier, Production of high-quality diesel from biomass waste products, Angew. Chem. 123 (2011) 2375−2378.
[89] A. Corma, O. de la Torre, M. Renz, High-quality diesel from hexose- and pentose-derived biomass platform molecules, ChemSusChem 4 (2011) 1574−1577.
[90] G. Li, N. Li, Z. Wang, C. Li, A. Wang, X. Wang, et al., Synthesis of high-quality diesel with furfural and 2-methylfuran from hemicellulose, ChemSusChem 5 (2012) 1958−1966.
[91] C. Zhu, T. Shen, D. Liu, J. Wu, Y. Chen, L. Wang, et al., Production of liquid hydrocarbon fuels with acetoin and platform molecules derived from lignocellulose, Green. Chem. 18 (2016) 2165−2174.
[92] S. Li, N. Li, G. Li, L. Li, A. Wang, Y. Cong, et al., Protonated titanate nanotubes as a highly active catalyst for the synthesis of renewable diesel and jet fuel range alkanes, Appl. Catal. B Env. 170 (2015) 124−134.

[93] W. Wang, N. Li, S. Li, G. Li, F. Chen, X. Sheng, et al., Synthesis of renewable diesel with 2-methylfuran and angelica lactone derived from carbohydrates, Green. Chem. 18 (2016) 1218–1223.
[94] G. Li, N. Li, J. Yang, A. Wang, X. Wang, Y. Cong, et al., Synthesis of renewable diesel with the 2-methylfuran, butanal and acetone derived from lignocellulose, Bioresour. Technol. 134 (2013) 66–72.
[95] D. Damodar, A. Kunamalla, M. Varkolu, S.K. Maity, A.S. Deshpande, Near-room-temperature synthesis of sulfonated carbon nanoplates and their catalytic application, ACS Sustain. Chem. Eng. 7 (2019) 12707–12717.
[96] Y.J. Luo, Y.H. Zhou, Y.B. Huang, A new Lewis acidic Zr catalyst for the synthesis of furanic diesel precursor from biomass derived furfural and 2-methylfuran, Catal. Lett. 149 (2019) 292–302.
[97] K.J. Zeitsch (Ed.), The chemistry and technology of furfural and its many by-products, **13**, 2000, 1–358.
[98] K. Yan, G. Wu, T. Lafleur, C. Jarvis, Production, properties and catalytic hydrogenation of furfural to fuel additives and value-added chemicals, Renew. Sustain. Energy Rev. 38 (2014) 663–676.
[99] I. Gandarias, S. García-Fernández, I. Obregón, I. Agirrezabal-Telleria, P.L. Arias, Production of 2-methylfuran from biomass through an integrated biorefinery approach, Fuel Process. Technol. 178 (2018) 336–343.
[100] M.K. Shukla, E. Singh, N. Singh, S.K. Singal, Prospects of 2,5-dimethylfuran as a fuel: physico-chemical and engine performance characteristics evaluation, J. Mater. Cycles Waste Manag. 17 (2015) 459–464.
[101] N. Viar, J.M. Requies, I. Agirre, A. Iriondo, P.L. Arias, Furanic biofuels production from biomass using Cu-based heterogeneous catalysts, Energy 172 (2019) 531–544.
[102] A. Bohre, S. Dutta, B. Saha, M.M. Abu-Omar, Upgrading furfurals to drop-in biofuels: an overview, ACS Sustain. Chem. Eng. 3 (2015) 1263–1277.
[103] P.P. Upare, J. Lee, D. Wang, S.B. Halligudi, Y.K. Hwang, J.-S. Chang, Selective hydrogenation of levulinic acid to γ-valerolactone over carbon-supported noble metal catalysts, Ind. Eng. Chem. Res. 17 (2011) 287–292.
[104] L.E. Manzer, Catalytic synthesis of α-methylene-γ-valerolactone: a biomass-derived acrylic monomer, Appl. Catal. 1 (2004) 249–256.
[105] K. Yan, J. Liao, X. Wu, X. Xie, A noble-metal free Cu-catalyst derived from hydrotalcite for highly efficient hydrogenation of biomass-derived furfural and levulinic acid, RSC Adv. 3 (2013) 3853–3856.
[106] A.M. Hengne, C.V. Rode, Cu-ZrO_2 nanocomposite catalyst for selective hydrogenation of levulinic acid and its ester to γ-valerolactone, Green. Chem. 14 (2012) 1064–1072.
[107] V. Mohan, V. Venkateshwarlu, C.V. Pramod, B.D. Raju, K.S.R. Rao, Vapour phase hydrocyclisation of levulinic acid to γ-valerolactone over supported Ni catalysts, Catal. Sci. Technol. 4 (2014) 1253–1259.
[108] L. Deng, Y. Zhao, J. Li, Y. Fu, B. Liao, Q.X. Guo, Conversion of levulinic acid and formic acid into γ-valerolactone over heterogeneous catalysts, ChemSusChem 3 (2010) 1172–1175.
[109] L. Hu, G. Zhao, W. Hao, X. Tang, Y. Sun, L. Lin, et al., Catalytic conversion of biomass-derived carbohydrates into fuels and chemicals via furanic aldehydes, RSC Adv. 2 (2012) 11184–11206.

[110] S. Chen, R. Wojcieszak, F. Dumeignil, E. Marceau, S. Royer, How catalysts and experimental conditions determine the selective hydroconversion of furfural and 5-hydroxymethylfurfural, Chem. Rev. 118 (2018) 11023–11117.
[111] M.J. Climent, A. Corma, S. Iborra, Conversion of biomass platform molecules into fuel additives and liquid hydrocarbon fuels, Green. Chem. 16 (2014) 516–547.
[112] S. Sitthisa, W. An, D.E. Resasco, Selective conversion of furfural to methylfuran over silica-supported NiFe bimetallic catalysts, J. Catal. 284 (2011) 90–101.
[113] H. Zheng, Y. Zhu, B. Teng, Z. Bai, C. Zhang, H. Xang, et al., Towards understanding the reaction pathway in vapour phse hydrogenation, Mol. Catal. 246 (2006) 18–23.
[114] D.M. Alonso, J.Q. Bond, J.A. Dumesic, Catalytic conversion of biomass to biofuels, Green. Chem. 12 (2010) 1493–1513.
[115] I. Obregõn, I. Gandarias, N. Miletic, A. Ocio, P.L. Arias, One-pot 2-methyltetrahydrofuran production from levulinic acid in green solvents using Ni-Cu/Al_2O_3 catalysts, ChemSusChem 8 (2015) 3483–3488.
[116] Elliott, Hydrogenated 5-carbon compound and method of making, US5883266A, 1999.
[117] P.P. Upare, J.M. Lee, Y.K. Hwang, D.W. Hwang, J.H. Lee, S.B. Halligudi, et al., Direct hydrocyclization of biomass-derived levulinic acid to 2-methyltetrahydrofuran over nanocomposite copper/silica catalysts, ChemSusChem 4 (2011) 1749–1752.
[118] X.L. Du, Q.Y. Bi, Y.M. Liu, Y. Cao, H.Y. He, K.N. Fan, Tunable copper-catalyzed chemoselective hydrogenolysis of biomass-derived γ-valerolactone into 1,4-pentanediol or 2-methyltetrahydrofuran, Green. Chem. 14 (2012) 935–939.
[119] X. Wang, X. Liang, J. Li, Q. Li, Catalytic hydrogenolysis of biomass-derived 5-hydroxymethylfurfural to biofuel 2,5-dimethylfuran, Appl. Catal. A Gen. 576 (2019) 85–95.
[120] F.K. Kazi, A.D. Patel, J.C. Serrano-Ruiz, J.A. Dumesic, R.P. Anex, Techno-economic analysis of dimethylfuran (DMF) and hydroxymethylfurfural (HMF) production from pure fructose in catalytic processes, Chem. Eng. J. 169 (2011) 329–338.
[121] W. Yang, A. Sen, One-Step catalytic transformation of carbohydrates and cellulosic biomass to 2,5-dimethyltetrahydrofuran for liquid fuels, ChemSusChem 3 (2010) 597–603.
[122] M.R. Grochowski, W. Yang, A. Sen, Mechanistic study of a one-step catalytic conversion of fructose to 2,5-dimethyltetrahydrofuran, Chem. A Eur. J. 18 (2012) 12363–12371.
[123] X. Kong, R. Zheng, Y. Zhu, G. Ding, Y. Zhu, Y.W. Li, Rational design of Ni-based catalysts derived from hydrotalcite for selective hydrogenation of 5-hydroxymethylfurfural, Green. Chem. 17 (2015) 2504–2514.
[124] Z. Gao, C. Li, G. Fan, L. Yang, F. Li, Nitrogen-doped carbon-decorated copper catalyst for highly efficient transfer hydrogenolysis of 5-hydroxymethylfurfural to convertibly produce 2,5-dimethylfuran or 2,5-dimethyltetrahydrofuran, Appl. Catal. B Env. 226 (2018) 523–533.
[125] E. Ahmad, M.I. Alam, K.K. Pant, M.A. Haider, Catalytic and mechanistic insights into the production of ethyl levulinate from biorenewable feedstocks, Green. Chem. 18 (2016) 4804–4823.
[126] K. Yan, G. Wu, J. Wen, A. Chen, One-step synthesis of mesoporous $H_4SiW_{12}O_{40}$-SiO_2 catalysts for the production of methyl and ethyl levulinate biodiesel, Catal. Commun. 34 (2013) 58–63.
[127] S.B. Onkarappa, M. Javoor, S.S. Mal, S. Dutta, Efficient and scalable production of alkyl levulinates from cellulose-derived levulinic acid using heteropolyacid catalysts, ChemistrySelect 4 (2019) 2501–2504.

[128] L. Peng, L. Lin, J. Zhu, J. Shi, S. Liu, Solid acid catalyzed glucose conversion to ethyl levulinate, Appl. Catal. A Gen. 397 (2011) 259–265.
[129] G. Zhao, L. Hu, Y. Sun, X. Zeng, L. Lin, Conversion of biomass-derived furfuryl alcohol into ethyl levulinate catalyzed by solid acid in ethanol, BioResources 9 (2014) 2634–2644.
[130] S.S. Enumula, K.S. Koppadi, V.R. Babu Gurram, D.R. Burri, S.R. Rao Kamaraju, Conversion of furfuryl alcohol to alkyl levulinate fuel additives over Al_2O_3/SBA-15 catalyst, Sustain. Energy Fuels 1 (2017) 644–651.
[131] E. García, M. Laca, E. Pérez, A. Garrido, J. Peinado, New class of acetal derived from glycerin as a biodiesel fuel component, Energy Fuels 22 (2008) 4274–4280.
[132] G. Sailaja, R.N. Pethan, G.S. Rao, V.R.C. Komandur, Acetalization of glycerol with acetone to bio fuel additives over supported molybdenum phosphate catalysts, J. Mol. Catal. A Chem. 410 (2015) 49–57.
[133] S. Ammaji, G.S. Rao, K.V.R. Chary, Acetalization of glycerol with acetone over various metal-modified SBA-15 catalysts, Appl. Petrochem. Res. 8 (2018) 107–118.
[134] J. Deutsh, A. Martin, H. Lieske, Investigations on heterogeneously catalysed condensations of glycerol to cyclic acetals, J. Catal. 245 (2007) 428–435.
[135] J. Kowalska-Kus, A. Held, M. Frankowski, K. Nowinska, Solketal formation from glycerol and acetone over hierarchical zeolites of different structure as catalysts, J. Mol. Catal. A Chem. 426 (2017) 205–212.
[136] S. Abd el-aziz, Ael-A. Mohammed, Effect of different metal sulfate precursors on structural and catalytic performance of zirconia in dehydration of methanol to dimethyl ether, J. Fuel Chem. Technol. 46 (2018) 67–74.
[137] F. Yaripour, F. Baghaei, I. Schmidt, J. Perregaard, Catalytic dehydration of methanol to dimethyl ether (DME) over solid-acid catalysts, Catal. Commun. 6 (2005) 147–152.
[138] V. Vishwanathan, K.W. Jun, J.W. Kim, H.S. Roh, Vapour phase dehydration of crude methanol to dimethyl ether over Na-modified H-ZSM-5 catalysts, Appl. Catal. A Gen. 276 (2004) 251–255.
[139] S. Kim, Y.T. Kim, C. Zhang, G. Kwak, K.W. Jun, Effect of reaction conditions on the catalytic dehydration of methanol to dimethyl ether over a K-modified HZSM-5 catalyst, Catal. Lett. 147 (2017) 792–801.
[140] W. Alharbi, E.F. Kozhevnikova, I.V. Kozhevnikov, Dehydration of methanol to dimethyl ether over heteropoly acid catalysts: the relationship between reaction rate and catalyst acid strength, ACS Catal. 5 (2015) 7186–7193.

Section III

Hydro Power Technology

Hydropower technology

4

Nitai Pal and Faizan A. Khan
Department of Electrical Engineering, Indian Institute of Technology (Indian School of Mines), Dhanbad, India

4.1 Introduction

A significant role is played by energy in the advancement of a nation and in facilitating the everyday comforts of people. Because of the fast development of the population and industrialization across the globe, global energy requirements are expanding rapidly. The rise in energy demand is speedily exhausting conventional resources. Traditional energy sources, including coal, oil, and nuclear, in addition to becoming restricted, are additionally answerable for contaminating the Earth and disturbing the natural balance. Due to the continuous consumption of fossils fuels like coal and petroleum, their reserves in the Earth are decreasing. So, there is focus on renewable sources of energy and how efficiently these renewable sources of energy can be used for the generation of power. Nonconventional sustainable power resources like wind energy, solar energy, small hydro, tidal and geothermal, are plentiful, endless, and noncontaminating to nature. In order to utilize these resources, the electricity generation process is shifting towards sustainable power sources from conventional system. Out of all the accessible sustainable power source assets, hydropower is an impeccable, productive, and trouble-free innovation for power generation [1]. The main sources of hydropower are the large river systems of the world along with their vast catchment areas. A survey of the characteristic features of some river systems shows that the usable potential increases more with a catchment area than with the mean altitude of the system. The usable potential is usually greater in downstream areas than in the upstream ones. In general, the potential does not only depend on such obvious figures as the total river length or the discharge at the river mouth. The topography of the continent together with the climate and the rainfall rate may favor the damming of a river even of low gradient provided there is a valley with sufficient depth. This and the collection of water creates head and discharge as the basic requirements of hydropower [2]. The theoretical potential of the world can be predicted from hydrographs, precipitation, and the gradient of rivers. It is always related to the usable potential. This depends on the theoretical potential and results from the technical possibilities available, from the feasibility of electric power transmission over a long distance, and from the power demand in the neighborhood of a certain scheme [3]. Any development requires a usable potential, a certain demand for power, and a reasonable electricity

Sustainable Fuel Technologies Handbook. DOI: https://doi.org/10.1016/B978-0-12-822989-7.00005-6

rate due to hydropower. Therefore the usable potential has to be considered in the context of the total power production, at least of the total electricity production, in the area of a certain site. For these reasons, the harnessable potential of river power does not only change with the system of rivers and their topography, but also differs from one country to another depending on their economies [4]. In this chapter, all the technologies used in hydropower plants, their basic constructions, and the structure of the turbines used in different types of hydropower plants will be discussed.

4.2 Hydropower technologies

The classification of hydropower plants is based on two factors, that is, technology and power scale. The main types hydropower plants are impoundment, run-of-river, gravitational vortex, pumped storage, etc. [5].

4.2.1 Impoundment hydropower

The most widely recognized type of hydroelectricity generation plant is the impoundment-type. Impoundment agency is an enormous hydropower framework, which stores the water of a river in a reservoir by utilizing a dam, as shown in Fig. 4.1. The reservoir discharges the water that flows through the turbine and rotates this, to generate electricity. This generated power might be utilized decently to meet the peak load and base-load requirements respectively. The discharge water may be utilized either in the generation of power or in maintaining the consistent water storage level. Fig. 4.1 shows the layout of the conventional impoundment

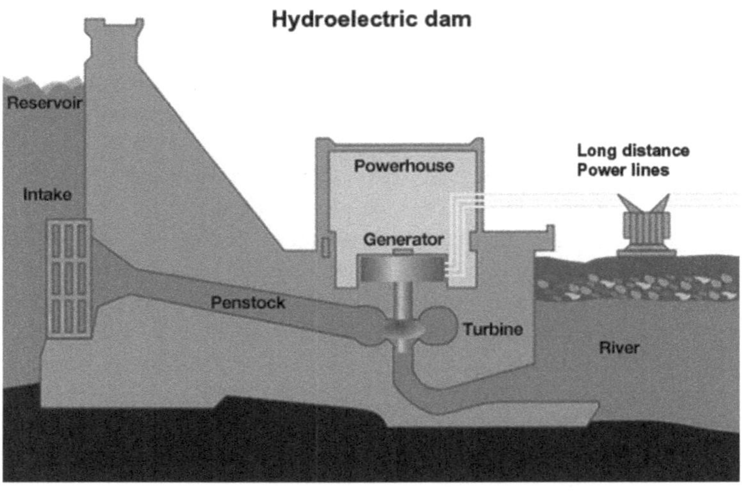

Figure 4.1 Conventional impoundment dam [6].

dam [6]. The Hoover Dam is the most acclaimed impoundment-type dam across the world, and it is shown in Fig. 4.2.

4.2.2 Run-of-river hydroelectricity plant

This is a kind of hydrogeneration plant where approximately no stockpiling of water or limited capacity of storage, known as poundage, is required. In the absence of poundage, a run-of-river (ROR) plant would be dependent on seasonal waterways, therefore, an ROR-based plant will work irregularly. A conventional hydropower system exploits water storages, which are also beneficial in flood control [7]. This type of hydropower system is viewed as perfect for a consistent river, which can upkeep the least possible stream or be controlled by reservoirs. A minute dam with a head pond is constructed in such a manner that there would be continuous availability of sufficient water for the penstock pipes and rotation of turbines. Ventures with poundage, rather than without poundage, can store water for day-to-day load requirements. These ventures are occupying the greater region of a river stream directly over the tunnel, which prompts the turbine to generate power. These ventures are amazing in appearance and design as compared to regular hydro system. These ventures are amazing in appearance and design compared to regular hydrosystems. Customary hydrosystems have huge reservoirs for water storage that may create flooding in vast tracts. Conversely, ROR ventures don't have reservoir-related disadvantages. This is the reason behind their lesser ecological effect. The expression "run-of-river" for hydropower ventures has different meanings across the world. In some places, it is considered to mean a plant with no storage, while restricted storing is supposed in other places. In order to get recognition, many developers try to present their similar projects as ROR hydro system. [7]. ROR hydroelectricity generation plants are standardized by the Bureau of Indian

Figure 4.2 The Hoover Dam [6].

Standards as generating stations using river streams for power production in a specified period with appropriate poundage and providing the water needed to meet diurnal or various demands. In these conditions, an ordinary river isn't materially adjusted. Several large ROR ventures are giving tough competition to customary hydro dams by their scales and specific designs. For instance, the Beauharnois hydropower generation venture at Quebec with a capacity of 1853 MW. Some ROR ventures are downstream of reservoirs and dams. An ROR venture doesn't construct a reservoir, however, it takes the freedom of the water supplied by it. The La Grande-1 generating station with a capacity of 1436 MW in 1995 is a suitable example. Former upstream dams and repositories are a fragment of the James Bay Project of 1980 [8]. There are some little and portable types of ROR power plants. For example, a little floating hydroelectric electrical plant. The energy of moving water rotates the power turbine of the generator that produces electric power [2].

4.2.2.1 Advantages

In consideration of the footprint size and region, ROR-based project produces sustainable energy, limiting effects on the environment conditions, and enclosed localities.

4.2.2.1.1 Inclusive advantages

The advantages of ROR hydroelectricity include cleaner power, less ozone harming substances, and the fact that ROR utilizes the potential energy of water like other hydroelectricity generated power sources, removing the need to consume coal/petroleum products to create the power required by purchasers and industries. In addition, ROR hydro plants don't have repositories, thereby wiping out the carbon dioxide and methane discharges brought by the disintegration of organic matters in the repositories of customary hydropower dams. It is a specific benefit in tropical nations where the generation of methane can be an issue.

4.2.2.1.2 Less flooding/store

Flooding in the upper portion of a river doesn't take place in a lacking reservoir. Therefore environments and individuals living close to the river are not affected. If the flooding pattern remains unaltered, thenthreat of flood hazard to the nearby territories always exists [2].

4.2.2.2 Disadvantages

The ROR system of power generation is viewed as an "unfirm" resource of intensity, that is, this venture has small energy-storing, therefore, the customer demand cannot be satisfied by the power generation. Considerably more power is produced when river streams run high, and depending on the region, less power is produced during summer or severe winter months.

4.2.2.2.1 Accessibility of site

At a specific site, the power potential is a consequence of the flow and head of the water. By blocking the river flow, the water head becomes more accessible to produce power at the dam front. The water reservoir developed by the dam may be several kilometers long. The large size of the reservoir also affects the water head. The water head created here is normally conveyed to the powerhouse by a passage developed on the upstream. Because of expenses in upstream development, a sheer drop is alluring, for example, rapids or falls.

4.2.2.2.2 Ecological effects

The small ROR ventures have insignificant natural effects as compared to the larger ventures that have higher ecological concerns. On account of fish-bearing streams, a ladder might be a requisite, and downstream disintegrated gases may influence the fish [7]. A large amount of water distraction from the river stream reduces the flow of the river, this event influences the water speed and diminishes the living quality for the fish and amphibian's life. The decreased water stream can prompt adverse conditions for salmon and other fishes in the summer. In case of remote location establishment of transmission system for new hydropower may cause disturbance for living system, while the absence of reservoir may create disturbance. The absence of reservoir storing may bring about discontinuous activity, lessening a venture's suitability.

4.2.3 Gravitational vortex turbine-based hydroelectricity plant

A gravitational vortex–based water turbine predominantly involves a tank and sprinter. It produces electrical energy by bringing a water progression into the tank using the gravitational vortex turbines. Power is created by the water flowing from the base of the tank. This type of turbine is fit for producing power utilizing a low stream and a low head with a moderately basic arrangement. The free stream surface makes water turbines action complicated. The water turbine performance can be anticipated by the stream field, estimations of the turbine output, and water head. The water turbine performance can be anticipated by this examination. It has been demonstrated that the rotating speed increments at the sprinter channel are extended to the forward stream zone. In any case, when the air zone diminishes, the backward stream territory grows [9]. Some large-scale customary hydraulic generators fundamentally utilize high or medium head and water turbines for channels, for example, the Francis turbine. Awareness regarding sustainable power sources has risen, and the scope of small-scale hydropower generation with open channels has been expanding. In this manner, attention is given to the water turbine utilized in gravitation water vortex (GWV) power plants, which produce power with low streams and low head. This GWV water turbine basically involves a tank and a sprinter [10]. On bringing a water flow into the tank, the turbine starts to produce power from the GWV by emptying the water out of the tank base. It is assumed that the air circulation capacity of a turbine is used for increasing the oxygen convergence in downstream water, by moving the air over the open surface of the runner. The stream field is incredibly complex on account of its free surface, except that the water turbine has a moderately straightforward structure. The detailed examination of the stream field is important for improvement in the turbine operation and performance. For improvement in the presentation of the turbine and its significance to the performance feature, it is imperative to detail examination of the stream field [11]. A numerical investigation is influential since this turbine works on the utilization of a gravitational energy vortex; therefore, it is important to perform a mathematical examination through an assumption of open surfaces.

4.2.3.1 Gravitational vortex type of water turbine

A sketch of the GWV type of turbine is shown in Fig. 4.3. This turbine involves a tank and a runner for generating power by the gravitational vortex in the tank. A diagram of a runner is shown in Fig. 4.4. Runners have differing structures that are unique in relation to the paddle runner that has been utilized in past investigations [9].

4.2.4 Pumped-storage hydroelectricity plant

Pumped storage hydroelectricity (PSH), or pumped hydropower energy storing (PHES), is a kind storing system used by electric power frameworks for load adjustment, as shown in Fig. 4.5. In this strategy, energy is stored in the form of the gravitational potential energy of water, by placing the entire amount at a higher escalation. Off-peak low-cost electric surplus power is regularly used to run the pumps. During peak demand, the stored water is discharged to create electricity through turbines. The losses of the draining process make such a plant an energy purchaser, while the income of the framework expands as more power is sold during the highest power demand [12]. Pump storage hydropower permits energy from irregular resources (sunlight and wind) and other renewables, or surplus power from constant base-load resources (coal or atomic), to be protected for periods of high demand. The pumped storage reservoirs are producing relatively compact hydropower as compared to the same capacity of regular hydropower plants, in

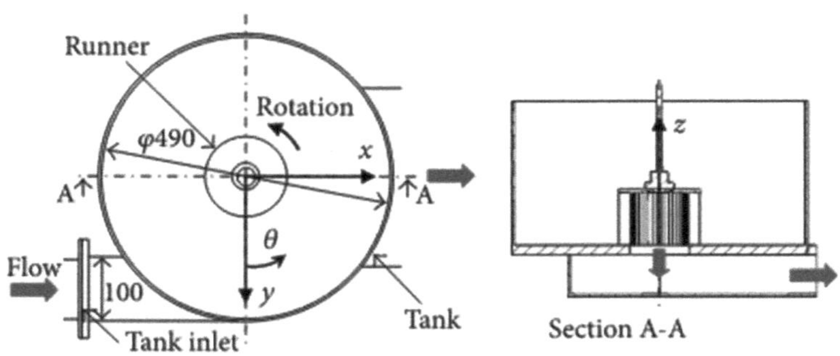

Figure 4.3 Sectional view of water turbine [9].

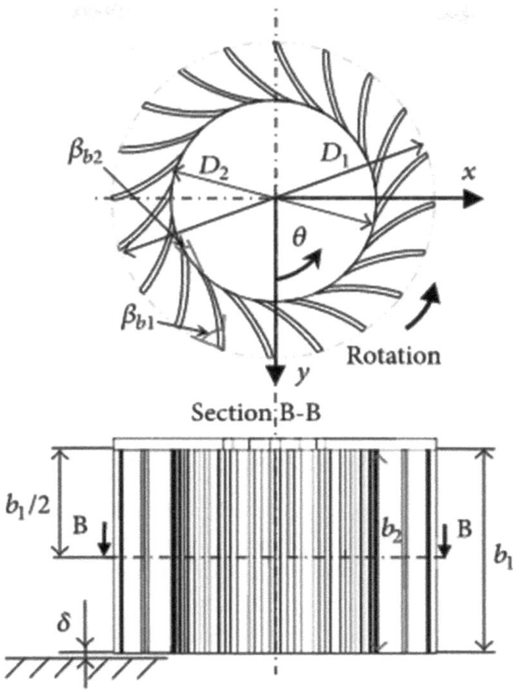

Figure 4.4 Sectional view of runner [9].

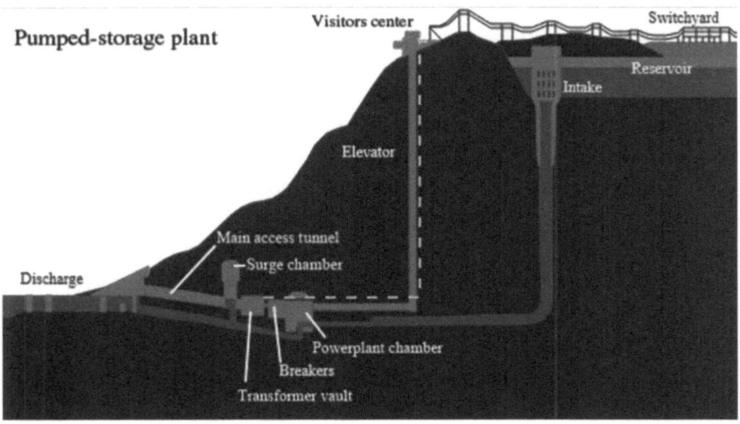

Figure 4.5 Pumped storage plant [6].

time periods of less than half-day. Pump storing is the biggest grid-accessible energy storing technology. The energy efficiency of PSH varies between 70% and 80%, with certain resources guaranteeing up to 87%. The primary hindrance to PSH is the requirement of a special landscape with access to great height and water.

In this manner, suitable areas are mountainous or hilly areas, with the possibility of ecological and social issues also existing. Numerous proposed activities avoid exceptionally sensitive or beautiful territories, for example, disused mines.

On the occasion of low electricity demand, the surplus generation limit is utilized in the pumping of water into the higher reservoir. In the case of a greater demand, water is discharged back to the lower reservoir through a water turbine, creating power. A generator and reverse turbine are used in combination with a turbine generator and consolidated pump as a single unit (the Francis turbine) [13]. For small-scale applications of PSH, a collection of pump-as-turbine (PAT) systems and pumps can be completed individually for pumping and creating power. A similar pump can be utilized in both modes by altering rotational speed and direction; the point of operation in pumping generally having differences is the action point of PAT mode.

4.2.4.1 Regular or artificial storages

In open-loop frameworks, the water is stored in the upper section of the untainted pump storing plant without regular influxes. The pump back plant utilizes a combination of pump storing and regular hydropower plants with an upper repository. The hydro plant that does not utilizes pump storage has alluded as a regular hydropower plant. The hydropower plant with sufficient storage capacity has an important role in the electricity generation system as the reservoir influences the power output according to requirement.

4.2.4.2 Economic efficiency

Considering evaporation damages from the uncovered surface of water and alteration adversities, an energy restoration of 70%–80% can be accomplished. This strategy is presently one of the savviest methods for storing a lot of electrical energy, however, capital expenses and the proximity of proper topography are basic choice factors in choosing sites for a pump capacity plant. Generally, the low energy density of pump storing frameworks requires enormous streams or huge gaps in elevation between repositories. The best method for the storage of a significant amount of energy is taking a huge amount of water, moderately close and as high as possible, from the second body of water. In certain places, this happens naturally, in others, one or two waterways are artificial. Ventures in which the repositories are artificial, and in which inflows are engaged naturally with reservoirs are referred to as "closed-loop" frameworks. These frameworks are more efficient as they are able to counter the load variations observed from the grid network. It is widely known that conventional thermal power plants are the best option for base-load electricity at the highest effectiveness. Numerous base-load thermal plants, however, have been intended for adaptability as opposed to maximal productivity. Thus pumped storing frameworks are vital when organizing huge gatherings of heterogeneous generators. The establishment of pump storage plants required large initial costs in general. In spite of the large establishment cost, It should be kept in the mind that the life of this system is multiple times longer than the small storage system. Alongside energy management, pumped capacity frameworks help to control electrical system frequency and give standby generations. Substantial thermal plants are less ready to react to abrupt changes in electricity demand, conceivably causing voltage and frequency unsteadiness. Pump storing plants, as another form of hydropower plant, can react to load variations in seconds. The most significant use for pump capacity plants has customarily been to modify base-load power, however, they may be used to lessen the fluctuation in yield of intermittent energy resources. Pumped storing system can afford a lesser load demand while the power production is higher, This situation empowers the framework with the additional peak capacity. In the case of more electrical generation than the load demand, the power costs might be zero or negative. Considering all things, pumped storage could turn out to be significant as an equalization for large-scale solar photovoltaic energy generation. The interconnection of transmission lines with the energy storage makes the standalone renewable system more feasible for large scale deployment [14].

4.2.4.3 Facilities at small scale

Small-scale pump storage has encountered an economic scale disadvantage. Proposals for small-scale pump storing plants have been observed, in spite of the financial unfeasibility of this scale. Likewise, a huge water-storing capacity is necessary, which is troublesome for urban zones. However, a few authors defend its innovative effortlessness and secure arrangement of water as important externalities [15].

4.2.4.4 Pump backed hydroelectric dams

A regular hydropower dam may utilize pumped storing in a hybrid framework so that both can produce power from streaming water normally into a reservoir while the stored water returns to the reservoir. The regular dams are empowered by associating it with the turbine system. A pumped back power plant is added to the existing structure for the expansion of the generation limit. The utilization of an upper reservoir and transmission framework in an existing dam may speed up the venture and lessen the cost [16].

4.2.4.5 Potential advancements

Pump storing plants can work with ocean water, despite the additional difficulties compared with utilizing freshwater.

4.2.4.6 Underground stores (reservoirs)

The utilization of underground stores is discussed here. It is proposed to utilize a vacant iron mine as the underground storage of water. The proposed energy storing would use the most profound base metal mine at 1450 m height gap. These proposed underground mines storage would be the most profound underground water reservoir at the depth of 1450 m. There are constrained opportunities, including reasonable underground space, yet the quantity of underground pumped storing opportunities may rise whenever deserted coal mine prove to be appropriate [17]. In Australia, the utilization of old gold mines has been proposed for pumped hydro energy storage. This location has the best centralization of deep hard-rock mines than any other place on the planet, with more than 5000 shafts. The most profound shaft broadens 1406 m straight underground [18].

4.2.4.7 Decentralized frameworks

Little (or small scale) applications for siphoned capacity could be based on streams and inside frameworks, for example, drinking water systems and counterfeit snowmaking foundations. Right now, the storm-water bowl has been solidly executed as a savvy answer for water supply in a small scale siphoned hydro vitality stockpiling. Such plants give dispersed energy storing and dispersed flexible electricity generation and can participate in the distributed assimilation of discontinuous renewable energy resources, like solar power and wind power. The reservoir can be used for small pump storing hydropower plants with the inclusion of natural or artificial reservoirs within other arrangements like irrigation, or unused portions of mines [19].

4.2.4.8 Underwater reservoirs

The research venture, Stored Energy at Sea (StEnSea), in March 2017, was declared an effective trial execution of a pump storing submerged reservoir. This submerged reservoir acts as lower reservoir that has existed in the shape of a sphere at incredible depth, while the encasing water stream behaves as an upper reservoir. The electrical power is generated by rotating the turbine with the coordination of the spherical water reservoir. During off-peak times, the turbine alters the progression and pumps the water outward once more, utilizing the "overflow" electrical power of the grid. The amount of power generated is proportional to the height of the water column with respect to the spherical lower reservoir. The more potential energy can be stored according to the depth of the lower reservoir. On the other side, The more water pumping at prominent depth consumes more power, since, the turbine–pumping should work upon a similar water column. In that capacity, the energy-storing capacity of submerged reservoirs is administered by gravitational energy in the conventional sense, or by the vertical variation of pressure [20]. The underwater storage requires an electrical connecting cable for transmission of power. This system is utilized at a depth of many hundred meter where the turbine can work, distinctly. The planning test of saltwater pumping capacity brings a scope of favorable circumstances:

- No requirement for land territory.
- Only an electrical connector other than mechanical structures is needed to traverse the separation of the potential energy distinction.

- In the vicinity of an adequate seabed region, numerous repositories can scale the capacity limit unbounded.
- Should a supply break down, the results would be restricted, separated from the loss of the repository itself.
- Vaporization from the higher surface has no impact on the energy change productivity.
- The transmission of power between the supply and the matrix can be built up from a close-by seaward wind ranch, limiting transmission errors and blocking the requirement for coastal cabling licenses [19].

4.3 DAMS

A dam is an obstacle that halts the movement of water or underground watercourses. The reservoir made by dams restricts flooding as well as offers stored water for useable works, for example, irrigation systems, industrial consumption, human uses, and aquaculture. In general, hydropower is utilized in association with dams to produce electrical power. A dam can likewise be utilized for water collection, which can be equally conveyed between nearby areas. The primary purpose of dams is to retain a huge amount of water [21]. Dams are categorized on the basis of structure and utilized material.

4.3.1 Arch dam

In this type of dam, the combination of gravity and an arc provides stability. For vertical upstream, the whole dam weight must be transferred to the foundation. The stiffness in horizontal and vertical directions would be responsible for the rotation of cantilevers. The distribution becomes more complicated for a sloped upstream face. The normal section of the arched load may be considered for this action and the normal hydrostatic pressure would be distributed in similar manner. These dams are required to have stable dependable supports at the projections. For an arc dam, a narrow canyon is the most alluring place with a sound rock-based steep sidewall. The security of such a dam is subject to the strength of the sidewall. Therefore the inspection of the characteristics of the material and the arc position on the sidewall should be cautiously examined. In general, constant-radius and constant-angle dams are used as singly-arced dams. In a constant-radius dam, a constant radial elevation is used, which implies that the angle created by the dam face decreases as the channel becomes smaller toward the base. In the case of a constant-angle (variable-radius) dam, the distance between different levels varies, thus, keeping the subtended angle constant. Constant-angle dams are substantially more popular than consistent-radius dams [16]. Another popular kind of dam is the thin-shell or double-curve dam, an optimized construction technique is used in double-curve dam construction; this type of technique requires less materials for huge loads. Double-curve dams are similar to single-arch dams in terms of appearance, yet the unique vertical shape and concave lending from downstream create differences. The multiple arched dams can be developed by adding supporting braces and projections in the single arched dams. The large establishment on the ground is required for the multiple curve dam that can support loads. Multiple curve dam doesn't require the multiple supports [22].

4.3.2 Gravity-based dam

In this type of dam, gravitational force is utilized as the main holding power against the thrust of water. The water of the reservoir tried to push the dam laterally downstream. This condition may overturn the dam about the base point but the weight of the dam counters the overturning tendency. Scholars states that the weight of the dam is sufficient to overcome the turning and damaging possibility of dams. It is possible due to the resultant of the gravitational force and water pressure of the dam acting in streamline.

Architects always attempt to shape dams in such a way that any section of a dam that is over a specific height is considered to be an entire dam itself. The highest point of the dam observed no pressure as compared to downward position of the dam. It is generally more realistic to construct a dam by accumulating the material upward than to keep the material composed alongside vertical strain. The upside pressure eliminating structure also reduced the downstream pressure on the dam, this makes the system extra economical. [23]. An impenetrable foundation with high bearing quality is required for this kind of dam. A porous foundation has a greater probability of generating upward pressures. These pressures are hydrostatic, generated by the water in the reservoir against the base of the dam. If the generated uplift pressure crosses a limit, then the danger of concrete structure destabilization increases. In order to stabilize the system, the gravity-based dam would become a greater alternative for odd conditions. Moreover, in the case of deliberately examined locations, a gravity-based dam becomes the best-created dam structure. In most flood-affected locales, gravity-based dams are the best option other than arc dams, which are economical indeed. Gravity-based dams are delegated as "hollow" or "solid" completed by either brickwork or concrete. The solid type is the most frequently utilized, however, the hollow type is more economically efficient [24].

4.3.3 Arc gravity dam

A gravity-based dam can be converted into an arc gravity dam by joining an arc in regions where few materials and a huge amount of water are accessible. The internal pressure of the water on the dam diminishes the horizontal forces. Therefore less gravitational power is required by the dam and the dam won't need to be so vast. This condition prefers thin dams that save resources [25].

4.3.4 Barrage dam

These dams are comprised of huge gates to control the flowing water through the dam. The gates are joined with flanking docks to support the water burden and settle the water stream for irrigation by controlling frameworks. Barrage dams that work at the entrances of rivers or lagoons to avert tidal attacks are known as tidal barrage dams [22].

4.3.5 Embankment dam

These are produced by using compacted earth; earth-fill and rock-fill dams are two fundamental categories. These dams are dependent on their weight to balance the water forces, similar to concrete structured gravity-based dams [23].

4.3.6 Rockfill dam

Rockfill dams have preferably constructed at the compressed, draining-free, depleting land situated in the impenetrable zone. The most regularly used soil is comprised of a high level of enormous particles, thus, the expression "rockfill" is justified. There might be an impenetrable region along the upstream face that is constructed by brickwork, plastic film, heaps, timber material, or solid steel sheets. This impenetrable region can be inside the embankment considered as the center; if clay is used as the material for the impenetrable region, then the structure is called a composite dam. The center is isolated by a filter, for avoiding the disintegration of the clay because of leakage. The filter is explicitly evaluated soil, anticipated to prevent the relocation of fine soil elements. In the case of the nearby availability of materials, the transport facility is minimally used, prompting a reduction in cost. These dams are impermeable to harm by seismic tremors. Nonetheless, a lack of quality control can prompt deprived compaction during the construction of the embankment, which may be the cause of liquefaction due to filling during a seismic tremor. The liquefying potential may be diminished by shielding vulnerable saturated material and sufficient compaction [25].

4.3.7 Concrete face rockfill dam

This is a stone-filled dam in which concrete slabs are placed on the upstream surface. Here, a concrete slab is used as an impenetrable partition to forestall spillage; furthermore, it is an arrangement without elevated pressure. The configuration is adaptable according to geography, can be rapidly constructed, and economic. This idea began in the 1860s during the California gold rush; at that time, excavators mainly built timber-face rockfill dams for drain activities. Later, concrete displaced timber as the most functional system for water systems for power generation. In the beginning, the concrete face rockfill (CFR) dam was constructed according to the height and the compacted filling of the CFR dam was used. The vertical and horizontal joints of the slabs were replaced by a vertical joint. Over the past couple of decades, the plans have become famous. The Shuibuya Dam is the tallest CFR dam in China at 233 m (764 ft) in height [26].

4.3.8 Earthfill dam

This type of dam is also known as an earthen dam or rolled-earth dam. It was developed in the form of a compacted earth embankment. Rolled-earth dams are completely developed by a single material, yet there may be a passage for the collection of leak water. This type of dam has various parts or sections of disparate material like waterproof clay and copious shell. A specific hydraulic fill strategy is used to produce a sealed center. A rolled-earth dam may utilize a sealed center similar to in rockfill dams. A frozen center dam is a different sort of earth dam utilized at high latitudes temporarily, where a coolant is streamed through funnels inside the dam. The Tarbela Dam is an example of this type of dam, located in Pakistan. It is estimated that 152.8 million cubic meters of rock and earth were utilized. Since earthfill dams are developed by materials available at nearby sites, they become economical in areas where delivering and processing concrete are restricted [27,28].

4.3.9 Fixed-crest dam

A fixed-crest dam is a simple concrete obstruction over a river. These dams are intended to maintain the depth in waterways for navigation and present dangers to boaters. They are helpful in identifying difficult spots and make traveling easy for boaters along incited currents. Presently, reservoirs are examined [13].

4.4 Reservoirs

A reservoir is an augmented natural/artificial lake, made by utilizing a dam or the lockage of water. Reservoirs may be prepared in various manners, including controlling a waterway that drains an existing stream or interfering with a waterway to frame an embayment inside it. It is also characterized as an extra room for liquids, which may hold any fluids including hydrocarbons. There are specific types of reservoirs like the tank-type reservoir situated at ground-level, raised, or covered. Cisterns are also a type of tank reservoir. Most subversive reservoirs are utilized to stock fluids, mostly oil or water, under the ground [29]. Reservoirs are categorized here.

4.4.1 Dammed valley

This type of reservoir is formed in a dam, developed in a valley. The natural geography of the valley provides the space for the basins of the reservoir. To obtain such structures, such dams are constructed at the narrowest part of the valley downstream. The sides of the valley work as natural fences of the dam situated at the tapered point to offer strength and economic construction. Rehabilitation of people has required environment safety in many cases of reservoir construction. In the development of a valley-based repository (reservoirs), the redirection of the waterway is needed during the construction of the dam. This frequent redirection is executed by an impermanent passage or bypass canal. Reservoirs can be frequently built by broadening prevailing lakes in hilly areas. In these cases, extra dam sides are needed to store water [29].

4.4.2 Coastal reservoirs

These reservoirs are constructed to store the floodwater primarily. The coastal reservoirs are situated on the ocean coast at the river gateway, to store the floodwater. The coastal reservoir has no land submergence as it occurs in the ground-based reservoir. The coastal reservoirs are financially preferred, it does not utilize the region of the fertile land. Numerous coastal storages were developed in Asia and Europe [30].

4.4.3 Bank-side reservoirs

Here, water is siphoned or guided by a variable river; bank-side repositories store the water. These are typically shaped by excavation and building a totally encompassing embankment or bund of approximately 6 km in perimeter. An impermeable covering or center is required for the floor of the repository and the bund. Originally, these were frequently prepared by puddled clay, but over time, the utilization of rolled clay has become preferred. Such repositories may contain water for numerous months; in this duration, numerous contaminants can diminish by natural procedures and any turbidity can nearly be disposed of. In the case of a river being unsuitably contaminated or extremely low because of drought, the utilization of a bank-side repository permits the water flow to be halted for a certain time-period. The water supply framework in London is an example of bank-side storing [30].

4.4.4 Service repositories

This type of storage completely treats consumable water near the place of appropriation. Service reservoirs are built as raised structures or water towers upon concrete columns, where the section is normally levelled. Other service reservoirs can be built underground, particularly in the case of sloping or uneven areas. The water ring reservoir constructed on the Thames river is one of the best example of service reservoir. At present, the upper surface of this reservoir has been grassed over and utilized by the golf club. Service reservoirs discharge various other duties. Many of them are mentioned here: Securing the efficient height of water head for the water circulation framework, ensure sufficient water capacity to handle the peak demand, and empowering the water treatment plant to run at optimal efficiency [28].

4.5 Small Scale hydropower plant

A small-scale hydropower "plant" is a kind of power plant that can produce 100 KW of power utilizing streaming steam or a water stream. The power generated by this framework is utilized to provide energy support for domestic consumers and the rural standalone system.

Spilling over into the national lattice when power creation is in surplus is additionally apparent at times. The small-scale hydropower system can be developed near to the consumers of the remote hilly stations with frequently available materials [15]. In

1995, the small-scale hydro limit on the planet was evaluated as 28 GW, providing around 115 TWh of power. Around 60% of this limit was in the developed nations, and the remaining 40% was in developing territories. Small-scale hydro plants are found in most of the hilly locales, comprising both electrical and mechanical power production. In the case of South America, various small-scale hydro schemes are executed across the continent along the Andes hills, for example, in Bolivia and Peru [16].

4.6 Micro-hydropower plant

The development subtleties of micro-hydropower plants are location dependent. There should be the presence of lagoon or other artificial sources that is accessible for power creation. All micro-hydropower frameworks are built by numerous segments. In the case of more consumption, Most of the hydropower significantly integrates more water either by redirected canal or by the river. An ingestion arrangement, for example, a catch box, is essential to screen out skimming fish and trash, utilizing a screen or bar-cluster to move out huge items. In mild atmospheres, this arrangement should oppose ice too. The intake may have a door to permit the examination and upkeep of the framework. The water intake is transported by a channel and forebay. This forebay is utilized for silt holding. At the framework base, water is passed by a conduit (penstock) toward the powerhouse and turbine. This penstock develops the pressure from the water in downward direction. In uneven zones, access to the penstock can give impressive difficulties. If the turbine and water resource are far away from each other, the development of the penstock might be the biggest section of the expenses [21]. A governing valve is introduced at the turbine, to manage the water stream and turbine speed. The turbine changes the water stream pressure into mechanical energy. The water moving outside upward out from the turbine comes back by a tailrace channel to the normal waterway. The turbine becomes a generator and is associated with the electrical load. This may be reasonably related to the power source of a lone structure in an exceptionally small establishment or may be connected with a network dissemination framework for residential buildings. Normally, a microscale hydrosystem does not have storage facilities or a dam, like huge hydroelectric plants, and is dependent on a minimal progression of water accessible over a year.

4.7 Benefits and shortcomings of micro hydropower plants

Small-scale hydropower stations and micro-hydro plants are related to the upsides of hydropower for power production, without the impediments of large-scale establishments. The most favorable characteristics are distinguished as:

In this case, a small quantity of water stream is sufficient to make it work. A small amount of two gallons of water may be able to create power with the micro-scale hydropower. The power generated can be utilized by the place located far away from the generation site.

Reliable power source: There is a steady and constant power supply from a hydro plant in contrast to small-scope sustainable power source advancements. There are high energy seasons (basically in winter) when huge amounts of power are required.

Reservoirs are not required: Micro-hydro plants are assumed to be ROR frameworks; this implies that the water goes through the turbines when it is used, but is later redirected to a waterway with a moderate effect on the encompassing ecology.

Power for developing nations: Having ease of flexibility and long-life expectancy, microscale hydro plants can be utilized by developing nations to provide power to small towns and linkages.

Environmental influence: The effect on nature is limited in contrast to the conventional energy generation plants that utilize petroleum derivatives [12].

4.7.1 Suitable conditions for micro hydropower plants

Perfect geological zones represent the preferential condition for utilizing small-scale hydropower plants like steep rivers that stream throughout the entire year. Islands with wet marine atmospheres are also reasonably suitable. For small-scale utilization, low-head turbines have been created for water channels having small head yet adequate stream. To recognize an appropriate potential location, the site hydrology should be recognized and a site study completed to decide the real stream and head information. Metrological departments or water system divisions provide the hydrological data of a particular location openly. Site studies ordinarily offer increasing point by point data of site conditions to permit power estimation to be accomplished and for the configuration work to start. Stream information ought to be gathered over a one-year period, where conceivable; this is to determine the vacillation of the river stream over the different seasons.

4.7.2 Turbines utilized in micro hydro plants

Various types of turbines are utilized in micro-hydro establishments. Thes turbines are preferred according to water head, stream volume, and accessibility of transportation. In a mountainous region (greater than 50 m height) Pelton turbines could be used. In the case of a low head establishment, Propeller, and Francis turbines are utilized. Archimedes wheel and propeller turbines may be utilized for fewer head establishments in exceptional cases. Small hydrosystems can be utilized in the industrial operation of the centrifugal pump effectively. For low head establishments, maintenance and component expenses may be generally high. The low head framework transfers a large amount of water and is bound to experience surface wreckage and debris. The Banki or Ossberger turbine, a pressurized self-cleaning cross-flow water wheel, is regularly favored for a low-head micro-hydrosystem. Except for the fact that it is less productive, its less complex assembly is more reasonable than other low-head turbines of analogous capacity. The water flowing stream cleans itself and is less inclined to stick with debris.

- Screw turbine (reverse Archimedes screw): In England, two low-head systems, namely Torrs Hydro and Settle Hydro, utilize Archimedes screw turbines, which are debris tolerant with a proficiency of 85%.
- It is streams free turbine can be used with or without a dam structure.
- Propeller and Francis turbines are used for low head establishment.
- Kaplan turbine is an enormous measurement, slow turning, perpetual magnet, slanted open stream very large head turbine with efficiencies of 90%.
- Water wheel: It is a hydraulic wheel. These hydraulic reaction turbines have efficiencies between 67% and 85%. The overshot proficiency of a water wheel is 85% and the undershot efficiency is underneath 30% when working with extremely low head.
- GWV power plant: In this case, the various water resources are channeled into a round-shaped basin with a vortex at the central bottom. A straightforward rotor is rotated by the kinetic energy with an efficiency of 64%–95% [32].

4.8 Small hydro

Small-scale hydro is the improved version of the hydropower generation system with a smaller rating. This small scale system is capable of satisfying the energy demand of nearby locality and industries. The meaning of a "small hydro" venture changes depending on what comprises the base and top cut-off points from nation to nation; yet the most basic range of "small hydro" stretches out in the range of 1–15 MW. Interestingly, numerous hydroelectric ventures are gigantic, for example, hydropower in India extends to 25 MW as the top limit for small hydropower (SHP) ventures [20]. SHP undertakings might be worked in segregated territories, which would be uneconomical for a national grid, or in regions where a national network doesn't exist. The hydropower system produces electricity by utilizing the energy of the water flow. A hydropower venture requires a reliable progression of water with a sensible elevation of waterfall. In a typical establishment, water is supplied from storage through a conduit into a turbine. The water moving past the turbine causes an electrical generator to change the movement into electrical energy. SHP can be created by building a new structure or through the readvancement of existing dams, whose main role is flood control or to provide water for the agricultural system. Old hydropower sites might be redeveloped with significant changes, like the improvement of penstock turbines. The economical advantages of the small hydropower are observed above discussion. It would be beneficial for any organization to invest in small hydro projects at a new site or in any existing sites.

The utilization of the expression "small hydro" fluctuates extensively across the globe; it typically refers to somewhere between 10 and 30 MW. Hydropower plants below 30 MW capacity are named as a small hydro plant. These small hydro plants are considered as renewable energy resources, while the hydroelectric stations with a higher limit are known as conventional energy generation system. Miniaturized hydropower is typically the utilization of hydropower estimated for small networks or small endeavors. The smallest establishments are Pico, hydro beneath 5 kW. The smallest hydro establishment is Pico-hydro power under the 5 kW generation capacity. The construction work related to these systems is also very small, therefore they have correspondingly little ecological effect contrasted with large hydropower [20]. The development in growth report by REN21 indicated that in 2008, SHP establishments developed by 28%, while in 2005, SHP ability was raised to 85 GW all over the world. 70% of this was executed by China (65 GW), trailed by Japan (3.5 GW), the United States (3 GW), and India (2 GW). Worldwide development is expected to be 2.8% every year up to the mid-2020s, when the limit will be around 150 GW [33].

4.9 Turbines

Two primary sorts of hydro turbines are reaction and impulse turbines. These sorts of hydropower turbines are chosen for a given venture based on the depth of the water (head) and the water stream volume. Additional dominant elements include how profoundly the turbine must be set, productivity, and cost [34].

4.9.1 Impulse turbines

These turbines utilize the water speed to rotate a runner and release the atmospheric pressures. Each container on a runner is attacked by the water stream. No suction would be happening on the downward back side of the turbine, and the water streams coming out from the base of the turbine hitting the runner. It is commonly reasonable for high-head and low-stream applications [31].

4.9.2 Pelton turbines

This type of turbine has one open jet for the release of water into circular air space; this water jet has a direct impact on the containers of the runner. There is no requirement of draft tubes in impulse turbines, therefore the runners are essentially located above the water-tail extreme tail to allow the operation of air pressure. Gilkes in England made the Turgo wheel turbine, which is quite different from the Pelton turbine, shown in Fig. 4.6. The cast type wheel is used as the Turgo runner whose structure seems like a fan blade that is shut on the outer sides. A water stream is applied on one side, goes over the blades, and exits on the opposite side.

Pelton is a Tangential flow impulse turbine, where the pressure vitality of water is converted into kinetic energy to produce fast water jet. After that, the jet attacks the wheel tangentially to rotate it.

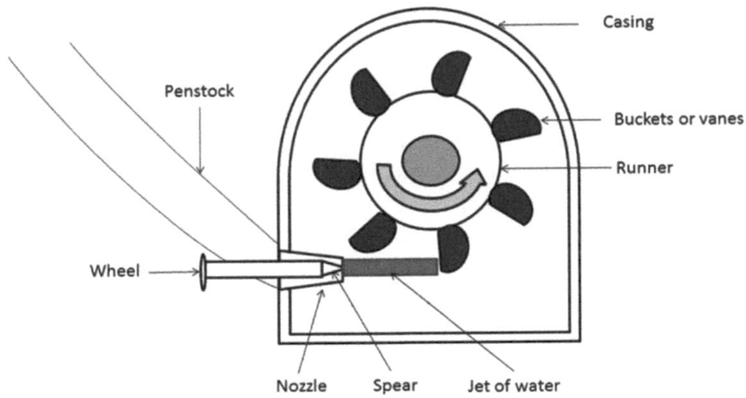

Figure 4.6 The Pelton turbine [31].

4.9.2.1 Nozzle and flow regulation

The movement of the water stream is through the penstock toward a nozzle is prearranged. By using a nozzle, a fast water stream can be created. To govern the nozzle water jet, a mobile spear needle is composed in the nozzle. The spear is allowed to move in reverse and in a forward axial motion. The stream will lessen when it is pushed ahead and will increment when in reverse [31].

4.9.2.2 Bucket and runner

The Pelton turbine is comprised of a runner; it is a rounded disc, on which, various buckets are mounted at the periphery with equal separation between them. The buckets are either in the shape of a twofold hemispherical or ellipsoidal mold. A partition called a splitter is used in each bucket for separating the containers into halves. The containers are made of solid metal, namely iron, steel, or bronze [31].

4.9.2.3 Casing

The entire system, including the runner and buckets, braking, and inlet jets, is covered by a casing. The Pelton turbine casing doesn't perform any hydraulic activities, however, it forestalls the water sprinkling during working and encourages the water release into the tailrace [31].

4.9.2.4 Braking jet

These braking jets are utilized to stop the running wheel those are not in working condition. This condition emerges when the inlet of the nozzle is closed by the spear, after that the water jet is stopped on the bucket. However, due to inertia, the runner won't quit rotating significantly after the complete conclusion of the nozzle inlet. To stop this, a brake nozzle is provided. The brake nozzle coordinates a stream of water on the rear of the buckets to halt the wheel. The jet stream coordinated by the nozzle brake is called the braking jet.

4.9.2.5 Working of Pelton turbine

The working of the Pelton turbine is described here:

The water is moved from the high head source through a long channel called the penstock.

The nozzle system of penstock helps the water to accelerate and flows out as high-speed jet and releases it at normal atmospheric pressure. The water jet will hit a splitter of the cans, which will circulate the jet stream into equal parts of the can and the wheel begins spinning [12]. The kinetic energy of the jet stream is diminished when it hits the can, and furthermore, because of the round state of the pails, the coordinated stream will alter its course, make a U-transform, and fall into the tailrace. The inlet angle of the stream is generally, observed between 1to3 degrees while the deflected angle of the water jet after hitting the bucket is observed between 165 and 170 degrees. The water gathered in the tailrace ought not to submerge the Pelton wheel. To create more force, two Pelton wheels can be organized on a solitary shaft or two water jets can be aimed at a solitary Pelton wheel (Fig. 4.6).

4.9.3 Cross-flow

This type of turbine seems to be in the shape of a drum. It uses a rectangular, long nozzle, synchronized with bent vanes on a circular runner. It seems like a "squirrel cage" blower (Fig. 4.7). This turbine permits water to flow twice along with the blades. In the primary step, water flows from the outside of the blades to the inner side while in the second step the water flows from inside to out. A guided vane at the passage to the turbine guides the stream to a restricted segment of the sprinter.

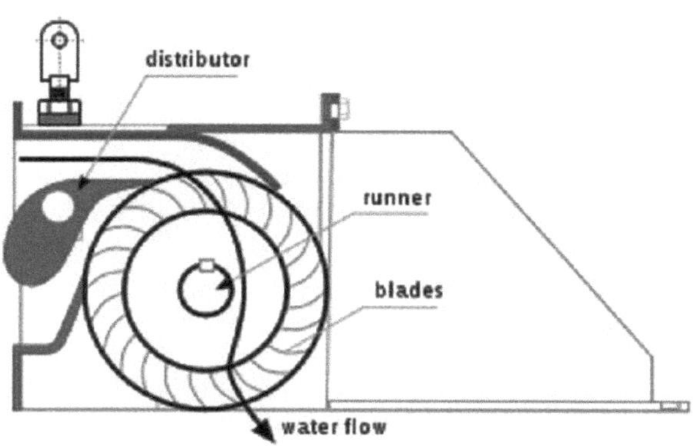

Figure 4.7 Cross-flow turbine [35].

A cross-flow is created to oblige bigger water streams and lower heads than the Pelton turbine [35].

This turbine is generally recognized as an Ossberger or Mitchell–Banki turbine. It is a kind of turbine that has the tendency to be used in minor hydropower sites with power outputs ranging from 5 to 100 kW. These turbines are beneficial for a big range of hydraulic heads, from 1.75–200 m; generally, cross-flow turbines are preferred for heads of below 40 m.

Not being utilized in small hydroelectric schemes, one advantage of these turbines is the requirement of relatively little maintenance to retain them in working condition. Along these lines, they are increasingly reasonable for use in remote networks. Albeit helpful for a wide scope of water-driven heads and power outputs, these turbines are generally effective for low heads and small power yields. Dissimilar turbines are likely progressively effective and helpful for huge scale applications.

4.9.3.1 Design structure

A cross-flow turbine is structured utilizing an enormous cylindrically composed mechanism made out of a rotor encompassed by a "cage" of sharp edges collected into a water wheel shape. These edges are commonly sharpened to build the effectiveness of the turbine by decreasing the resistance from the water stream. Water is coordinated onto the turbine through a jet that makes a level sheet of water, and afterward it is coordinated onto the cutting edges utilizing a guide vane. The water first hits the sharp edges and moves to within the turbine, with the water hitting the edges once again as the water leaves the center point of the turbine [30].

4.9.3.2 Operational activity

In the present turbine system, the water stream enters as a flatten sheet instead of a round jet. The sheet of water is guided onto the edges of the turbine by a guided inlet vane, guaranteeing that the water hits the blades at the best possible angle to amplify productivity. The water runs over the blades making a torque on the edges. In the wake of hitting the sharp edges, the sheet of water travels through the turbine and hits the blades again as it leaves, creating more torque. The primary effect of the water hitting the sharp edges is that it delivers more power than the subsequent hit, since this kind of turbine utilizes water jets that make an impact on the turbine. The best example of a drive turbine is the cross-flow turbine like the Pelton turbine.

4.9.4 Reaction turbine

Power is generated by a reaction turbine from the combined action of moving water and pressure. The runner is set straight forward in the water stream flowing above the blades instead of separately striking each. Low head and high streams are preferred for the common utilization of reaction turbines [14].

4.9.5 Propeller turbine

The configuration of the propeller turbine is based on three to six blades associated with the runner; here, the blades are continuously in contact with the water. A boat propeller runs in a channel. Through a funnel, the pressure is steady; but in the event that it isn't, the runner would be out of parity. Variations in blade pitch are also allowed as they may be fixed or flexible. Other important parts of this system are draft tubes, scroll cases, and wicket entryways other than the sprinter. A few distinct kinds of propeller turbines are described here [17].

4.9.6 Bulb turbine

In this case, a combined sealed unit, including a generator and turbine, is placed in the water stream as shown in Fig. 4.8.

4.9.7 Straflo turbine

In the Straflo turbine (Fig. 4.9), a generator is directly combined at the perimeter of the turbine.

4.9.8 Tube turbines

The penstock bends just before or after the runner, allowing a direct connection to the generator (Fig. 4.10).

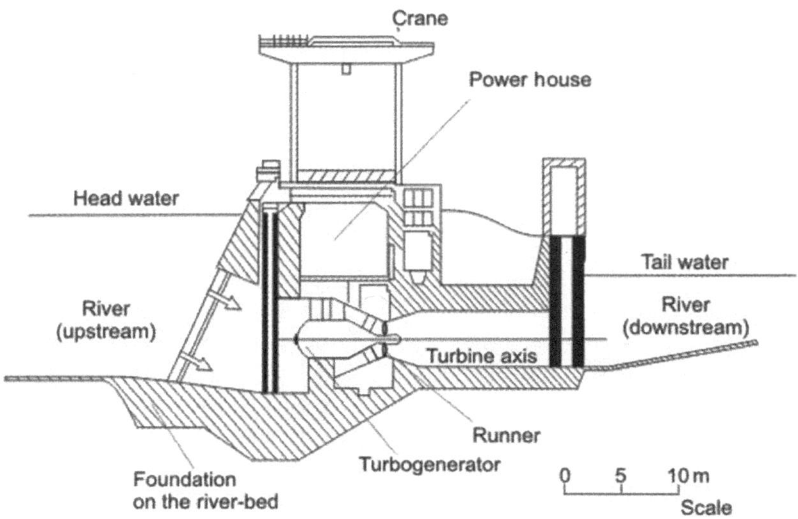

Figure 4.8 The Bulb turbine [14].

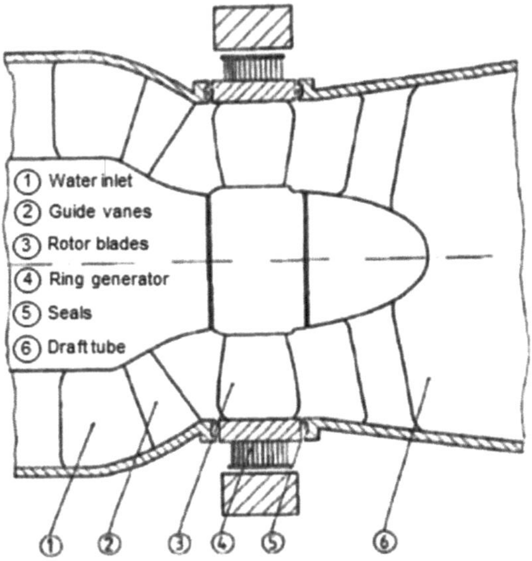

Figure 4.9 The Straflo turbine [17].

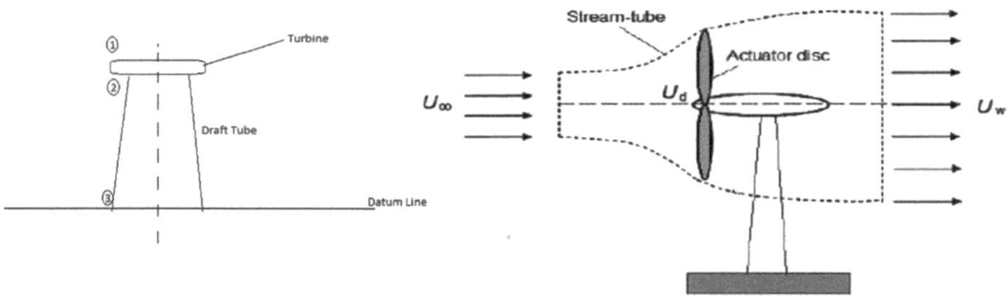

Figure 4.10 Tube turbines [20].

4.9.9 Kaplan turbine

This turbine is basically classified as a propeller-type turbine with movable blades. Viktor Kaplan introduced this turbine in 1913. In this turbine, propeller blades and wicket entryways are combined and operated automatically to accomplish a high efficiency over a wide water level and flow (Fig. 4.11). It is an enhanced version of the Francis turbines that permits the proficient creation of power in low-head conditions. The operative head range lies between 10 and 70 m with an output of 5–200 MW. A range of 2–11 m runner widths are considered here. This turbine rotates at a consistent speed, differing from place to place in the range of 54.5–450 rpm. Generally, these turbines are utilized across the world for low-head and high-stream power creation.

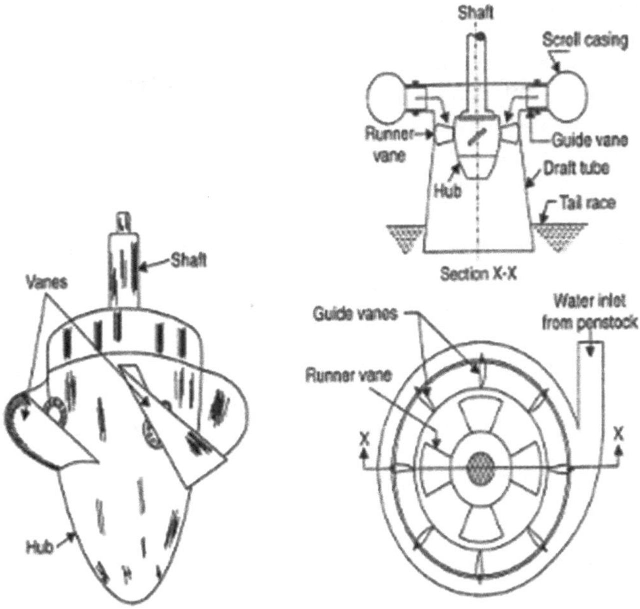

Figure 4.11 The Kaplan turbine [12].

4.9.9.1 Operational activity

The Kaplan turbine is an internal stream-type reaction turbine; the working fluid changes its pressure on traveling through the whole turbine and releases energy. The hydrostatic head and kinetic energy recuperate power from streaming water. The combined features of axial and radial turbines can be observed in the Kaplan turbine. A scrolled-type tube is wrapped on the wicket inlet of the turbine. For rotating the propeller runner, a water jet is tangentially applied onto the wicket entryway and spirals. The extraordinarily molded draft outlet tube slows down the water and recuperates kinetic energy. The draft tube is required to stay full of water, and for this reason, the turbine shouldn't be placed at the bottom of a water stream. A higher latitude of the turbine is helpful for the suction process offered by the draft tube on the turbine blades. The subsequent drop in pressure may prompt cavitation. Productive operations for long-range flow conditions are permitted by the variable geometry of the turbine blades and wicket door. Commonly, over 90% efficiencies are observed in Kaplan turbines, however, it might be lesser in low-head uses. Hydraulic high-pressure oil bearings are used for pivoting the propeller blades; therefore, a basic component is used to keep up a continuous discharge of oil into the system [14].

4.9.9.2 Applications

Kaplan turbines are generally utilized across the world for power generation. These turbines are used for low head and high flow conditions. The cost-effective, small Kaplan turbines are fabricated for the individual uses. These micro-level turbines can work for the range of 3 to 0.3 m head with sufficient water stream. Enormous Kaplan turbines are separately intended for each site to work at the most elevated conceivable productivity, ordinarily over 90%. They are extremely expensive to configure, fabricate, and introduce, however, they work for a considerable length of time.

4.9.10 Francis turbine

The structure of a Francis turbine consists of a runner having buckets fixed onto its periphery (at least nine). A water stream is thrown on the sprinter and afterward falls to make it turn. In addition to the sprinter, the other significant segments are the wicket gates, scroll case, and draft tube.

This turbine was created by James B. Francis in Massachusetts, USA, and is shown in Fig. 4.12. It is based on the concept of a combined axial and radial flow and inward-flow reactions. It is the most widely recognized water turbine used today. It works in the range of 40–600 m head for electrical power generation. The power output of electric generators using this turbine lies in the range of a few kilowatts up to 800 MW. The diameter of the penstock is in the range of 0.91–10 m. The turbine speed is in the range of 75–1000 rpm. A wicket door on the rotating runner controls the water movement through the turbine. These turbines are always mounted with a vertical shaft to confine water from the generator. The components that comprise a Francis turbine are described here.

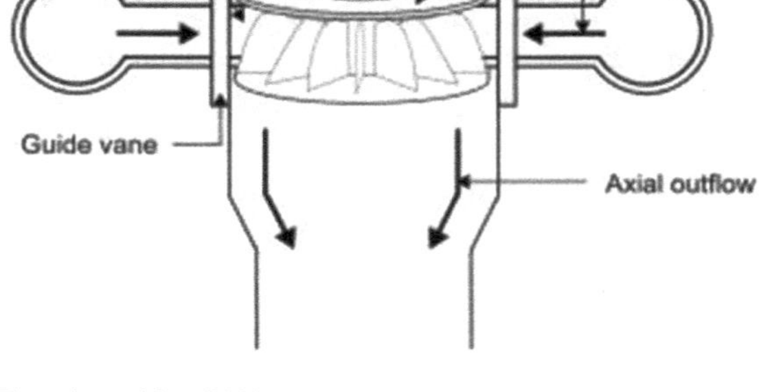

Figure 4.12 The Francis turbine [12].

4.9.10.1 Spiral packaging

This packaging exists around the sprinter of the turbine and is recognized as a scroll case or volute packaging. It has various openings through its entire length at specified intervals to permit the fluid to encroach the blades of the runner. The pressure of the liquid is converted into kinetic energy by the openings just before the fluid encroaches the blades. This system stabilizes a steady speed of fluid regardless that various openings. These openings have been accommodated for entering the fluid as the cross-sectional region of this casing continuously decreases along the outer periphery [36].

4.9.10.2 Guide and stay vanes

Guide and stay vanes are used to change the pressure of the fluid into kinetic energy. It also provides direction to the flow at the certain designed angles of runner's blades.

4.9.10.3 Runner blades

These are the core of turbines. Runner blades are the main target of the striking fluids, which generates tangential power to rotate the turbine and creating torque. The design of inlet and outlet blade angles are significant parameters for influencing power generation.

4.9.10.4 Draft tube

The tube interface between the runner outlet and the tailrace is known as the draft tube, which is used for releasing the turbine water. It is used for diminishing the speed of the released water to limit the loss of motor vitality at the outlet. This allows the turbine to be set over the tailwater without a calculable drop in the accessible head.

4.9.10.5 Operation activity

The Francis turbine is a sort of response turbine, a classification of turbines wherein working liquid goes into the turbine under enormous tension and energy is extricated by the turbine blades from the working liquid. A part of the fluid vitality is transferred due to a change in pressure, while the remaining energy is extracted by the volute turbine casing. The water leaving the outlet at a low speed and low whirl has only minor kinetic and potential vitality left. The outlet tube of the turbine is molded to decelerate the water stream and regain the pressure.

4.9.10.6 Applications

This turbine is designed for being used in a wide range of streams and heads. This, alongside its high effectiveness and flexibility, has made the Francis turbine the most utilized turbine. A head extends from 40 to 600 m and the associated generated power yields vary from a few kilowatts to 800 MW. Enormous Francis turbines are exclusively designed for sites to synchronize with any water head at maximum efficiency, commonly over 90%. The Francis turbine works at its best when totally filled with water, unlike the Pelton turbine. The outlet channel and the turbine itself might be set lower than the lake or ocean level used, diminishing the inclination for cavitation.

In addition to electricity generation, this system may be utilized for the pump storage. This pump reservoir is acting as a huge vitality stocking source to store the excess electrical vitality in the form of water at elevated storages. This is one of a couple of strategies that permit excess electricity to be stored for later usage [36].

4.9.11 Kinetic turbine

This is also called a free-stream turbine. The kinetic energy of flowing water is utilized for the generation of electricity instead of the potential energy. This framework can be used in artificial canals, rivers, ocean flows, and tidal waters. The kinetic framework uses the specific way of the water stream. It does not require the redirection of water in artificial conduits or rivers. These frameworks don't require enormous civil work; rather, they can utilize existing structures [32].

4.10 Conclusion

This chapter presents a brief description of hydropower technologies. The approach follows a basic- to high-level explanation of hydropower plants. It has been discovered that several parameters influence the activity of hydropower plants, including water head, dams, turbines, discharge, and generators. Various types of dams are discussed with their specifications and shortcomings. Turbines are one of the most important segments of hydropower plants. Most types of hydro turbines are presented with diagrams and specifications. It has been also found that different parameters, like penstock, type of turbine, and the number of turbines, were not used in earlier works. The small hydropower plants are generally built as ROR system. In order to improve the small hydropower system further researches are required. The various aspects of hydropower plants discussed in this study can be useful for scholars working in the same field.

References

[1] V.K. Singh, S.K. Singal, Operation of hydro power plants-a review, Renew. Sustain. Energy Rev. 69 (2017) 610–619March. Available from: https://doi.org/10.1016/j.rser.2016.11.169.

[2] J.S. Anagnostopoulos, D.E. Papantonis, Optimal sizing of a run-of-river small hydropower plant, Energy Convers. Manag. 48 (10) (2007) 2663–2670.

[3] M. Nakajima, S. Iio, T. Ikeda, Performance of savonius rotor for environmentally friendly hydraulic turbine, J. Fluid Sci. Technol. 3 (3) (2008) 420–429.

[4] N. Kishor, R.P. Saini, S.P. Singh, A review on hydropower plant models and control, Renew. Sustain. Energy Rev. 11 (5) (2007) 776–796.

[5] C. Tong, Chapter 7 Advanced Materials and Devices for Hydropower and Ocean Energy, Springer Science and Business Media LLC, 2019.

[6] <https://www.e-education.psu.edu/earth104/node/1067> (accessed 23.04.20).

[7] V. Yildiz, J.A. Vrugt, A toolbox for the optimal design of run-of-river hydropower plants, Environ. Model. Softw. 111 (2019) 134–152. January.

[8] <http://www.hydroquebec.com/visit/monteregie/beauharnois.html> (accesed 26.04.20).

[9] Y. Nishi, T. Inagaki, Performance and flow field of a gravitation vortex type water turbine, Int. J. Rotating Mach. 2017 (2017)Article ID 2610508, 11. Available from: https://doi.org/10.1155/2017/2610508.

[10] Y. Nishi, T. Inagaki, K. Okubo, N. Kikuchi, Study on an axial flow hydraulic turbine with collection device, Int. J. Rotating Mach. 2014 (2014). Article ID 308058, 11.

[11] N. Kolekar, A. Banerjee, Performance characterization and placement of a marine hydrokinetic turbine in a tidal channel under boundary proximity and blockage effects, Appl. Energy 148 (2015) 121–133.

[12] C.P. Jawahar, P.A. Michael, A review on turbine for micro hydro power plant, Renew. Sustain. Energy Rev. 72 (2017) 882–887. May.

[13] <https://www.energy.gov/eere/water/pumped-storage-hydropower>.

[14] S. Nash, A. Phoenix, A review of the current understanding of the hydro environmental impacts of energy removal by tidal turbines, Renew. Sustain. Energy Rev. 80 (2017) 648–662. December.

[15] C. Cheng, A. Blakers, M. Stocks, B. Lu, Pumped hydro energy storage and 100% renewable electricity for East Asia, Glob. Energy Inter. Connect. 2 (5) (2019) 386–392.

[16] F.J. Peñas, J. Barquín, Assessment of large-scale patterns of hydrological alteration caused by dams", J. Hydrol. 572 (2019) 706–718. May.

[17] N.F. Yah, A.N. Oumer, M.S. Idris, Small scale hydro-power as a source of renewable energy in Malaysia: a review, Renew. Sustain. Energy Rev. 72 (2017) 228–239. May.

[18] https://reneweconomy.com.au/victoria-plan-use-old-mines-pumped-hydro-heads-next-stage-33901/ (accessed 2.05.20).

[19] P. Orban, S. Bodeux, P. Archambeau, S. Erpicum, A. Dassargues, Underground pumped storage hydropower plants using open pit mines: How do groundwater exchanges influence the efficiency? Appl. Energy 190 (2017) 135–146. 15 March.

[20] S. Saad, A.Y. Hatata, M.M. El-Saadawi, A feasibility study of small hydro power for selected locations in Egypt, Energy Strategy Rev. 24 (2019) 300–313. April.

[21] D. Burchsted, M.D. Daniels, Classification of the alterations of beaver dams to headwater streams in north-eastern Connecticut, U.S.A, Geomorphology 20515 (2014) 36–50. January.

[22] Z. Tao, Y. Liu, Q. Yang, S. Wang, Study on the nonlinear deformation and failure mechanism of a high arch dam and foundation based on geo mechanical model test, Eng. Struct. 20715 (2020). March Article 110287.

[23] S. Anders Brandt, Classification of geomorphological effects downstream of dams, CATENA 40 (4) (2000) 375–401. August.

[24] W. Ge, H. Sun, H. Zhang, Z. Li, P. van Gelder, Economic risk criteria for dams considering the relative level of economy and industrial economic contribution, Sci. Total. Environ. 72510 (2020). July.

[25] H.A. Nefeslioglu, T. Gorum, The use of landslide hazard maps to determine mitigation priorities in a dam reservoir and its protection area.", Land. Use Policy 91 (2020). February.
[26] N.L.S. Pinto, Concrete face rockfill dams, in: E.M. das Neves (Ed.), Advances in Rockfill Structures. NATO ASI Series (Series E: Applied Sciences), vol 200, Springer, Dordrecht, 1991, pp. 341–373.
[27] G. Blöschl, A. Viglione, A. Montanari, Emerging approaches to hydrological risk management in a changing, Climate Vulnerability: Understanding and Addressing Threats to Essential Resources, 5, 2013, pp. 3–10.
[28] <https://www.mottmac.com/water-and-wastewater/dams-and-reservoirs> (accesed 25.04.20).
[29] L. Bengtsson, R.W. Herschy, R.W. Fairbridge, Encyclopedia of Lakes and Reservoirs, Springer Science and Business Media LLC, Netherlands, 2012.
[30] R.K. Ranjan, N. Alom, J. Singh, B.K. Sarkar, Performance investigations of cross flow hydro turbine with the variation of blade and nozzle entry arc angle, Energy Convers. Manag. 18215 (2019) 41–50. February.
[31] J. Ma, Y. Wang, X. Feng, Energy recovery in cooling water system by hydro turbines, Energy 13915 (2017) 329–340. November.
[32] J. Riglin, W. Chris Schleicher, I.-H. Liu, A. Oztekin, Characterization of a micro-hydrokinetic turbine in close proximity to the free surface, Ocean. Eng. 110 (2015) 270–280.
[33] <https://www.ren21.net/wp-content/uploads/2019/05/gsr_2019_full_report_en.pdf> (accessed 28.04.20).
[34] R. Uhunmwangho, M. Odje, K.E. Okedu, Comparative analysis of mini hydro turbines for Bumaji Stream, Boki, Cross River State, Nigeria, Sustain. Energy Technol. Assess. 27 (2018) 102–108. June.
[35] A. Georgescu, S. Georgescu, C.I. Cosoiu, N. Alboiu, Efficiency of marine hydropower farms consisting of multiplevertical axis cross-flow turbines, Int. J. Fluid Mach. Syst. 4 (1) (2011) 150–160.
[36] M. Anyi, B. Kirke, Evaluation of small axial flow hydrokinetic turbines for remote communities, Energy Sustain. Dev. 14 (2) (2010) 110–116.

Section IV

Wind Power Technology

Wind power technology

5

Subhadeep Bhattacharjee
Department of Electrical Engineering, National Institute of Technology (NIT), Agartala, India

5.1 Historical background

Wind is one of the primordial forms of energy known to man and has been playing a significant role in the history of human civilization [1,2]. Utilizing wind as an energy resource started around 7000 years ago [3]. More than 5000 years back in ancient Egypt, sailboats powered mainly by the wind were used for water transportation and international trade. However, no one has precise knowledge when the first windmill was constructed. Archaeologists found remains of a primitive Chinese vase dating back to 3000 BCE with a painting resembles a windmill. By 2000 BCE Babylonians used windmills extensively for irrigation. These windmills were alike a revolving door system and similar to the modern vertical axis wind machine. Wind machines were used in Persia in 200 BCE. The wind-wheel of Heron of Alexandria is one of the first known instances of a wind power–driven machine in olden times [4]. From the 7th century onwards, practical windmills of the earliest design were built in the Sistan region of Persia [5]. By the 10th century, windmills were set up for grinding grains in eastern Iran and Afghanistan. During the 12th century, windmills were established in Europe to grind grains and pump water. Such windmills were also used extensively in The Netherlands to drain areas of the Rhine River delta [2,4].

Working in windmill plants during earlier days was a difficult task. Workers were often wounded as the windmills were constructed of a huge rotating mass with little or no control of its revolution. Several workers became deaf due to the loud grinding and hammering noises. Also, the grinding dust of some materials such as wood used to cause respiratory problems, and the process of grinding stones frequently resulted in sparks and fires.

By the 19th century, many windmills were commissioned in Denmark for serving mechanical loads such as pumping and grinding purpose. Also, by the end of the 19th century, many windmills were constructed in the United States. Until the use of diesel engine in rail, various US rail routes were dependent on large multivane windmills to pump water for steam locomotives. Such windmills are still being used in small numbers and are mostly suited for pumping ground water in small amounts to farm animals' water tanks (i.e., a farm windmill).

Sustainable Fuel Technologies Handbook. DOI: https://doi.org/10.1016/B978-0-12-822989-7.00006-8

Besides performing mechanical work such as grinding grains and pumping water, windmills were also used to communicate with neighbors by locking the windmill sails (or vanes or blades) in a certain arrangement. In The Netherlands, windmill sails were set in certain positions to alert the community of a likely attack during World War II.

In this day and age, windmills are mainly used to generate electricity by converting kinetic energy (KE) into mechanical energy and then into electrical energy. These types of windmills are known as wind turbines. The first known electricity-generating wind turbine was installed in 1887 by James Blyth in Scotland (Fig. 5.1) with a view to charging a battery [4]. The first wind turbine for electricity production in America was built by Charles F. Brush in 1888 (Fig. 5.2). In the 1930s and 1940s, numerous wind turbine projects were developed in the United States for electricity supply in off-grid areas. They had two or three thin blades which rotated at high speeds to drive electrical generators. However, the advent of the central power grid line extension in nearly all parts of the United States eliminated the requirement of such wind turbine projects and consequently wind turbine development remained largely dormant for 20 years until humanity was faced with the 1970s energy crisis due to interruptions in Middle Eastern oil exports for major industrial countries which caused substantial petroleum shortages [8].

Over the past few decades, global warming, climate change, and depleting fossil fuel reserves have caused a resurgence in interest for wind energy systems. Wind renewable power is currently the world's fastest growing energy source and has great potential to save a substantial amount of the Earth's fossil fuel resources. The design and operation of wind turbines have been improved considerably making them a highly viable technology. The modern wind turbine's size ranges from a few kW machines to up to 5 MW single unit systems.

Figure 5.1 Wind turbine of James Blyth [6].

Figure 5.2 Wind turbine of Charles F. Brush [7].

5.2 Energy and power from the wind

5.2.1 Available power in wind

We know that power is equal to energy per unit time. The energy available is the KE of wind. The KE of any particle is:

$$KE = \frac{1}{2}mv^2 \quad (5.1)$$

where m = mass of the particle (kg), v = velocity of the particle (m/s).

The amount of air passing for t sec through an area A, with velocity v, is Avt and its mass (m) is equal to its volume multiplied by its density ρ of air,

$$m = \rho Avt \quad (5.2)$$

Thus m is the mass of air traversing the area A swept by the rotating blades of a wind turbine.

Substituting this value of mass m from the Eq. (5.2) in Eq. (5.1), the following equation is obtained

$$KE = \frac{1}{2}\rho Atv^3 \quad (5.3)$$

Since energy is power multiplied by time, the wind power P_{wind} is

$$P_{wind} = \frac{KE}{t} = \frac{1}{2}\rho A v^3 \tag{5.4}$$

5.2.2 Modern wind turbine

Wind turbine is a generic name for the wind energy system as a whole that converts the KE in wind into electrical energy. Modern wind turbine sizes range from 50 kW to more than 5 MW. Wind turbines are available in a variety of designs and power ratings, but the most familiar design is the horizontal axis propeller type wind turbine which is shown in Fig. 5.3 in a simplified form. The basic components of the wind turbine are:

1. *Tower*: A tower carries the rotating blades at a satisfactory height to increase the exposure of the blades to wind. Large wind turbines (in the MW range) have towers as high as 250 m above the base.
2. *Rotating blades*: Rotating blades intercept the KE of wind. Blades are generally made of fiberglass-reinforced polyester or wood-epoxy material. The length of the rotating blades ranges from 5 m to over 60 m. The majority of wind turbines have three rotor blades. The pitch angle of the blades is changeable by advanced blade systems in modern wind turbines in order to maximize the capture of the wind's KE.

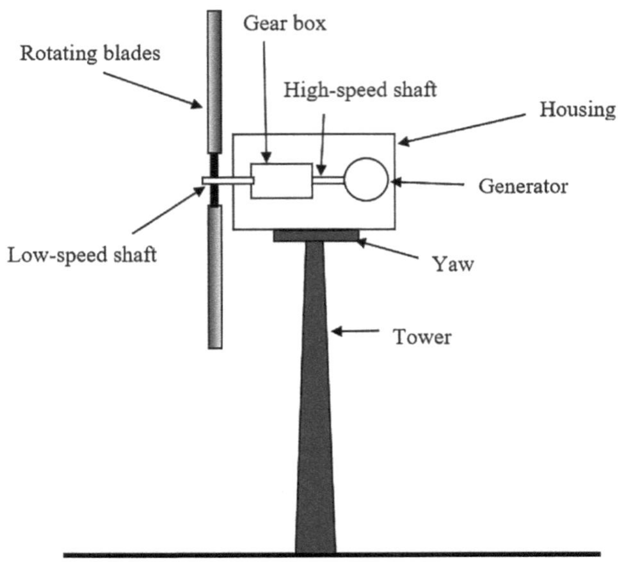

Figure 5.3 Basic parts of modern wind turbine system.

3. *Yaw mechanism*: A yaw (the left–right motion) mechanism allows the housing box to turn and maintains the blades perpendicular to the wind speed, thereby increasing the exposure of the blades to the wind.
4. *Gearbox*: A gearbox is used to connect the low-speed rotating blades to the high-speed generator. It also serves as a clutch.
5. *Generator*: A generator is connected to the high-speed shaft of the gearbox to transform the mechanical energy of the rotating blades into electrical energy.
6. *Controller*: A controller connects the utility system to the wind generator and locks the blades when the wind speed is less than the minimum generation limit or when the wind speed surpasses the design restrictions of the system [2].

5.2.3 Power output and efficiency considerations

Fig. 5.4 illustrates that all of the wind energy is not converted into electrical energy due to various losses in the system. Only a fraction of the available wind energy is captured by the rotating blades. The ratio of the power captured by the blades P_{blade} to the available power from the wind P_{wind} is defined as power coefficient C_p. The maximum theoretical value of the power coefficient is 0.5926, which is the Betz limit [2]. For the best practical wind turbine rotors, the power coefficient is 0.4–0.45, which means that the rotors cannot exploit more than 40%–45% of the available wind power.

$$C_p = \frac{P_{blade}}{P_{wind}} \tag{5.5}$$

A portion of the blade power inflowing the turbine is dissipated as rotational losses (frictional and windage) in the rotating blades and rotor mechanism. The remaining power P_{gear} enters the gearbox, where part of it is wasted as gearbox rotational loss. The rest of the power P_g goes into the generator where some gets wasted in the form of electrical losses in the generator's windings and core in addition to the generator's own rotational losses. The residual power P_{out} is the output

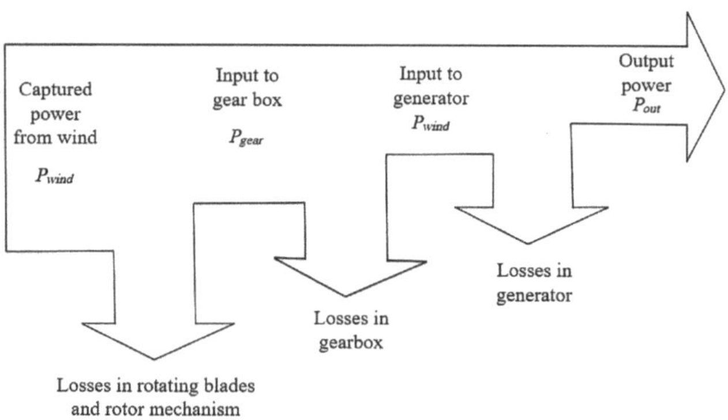

Figure 5.4 Power flow of a wind turbine.

electric power supplied to the load [2]. Thus the efficiency of the rotating blade (η_{blade}), the gearbox efficiency (η_{gear}), and the generator efficiency (η_g) can be expressed as:

$$\eta_{blade} = \frac{P_{gear}}{P_{blade}} \tag{5.6}$$

$$\eta_{gear} = \frac{P_g}{P_{gear}} \tag{5.7}$$

$$\eta_g = \frac{P_{out}}{P_g} \tag{5.8}$$

Therefore the total efficiency is:

$$\eta_{total} = C_p \eta_{blade} \eta_{gear} \eta_g \tag{5.9}$$

The output power of the wind turbine is:

$$P_{out} = \eta_{total} P_{wind} = (C_p \eta_{blade} \eta_{gear} \eta_g) P_{wind} \tag{5.10}$$

5.2.4 Power versus wind speed characteristics

The correlation between wind speed and generator electrical output is described by the wind turbine power curve, which is technically informative for a specific wind turbine. Fig. 5.5 shows the typical power curve for a wind turbine indicating the cut-in wind speed, rated wind speed, and cut-out wind speed. These are explained through the power–speed characteristics of the turbine.

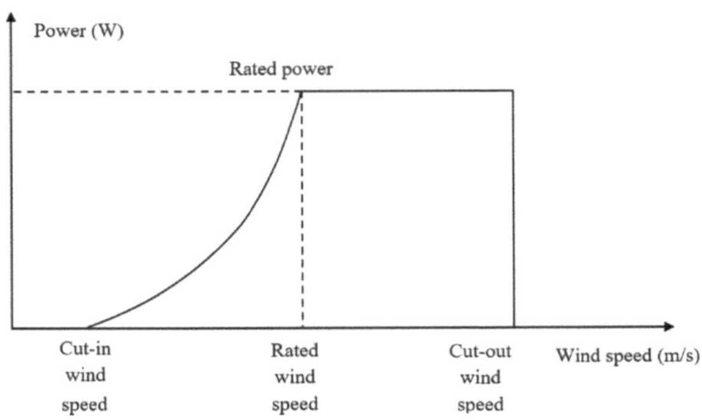

Figure 5.5 Power–speed characteristics of wind turbine.

1. *Cut-in wind speed*: Wind at low speed may not have sufficient power to overcome friction in the drive train of the turbine. Even if it does and the generator rotates, the electrical power generated may not be sufficient to offset the power needed by the generator's field windings. Thus the cut-in wind speed (V_c) is the minimum wind speed required to generate net power. Since no power is generated at wind speeds below V_c, the portion of the wind's energy below V_c is practically lost. However, the amount of wind energy in low-speed winds are not very significant and as a result generally not much wind energy is lost below V_c.
2. *Rated wind speed*: As wind velocity exceeds the cut-in wind speed, the power delivered by the generator rises in accordance with the cube of wind speed. When wind attains the rated wind speed (V_r), the generator produces as much power as it is designed for. Beyond V_r, there must be some way to shed some of the wind's power to avoid the generator being damaged.
3. *Cut-out or furling wind speed*: When the wind is very strong (stormy) that presents danger to the wind turbine, the machine must be shut down. This wind speed is called the cut-out wind speed or the furling wind speed (V_f) ("furling" is the term used in sailing to describe the practice of folding up the sails when winds are too strong). Therefore above V_f, power generation is stopped and output power becomes zero [9].

5.2.5 Extractable limits of wind power

The physical conditions in a wind turbine are such that only a fraction of the available wind power can be converted into useful power. As the free wind stream encounters and passes through a rotor, the wind transfers some of its energy to the rotor and its speed decreases to a minimum in the rotor wake. Subsequently, the wind stream regains energy from the surrounding air and at a sufficient distance from the rotor, the free wind speed is restored. While the wind speed is decreasing, the air pressure in the wind stream changes in a different manner. It first increases as the wind approaches the rotor, and then drops sharply as it passes through and energy is transferred to the rotor. Finally the pressure increases to the ambient atmospheric pressure.

The most common wind turbine today is the horizontal axis, propeller type. Assume that the wheel of such a turbine has thickness "*ab*" as shown in Fig. 5.6.

Let a = Inlet plane
b = Exit plane
P_i = Incoming (upstream) wind pressure
V_i = Incoming (upstream) wind velocity
P_e = Wind pressure at exit from blade (downstream of the turbine)
V_e = Wind velocity at exit from blades (downstream of the turbine)
ρ = Air density
v = Specific volume = $1/\rho$

V_e is less than V_i because KE extracted by the turbine. Assuming no energy loss and no change in air density. The velocity and pressure changes are plotted as shown in Fig. 5.6A and B respectively.

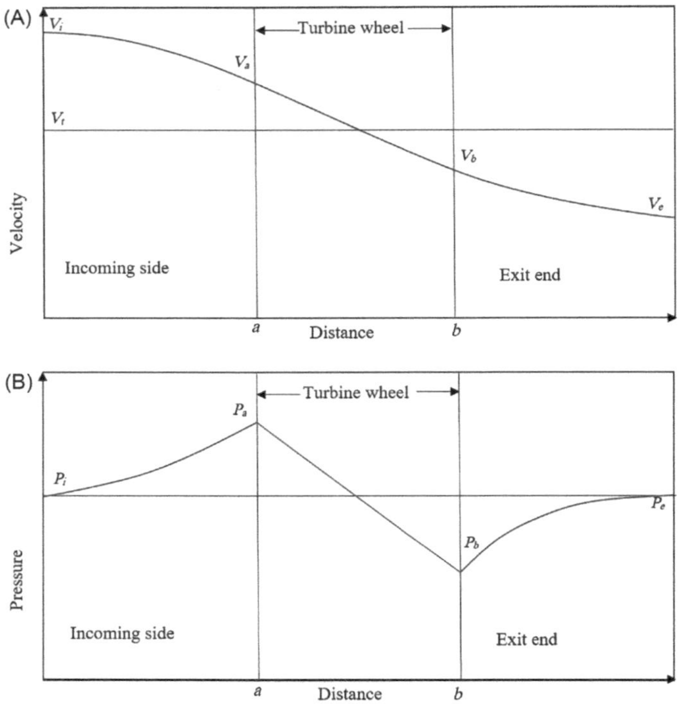

Figure 5.6 (A) Velocity and (B) pressure profiles while traversing a wind turbine rotor.

Applying the total energy equation, we get:

$$P_i v + \frac{V_i^2}{2} = P_a v + \frac{V_a^2}{2} \tag{5.11}$$

$$P_i + \frac{V_i^2}{2v} = P_a + \frac{V_a^2}{2v}$$

$$P_i + \frac{\rho V_i^2}{2} = P_a + \frac{\rho V_a^2}{2} \tag{5.12}$$

Similarly, for exit area:

$$P_e + \frac{\rho V_e^2}{2} = P_b + \frac{\rho V_b^2}{2} \tag{5.13}$$

The wind velocity across the turbine decreases from a to b, because KE is converted to mechanical work. The incoming velocity V_i does not decrease abruptly, but gradually as it approaches the turbine.

$$V_i > V_a$$

$$V_b > V_e$$

Therefore from Eqs. (5.12) and (5.13), it is apparent that

$$P_a > P_i$$

$$P_b < P_e$$

That is, the wind pressure rises as it approaches, then as it leaves the wheel. From Eqs. (5.12) and (5.13)

$$P_a - P_b = \left(P_i + \rho\frac{V_i^2 - V_a^2}{2}\right) - \left(P_e + \rho\frac{V_e^2 - V_b^2}{2}\right) \tag{5.14}$$

It can be assumed that wind pressure at exit end (away from the turbine) is equal to ambient pressure, that is:

$$P_e = P_i \tag{5.15}$$

As the blade width "*ab*" is very thin compared to total distance considered, it can be assumed that velocity within the turbine does not change much.

$$V_t \approx V_a \approx V_b \tag{5.16}$$

Considering the assumptions in Eqs. (5.15) and (5.16), Eq. (5.14) reduces to

$$P_a - P_b = \rho\left(\frac{V_i^2 - V_e^2}{2}\right) \tag{5.17}$$

If A is the projected area of windmill perpendicular to the wind stream, the axial force F_x in the direction of wind stream is given by

$$F_x = (P_a - P_b)A = \rho A\left(\frac{V_i^2 - V_e^2}{2}\right) \tag{5.18}$$

From Newton's second law:

$$F_x = mf = m\frac{dV}{dt} = \dot{m}\Delta V$$

where m = mass, f = acceleration = dV/dt, $\dot{m}$ = mass flow rate = $\rho A V t$.

Thus axial force can also be expressed as:

$$F_x = \rho A V_t (V_i - V_e) \tag{5.19}$$

Equating the Eqs. (5.18) and (5.19)

$$\rho A \frac{V_i^2 - V_e^2}{2} = \rho A V_t (V_i - V_e)$$

$$V_t = \frac{V_i - V_e}{2} \tag{5.20}$$

Considering the total thermodynamic system bounded by i and e, the general energy equation is now reduced to the steady flow work, W, which is expressed in terms of KE:

$$W = KE_i - KE_e = \frac{V_i^2 - V_e^2}{2} \tag{5.21}$$

Power, P, is defined as rate of work and can be expressed in terms of mass flow rate:

$$P = \dot{m}\left(\frac{V_i^2 - V_e^2}{2}\right) \tag{5.22}$$

$$P = \rho A V_t \left(\frac{V_i^2 - V_e^2}{2}\right)$$

$$= \rho A \left(\frac{V_i + V_e}{2}\right)\left(\frac{V_i^2 - V_e^2}{2}\right)$$

$$P = \frac{1}{4}\rho A (V_i + V_e)\left(V_i^2 - V_e^2\right) \tag{5.23}$$

It is evident from Eq. (5.23) that V_e is positive in one term and negative in the other—that too low or too high a value for V_e results in reduced power. Therefore there is an optimum exit velocity $V_{e_{opt}}$ that results in maximum power P_{max} which can be obtained by differentiating P, and equating the derivative to zero.

For maximum power:

$$\frac{dP}{dV_e} = 0$$

$$3V_e^2 + 2V_i V_e - V_i^2 = 0 \tag{5.24}$$

The quadratic Eq. (5.24) has two solutions, namely:

$$V_e = -V_i \text{ and } V_e = \frac{1}{3}V_i$$

But only the second solution is physically acceptable

$$\text{Thus, } V_{e_{opt}} = \frac{1}{3}V_i \tag{5.25}$$

Using Eq. (5.23) and substituting $V_e = \frac{1}{3}V_i$, maximum power for an ideal horizontal axis propeller type wind turbine is derived as:

$$P_{max} = \frac{1}{4}\rho A\left(V_i + \frac{1}{3}V_e\right)\left(V_i^2 - \frac{1}{9}V_i^2\right)$$

$$P_{max} = \frac{8}{27}\rho A V_i^3 = \frac{16}{27}\left(\frac{1}{2}\rho A V_i^3\right) = 0.593\left(\frac{1}{2}\rho A V_i^3\right) = 0.593 P_{wind} \tag{5.26}$$

The ideal or maximum theoretical efficiency η_{max} (also called the power coefficient C_p) of a wind turbine is the ratio of the maximum power obtained from the wind, to the total power available in the wind. The factor 0.593 is known as the Betz coefficient (from the name of the man who first derived it). It is the maximum fraction of the power in a wind stream that can be extracted.

5.2.6 Axial thrust and torque on blades

Here blades of propeller type wind turbines are considered. There are two types of forces acting on the blades. One is a circumferential force acting in the direction of wheel rotation that provides the torque and other is the axial force acting in the direction of the wind stream that provides an axial thrust and must be counteracted by proper mechanical design.

The circumferential force, or torque, T, can be obtained from:

$$T = \frac{P}{\omega} = \frac{P}{\pi DN} \tag{5.27}$$

where T = torque (kgf) or Newton (N), ω = angular velocity of turbine wheel, m/s, $D = \sqrt{\frac{4}{\pi}}\ A$ = diameter of the turbine wheel, m, N = wheel revolutions, s^{-1}, A = rotor swept area, m^2.The real efficiency

$$\eta = \frac{P}{P_{wind}}$$

Hence,

$$P = \eta \times P_{wind} = \eta \times \frac{1}{2}\rho A V_i^3$$

For a turbine operating at power P, the expression for torque becomes:

$$T = \eta\frac{1}{2}\frac{\rho A V_i^3}{\pi D N} = \eta\frac{1}{2}\frac{\rho V_i^3}{\pi D N}\frac{\pi D^2}{4} = \eta\frac{1}{8}\frac{\rho D V_i^3}{N} \tag{5.28}$$

At maximum efficiency $\left(\eta_{\max} = \frac{16}{27}\right)$, the torque has maximum value $T_{\max}$ which is equal to:

$$T_{\max} = \frac{16}{27}\frac{1}{8}\frac{\rho D V_i^3}{N} = \frac{2}{27}\frac{\rho D V_i^3}{N} \tag{5.29}$$

The axial force or thrust by Eq. (5.18)

$$F_x = \frac{1}{2}\rho A\left(V_i^2 - V_e^2\right) = \frac{1}{2}\rho\frac{\pi D^2}{4}\left(V_i^2 - V_e^2\right) = \frac{\pi \rho D^2\left(V_i^2 - V_e^2\right)}{8} \tag{5.30}$$

The axial force on a turbine wheel operating at maximum efficiency where $V_e = \frac{1}{3}V_i$ is given by:

$$F_{x_{\max}} = \frac{\pi \rho D^2\left(V_i^2 - \frac{1}{9}V_i^2\right)}{8} = \frac{\pi \rho D^2 V_i^2}{9} \tag{5.31}$$

Thus it is evident that axial forces are perpendicular to the square of the diameter of the turbine wheel, this limits the turbine wheel diameter of large size.

5.2.7 Concept of solidity

Solidity is defined as the percentage of the circumference of the rotor which is filled by rotor blades. It is actually the portion of the swept area of the rotor which is occupied with metal. The general equation representing the solidity is

$$\text{Solidity}(\%) = \frac{\text{Numberofblades} \times \text{Bladewidth}}{\pi \times \text{Rotordiameter}} \times 100 \tag{5.32}$$

For illustration, if a 6 m diameter rotor has 24 blades, each 0.35 m wide, the solidity is calculated as 45%. The greater the solidity of a rotor, the slower it needs to turn to capture the wind. A two- or three-bladed wind turbine has a very low solidity and therefore must rotate quickly to capture the wind or else a lot of wind energy would be lost through the large gaps between the blades [10].

5.2.8 Concept of tip speed ratio

Fig. 5.7 shows a front view of the rotating blades of a three-bladed, propeller type, horizontal axis, wind turbine. The linear velocity of the blade is known as the tip velocity v_{tip} and is a function of the rotating speed of the blade ω (rad/s) and the length of the blade r (m).

$$v_{tip} = \omega r = 2\pi n r \tag{5.33}$$

where n is number of revolutions the blade makes per second:

$$n = \frac{v_{tip}}{2\pi r} \tag{5.34}$$

A wind turbine is usually designed in such a way that its tip velocity should be faster than the wind speed. This allows the turbine to generate electricity at low wind speeds. The ratio of the tip velocity v_{tip} to the wind speed v is known as tip speed ratio (TSR).

$$\text{TSR} = \frac{v_{tip}}{v} \tag{5.35}$$

The TSR is a significant factor for designing of wind turbines. If the TSR is very small, most of the wind passes through the open sections between the rotor blades and a small energy is intercepted by the blades. If the TSR is very large, the fast-

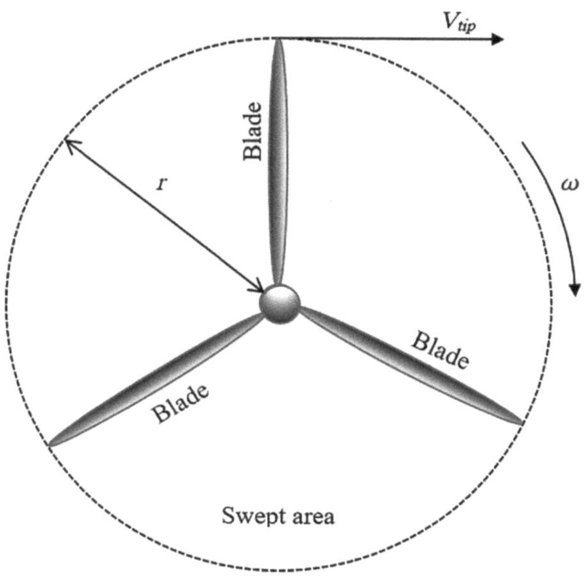

Figure 5.7 Front view of the rotating blades of wind turbine.

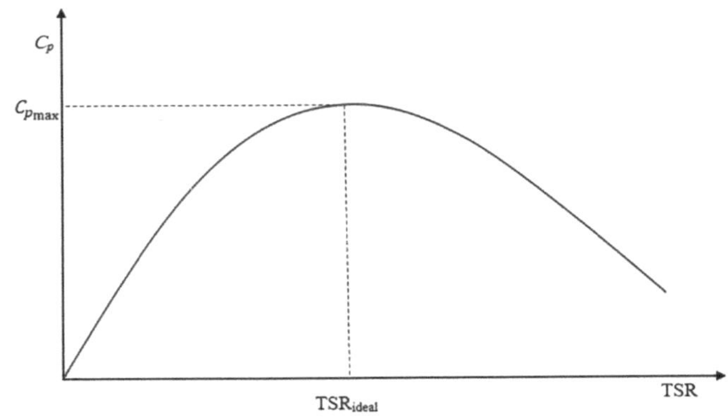

Figure 5.8 Relationship between power coefficient and TSR.

moving blades act similarly to a solid wall to the direction of the wind flow. When a rotor blade moves through the wind it causes a wake of air turbulence that is normally diminished quickly. But if the TSR is high, the next blade may reach the region swept by the earlier blade before the turbulence is diminished, thereby the efficiency of the blade is largely reduced as blades intercept a lesser amount of energy from turbulent wind. Fig. 5.8 portrays the correlation between the TSR and the efficiency of the blades (i.e., power coefficient). It is evident from the illustration that the power coefficient is considerably reduced when there is a deviation from an ideal value of the TSR (TSR_{ideal}).

In advanced wind turbines, the value of TSR can be corrected by varying the pitch angle of the blades. At light windy conditions, the pitch angle is adjusted to increase the TSR and enables more energy to be captured from the wind. At high wind speeds, the pitch angle is set to reduce the TSR, which helps to maintain the rotor speed of the generator within its design limits. In some wind turbines the TSR can be adjusted to nearly zero in order to lock the blades during extremely windy conditions. Variable TSR wind turbines generate more energy as they operate for wider ranges of wind speeds and for a longer time than fixed pitch turbines [2].

5.3 Wind resource feasibility

The accurate determination of the wind energy resource at a specific location is one of the most essential considerations for the development of a wind energy project as the available wind resource at the project site significantly influences the cost of wind energy generation. When a prefeasibility study indicates that a proposed wind energy project could be economically viable, then it is also usually suggested that a project developer should take at least a full year of wind measurement at the exact site where the wind energy project is going to be implemented. However, for very small projects (e.g., off-grid battery charging and water pumping), the price of wind monitoring could be higher than the cost of purchase and installation of a small wind turbine. In these types of minor projects, a detailed wind resource measurement would generally be discarded [11].

5.3.1 Wind power density

Wind resource is expressed in terms of power per unit area and is known as wind power density (WPD). Its unit is W/m^2.

$$\mathrm{WPD} = \frac{P_{wind}}{A} = \frac{1}{2}\rho V^3 \tag{5.36}$$

5.3.2 Variation of wind speed with height

Wind speed generally increases with height above the ground as wind moving across the Earth's surface encounters friction caused by the turbulent flow over and around mountains, hills, trees, buildings, and other obstructions in its path. However, there could be exceptional cases where wind speed decreases with height due to special geographical features. The variation of wind speed with variation of height is given by the power law that can be described by the following relation:

$$\left(\frac{V_2}{V_1}\right) = \left(\frac{H_2}{H_1}\right)^{\alpha} \tag{5.37}$$

where V_2 and V_1 are the mean speed at height H_2 and H_1 respectively and α is known as the *friction coefficient* or *power law index (PLI)*. The friction coefficient is a function of the terrain over which the wind blows and represents the roughness of the surface. The wind speed increases significantly with the height above the ground, with the speed zero at the surface. Its rate of change decreases with height as the frictional forces decrease, as is apparent in Fig. 5.9. Table 5.1 gives representative values of friction coefficient for some different terrain types [5,9].

The first fundamental guideline is that roughness should be as low as possible, since friction against the terrain will slow down the wind and create turbulence. Considering this aspect, it is always best to locate sites with low roughness. The second guideline is to increase the hub height in regions with high roughness, because it will help for enhanced power production [12].

5.3.3 Wind measurement

Wind speed is measured through anemometers. Three major types of anemometers that are principally used are:

1. *Cup anemometer*: This is the most widely used anemometer for wind speed measurement. The majority of present anemometers have three cups with a vertical axis of rotation.

The rotational speed of the cups is proportional to the wind speed. The output of an anemometer is a low-level sinusoidal AC waveform and the frequency of the sine wave is proportional to the wind speed [13]. Fig. 5.10 shows a cup anemometer.

2. *Propeller anemometer*: In propeller anemometer, a propeller is used to measure the wind speed. The axis of rotation of propeller anemometer is horizontal. This type of anemometer also contains a wind vane in order to align the axis of rotation with the direction of

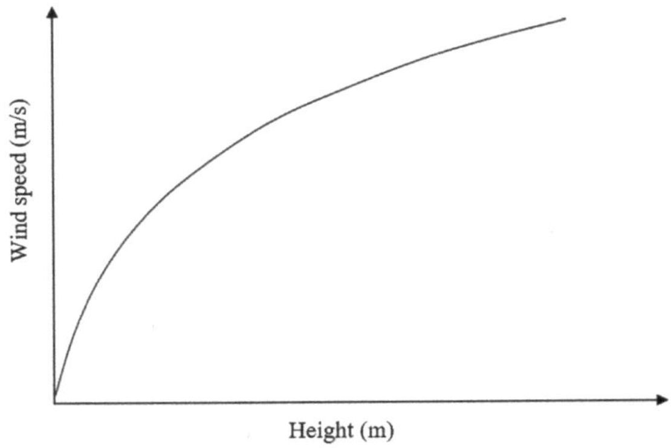

Figure 5.9 Variation of wind speed with height.

Table 5.1 Friction coefficient for different terrain types.

Terrain characteristics	Friction coefficient (α)
Smooth hard ground, calm water	0.10
Tall grass on level ground	0.15
High crops, hedges and shrubs	0.20
Wooded countryside, many trees	0.25
Small town with trees and shrubs	0.30
Large city with tall buildings	0.40

Figure 5.10 Three-cup anemometer.

wind. Both wind speed and wind direction measurement are possible with propeller anemometer. A special type of propeller anemometer is used to measure the vertical component of wind speed where the axis of rotation is vertical [13]. Fig. 5.11 shows a propeller anemometer.

3. *Sonic anemometer*: In an ultrasonic flowmeter, a beam of ultrasonic energy is used to measure the velocity of flowing fluid. In an ultrasonic flowmeter a pulsed beam is directed into the fluid at a shallow angle and its transit time is then measured. When the fluid flows in the direction of the energy transmission, the transit time is shortened. If it flows in the opposite direction, the transit time is lengthened. In this design the time of flight of the ultrasonic signal is measured between two transducers —upstream and downstream. The difference in elapsed time going with or against the flow determines the fluid velocity. When the flow is zero, the time for the signal T_1 to get T_2 is the same as that required to get from T_2 to T_1. When there is flow, the effect is to boost the speed of the signal in the downstream direction, while decreasing it in the upstream direction.

The ultrasonic anemometer shown in Fig. 5.12 measures the wind speed by the time taken for an ultrasonic pulse of sound to travel from the north (N) transducer to the south (S) transducer, and compares it with the time for a pulse to travel from the S to N transducer. Likewise times are compared between west (W) and east (E), and E and W transducers (Fig. 5.13). If, for example, a north wind is blowing, then the time taken for the pulse to travel from N to S will be faster than from S to N, whereas the W to E, and E to W times will be the same. The wind speed and direction can then be calculated from the differences in the times of flight on each axis (Fig. 5.14). This calculation is independent of factors such as temperature.

$$T_2 = \frac{L}{C - V} \quad \text{and} \quad T_1 = \frac{L}{C + V} \tag{5.38}$$

Figure 5.11 Propeller type anemometer.

Figure 5.12 Two-axis ultrasonic anemometer.

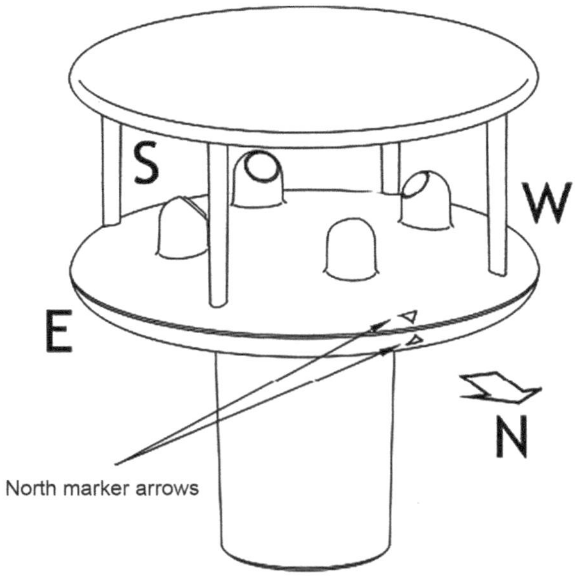

Figure 5.13 Compass points.

Therefore

$$V = \frac{L}{2}\left\{\frac{1}{T_1} - \frac{1}{T_2}\right\} C = \frac{L}{2}\left\{\frac{1}{T_1} + \frac{1}{T_2}\right\} \tag{5.39}$$

where L = distance between transducer faces, C = speed of sound, V = velocity of gas flow, T_1 and T_2 = transit time of ultra sound.

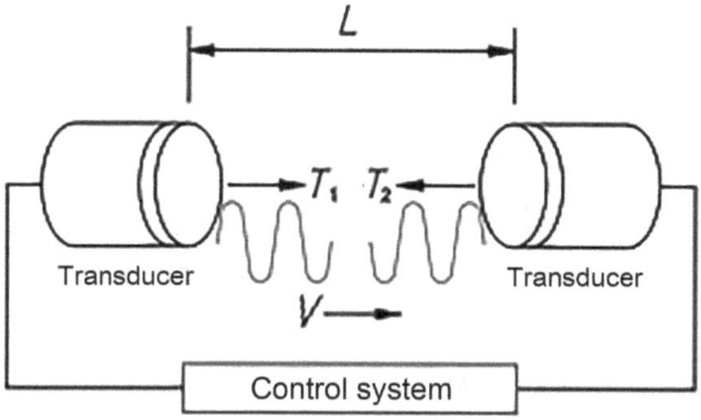

Figure 5.14 Time of flight calculation of ultra sound.

5.3.4 Evaluation of suitable sites

The selection of a site is one of the most vital aspects of planning a proposed wind turbine system. The major factors governing the selection of site are [14]:

1. *Wind availability*: As wind velocity is not constant, the average velocity should be calculated hourly, daily, weekly, and monthly basis from the readings taken over a period of 1 year. The site should have satisfactory and uniform average wind velocity over the year. The useful operating range for wind speed is 4–30 m/s.
2. *Availability of land*: Installation of 1 MW wind turbine plant necessitates around 25 acres. The actual land used directly for wind turbine installation is only 10% of this value. The remaining land may be utilized for agricultural farming of low-height vegetation.
3. *Access to land*: The land should be linked with roadways for transportation of machinery and construction material.
4. *Grid stability*: The wind turbine generator system generally feeds the power to the grid and is operated in tandem with an electric utility network. Therefore the grid should be free from voltage and frequency fluctuations as far as possible

5.3.5 Arrangement of turbines in a wind farm

Unless it is a single wind turbine for a particular site, for example, an off-grid home system, a group of wind turbines on the same site is generally called a *wind farm, wind array*, or *wind park*. [9]. When installing a group of machines in a wind farm, a certain distance between the wind towers must be kept to optimize capturing the wind power. The spacing depends on the topography, wind direction and speed, and the turbine size [15]. We know that wind velocity is reduced at the exit side of the wind turbine due to the extraction of some of wind energy by a wind turbine rotor. As a result, reduced wind power is available for downstream wind machines. However, at some distance downstream, the wind speed is regained [9]. Evidently due to the rotation of a wind turbine, the wind velocity and energy at downstream are less compared with the upstream wind velocity and energy. Inadequate wind turbine spacing on a site causes fluctuation of the power output

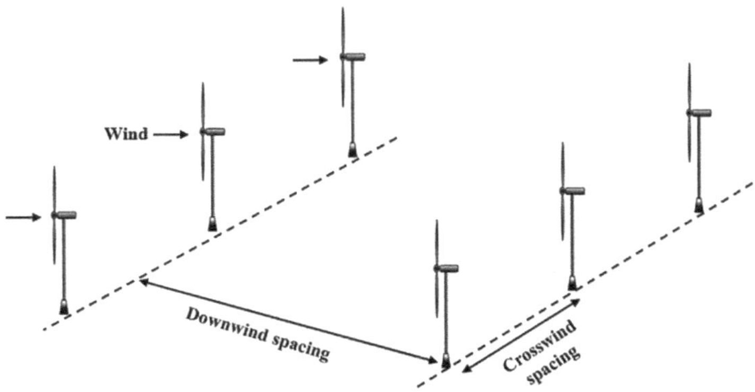

Figure 5.15 Wind turbine spacing.

that may further lead to instability in the electrical network to which it is connected. There is no scientific and systematic analytical method for computing the wind turbine spacing in a wind farm. This is a multivariable problem and no plain formula solution is available. Therefore the design of wind turbine spacing is on empirical basis, based on the practical knowledge and on the different theoretical model solutions available in the technical literature [16]. Practical experience has recommended some rough rules-of-thumb for tower spacing. Recommended spacing is 3–5 rotor diameters separating towers within a row (crosswind spacing) and 5–9 diameters between rows (downwind spacing). Fig. 5.15 shows the downwind and crosswind spacing in a wind farm.

5.4 Wind energy conversion devices

5.4.1 Types of wind energy conversion devices

Depending on the axis of rotation, wind energy conversion devices such as wind turbines are categorized as horizontal axis wind turbine (HAWT) and vertical axis wind turbine (VAWT). HAWTs are the most common type of wind turbine in present days. In HAWTs, the axis of rotation of the turbine blades (rotor) is horizontal or parallel to the ground (Fig. 5.16) while that of in VAWTs, the axis of rotation of the blades is vertical with respect to the ground (Fig. 5.17). Classification of different types of wind turbines are shown in Fig. 5.18.

A brief introduction of various types of wind turbines are given next.

5.4.2 Dutch type wind turbine

There are usually four blades in Dutch type wind turbine (Fig. 5.19) and they are inclined at an angle to the plane of rotation. These windmills are operated on the thrust force of the wind. The blades are made of sails or wooden slats. The

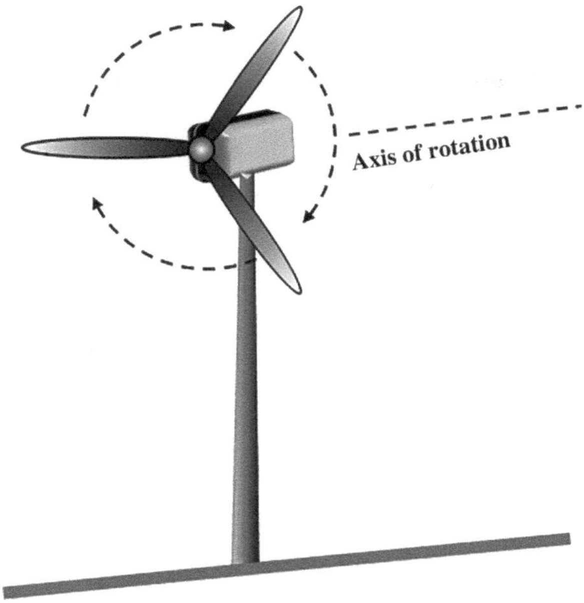

Figure 5.16 Horizontal axis wind turbine.

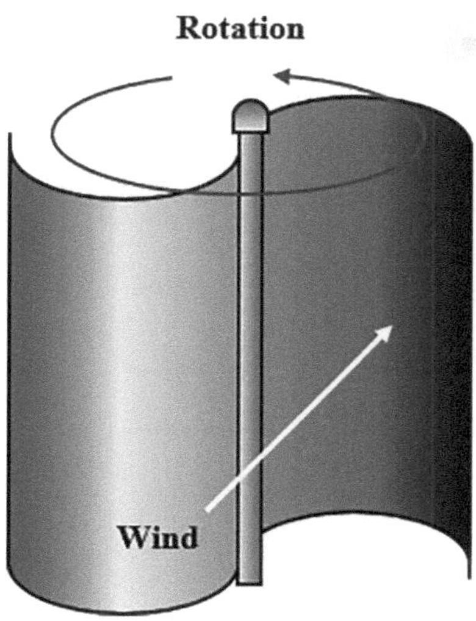

Figure 5.17 Vertical axis wind turbine.

operation of Dutch windmill needs highly skilled workers. The Dutch windmill became obsolete with the advent of cheaper fossil fuel. It was found that certain modifications in the earlier design of windmills made them very useful for pumping water.

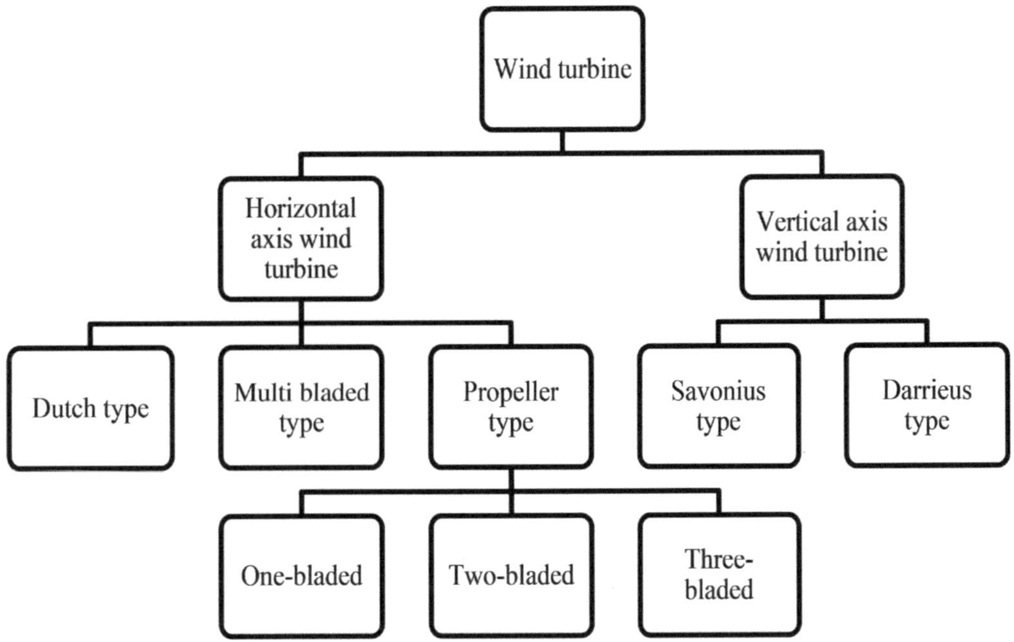

Figure 5.18 Classification of wind turbines.

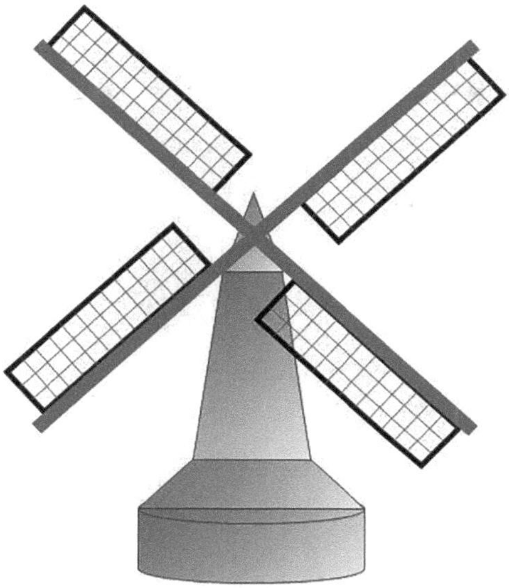

Figure 5.19 Dutch type wind turbine.

5.4.3 Multibladed type wind turbine

Wind turbines designed with a large number of blades (Fig. 5.20) are used for pumping water or other mechanical works that require a high starting torque. Modern water pumping windmills have a large number of blades—generally wooden or metallic slats—driving a pump. The large number of blades means high solidity, and therefore high torque and low speed [17] which are necessary for driving a pump, even at low winds. As the mill has to be placed directly over the well for pumping purpose, the condition for site selection is governed by water availability and not windiness. Multibladed wind turbines are not used for electric power production [12].

5.4.4 Propeller type wind turbine

High-speed propeller type HAWT are efficient and very suitable for electrical power generation. The vast majority of current designs are horizontal axis propeller type wind turbine [5]. Modern wind power plants have three rotor blades. The constructional features of the modern three-bladed propeller type HAWT is described in Section 5.2.2. The efficiency of power extraction from wind depends on the proper choice of the number of blades. If the blades are very close to each other or move very fast, then every blade passes through the turbulent air produced by the previous blade and eventually results in little power extraction. On the other hand if the blades are at a distance from each other or move very slowly than most of the air stream will pass through the spacing between the blades (i.e., no power will be

Figure 5.20 Multibladed type wind turbine.

intercepted by the blades). Thus the choice of the number of blades should depend on the TSR [17]. A small number of blades imply low torque and high speed. For modern electricity-generating wind turbines, the requirement of high TSR necessitates a small number of blades, generally two or three.

5.4.4.1 One-bladed wind turbine

Single-bladed wind turbines (Fig. 5.21) are not commonly used although they save the cost of another rotor blade. As stability is a problem, one-bladed wind turbines require a counterweight on the opposite side of the hub in order to balance the weight of the single blade. However, this counterweight cancels out the savings on weight compared to a two-bladed design. One-bladed turbines have higher rotational speeds and thus produce more sound. The visual impact due to asymmetrical

Figure 5.21 One-bladed wind turbine.

Figure 5.22 Two-bladed wind turbine.

rotation creates annoyance. Besides these disadvantages, one-bladed wind turbines require comparatively higher wind speeds than that needed by three-bladed wind turbines to generate the same power output.

5.4.4.2 Two-bladed wind turbine

Two-bladed wind turbines (Fig. 5.22) imply less material is used than three-bladed turbines. A rotor with two blades has to move faster to perform the same amount of work as a three-bladed rotor. Higher rotor speeds transform into less torque and is the reason for smaller sized and lower cost gearboxes. Two-bladed rotors intercept somewhat less energy, about 97% of a three-bladed design with the same rotor

Figure 5.23 Three-bladed wind turbine.

diameter. However, this can be easily compensated by employing a slightly larger diameter two-bladed wind turbine rotor.

5.4.4.3 Three-bladed wind turbine

Most modern wind turbines are three-bladed designs (Fig. 5.23) with a upwind rotor and electrical motors to operate the yaw mechanism. The three-bladed design yields steady power output and a more balanced gyroscopic force and consequently less blade fatigue and possibilities of failure [17]. In three-bladed wind turbines, noise and wear are in general lower, and efficiency is higher in comparison to two-bladed wind turbines [12].

5.4.5 Savonius type wind turbine

This is a simple vertical axis wind turbine that functions completely because of the thrust force of wind (Fig. 5.24). The basic equipment is a drum cut into two halves vertically. The two parts are joined to the two opposite sides of a vertical shaft. The Savonius rotor is cheap and simple, and the material required for it is normally available in any rural area. However, its utility is restricted to pumping water because of its comparatively low efficiency.

5.4.6 Darrieus type wind turbine

Two or more flexible blades are attached to a vertical shaft (Fig. 5.25). The blades bend outward, taking approximately the shape of a parabola, and are symmetrical to the aerofoil section. Torque is zero when the rotor is stationary. A positive torque is developed by it only when it is already rotating. This implies that such a rotor has no starting torque and has to be started with some auxiliary means. The Darrieus rotor, with its high efficiency and high speed, is suitable for electrical power production. The cost of construction is low because the generator and the

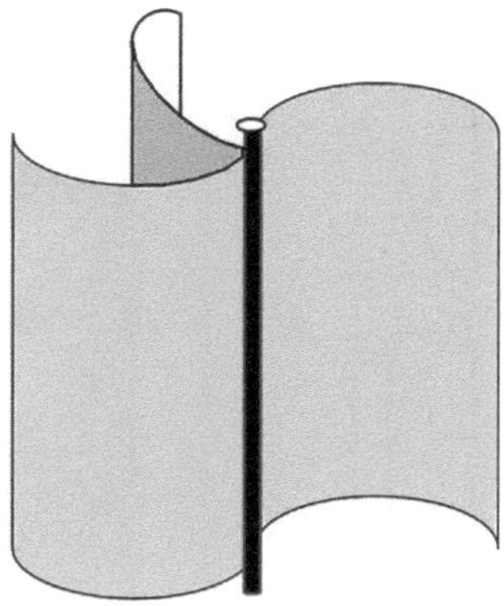

Figure 5.24 Savonius type wind turbine.

Figure 5.25 Darrieus type wind turbine.

gear assembly can be located at ground level, which substantially reduces the cost of the tower. However, it is unable to take advantage of the high wind speed available at a higher height.

5.4.7 Upwind and downwind wind turbine concept

Depending on the position of the wind turbine rotor with reference to the wind direction and tower, HAWT can be grouped into upwind and downwind wind turbines. The term "upwind wind turbines" signifies that the wind hits the blades first

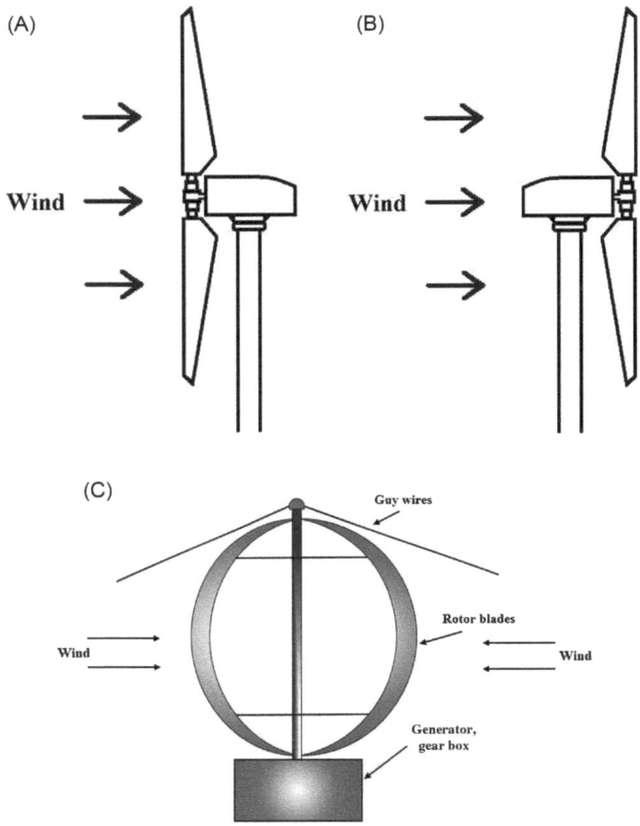

Figure 5.26 Horizontal axis wind turbines (HAWT) are either upwind machines (A) or downwind machines (B). Vertical axis wind turbines (VAWT) accept the wind from any direction (C).

before it hits the tower (Fig. 5.26A). In "downwind wind turbines" (Fig. 5.26B), the wind has to pass the tower before it reaches the rotor blades [12]. Despite the fact that almost all wind turbines presently are of the horizontal axis type, there is still some disagreement about superiority between upwind and downwind turbines [9].

The advantage of the upwind design is that the blades work in an uninterrupted airflow. The bottleneck of upwind designs is that they are inherently unstable, so they require some kind of mechanism for always keeping the rotor blades (using yaw system) upwind of the tower. In upwind wind turbines, the air turbulence due to the "tower shadow effect" is greatly reduced for the same blade-tower spacing, thereby decreasing the dynamic loads on the blades and the rhythmic noise effects. To avoid blades from hitting the tower, the blades spin at a sufficient distance before the tower. In addition, the blades are relatively stiff to prevent them from being pushed into the tower by high winds.

In contrast, downwind turbines do not necessitate a tail or yaw drive as the wind itself keeps the rotor downwind [12]. A downwind turbine has the advantage of allowing the wind itself control the yaw (the left–right motion), therefore it naturally orients itself properly with respect to wind direction [9].This makes the downwind turbine system simpler and more economic than yaw-driven upwind turbines. In high winds the blades can bend or cone, reducing their swept area and hence the power extracted. This design approach also reduces the load on the tower. However, when the wind direction changes quickly, the downwind turbine will not yaw fast enough. The limitation of downwind turbines is that the tower perturbs the airflow resulting turbulence as each blade has to move through this disturbed air stream once every rotation. This disturbance is known as "tower shadow." The disturbed air stream can produce cyclic forces that in rigidly mounted blades can cause fatigue if not the blades are properly designed. The blades can also produce a thumping sound as every blade moves through the tower shadow [12].

5.4.8 Horizontal axis versus vertical axis wind turbine [18]

5.4.8.1 Advantages of horizontal axis wind turbine

- The tall tower base allows access to stronger wind. In some wind shear sites, every 10 m height the wind speed can increase by 20% and the power output by 34%.
- High efficiency, as the blades always rotates perpendicularly to the wind, capturing power through the whole rotation.
- The ability to pitch the rotor blades in a storm so that damage is minimized.
- HAWT develops a high starting torque.

5.4.8.2 Disadvantages of horizontal axis wind turbine

- Massive tower construction is required to support the heavy blades, gearbox, and generator.
- Components of a HAWT (gearbox, rotor shaft and brake assembly) being lifted into position.
- Downwind variants suffer from fatigue and structural failure caused by turbulence when a blade passes through the tower's wind shadow (for this reason, the majority of HAWTs use an upwind design, with the rotor facing the wind in front of the tower).
- HAWTs require an additional yaw control mechanism to turn the blades toward the wind.
- HAWTs generally require a braking or yawing device in high winds to stop the turbine from spinning and damaging itself.

5.4.8.3 Advantages of vertical axis wind turbine

- VAWT does not need a yaw mechanism, because it can extract the wind from all directions (Fig. 5.26C).
- A strong supporting tower is not required since generator, gearbox, and other components are placed on the ground.
- Low production and maintenance cost in comparison to HAWT.
- Easy installation as compared to other wind turbine.
- Low risk for human and birds because blades move at relatively low speeds.

- VAWTs are quieter than the HAWTs.

5.4.8.4 Disadvantages of vertical axis wind turbine

- They need an initial push to start for which a small motor is required.
- Compared to HAWT they are much less efficient.
- They have relatively high vibration because the airflow near the ground creates turbulent flow.
- VAWTs may require guy wires to hold it up.

5.4.9 Concepts of aerodynamic forces

All wind turbines work on basically two principles by which the energy is extracted from the wind. These principles are:

- *drag force* which is in the direction of the wind flow and
- *lift force* which is perpendicular to the wind flow.

When we stretch our arms out through the window of a moving car, we sense our arms being pushed backwards. This is because of the drag force. If we hold our arms straight with our hands parallel to the road, and change the angle a little, we can suddenly feel that it is drawn upward. The hand and arm function like the wing of an airplane, and with the appropriate angle we can feel a strong lift force. Therefore both lift and drag forces act on a body immersed in a mass of moving fluid (air). These two forces determine the properties of an aerodynamic blade profile (or air foil). One force pulls backward drag, and another force pulls upward lift [12,19]. Wind turbines make use of either lift force or drag force predominantly to cause motion and are known as lift- or drag-type wind turbines [20].

5.4.9.1 Drag-type wind turbine

In drag-type wind turbines, the wind *drags* the blades along the direction in which it is flowing, but as the blades are so fixed that it cannot move in the wind's direction, it moves in the direction in which it can and that results in the rotational movement [12]. Drag devices are less efficient and turn slower than wind. They produce high torque and thus are suitable for pumping applications. At high wind speeds, they spill wind instead of producing more energy. Hence they do not benefit from the high energy density available in wind [20,21].

5.4.9.2 Lift-type wind turbine

In lift devices, the ratio of lift-to-drag forces may be as high as 30:1. The lift-blade design adopts the same principle that enables airplanes, kites, and birds to fly. The blade is fundamentally an airfoil or wing. When air flows past the blade, a wind speed and pressure differential is created between the upper and lower blade surfaces. The pressure at the lower surface is greater and thus acts to *lift* the blade. When blades are attached to a central axis, as in a wind turbine rotor, the lift is transformed into rotational motion [20]. The forces that can be developed in lift machines are more than those achievable with a drag machine with the same surface area [12]. Lift-powered wind turbines have much higher rotational speeds than drag types and hence are suitable for electricity production. Lift devices are more efficient and turn faster than wind. They are able to benefit from high-power densities available in strong winds. The power captured from wind by lift device to that of a drag device is greater for the same swept area [20]. Typically lift devices can intercept 50–100 times more power per unit of the projected blade area than drag devices [12]. The types and characteristics of various lift and drag variety machines are listed in Table 5.2 [20].

5.5 Wind electrical generators

5.5.1 Working principle of generators

While blowing through the windmill, the wind translates its KE by means of wind turbine blades into mechanical power and that mechanical power is converted into electrical power with the help of a generator. Generators include a rotor that rotates inside a stationary housing called a stator. Electricity is generated when conductors move through a magnetic field, cutting lines of flux, and producing voltage and current [9]. The principle of operation of all electrical machines is based on Faraday's law of electromagnetic induction and the interaction of magnetic flux and electric current. As the operation of the electrical machines can be made reversible, the same machine can be used as a conventional electrical motor for converting the electrical power into mechanical power, or as a generator for converting the mechanical power back into the electrical power. There are three main prerequisites for electrical generation: (1) a coil or set of conductors, (2) a magnetic field system, and (3) relative motion between the conductors and field [22]. Fig. 5.27 illustrates the generator working principle in a simple way.

5.5.2 Classification of wind electrical generators

Predominantly there are two types of AC generators that find applications in wind power plants — synchronous generators and induction generators (asynchronous generators). The induction generators constitute the largest segment in the wind power industry and about 85% of the wind generators today are induction generators. The type of generators used for wind energy conversion system mostly of either doubly-fed induction generator (DFIG) or permanent magnet synchronous generator (PMSG) [23].On the other hand, DC generators are used in small wind turbines that are used for battery charging purposes. The classification of wind turbine electric generators is shown in Fig. 5.28 and a brief overview on prominent electrical generators for modern wind power plants is given next.

Table 5.2 Types and characteristics of various lift and drag variety machines.

Sl. no.	Type of wind turbine	Axis of rotation	Aerodynamic principle for working	TSR	RPM	Torque	Purpose
1	Propeller type (1–3 blade)	Horizontal	Lift	6–20	High	Low	Electricity generation
2	Multibladed wind turbine	Horizontal	Drag	1	Low	High	Pumping
3	Dutch type	Horizontal	Drag	2–3	Low	High	Pumping
4	Savonious type	Vertical	Drag	1	Low	High	Pumping
5	Darrieus type	Vertical	Lift	5–6	High	Low	Electricity generation

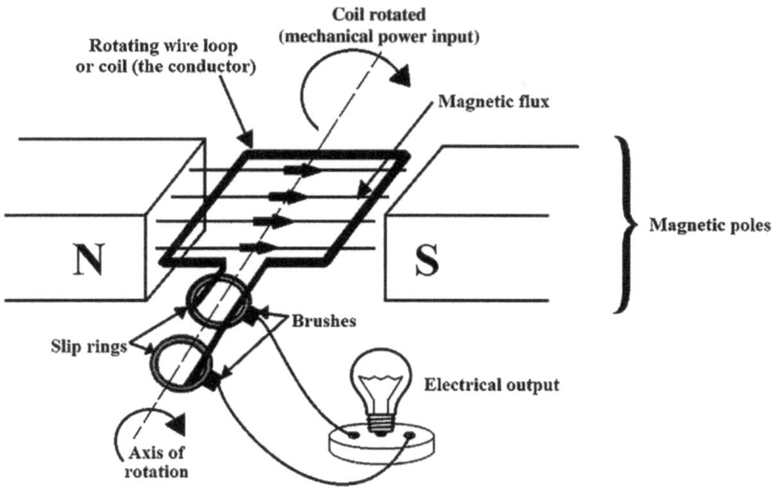

Figure 5.27 Simple generator with magnetic induction.

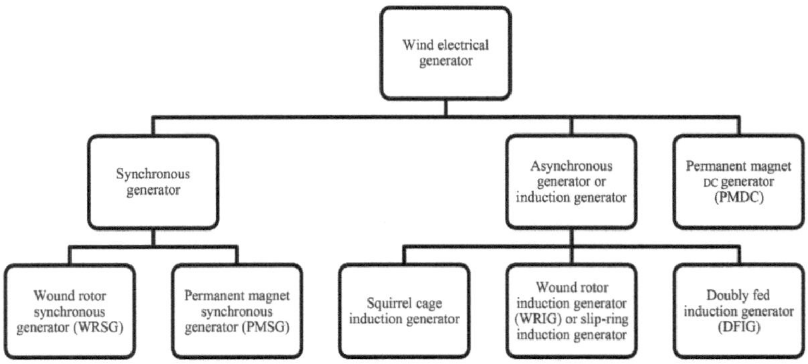

Figure 5.28 Wind turbine electric generators' classification.

5.5.3 DC generator

Though there are many different types of electrical generators, all electrical generators are in fact internally alternating current (AC) machines due to the conductor rotation in the magnetic flux of alternate north and south polarity. A DC generator is normally self-excited either by shunt or series coils carrying dc current in order to produce a magnetic field. The DC generator converts AC into DC using the commutator that is basically a mechanical rectifier that converts the AC voltage generated in the armature winding into DC voltage across the sliding carbon brushes on a series of copper segments. The sliding contacts have low reliability and high-maintenance costs. However, DC generators of the present day are often designed with permanent magnets and known as permanent magnet DC (PMDC) generators that eliminate the field current requirement, and the commutator and the brushes, and hence the reliability is greatly improved (Fig. 5.29). The AC is then rectified using the solid state rectifiers [15]. The PMDC generator is popular for small-scale wind turbine systems as they are reliable, can operate at low-rotational speeds, and provide good efficiency particularly in light wind conditions as their cut-in points are quite low [24]. However, the size of the brushless PMDC machine is limited due to the limitation of the permanent magnet capacity and strength [15].

5.5.4 Synchronous generator

Most of the electrical power consumed in the world is supplied by synchronous generators (alternator). A synchronous generator operates at constant speed related to the fixed frequency [15]. Fig. 5.30 shows the basic system for a wind turbine with a synchronous generator where the generator and wind turbine blades are coupled through a gearbox to match the speed required of each [9]. Basically the synchronous generator comprises a magnetic field on the rotor that spins and a stationary stator consists of multiple windings that provide the generated power. The synchronous generator needs DC to excite its rotor field winding. In conventional synchronous generator, the rotor winding is injected by DC from a separate

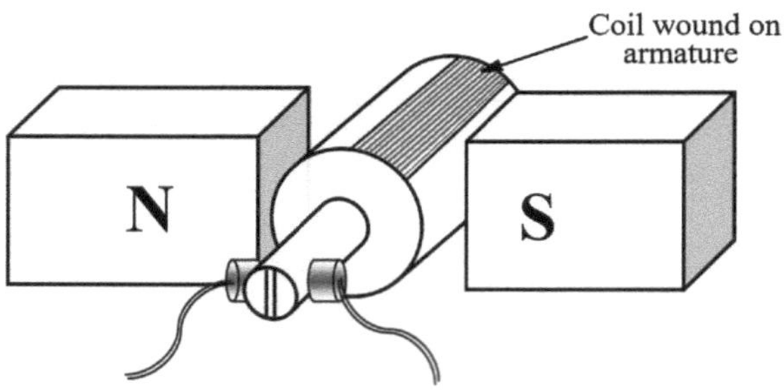

Figure 5.29 Permanent magnet DC generator.

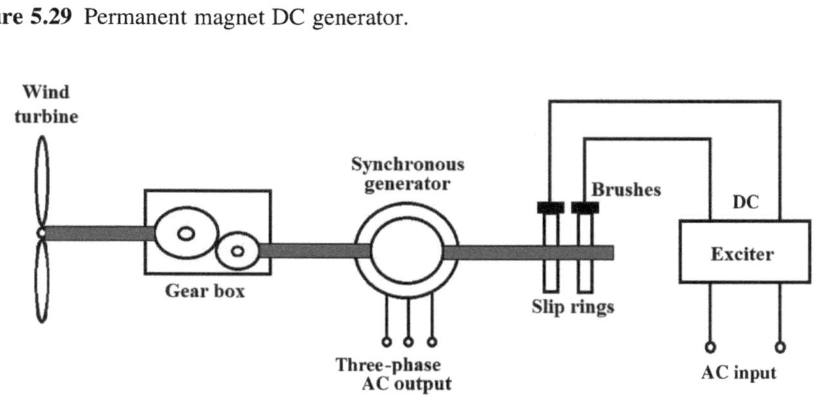

Figure 5.30 Schematic of synchronous generator-based wind turbine power plant.

rectifying circuit known as *exciter* which converts AC from the grid into DC for the rotor. This kind of synchronous generator is called electrically excited synchronous generator, or wound field or wound rotor synchronous generator (WRSG) [25]. This excitation is done through the use of brushes and slip rings on the generator shaft that poses several limitations such as regular maintenance, cleaning of the carbon dust, and so forth. An alternative approach is to use brushless excitation that employs permanent magnets instead of electromagnets. As its name implies, in a PMSG the excitation field is created using permanent magnets. The permanent magnets are installed in the rotor [26].

5.5.5 Induction generator

The major advantage of the induction generator is the rugged brushless construction and no requirement for separate DC field power. The limitations of the DC generator and the synchronous generator are removed in the induction generator, resulting in low-capital cost, low maintenance, and better transient performance. As a result the induction generator is widely used in small and large wind farms and small hydroelectric power plants. There are two categories of induction generators—squirrel cage and wound rotor (slip ring). Squirrel cage induction generators are found in a variety of types of small generating plants in the kW range, while wound rotor induction generators are preferred for wind electricity generation in the MW range. The machine is available in numerous power ratings up to several MW capacity, and even larger. The induction machine requires AC excitation current. The machine is either self-excited or externally excited. Since the excitation current is mainly reactive, a standalone system is self-excited by shunt capacitors. The induction generator when connected to the grid draws the excitation power (reactive power) from the utility grid [15]. As an induction generator cannot be functional without an electricity source it normally operates with the help of the grid only. Hence usually induction generator driven windmills are interactive to the electric utility grid and power generated is fed to the electric utility network. Induction generators are also self-synchronized with the power grid witout the need for auxiliary synchronization apparatus [2]. Fig. 5.31 shows the schematic of an induction generator-based wind turbine power plant.

5.5.6 Constant speed and variable speed operations of wind turbines

Constant speed generation system cannot maximize the extraction of the power contained in wind. For every type of wind turbine, power coefficient reaches a maximum at a particular value of TSR. Hence to extract the maximum amount of power from wind, the turbine should function at a constant TSR, which implies that the rotational speed should be proportional to the wind speed. Therefore harnessing of maximum power needs a variable speed generation system with the speed control aimed at maintaining a constant TSR. Such a system can provide 20%–30% more

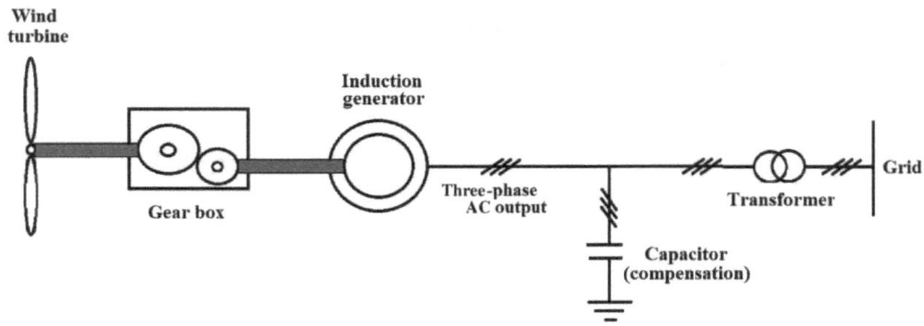

Figure 5.31 Schematic of induction generator-based wind turbine power plant.

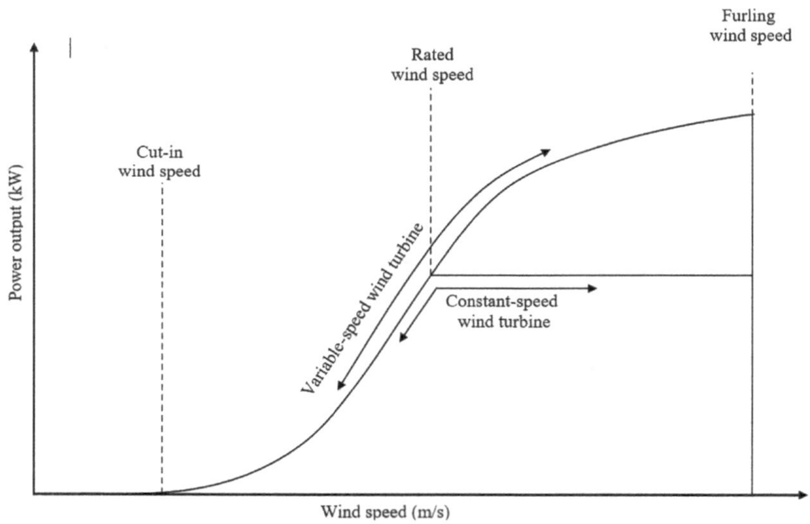

Figure 5.32 Power–speed characteristics of constant speed and variable speed wind turbine.

power than constant speed generation systems. Previously the wind industry was based on constant speed wind turbines. But with the scientific and technological developments in electrical, electronics, computers, and instrumentation engineering, modern wind industries are preferring variable speed wind power technology. Fig. 5.32 compares the power–speed characteristics of constant speed and variable speed wind turbines.

5.5.7 Common strategies for constant speed and variable speed operations

Constant wind speed turbines employ different types of asynchronous (induction) generators. On the other hand, variable speed wind turbines employ a DFIG, WRSG, or PMSG [12].

5.5.7.1 Fixed speed system ("Danish" concept)

In this concept (Fig. 5.33), normally three-phase induction generator (asynchronous generator) is directly connected to the electric utility grid (local AC power network) [27]. The rotor of the wind turbine is coupled to the generator shaft with a fixed-ratio gearbox. As the electric generator output is connected directly to the grid, the rotating speed of the generator is fixed (they usually vary a little as the slip is allowed to vary over a range of typically 2%–3%). The frequency of the grid decides the rotating speed of the generator and in consequence the speed of the wind turbine rotor [28]. The wind turbine with induction generators is advantageous due to the simple, rugged, and cheap constructional features of induction generators. Besides these, no synchronization equipment is required [27]. However, any variation in wind speed naturally causes the mechanical power at the wind turbine rotor to fluctuate and since the rotating speed is fixed, this results in torque variation at the wind turbine rotor. When a wind gust occurs, the torque at the wind turbine rotor therefore increases considerably while the rotor speed varies little. Thus every wind gust stresses the mechanical components (especially the gearbox) in the wind turbine and leads to a sudden increase in rotor torque and fluctuation in the output power that causes instability in the power network to which it is connected.

5.5.7.2 Variable speed (adjustable speed) generators

The generator in variable speed systems is usually connected to the grid by a power electronic interface. Power electronic converters are used for the control of variable speed wind turbines. One such arrangement is shown in Fig. 5.34. As the wind velocity varies, the turbine speed and in turn the generator rotor speed varies. In this configuration, the generator rotates at a speed that is proportional to the wind speed. This gives a variable output at the generator terminals. The output of the generator is therefore first rectified (from variable AC to fixed DC) and then inverted (from fixed DC to fixed AC which is synchronous with the ac power network) using a voltage source inverter. Thus the power electronic converters are connected in series with the generator to have the size and capacity to process 100% of the generator output power [29]. Operation of diode bridge rectifiers and

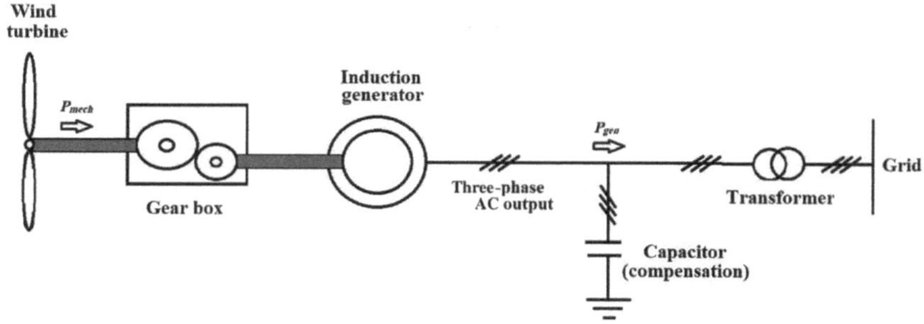

Figure 5.33 Fixed speed "Danish" concept.

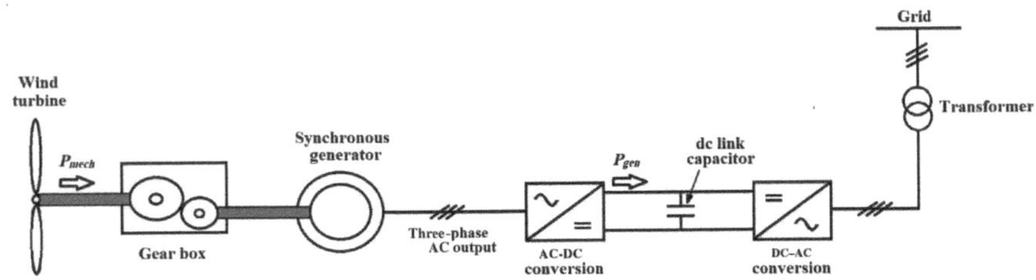

Figure 5.34 Variable speed wind generator system.

thiristorised inverters generate harmonic currents and voltages, and their interaction results in considerable power fluctuation through the DC link. This interaction can be considerably reduced by means of the capacitor filter at the output of the diode bridge. Therefore a DC link capacitor is connected at the output of the rectifier that acts as a buffer between the rectifier and inverter. Independent control of rectifier and inverter is possible by introducing this capacitor.

The key advantages of adjustable or variable speed generators in comparison to fixed speed generators are the reduction of mechanical stresses; absorption of wind gusts, dynamic compensation of torque and power pulsations caused by back pressure of the tower, improvement in power quality; torque pulsation can be mitigated due to the elasticity of the wind turbine system that reduces electrical power fluctuations, that is, less flicker, improvement in system efficiency (energy efficiency improvement up to 10% is possible), and reduction in noise as low-speed operation is feasible at low power conditions [28].

5.5.7.3 DFIG for variable speed operation

The majority of DFIGs nowadays are three-phase wound rotor induction generators that are fed with AC into the stator and rotor windings [27]. The schematic arrangement of a DFIG-based variable speed wind turbine system is shown in Fig. 5.35. The current in DFIG's rotor is controlled by a power electronic converter [30]. Using a DFIG in wind turbines offers the same advantages as when a synchronous generator is used as a variable speed generator. Both types of schemes require two AC/DC converters. But, the two converters in DFIGs are much smaller than those in synchronous generators of similar output power. This is because the power electronics devices employed in DFIGs are required to process only a small portion of the generator output power, that is, the power which is supplied to or from the generator rotor windings and which is normally about 25%–30% of the generator rated power. Therefore the power electronic devices in variable speed wind turbines using DFIGs are typically only around 25%–30% of the size of the power electronics devices employed for comparable sized three-phase synchronous generators as shown in Fig. 5.34. This decreases the cost and power losses associated with the power electronics devices [29]. Also the response to grid requirement is improved. The converter has to provide only excitation energy [28], while the AC/DC

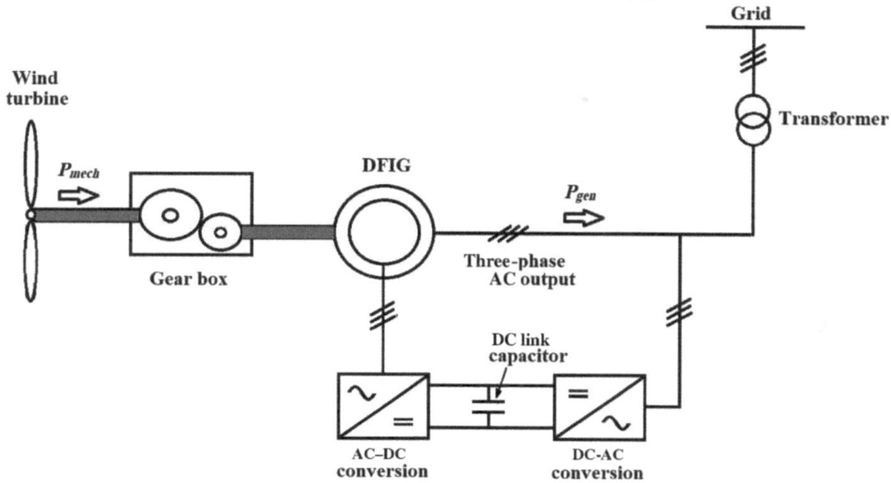

Figure 5.35 Doubly-fed induction generator wind turbine system.

converters in synchronous generators have to transmit 100% of the nominal generator output power [29]. With a generator side converter it is possible to control the torque or speed of the DFIG as well as the power factor of the stator terminals, while using the grid side (or network side) converter the DC link voltage is kept constant [27].

5.5.8 Classification of wind power plants on the basis of generators

The electrical generators used in wind turbine power plants have undergone continuous changes. All the different types of electric generator systems have their specific pros and cons when used in wind power plants [12]. Depending on the type of electrical generator employed, the classification of wind turbine power plants is shown in Fig. 5.36.

Table 5.3 shows the market penetration and development of Type-A, Type-B, Type-C, and Type-D wind turbine power plants [31].

5.6 Speed control strategies for wind turbines

There are various options for speed control of wind turbines. The type of control strategy depends on the size of the turbine. Small wind turbines use simple, low-cost methods while large machines use more sophisticated methods incorporating pitch control along with power electronic devices [20]. For large wind turbines the three common speed control approaches are active pitch-control system, passive

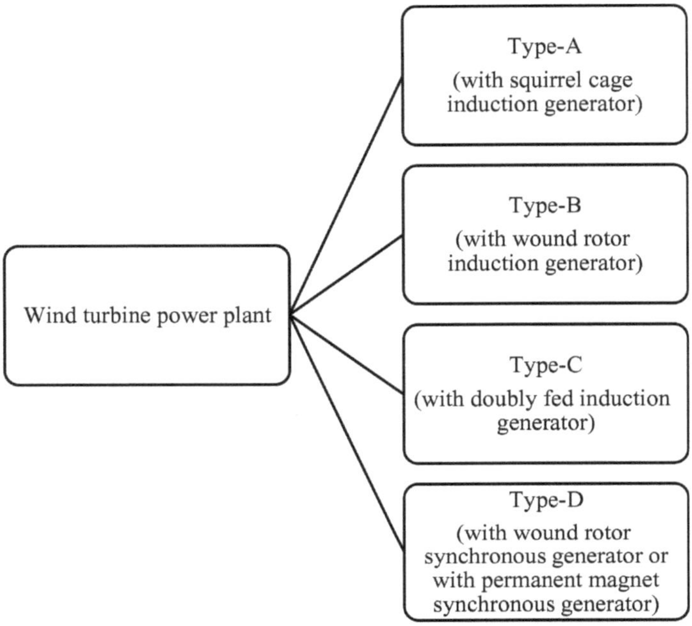

Figure 5.36 Category of wind turbine power plant on the basis of generator used.

stall-control design, and a combination of the two. However, once a rotor has been stopped by any control mechanism, a mechanical brake locks the rotor shaft in place, which is essential from a safety point of view during maintenance [9]. The different speed control schemes are:

1. *Pitch control*: In this speed control approach the pitch of the rotor blades is controlled to regulate the speed [20]. With pitch control, the TSR can be adjusted to produce power at a diverse range of wind speeds. But controlling the pitch angle increases the complexity of the system and augments the maintenance cost, so it is more suitable for large turbines [2]. For *pitch-controlled* turbines an electronic system supervises the generator output power; if it exceeds the limits, the pitch of the turbine blades is adjusted to spill some of the wind. In practice, a hydraulic system gradually moves the blades about their axes, turning them a few degrees at a time to decrease or increase their efficiency as conditions demand. The approach is to decrease the blade's angle of attack (Fig. 5.37) when wind speeds are high [9].
2. *Stall control*: For stall control, when the wind speed exceeds the safe limit blades are moved to a position such that they stall (stop) [20]. For *stall-controlled* machines, the blades are critically designed to automatically decrease efficiency when wind speeds are very high. Unlike the pitch-controlled scheme, in the case of stall-controlled scheme nothing rotates and there are no moving parts, so this is called passive control. The aerodynamic design of the blades, particularly their twist as a function of distance from the hub, must be very cautiously done so that a gradual reduction in lift occurs as the blades revolve faster. Most modern, large wind turbines apply this passive, stall-controlled approach.

Table 5.3 Trend of wind power generation strategies.

Wind energy plant type	**World market share of yearly installed wind power (%)**									
	2004	**2005**	**2006**	**2007**	**2008**	**2009**	**2010**	**2011**	**2012**	**2013**
Type-A	69.5	62.7	53.5	39.6	40.8	39.0	30.6	27.8	19.2	24.7
Type-B	16.2	23.4	27.0	17.8	17.1	17.2	15.2	5.2	3.1	2.2
Type-C	0.0	0.1	3.1	26.5	28.1	28.2	37.3	46.7	59.8	54.8
Type-D	14.3	13.7	16.3	16.1	14.0	15.6	16.9	20.3	17.9	18.3

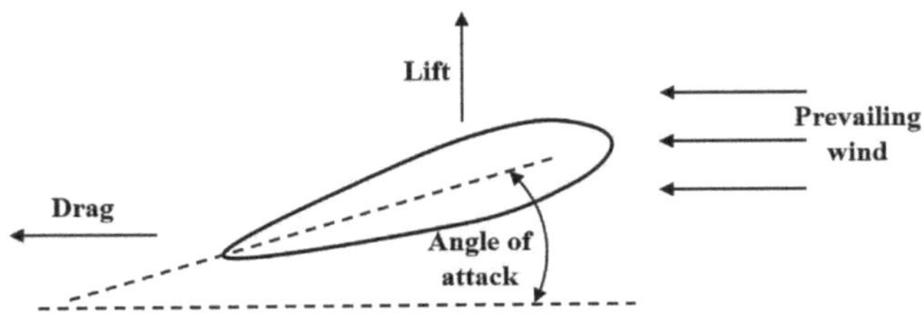

Figure 5.37 Concept of angle of attack.

3. *Active stall control*: For very large machines, typically above 1 MW, an *active stall control* scheme may be found suitable. In such large machines, the blades rotate in the same way as they rotate in the active, pitch-control approach. But the difference is that when winds go beyond the rated wind speed, rather than reducing the angle of attack of the blades, it is increased to induce stall [9].
4. *Yaw and tilt control*: Small, kW-size wind turbines can have any of a variety of approaches to shed wind. Passive yaw controls are common that cause the axis of the turbine to shift more and more off the wind as wind speeds increase [9] through which the rotor axis is moved out of wind direction, either by yaw control or by tilting the rotor plane with respect to normal vertical plane when the wind goes above the design limit [20]. Most wind turbines have a yaw mechanism to keep the blades facing into the wind as the wind direction changes. Some turbines are designed to operate on downwinds; these turbines do not require yaw mechanisms as the wind aligns these turbines [2]. In another simple technique, a wind vane is mounted parallel to the plane of the blades. When the wind is very strong, wind pressure on the vane rotates the machine away from the wind [9].
5. *No speed control at all*: Different components of the whole system are designed to survive the extreme speed during gusty winds [20].

Table 5.4 summarizes speed control strategies for various types of wind power plants.

5.7 Environmental impact and public perception

Despite the fact that wind energy technology is highly popular, there are certain issues with regard to the environment and public perception [2].

Noise pollution: Noise is disturbing for some people. There are two main sources of noise from a wind turbine — mechanical noise from the gearbox and generator, and aerodynamic noise from the rotor blades. Aerodynamic noise is the major component of the total noise and resembles the swish sound created by a helicopter. At 400 m downwind, the turbine noise can be as high as 60 dB, which is comparable to the noise level from an air conditioner or dishwasher. However, the noise from wind farms is far below the threshold of pain, which is 140 dB. Furthermore noise has been reduced substantially with the advent of new and improved designed wind turbines.

Aesthetics: Some people perceive wind farms as ugly for the landscape, forcing developers to install them in remote areas, including offshore as shown in Fig. 5.38.

Table 5.4 Speed control strategies and important attributes for different types of wind power plants.

Wind power plant category	Wind electric generator used	Blade angle control	Speed control	Drive train	Grid connection
Type-A	SCIG	Stall control	Fixed speed	Gear	Direct connection
Type-B	WRIG	Pitch control	Variable speed	Gear	Direct connection
Type-C	DFIG	Pitch control	Variable speed	Gear	Grid connection with partial scale converter
Type-D	PMSG, WRSG	Pitch control	Variable speed	Gear or gearless (for multipole machines)	Grid connection with full scale converter

Figure 5.38 Offshore wind farm.

Bird collisions: Some instances of birds colliding with the rotating blades are reported. However, the problem can be reduced to a great extent with newer turbines being designed with lower blade speeds.

Ice accumulation: In cold weather regions, ice can accumulate on the blades. When the turbine spins, the ice can be dislodged and such projectiles endanger the public and structures in the vicinity. In some models the blades are heated to avoid the ice from accumulating on them.

Sunlight flicker: Some blades reflect sunlight. When these blades move, the reflections of the sunlight on the blades flicker which creates an annoyance for some people.

5.8 Offshore wind energy

HAWT are commonly used for generating electricity offshore [32]. The installations of large offshore wind farms are possibly going to be the major development in wind energy sector in near future. There are already some small offshore wind farms located in shallow waters off Denmark and The Netherlands. However, future wind farms will be expected to be large, typically 50–100 MW, and may be located many kilometers offshore. The integration of offshore wind farms with distribution networks may cause a number of new challenges, primarily due to their size and remote location. Emphasis is to be given for using voltage source HVDC transmission to bring the power onshore and so avoid the problems and expenditure related with long AC high-voltage submarine cables.

The advantages of offshore wind farms
- Reduced visual impact
- Higher mean wind speed
- Reduced wind turbulence
- Low wind shear leading to lower towers

The disadvantages of offshore wind farms
- Higher capital cost
- Access restrictions in poor weather
- Requirement of submarine cables [33]

5.9 Wind energy applications

Wind energy projects are common for off-grid applications. However, the largest market potential for wind energy projects is with on-grid (or grid-connected) applications.

Off-grid application: Traditionally wind energy was most competitive in remote sites, far from the electric grid and requiring relatively small amounts of power, typically less than 10 kW. In these off-grid applications, wind energy is usually used for charging batteries in order to store energy captured by the wind turbines and provides the user with electrical energy on demand.

Hybrid system: Electricity generation is often relatively expensive due to the high cost of transporting diesel to remote or isolated areas. However, if the site has good local winds, a small wind energy project could be integrated to help supply a portion of the electricity requirements. These wind energy projects are normally referred to as wind–diesel hybrid systems. The wind energy system's primary role is to help reduce the amount of diesel consumption [11]. Also PV-wind hybrid configuration is a popular combination across the globe. It is observed that there is a good monthly complementary characteristic between solar and wind power production. When solar irradiance is high during that time, wind velocity is generally found to be less. Conversely, during the period when wind velocity is found to be significant then solar irradiance is less compared to other months [34].

On-grid application: Central-grid applications for wind energy projects are becoming more common. In relatively windy areas, larger-scale wind turbines are grouped together to develop a wind farm with capacities in the MW range. The land within the wind farm is normally used for other purposes, such as agriculture or forestry.

Wind pumps: Wind energy is also competitive in water pumping applications, which is also a key historical application of wind energy. In water pumping, water rather than energy can be stored for future use [11].

5.10 Wind turbine economics

Wind energy is one of the cheapest sources of electrical energy. Although the price of wind power has decreased noticeably over the past few years, the technology involves a higher initial investment than fossil fueled generators. Almost 80% of the capital cost is associated with the machinery, while the remaining expenditure includes costs related to site preparation, installation, and grid connections. The wind energy market trend shows that the rated power of new machines has increased year by year and the corresponding capital cost per kW dropped. The labor required to erect a larger machine is not that much higher than for a smaller one; the cost of electronics is only somewhat different. Wind energy projects are much more economical compared to other generating technologies since there is no fuel to purchase and nominal operation and maintenance expenses [4,12]. A coal mine and power plant require more than five times the amount of land to a comparable wind farm to produce the same amount of electricity during a 20-year period. Running a wind energy project in tandem with grid power may result in stable, uninterrupted, and cheaper power for manufacturing units. Nevertheless, to make a good decision about a wind power project a through economic analysis is necessary [12].

5.11 Conclusion

Global warming, climate change, and fossil fuel crisis compel the world for a paradigm shift from non-renewable resources to promoting renewable energy resources. Currently, wind renewable power is the world's fastest growing energy source and has great potential to save substantial amount of fossil fuel resources of the earth. The present chapter is conceived to describe the rudiments of wind power and includes most of the major aspects of wind energy and various relatable issues for development of wind power technology. The contents are organized in a cohesive manner for an uncomplicated comprehension of stakeholders. The chapter is initiated with a discussion of historical background of wind power technology. Subsequently, the principle of aerodynamic forces behind the working of modern wind turbines are explained and various wind energy conversion devices (wind turbines) are introduced with a focus on most popular contemporary horizontal axis wind turbine. Basics of wind energy and power, power output and efficiency considerations, extractable limits of wind power, axial thrust and torque on blades, concepts of tip speed ratio, and various operational characteristics are made clear in the chapter. The chapter also discusses on wind resource assessment, wind measurement techniques, evaluation of suitable sites, and wind tower spacing for development of wind power projects. Besides, different types of wind electric generators, constant speed and variable speed operation, and speed control strategies of wind power plant have been put in plain words. In addition, the chapter focuses on environmental impact and public perception on wind power technology and also contains the topics like wind energy applications, offshore wind energy, and wind turbine economics.

References

[1] Wind generation. <https://nptel.ac.in/content/storage2/courses/108108078/pdf/chap6/teach_slides06.pdf> (accessed 27.10.19).
[2] M.A. El-Sharkawi, Electric Energy: An Introduction, second ed., Taylor & Francis, Boca Raton, FL, 2009.
[3] Renewable energy, facts and information. <https://www.nationalgeographic.com/environment/energy/reference/renewable-energy/> (accessed 15.08.19).
[4] U. Rathore, Energy Management, second ed., S.K. Kataria & Sons, New Delhi, 2014.
[5] J. Andrews, N. Jelley, Energy Science: Principles, Technologies, and Impacts, Oxford University Press, New York, 2007.
[6] James Blyth's wind turbine. <https://commons.wikimedia.org/wiki/File:James_blythe_turbine.jpg> (accessed 17.08.19).

[7] History of wind power. <https://en.wikipedia.org/wiki/History_of_wind_power> (accessed 17.02.20).
[8] 1970s energy crisis. <https://en.wikipedia.org/wiki/1970s_energy_crisis> (accessed 29.09.19).
[9] G.M. Masters, Renewable and Efficient Electric Power Systems, John Wiley & Sons, Hoboken, NJ, 2004.
[10] D.S. Chauhan, S.K. Srivastava, Non-Conventional Energy Resources, New Age International (P) Ltd., New Delhi, 2006.
[11] Clean Energy Project Analysis: RETScreen® Engineering & Cases Text Book, Ministry of Natural Resources, Canada, 2005.
[12] J. Earnest, T. Wizelius, Wind Power Plants and Project Development, PHI Learning Private Ltd., New Delhi, 2011.
[13] P. Jain, Wind Energy Engineering, McGraw-Hill, 2010.
[14] B.R. Gupta, Generation of Electrical Energy, Eurasia Publishing House (Pvt.) Ltd., New Delhi, 2007.
[15] M.R. Patel, Wind and Solar Power Systems, CRC Press, Boca Raton, FL, 1999.
[16] W. Shephers, L. Zhang, Electricity Generation Using Wind Power, World Scientific Publishing Co. Pte. Ltd., Singapore, 2011.
[17] S.N. Bhadra, D. Kastha, S. Banerjee, Wind Electrical Systems, Oxford University Press, New Delhi, 2005.
[18] <https://slideplayer.com/slide/10537842> (accessed 18.08.19).
[19] G.N. Tiwari, M.K. Ghosal, Renewable Energy Resources: Basic Principles and Applications, Narosa Publishing House, New Delhi, 2005.
[20] B.H. Khan, Non-Conventional Energy Resources (second ed.), Tata McGraw Hill Education Private Ltd., New Delhi, 2009.
[21] Wind energy. <https://quizlet.com/80781637/wind-energy-flash-cards/> (accessed 26.10.19).
[22] Wind turbine generators used for wind power generation. <http://www.alternative-energy-tutorials.com/wind-energy/wind-turbine-generator.html> (accessed 27.10.19).
[23] R. Sharma, V.S. Chauhan, H. Shah, Doubly fed induction generator for wind energy system, Int. J. Adv. Res. Elect. Electron. Instrum. Eng. 6 (2) (2017) 3208–3213.
[24] Permanent magnet DC generator as a wind power generator. <www.alternative-energy-tutorials.com/wind-energy/pmdc-generator.html> (accessed 27.10.19).
[25] N. Goudarzi, W.D. Zhu, A review on the development of wind turbine generators across the world, Int. J. Dynam. Control. (1)(2013) 192–202.
[26] Synchronous generator as a wind power generator. <www.alternative-energy-tutorials.com/wind-energy/synchronous-generator.html> (accessed 27.1019).
[27] S. Hlaing, Basic concepts of doubly fed induction generator driven by wind energy conversion system, Int. J. Sci. Eng. Technol. Res. 3 (15) (2014) 3242–3246.
[28] S. Muller, M. Deicke, R.W.D. Doncker, Doubly fed induction generator systems for wind turbines—a variable alternatives to adjust speed over a wide range at minimal cost, IEEE Industrial Applications Magazine, May–June 2002, pp. 26–33.
[29] Principles of Doubly-Fed Induction Generators, Lab-Volt Ltd., 2011. <http://www.kves.uniza.sk/kvesnew/dokumenty/rep1/doubly-fed_induction_generators.pdf> (accessed 27.10.19).
[30] C.V. Hernández, J.S. González, G. Centeno, A market-based analysis on the main characteristics of gearboxes used in onshore wind turbines, Energies 10 (1686) (2017) 1–17.

[31] S. Chandrasekaran, Grid connected doubly fed induction generator based wind turbine under LVRT. <http://www.amsdottorato.unibo.it/6243/1/Grid_Connected_Doubly_Fed_Induction_Generator_Based_Wind_Turbine_under_LVRT.pdf> (accessed 27.10.19).

[32] S.P. Neill, M.R. Hashemi, Fundamentals of ocean renewable energy. <sciencedirect.com/topics/engineering/horizontal-axis-wind-turbine>, 2018 (accessed 26.10.19).

[33] N. Jenkins, R. Allan, P. Crossley, D. Kirschen, G. Strbac, Embedded Generation, IEE, London, 2000.

[34] S. Bhattacharjee, S. Acharya, PV-wind hybrid option for a low wind topography, Energy Convers. Manag. 89 (2015) 942–954.

Section V

Solar Power Technology

Enabling solar photovoltaics penetration in highly dependent African fossil fuel markets

6

Ogechi Vivian Nwadiaru[1,2], Tonny Kukeera[1,2] and David Ebanehita[3]
[1]Microenergy Systems Research Group, Technical University of Berlin, Berlin, Germany, [2]Working Group of Economic and Infrastructure Policy, Technical University of Berlin, Berlin, Germany, [3]Department of Energy Management, ESCP Europe Business School Paris, Paris, France

6.1 Introduction

The great acceleration [1,2] is an indication of the effects of humanity's continuous indiscriminate exploration and use of natural resources. While predictions indicate that some aspects of the great acceleration such as population will level off, other factors such as primary-energy use show no sign of slowing down. At this rate, increases in the extraction and utilization of the Earth's resources are predicted to continue, regardless of a leveled-off population growth [3]. Unabated use of fossil fuels has contributed significantly to the production of anthropogenic CO_2, contributing to climate change [4–6]. Some of the remarkable effects include melting ice in the Arctic, rising sea levels, desertification, poor crop yields, and global warming, which have been reported by several studies and most prominently in the IPCC report [7]. This growing concern on the causes and effects of climate change together with the need to ensure sustainable economic growth is putting pressure on global economies to rethink their development agendas [8–11]. Energy being the foundation of development and hence a determinant of the level of economic prosperity of a country, any shock to the sector can have severe effects on the global economy. The challenges such as the oil price crises in 1973, 1979, and 2008, and climate change and the inevitable depletion of fossil-energy sources present an opportunity for reflection and seeking alternatives [12].

Global estimates place the true annual cost of burning fossil fuels at US$5.3 trillion inclusive of costs to the fossil fuel expenditures, and environmental and healthcare costs.[1] According to IRENA's global energy transformation report, renewable energy (RE) scale-up combined with electrification could deliver more than three quarters of the energy-related emission reductions needed to meet global climate goals [13]. Key insights from the report indicate, that generation from renewables

[1] The Economics Of Solar Power https://www.investopedia.com/articles/investing/061115/economics-solar-power.asp.

Sustainable Fuel Technologies Handbook. DOI: https://doi.org/10.1016/B978-0-12-822989-7.00007-X

has the potential to achieve desired environmental targets as well as result in US $160 trillion worth of cumulative saving over the next 30 years in avoided health costs, energy subsidies, and climate damages and each energy transition investment dollar generating seven-fold return on investment. In West Africa alone, reports from World Bank predict the realization of cost savings of US$5−8 billion/year by enabling countries to import cheaper sources of electricity through integrated power trade that could be achieved by the implementation of largescale, cross border, renewable-energy projects leading to, increased access to affordable, reliable and modern energy, and reduced CO_2 emission intensity.[2]

6.1.1 Complexity of Africa's energy landscape

The global energy transition is quite a complex process with multiple interacting layers. Therefore the task posed to countries with a low human development index is not an easy one—ensuring significant improvement in the standard of living of their citizens alongside resource conservation. In this regard, the African continent is put on the spotlight. Of the 840 million people without electricity access globally approximately 600 million are in Africa [14]. Prior to the turn of the century, several authors argued the lack of adequate alternate resource and technology to replace oil as the backbone resource of industrial society, which could pose a challenge for Africa in its attempts at industrialization [15]. This has been disputed with several scenarios for a 100% RE future and though fossil fuel still plays a significant role in the middle term, there is significant room for fuels from renewable sources to replace unsustainable fuels [16,17]. Though this could pose a tremendous challenge, it is laced with an opportunity for Africa to leapfrog current energy-intensive and high CO_2 energy sources. The energy access problem in Africa is multidimensional; from access to clean cooking fuels, to electricity for lighting and productive use as well as fuel for transportation, cooling and heating needs. The current energy mix of the continent reflects the rate of development and advancement of its energy sector. While the global energy landscape is experiencing a rapid transition, Africa's energy landscape is not gaining enough momentum to satisfy the current and future needs of its population. According to the Energy Sector Management Assistance Program (ESMAP) report on tracking SDG 7 [18], 90% of the 8% of global population without electricity access in 2030 will be in sub-Saharan Africa under the current policy situation.

6.1.2 Research contribution

Though most of the research and attention in current energy studies on the continent focuses on sub-Saharan Africa, due to the region's peculiar statistics this chapter examines the green energy transition from the perspective of the top oil-producing countries in Africa (Table 6.1). The aim of this approach is to highlight the complexity of the transition process in such a context, their role in shaping African

[2] Large-scale Solar Energy Gets a Boost in West Africa https://www.esmap.org/node/170463.

Table 6.1 Proven oil reserves for select African countries.

Country	Reserves (millions of barrels)	R/P ratio[a]	World/continental ranking
Libya	48,363	127	9/1
Nigeria	36,182	49	11/2
Algeria	12,200	26.55	16/3
Angola	8384	14.42	17/4
Kenya	0	0	N/A

[a]R/P ratios are estimations from calculations using the proven reserves of 2019 and the average daily production of 2018.
Source: Author's estimation/data from EIA, International energy statistics. <https://www.eia.gov/beta/international/data/browser/#/?pa = 000000000000000000000080000000000002&c = ruvvvvvfvtvnvv1urvvvvfvvvvvvfvvvou20evvvvvvvvvnvvuvo&ct = 0&tl_id = 5-A&vs = INTL.57-6-AFG-BB.A&cy = 2018&vo = 0&v = H&s = INTL.57-6-NGA-BB.A~ ~INTL.57-6-DZA-BB.A~ ~INTL.57-6-AGO-BB.A~ ~INTL.57-6-LBY-BB.A~ ~INTL.57-1-DZA-TBPD.A~ ~INTL.57-1-AGO-TBPD.A~ ~INTL.57-1-NGA-TBPD.A~ ~INTL.57-1-LBY-TBPD.A>, 2019 (accessed 14.11.19) [19].

energy markets, and the implications on their evolving political and economic landscapes. Generally taking perspectives from Algeria which has achieved 99% electrification, to evaluating Nigeria's electrification scenario which is dominated by backup generators (BUGs), Angola's largely hydro-powered grid, and finally, exploring Kenya with a dual status as a nonoil-producing country and new entrant into the oil market, provides a diverse set of scenarios that could offer directions to future frameworks for efficient green transformations.

In the first section, a review of the energy transition in Africa is presented with a look into electricity access rates and the overall energy mix. A deep dive into the energy scenario of the different countries gives an argument for the need to integrate more RE in the mix for oil-producing countries. As a second contribution, a review of the status of policies based on the Regulatory Indicators for Sustainable Energy (RISE) developed by the World Bank is presented. This helps in identifying leaders and laggards in the selected countries. A compelling case is then made for solar as a prime mover for improving electricity access in Africa, and diversifying fossil-dependent economies. Finally, we identify interventions to promote the growth of the solar industry in Africa. We discuss the necessary policies, private sector participation, and technology combinations based on historical evidence from literature.

6.2 Energy transition in Africa

Africa's population makes up 16% of the global population and accounts for only 4% and 3% of production and consumption-based CO_2, notwithstanding the continent remains the most vulnerable from the effects of climate change [20,21]. Like elsewhere, Africa's economic development is linked to its energy consumption which makes considerations for energy-related emissions a priority in new infrastructure plans. The challenge faced by the energy transition process in Africa is one of modernization and expansion—in utilizing the vast RE potential, moving away from harmful and wasteful conversion technologies, and scaling up deployment of new technologies to meet the growing demand [22]. The path toward this transformation process is often blurry with no end game in mind except an assured economic and environmental prosperity for all. To demystify the transition process, research efforts are being made to understand the drivers of RE diffusion in Africa. Ackah et al. [23] analyzed the efficiency of RE consumption and conclude that RE sectoral growth is tied to citizen welfare. Their argument buttresses the point that poverty results in limitations in choice of energy access technologies. In a study by Aliyu et al. [24], where policies for RE deployment were explored, an emphasis is made on the need for proper technology, awareness and skills to fully harness the RE potential. Schwerhoff and Sy [25] analyzed the optimal energy mix required for fueling rapid growth in Africa within the 2°C global warming limits, and found that though intermittency and higher capital intensity of RE are big challenges, by 2050, solar, wind and hydropower could make up a significant portion of electricity

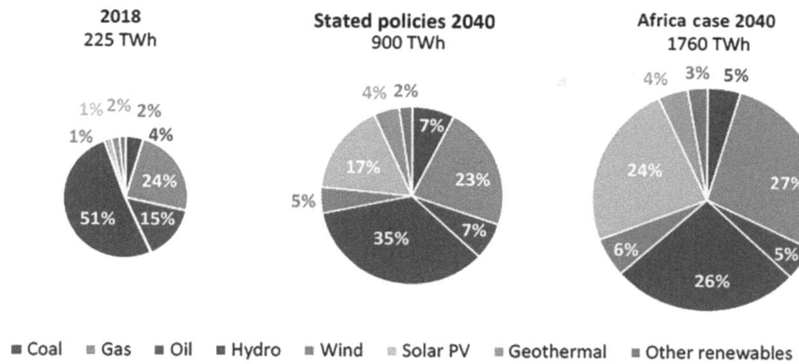

Figure 6.1 Electricity supply by type, source, and scenario in sub-Saharan Africa (excluding South Africa) in 2018 and 2040.
Source: IEA, Africa Energy Outlook 2019, International Energy Agency, November 2019.

generation. Trotter et al. [26] reviewed a strand of literature about electricity planning and implementation in SSA and found out that 63% of the past 306 studies in the region, pointed at RET (RE technologies) as the preferred energy source. More research work indicates that sufficient investment in RE could free up short-term revenues as it frees up petroleum resources for export in oil-producing countries [27]. Further studies on the growth of RE in oil-producing countries in Africa indicate that the main drivers of RE consumption are real income per capita, energy resource depletion per capita, carbon emissions per capita and energy prices; hence, providing recommendations for the encouraged consumption of commercial sources of RE to attract the needed investments [12]. Given the stated literature review and projections from IEA [28] as seen Fig. 6.1, it is evident that the future of Africa's energy sector will be greatly influenced by RE for improving the quality of life on the continent as well as for economic diversification. Therefore the top oil producers in the region, considering their large economies, will have to equally adapt.

6.2.1 Country insights

According to IEA's African Energy Outlook [28], one of two people in Africa still lack access to electricity and only six countries in SSA will achieve universal access by 2040. SSA (excluding South Africa) is expected to see an income and industrialization driven demand increase by a factor of four. In its two different scenarios (Fig. 6.1) the share of oil in the electricity mix is expected to drop, while solar energy will see an increase in multiple degrees [28]. The energy sector challenges, and weak grid situation costs the region 2%–4% of its gross domestic product (GDP) annually, undermining sustainable economic growth, jobs, and investment [29]. Though carbon emissions from Africa are currently quite low when compared to the global total, given the demand and population projections, the carbon intensity from energy generation in Africa is poised to become

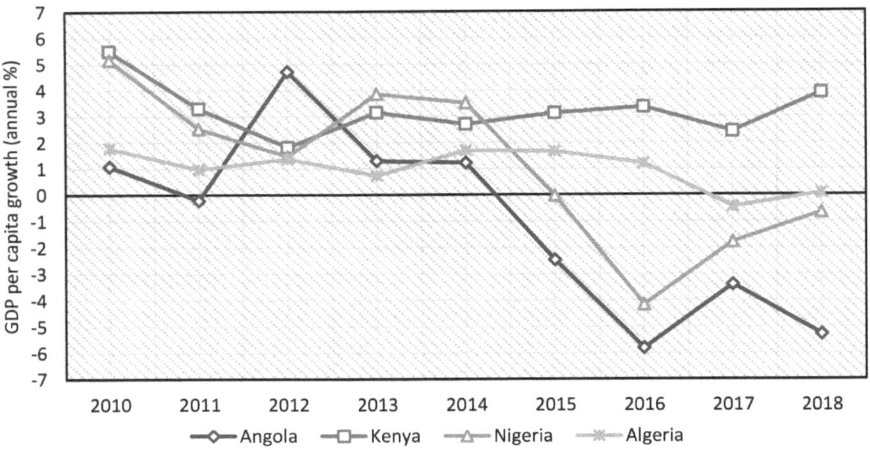

Figure 6.2 Carbon intensity (CO_2 emissions to total primary energy supply).
Source: CO_2 Emissions from Fuel Combustion IEA, CO2 Emissions from Fuel Combustion, (2018 ed.), OECD/IEA, Paris, 2018 [30].

increasingly significant [25]. In addition to the challenge of energy access and environmental sustainability (Fig. 6.2), the economies of top oil producing are under constant threat due to fluctuating oil prices. Major oil-producing countries in Africa took a major hit following the 2010s oil glut that occurred between 2014 - 2016 leading to crash in oil prices. The decline cut deep into the pockets of these countries. For Nigeria and Angola which rely on crude oil for over 80% of their exports,[3] the past few years meant tough economic recession (Fig. 6.3). Furthermore, IEA projects funding needs of around US$25 billion/year above the level mobilized in the Stated Policies Scenario (Fig. 6.1) over the period to 2030 to ensure electricity access for all. This section presents an overview of the socioeconomic context in the countries, including discussions on subsidies and climate commitments.

6.2.1.1 *Algeria*

Algeria is the leading natural gas producer and one of the top three oil producers in Africa (Table 6.1). Algeria currently has an electricity access rate of 99% that was achieved through the implementation of strategies and policies to promote the economic and social utilization of energy resources [31]. Though achieved through its hydrocarbon-based growth model [32], Algeria is yet to adopt strategies for proper economic diversity. Revenues from oil and gas constitute the backbone of the Algerian economy evident in its growth model. The electricity sector of Algeria is dominated by fossil energy due to the availability and low cost of natural gas as

[3] https://oilprice.com/Energy/Crude-Oil/The-Biggest-Losers-Of-The-Current-Oil-Price-Slump.html.

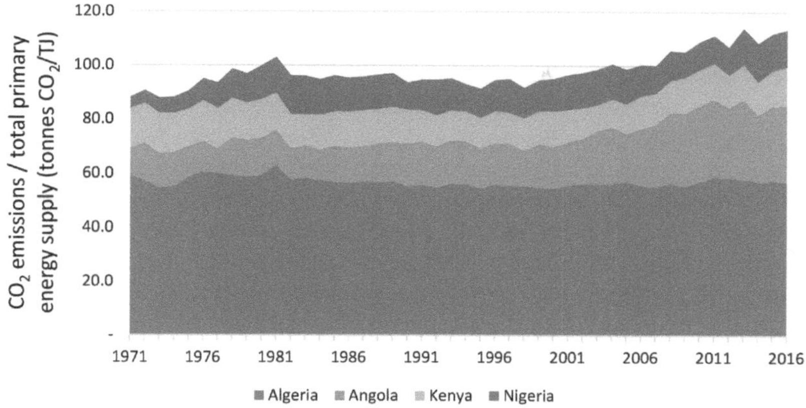

Figure 6.3 GDP per capita growth for selected African countries.
Source: World Development Indicators.

most generating capacity comes from natural gas-fired and combined-cycle power plants [33].

According to IMF [34], from 2002 until 2014 (the period of rising oil prices) hydrocarbons on average accounted for 98% of exports earnings, 69% of fiscal revenues, and 36% of the country's GDP. Consequently in 2015, the oil price slump strongly hit the economy as the oil and natural gas export revenues reduced by 41%. The foreign exchange reserves helped softening the extent of the economic shock, going down from $194 billion in December 2013 to $153 billion in late 2015. The continuous dependence on crude oil revenue for economic growth increases Algeria's vulnerability to oil price volatility, as such, the 2016–19 economic plan was designed to diversify the country's exports and reduce its hydrocarbon dependent domestic energy consumption growth to 3% in 2030.

For Algeria, the energy transition implies confronting an old challenge with new urgency—how to diversify the economy away from hydrocarbons. Moreover, low-energy efficiency and fossil fuel subsidies strongly contribute to the unsustainable use of energy in Algeria.[4] Hence RE could be the answer to the challenge of finding new sources of growth that will create jobs for a youthful and fast-growing population. With an average sunshine hour of 2000 h, the country's solar power is estimated to range from 1700 kWh/m^2 in the north to 2650 kWh/m^2 per year in the south [31].

6.2.1.2 Kenya

With an approximate population of 52 million[5] and economic growth of 6% per annum [35], Kenya has a GDP per capita of US$2010, making it the highest among

[4] https://www.forbes.com/sites/judeclemente/2016/05/04/will-algeria-be-able-to-export-more-natural-gas-and-lng/.
[5] http://worldpopulationreview.com/countries/kenya-population/.

East African countries. In comparison to the countries under study, Kenya has the lowest GDP per capita; however, it has experienced a stable growth at approximately 3% per annum (Fig. 6.3). In the past two decades, the country's power sector has seen an increased pressure from the unprecedented economic growth. An 18.9% estimated growth in energy demand was observed between 2004 and 2013. This saw a transformational change in Kenya's energy policies and infrastructure which has supported an astronomical growth in the electrification rates from 19% in 2010 to 64% in 2018 [36]. According to the Energy Act of 2006, the energy regulatory commission—the responsible body in Kenya—was mandated to create short-term and long-term least cost–power development plans that saw a focus on the development of renewables ranging from geothermal, photovoltaics (PV) and hydro on the forefront [37]. Currently the total installed capacity is 2234.83 MW [37], with a 71.95% RE share consumption [36].

However, the country still suffers from insufficient supply as the existing demand is not met. Considering hydro contributes the highest percentage of electricity, the power industry is always affected during summer periods when water levels are low. This gap in supply is normally catered for by the more costly fossil fuel-based thermal generation. The country spends a substantial amount of foreign reserves to import fossil-based fuels. According to Kiplagat et al. [38], 55% of foreign exchange earnings were consumed by the oil import bill of 2008.

This is supplemented with electricity imports from neighboring countries like Uganda, Ethiopia, and Tanzania. The exorbitant costs involved in energy imports coupled with unreliable energy infrastructure have been reported as one of the issues that continue to contribute to the sluggish growth of the nation's economy [35]. Kenyan companies experience production losses of close to 10% due to power instabilities and outages [39]. The Government of Kenya (GoK) set ambitious targets of achieving universal access by the end of 2020 with renewables at the forefront.[6] By 2010, Kenya had 320,000 solar home systems installed and accounting for over 20% of electricity grid consumers [40]. While Kenya continues to grow its RE industry, there are concerns that the discovery of oil in the country will pose a threat to its commitment to RE. Though, commercial production is still quite a ways. this changes the landscape in Kenya and needs to be considered when analysing the government's priorities and energy access targets.

6.2.1.3 Nigeria

Nigeria is one of the world's largest producers of oil and gas. The electricity grid in the country barely functions despite the 23 on-grid power plants concentrated in the southern part of the country. Oil accounts for more than 80% of Nigeria's export revenues, making it riskily reliant on a single industry already stuck in corruption [41–43]. Notwithstanding its status as a top global oil producer, Nigeria still imports around 85% of its petroleum products due to ill-managed refineries and

[6] https://ec.europa.eu/europeaid/blending/rural-electrification-kenya-last-mile-connectivity_en.

insufficient refining capacity. These imports are used to meet the increasing demand gap left by domestic production [44].

In 2016 a drop in global oil prices pushed Nigeria into a recession[7] (Fig. 6.3) that made diversifying energy resources and the economy a priority. To stimulate other industries such as agriculture, food processing, construction, and technology, Nigeria needs electricity. Nigeria has the largest economy in Africa and is projected to be among the top 10 global economies by 2050. Currently the country ranks 131, which is considerable development after it dropped drastically from 102 in 2009 to 169 in 2016.[8] In addition to its reliance on oil revenues, Nigeria's embattled economy is alsoburdened by its weak on-grid electrification. On-grid consumers experienced approximately 14.8 blackout hours daily for the first half of 2019.[9] One of the challenges to investments identified by Nigerian businesses is the inadequacy of power supply, forcing them to spend billions annually on self-generation. IMF [45] estimates the annual cost to the economy of the inadequate power supply to be US $29 billion. Given the new United Nations rules aimed at cleaning up international shipping which come into effect at the beginning of 2020, with many ships expected to burn distillates instead of dirtier fuel oil, Nigeria's heavily reliant business landscape will need to adapt rapidly [46].

This implies higher costs for operating generators that facilitate Nigeria's economy, leading to impaired growth in GDP at a time of high inflation rates. With the population growing at 2.6% yearly, poverty levels are getting higher. Increasing poverty levels and income disparity also make it difficult for end-users to have control over their energy options, creating an energy poverty cycle and a general unwillingness to pay. However, reliability studies [47] indicate that consumers are willing to pay for improved reliability and opt out of self-generation for an improved service quality.

The share of RE in the total final energy consumption (TFEC) has been almost constant since the 1990s at around 88%. Traditional solid biofuels form the biggest share of renewable sources at 77.1% of TFEC in 2012, while the modern solid biofuels contributed 9% and hydro only 0.4% [48]. Solar abundance in the country creates an opportunity for addressing some of the discussed challenges, on average, the country receives solar radiation of about 19.8 $MJ/m^2/day$ with an estimate of six sunshine hours daily. At current solar efficiency standards, Nigeria could generate 190,550 GWh/year from 1% of its land area, which exceeds the current grid capacity [49]. This resource has remained largely untapped and the number of solar PV installations in Nigeria have been almost negligible. The RE Agency through the Nigeria Electrification Project (NEP) is taking a private sector–driven approach to provide electricity access to households, micro, small, and medium enterprises through renewable power sources. The program, though robust and ambitious is still

[7] https://www.export.gov/apex/article2?id = Nigeria-Market-Overview.

[8] Ease of doing business ranking by World Bank https://openknowledge.worldbank.org/bitstream/handle/10986/32436/9781464814402.pdf?sequence = 24.

[9] https://www.bloomberg.com/news/articles/2019-09-23/nigeria-runs-on-generators-and-nine-hours-of-power-a-day.

at its inception, addresses off-grid communities and the long-term impact is yet to be assessed. Furthermore there are still fiscal barriers to accessing solar household devices as small solar lanterns, and charging systems attracts an import tax of up to 30% [50].

6.2.1.4 Angola

Angola is the second largest producer of oil in Africa after Nigeria, with proven reserves of approximately 8.3 billion barrels (Table 6.1). Oil currently accounts for 90% of its exports. The post-civilwar economic boom saw an increase in incomes and public revenue without a commensurate oversight on public spending. The GDP per capita soared from US$711 in 2002 to US$4804 in 2013, which led to its ascent from the level of low-income country to that of upper-middle-income country. Despite this Angola shares similar characteristics with impoverished countries and is still ranked by the UN as one of the least-developed countries (LDC). Due to the economic crisis that commenced in 2014, Angola's GDP saw a steep fall (Fig. 6.3) and reduced its ranking to lower-middle income country in 2017 [51].

Electricity production in Angola as at 2015 was 617 ktoe with a 73.2% share from hydro and the other 24.7% from fossil fuels [52]. There is a huge potential for hydropower, solar power, and natural gas-based generation, and expanding production has been one of the biggest priorities of the government over the past decade [51]. Generation capacity and production have both increased massively (Fig. 6.4) thanks to large infrastructure investments, but still falls below national targets, primarily due to a delay in the 2170 MW Caculo Cabaca hydropower project. Regardless of these shortcomings, oil production has allowed for a remarkable economic growth, acting as a buffer for the impacts of civil war on the economy

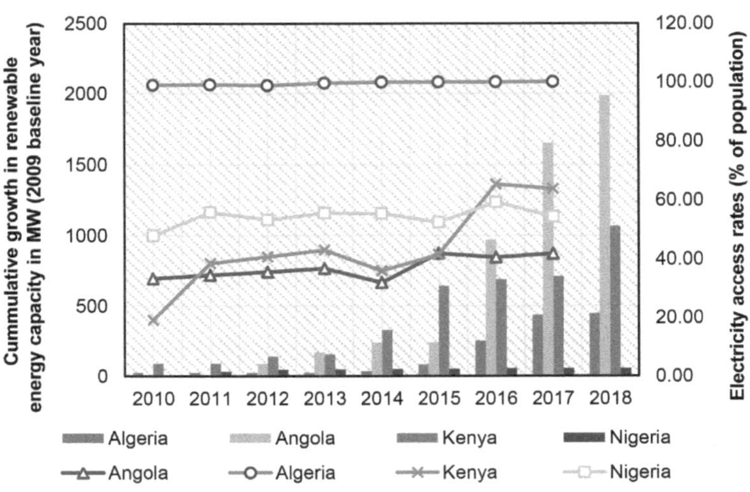

Figure 6.4 Electricity access rates and renewable energy capacity added from 2010 to 2017. *Source*: World Bank, IRENA.

and electricity demand. Like in Nigeria, the presence of local refining is almost nonexistent due to the inefficient and underdimensioned refining capacity [53]. This has resulted in a high dependence on imports of oil-derived products such as diesel and gasoline, thereby increasing dramatically the cost of the associated subsidies. Repressed demand has been predominately compensated by diesel-based generation [53]. There is a high disparity in the electricity access rates between urban and rural populations with an overall access rate of approximately 42%, 72% in urban areas, and a meager 16% in rural areas [53]. This has led to the widespread use of diesel engines for self-generation, with negative impacts on emissions (Fig. 6.2) and end-user energy costs. In 2017 the country ranked 166 out of 189 and obtained a score of zero out of eight on the "getting electricity" and "reliability of supply and transparency of tariffs index" indicators, respectively.[10] According to the IFC country diagnostic report, electricity hampers investment in manufacturing and hinders growth in the agricultural, ICT, and health sectors [54].

The government of Angola through its 2025 strategy indicates an interest to deploy an estimated 99.9 MW of solar PV at a total investment cost of US$267.8 million among other improved generation capacity improvements. The Angolan NDP stipulates that 10% of infrastructure needs of $47 billion could be met through public–private partnerships (PPPs) with 30% of the needs being in the energy sector [55].

These planned improvements, however, will not translate to significant improvement in electricity access if not paired with transmission infrastructure [51]. The major barriers to growth in the electricity sector include governance, a lack of creditworthy utilities, and a cost-reflective tariff.

6.2.2 Subsidies as enablers of fossil fuel dependency

Ensuring that global temperature rise is limited to well below the 1.5 - 2 degrees mark is achievable, however current policies need to reflect these targets. Policies such as fossil fuel subsidies that favor fossil fuel use and deployment, counteracts and even negates the efforts to mitigate climate change.

Subsidies have been used globally by governments as an incentive for promoting specific economic and social policies. This can take any form as price reductions, tax cuts, or cash payments. In the energy sector they are provided as consumer and producer subsidies to reduce the end-user cost of electricity and petroleum products. Current global fossil fuel subsidies are so large that if redirected enough money would be available for investments in clean energy and energy efficiency needed to meet the Paris Climate Agreement targets [56]. Estimates from the IMF [57] highlight that eliminating fossil fuel subsidies could free up US$2.9 trillion in government revenue annually. That amount is more than double the annual investment of US$1.25 trillion the International Energy Agency estimates is needed by 2035 in clean energy and energy efficiency to stop the world from warming up by 2°C.

[10] World Bank Doing Business https://www.doingbusiness.org/en/reports/global-reports/doing-business-2016.

Angola, Nigeria, and South Africa were identified as some of the SSA countries that provided fossil fuel subsidies in excess of US$1 billion in 2015. Adding the costs of externalities—such as local pollution, impacts on climate change, road accidents, and congestion—the estimated costs related with fossil fuel subsidies increased to US$75 billion in 2015 for the region [57]. In 2018 according to the IEA [58] Algeria and Nigeria spent a total of US$20.1 billion on subsidies accounting for 9% and 1% of GDP, respectively. Though the motivation for providing subsidies is consumer protection, the rebound effect is a distortion in resource allocation, driving down priority public spending, and suppressing private investment [59]. These effects have been summarized by the IMF as retrogressive with impacts on economic growth. Subsidies do not necessary address the concerns of the poor, as the richest 20% are estimated to benefit six-times more than the poorest 20% [60]. Attempts to remove subsidies from fossil fuel products have been met with displeasure, as was the case in Nigeria in 2012, due to the corruption that has ravished the sector.[11] The 112% increase in gasoline prices was eventually scaled back to 49% after strong protests. Hence a critical success factor for the removal of subsidies will lie in the state's ability to regain the trust of the people through transitional policies that support and provide a soft landing for the poor [57,59,61,62]. In some cases removal of a subsidy does not necessitate an increase in the end-user price, and can be achieved by improving the overall system efficiency [61]. At the beginning of 2016 the Algerian government raised the price of gasoline and other energy products for the first time since 2005, yet despite a 34% increase, the gasoline remains among the least expensive in the world and cheap gasoline mainly benefits the well-to-do.[12] There is no easy path to the removal of subsidies; however, political credibility, will, and diversification can support a gradual phase-out of fossil fuel subsidies. Lessons from history indicate that subsidy reforms are only as successful as a citizen-participation process—people are more likely to accept a program of subsidy reinvestment if they are involved in the decision-making process.

6.2.3 Carbon intensity and Paris agreement

Through the continued exploration of oil, overall carbon intensity is on a rise, further strengthening the case for RE especially solar. At the current predicted levels of population growth rates, energy demand will see tremendous growth. The energy-related emissions will be defined by the continents' ability and commitment to reducing greenhouse gas emissions.

Despite the commitment to the Paris agreement and several RE targets set out in countrywide policies (Table 6.2) there is no drop in carbon intensity as seen from the time series in Fig. 6.2. The graph indicates that Angola ranks top in the rate of increase in carbon intensity for the period under review.

[11] https://www.un.org/africarenewal/magazine/april-2012/bid-end-subsidy-stirs-protest-nigeria.

[12] The Need for Subsidy Reform in Algeria (IMF).

Table 6.2 Breakdown of emissions and climate commitments.

Country	Intended nationally determined contributions [48,62]	Targets and supporting policies [28]
Algeria	• 27% of its electricity for domestic consumption from renewable energy sources by 2030. • 7% below BAU levels by 2030. • Additional 15% reduction (conditional)	• National Program for the Development of Renewable Energies (NPDRE)
Angola	• Reach 27% of electricity generated from renewable sources of energy by 2030; • Up to 35% (unconditional) to 50% (conditional) reduction in GHG emissions by 2030.	• Install about 30,000 solar photovoltaic systems in rural areas. • 800 MW of clean energy capacity installed by 2025. • National Development Plan (NDP) 2018–22: Lessen economic dependence on fossil fuel revenues. • Achieve middle-income status by 2022.
Kenya	• Enhance energy and resource efficiency across the different sectors. • Adopt clean energy technologies to reduce over reliance on wood fuels. • Develop low carbon and efficient transportation systems. • 30% below BAU by 2030 (Conditional)	• National Electrification Strategy (2018): investment of $2.8 billion from 2018–22. • Distribute 250,000 solar home systems to power households, schools, health facilities, and agriculture by 2030. • Produce 100,000 barrels of oil per day from 2022. • 2275 MW of geothermal capacity by 2030. • Achieve middle-income status by 2030.
Nigeria	• Work towards ending gas flaring by 2030. • Work towards off-grid solar PV of 13 GW. • 20% unconditional reduction below BAU by 2030 (Additional 25% is conditional)	• Increase oil production to 2.5 mb/d and become a net exporter by 2020. • Dedicate at least 30% of the federal budget to capital expenditure. • Achieve GDP growth of 7% and create over 15 million jobs by 2020 and double manufacturing output to 20% of GDP by 2025.

6.3 Solar photovoltaics deployment in oil producing African countries

The African continent, besides being endowed with fossil fuel reserves, boasts of an abundance of RE resources that could potentially support in illuminating the homes of hundreds of millions if deployed correctly. RE are particularly well-suited for Africa due to the favorable natural conditions. Africa's RE resources are varied, unequally spread and vast—almost unlimited solar potential (10 TW), abundant hydro (350 GW), wind (110 GW), and geothermal energy sources (15 GW) [29,48]. Currently, heavily reliant on hydropower for most of the renewable electricity generated, the continent is gradually turning to solar mainly due to its flexibility of deployment [48]. In 2015 the total installed solar PV capacity on the continent stood at 2.1 GW with South Africa accounting for 65% and Algeria being the next largest shareholder at 13% [64]. This figure is equivalent to only 3% of IRENA's 2030 projected PV capacity installations in Africa. Solar PV can be developed under two main paths—as utility scale or off-grid.

6.3.1 Rationale for a solar-powered Africa

Next to the Middle East and North Africa, SSA is estimated to have the second highest technical potential for solar energy ranging from 8860 to 227,529 Mtoe; taking into consideration different assumptions [65]. Solar PV serves a unique role for developing countries by providing opportunities for electrifying rural and periurban areas [65]. In Africa there has been a growth in the number of decentralized energy systems with the continent having the most significant share of the planned 210,000 minigrids by 2030 [66]. The inclusion of solar PV electricity generation in the global energy mix might be gaining momentum but the rates of adoption in Africa are still low especially at utility scale. This is in part due to the historical role of governments in energy production for largescale utilities and, on the other hand, the high initial capital required for standalone systems that is still out of reach for a reasonable percentage of people on the continent. Nonetheless the falling cost of solar panels together with increased efficiencies of solar cells makes it attractive and competitive for electricity generation. A significant percentage of the population would afford the technology as time elapses. The cost of electricity generation from low-carbon energy technologies has seen an incredible decline in the past decade with wind and solar seeing a tremendous drop of 34% and 76% respectively between 2009 and 2017. The average cost of developing new solar PV projects has been estimated by IRENA [13] at $0.10/kWh and though this is given as the global average, the variation for individual projects can be significantly wide. Despite this variation, the prices are an indication of the competitiveness of new PV power with flexibility of deployment making an additional case for its use in low access and hard to reach regions.

This technology currently presents the fastest installation time (less than 1 year) and lowest maintenance costs. Minigrid construction can take less than 1 month—far less time than it takes for building a hydropower plant. Furthermore, the modularity of solar PV implies that small businesses and stakeholders can partake in the boom.

6.3.2 Climate resilience

Compared to large hydro, solar PV is resilient to the effects of climate change. As the amount of annual rainfall continues to vary and seasonal patterns deviate from weather predictions, it will become increasingly difficult to estimate the electricity production from large hydropower dams, making PV a more economical solution for climate resilience, especially for new installations.

6.3.3 Boost to businesses and rural electrification

By the sheer delocalization of the resource, the use of solar PV offers the possibility of citing generating plants in locations where the demand is, cutting high costs for long transmission and distribution lines. This also promotes the growth of local businesses.

6.4 Institutional and economic barriers to solar deployment

On the pathway to promoting sustainable energy and climate change mitigation polices there are several choices to be pursued. (Re-)allocating resources, creating institutions, incentivizing investments, and influencing behavior or culture has a subtle but critical impact on the patterns of politics and power relations, in turn shaping future landscapes [67]. The right combination of instruments depends largely on the country, political dynamics, commitments, and its emissions targets. Research by MIT analysts [68] has found that government policy to help grow markets around the world has played a critical role in reducing the technology's costs. An example of such policies that contributed in stimulating market growth globally include measures such as renewable portfolio standards, feed-in tariffs, and a wide range of subsidies [65]. Furthermore, while attracting private sector has an effective role to play, the public sector still plays a steering role, effective governments have been shown to deliver better quality infrastructure, especially in the electricity sector [69].

In this section the Regulatory Indicators for Sustainable Energy (RISE) framework developed by the World Bank is adopted to evaluate current policies implemented by the countries of study. RISE is a global inventory of policies and regulations that support the achievement of SDG 7 [70]. RISE scores reflect a snapshot of a country's policies and regulations in the energy sector, organized by the three pillars of the SEforAll[13] initiative: energy access, energy efficiency, and renewable energy. Arguably the RISE framework employs a one-size-fits-all approach [71]; however, comparing different African countries with similar

[13] https://www.seforall.org/.

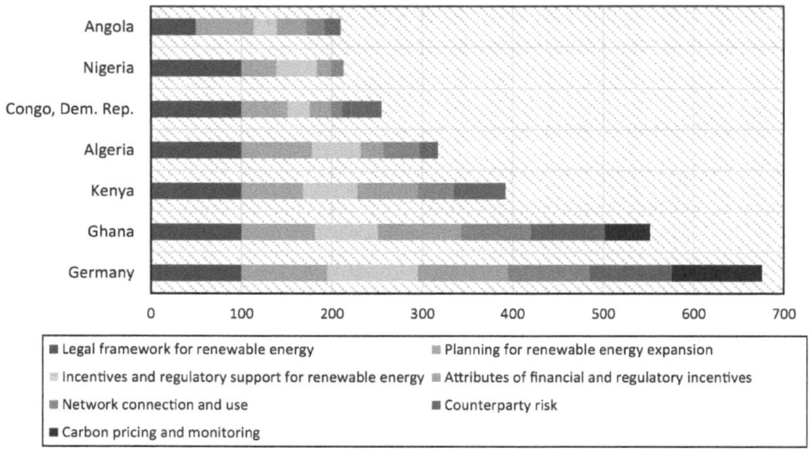

Figure 6.5 2017 RISE (renewable energy indicator) scores for select countries.

microeconomics and challenges as shown in Fig. 6.5 provides better insights on how the future of RE can be shaped especially solar.RISE uses a variety of indicators for renewable energy, energy access and energy efficiency offering a deep dive into several countries. Our focus in this chapter is to review the indicators for renewable energy for the countries of focus and setting as a benchmark lighthouse countries in this sector. In using this framework we reveal the bottom four indicators for the countries of study; carbon pricing and monitoring, network connection and use, attributes of financial and regulatory incentives, and counterparty risk (Fig. 6.5). These indicators and their impact on the rate of deployment of solar PV are discussed in the subsequent sub chapters (section 6.4.1 through 6.4.4).

6.4.1 Carbon pricing and monitoring

Carbon pricing is emerging as one way of internalizing the societal cost of emissions. This works based on the rationale that pricing carbon incentivizes various economic actors to seek alternative options [71]. Such policies applied to economies like Algeria or Angola with high and increasing carbon intensity, respectively, could serve as the needed push to rapidly deploy low-carbon alternatives like solar. However, the four countries presented in this study are yet to implement any policy framework in support of carbon tax or emissions trading, and lack the requisite monitoring and verification system for greenhouse gas emissions. Carbon pricing functions based on the assumption that consumers respond to price signals, which may not necessarily be the case, although there can be no certainty in these oil dependent economies except some measures are taken. In addition, such a policy rides on the back of high political trust and low corruption levels.

6.4.2 Network connection and use

This indicator measures the characteristics of connection and cost allocation. A lack of investment in grid infrastructure can limit solar development. Grid integration and variable RE dispatch capability are important considerations for commercial investors regardless of the market size. Support schemes that establish quality assurance standards for forecasting and measuring the variability of solar energy are needed to support industry growth and boost investor confidence. For example, the Kenyan solar PV and Solar Thermal Regulations Act 2012 mandates the energy regulatory commission to regulate the production, distribution, supply, and use of RE. The policy further details the requirements for authorizing and registering solar PV systems technicians, manufacturers, importers, vendors, and contractors, including the design, installation, and setup as well as repair and maintenance procedures. This helped to streamline different processes in the use of PV technologies, setting qualifications for jobs, standard setting, and making PV technologies safer for everyone thus giving it a score at par with Algeria, while Angola and Nigeria have scores below 20 according to the RISE framework depicted in Fig. 6.5.

6.4.3 Attributes of financial and regulatory incentives

Reforms that create strong regulatory and legal frameworks and financially sustainable power utilities are essential. Financial incentives for deploying solar technology could come in the form of income tax incentive, cash, sales tax incentive, or property tax incentives. These incentives target different segments of the value chain, from manufacturing to consumer, and are generally administered differently. Some are upfront and others are accrued over a period. A study conducted in the United States found that provision of cash incentives like rebates and rants increased the rate of deployment of grid-tied PV [73].

6.4.3.1 Fixed tariffs for small producers

This policy instrument is used to catalyze the generation and integration of renewables in the country's total energy consumption, encourage local distributed generation thereby reducing pressure on the network and losses, and reduce GHG emissions through the reduction of over reliance on unclean energy sources. In the absence of fixed feed-in tariffs for small producers, there is no incentive to generate electricity from solar modules. Angola has set ambitious plans for 2025 to reach 800 MW RE generation capacity (excluding hydropower) to achieve this, an appropriate feed-in mechanism needs to be explored, including ensuring the legal right for independent power production and sale, establishing fast, low cost, and uniform licensing and permission for minigrids. In Nigeria the feed-in tariff is not underwritten by the government-backed bulk electricity trader (NBET) but is directly negotiated with, and paid by, one of Nigeria's 11 distribution companies, depending on the project's location. This is offered only to generation in excess of 1 MW and

Table 6.3 Kenya's FiT structure for projects below 10 MW [74].

Technology	Installed capacity (MW)	Standard fit (US$/kWh)	Minimum capacity (MW)	Maximum capacity (MW)
Solar (grid)	0.5–10	0.10	0.5	10
Solar (grid)	10.1–40	0.12	10.1	40
Solar (off-grid)	0.5–10	0.12	0.5	10

with a limit of 5 MW for PV projects.[14] Kenya is the only country of the four evaluated countries which specifies FiTs (Table 6.3) for capacities less than 1 MW.

6.4.3.2 Auctions

African countries face several challenges with attracting investment, while feed-in tariffs have proven successful in some countries, another emerging tool for pooling in cost-reflective investment in the solar industry is auctions. The use of auctions can provide evidence for the true cost of electricity through solar to help stimulate further deployment. Governments (especially in the Middle East) have been holding competitive bidding processes when launching contracts to develop new power plants as seen in Saudi Arabia and UAE[15] for reducing costs of utility-scale PV power generation. An efficient bidding process and high-quality contractual documents that reflect a deep understanding of the private sector's perspective helped ensure a high-response rate to the tenders.

Regardless of historical challenges to its implementation in Africa, evidence from South Africa and now Zambia[16] reflect this as an effective policy tool for creating price competition between bidders.[17] Achieving such competitive processes requires trust and transparency. Establishing bankable standardized transaction agreements, effective tender implementation, and ensuring steadfast commitment of the national government and utility can yield low tariffs with both large corporations and small businesses emerging as successful bidders in the process. Several attempts have been made to have auctions in Nigeria, but internal disagreements on the price between different stakeholders have hindered the process.

6.4.4 Counterparty risk

This indicator is used to address factors such as utility, transparency, creditworthiness, and payment-risk mitigation. Due to poor network connections, RE

[14] Nigeria's FiT http://global-climatescope.org/policies/4097.

[15] https://www.pv-magazine.com/2019/10/10/dubais-900-mw-solar-tender-sees-lowest-bid-of-0-0169-kwh/.

[16] https://www.scalingsolar.org/active-engagements/zambia/.

[17] http://taiyangnews.info/markets/zambia-awards-120-mw-under-get-fit-auction/.

independent power producers (IPPs) often face the need to assess whether they can operate at partial capacity or bear the full cost of grid expansion and interconnection. A delay in this process prevents RE power generators from interconnecting with the national grid.[18] Instruments such as a risk guarantee that ensures off-taker's payment obligation or the partial credit guarantee to tackle currency transfer and convertibility risk caused by host government action can be used. The upfront capital-intensive nature of solar PV projects, especially at utility scale, makes it very sensitive to high interest rates. The interest rates demanded by lenders can significantly increase the overall cost. Hence, with the support of governments, development financing institutions can offer guarantees depending on the details of each project.

The other challenge to solar development is the transparency of utility and revenue collection rates. Governments are often challenged with difficult choices between how best to balance strategies that promote investment and energy access, while also ensuring that electricity is affordable. Until recently Angola had the lowest electricity cost which significantly increased public expenditure; however, the adoption of 14% VAT tax to replace the Consumption Tax in October 2019 and a cut on subsidies has led to a readjustment of power prices, the increment varies for different categories of consumers.[19]

6.5 Outlook and recommendations

Based on the assessment of electricity access improvements, carbon emissions growth, and regulatory indicators for sustainable energy growth, Angola show the least signs of growth, with Nigeria following closely. While Algeria may not have electricity access concerns, there are strong questions about its carbon intensive economy, which can be significantly lowered by utilizing the solar potential. These countries must build the capacity of key institutional stakeholders to face challenges associated with the scale-up of solar power to facilitate the shaping of the future energy landscape. Through an extensive literature review and data analyses presented in this work, three areas that can facilitate economic growth through solar PV deployment are summarized next.

6.5.1 Off-Grid photovoltaics for improved electricity access

The off-grid industry as proven by Kenya in Fig. 6.6 holds one of the keys for improving electricity access. Despite the significant addition to the RE capacity of Angola. The massive gains in electricity access by Kenya was achieved by tapping into decentralized energy systems in low-demand areas. This could prove to be a

[18] https://www.reuters.com/article/safrica-renewables/south-africas-power-crisis-risks-choking-green-energy-drive-idUSL5N0W61SY20150305.

[19] New electricity tariff Angola https://macauhub.com.mo/2019/07/04/pt-preco-da-energia-electrica-para-a-industria-em-angola-aumenta-8199/.

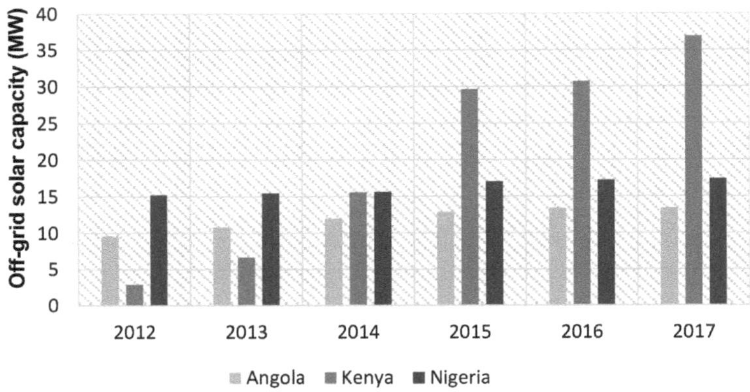

Figure 6.6 Off-grid solar capacity in select African countries.

beneficial strategy for Angola, considering a majority of the unelectrified population are in rural areas. Strategies to increase electrification and economic diversification can be enforced by the growth of the off-grid solar industry. The off-grid market growth in Kenya, like most of East Africa is borne out of creative mobile solutions and driven by the lack of subsidies. Kenya has been able to grow off-grid solar access due to simplification of payment services through mobile money and the integration of smart meters. In Nigeria poor mobile coverage in remote areas is a barrier to the growth of this industry as it hinders the penetration of mobile financing platforms.

Governments of fossil-dependent countries can leverage this industry by channeling in-kind transfers [61] of prior fuel subsidy regimes to the development of off-grid PV. These in-kind transfers can take different forms but are targeted toward the most vulnerable[20] in the society. By communicating these initiatives the government can garner some goodwill from its citizens. Examples include the distribution of energy-efficient bulbs by Ghana with compact fluorescent lamps (CFL) at the time or instituting a tax-free repayment on purchased solar home systems [62] thereby guaranteeing households a fixed-energy cost within the payment period as a substitute for kerosene lamps or diesel generators.

Further growth of off-grid PV can be ensured through provision of information and increased consumer awareness. People are more willing to pay for services they understand—a typical example is the rate at which clean cook stoves are abandoned for the more harmful alternatives [75,76].

6.5.2 Financing solar projects and private sector participation

Financing solar projects is a major hindrance to its growth especially in the unstable political environments [77] of Algeria, Angola, and Nigeria. These

[20] "Groups whose access to energy services needs to be preserved through appropriate transitional policies"—G20 St. Petersburg.

countries need to build the capacity of key institutional stakeholders to effectively face challenges associated with the scale-up of solar power. In general, predictable policies and a clear indication of stable government strategies and vision build market confidence. Financing plays a much larger role in solar projects than in fossil fuel projects because of the high initial costs [71].

Nigeria's NEP looks at crowding in private sector investment, that is to increase private sector investment by reducing risks, an initial assessment indicates a segmentation of interests from different private sector actors. First, historical challenges of low revenue collection rates that has plagued the electricity sector, engenders a level of cautiousness, with some financing institutions willing to enter the market at scale, while others take the wait-and-see approach.[21]

Second, the market price of solar products is still high for the most energy-disadvantaged; hence, official development assistance needs to be channeled to provision of energy services and market creation. Third, to improve private sector participation investment policies such as the 51/49 ownership policy of Algeria [78] need to be revised to allow for foreign direct investments. Though there are efforts being made to remove this policy, it may only apply to "nonstrategic sectors" in the country.[22]

6.5.3 Market creation

Internal markets can help stimulate economic growth and speed up the adoption of PV technologies. In the absence of markets, there is no real perceived benefit for new electricity connections. Policy measures such as fair market pricing, that are market enabling and inclusive can play a significant role in encouraging investment in the local production and distribution of electricity. The presence of international energy companies has historically enabled export value chains but adopting policies for PPPs can stimulate the growth of profit oriented local markets. Furthermore, adopting a wide range of business models such as pay-as-you-go model and fee-for-service models can support economic growth by opening new markets in hard-to-reach areas as they combine with more services in their delivery model.

6.6 Conclusion

Diversification of the energy mix of a country is important not only in ensuring sustained economic growth but also to build resilience and efficiently harnessing the resources of the sun offers these benefits.There are several considerations that shape a country's energy priorities, including its climate ambitions, energy security need, or improving the lives of its citizens economically. Despite promising technology and market trends, today's policies are slow in accounting for the transformative

[21] https://blogs.worldbank.org/ppps/what-public-sectors-role-mobilizing-private-finance-solar-energy.

[22] https://www.mees.com/2019/9/13/geopolitical-risk/algeria-looks-to-ease-5149-rule/543d9780-d63d-11e9-b01c-b91c29b53db9.

potential of solar off-grid and minigrid solutions to deliver clean energy access. They do not incorporate the possibly huge social and economic benefits to be derived from renewable electricity access and clean cooking. Domestic leadership, policy reforms, and capacity must lay the foundation for more effective public investment and to facilitate private investment. In addition to the economic gains to be made from improving electricity access, energy access, and investing in sustainable fuel options. There are also cobenefits for health and culture. The arguments presented in this chapter show that the significantly large economies of oil producing countries in Africa have been fueled by the oil boom and will be faced by a struggle-or-adapt scenario. This scenario would be determined by the ability to adopt progressive reform policies such as cost-reflective tariffing, removal of subsidies, and liberalized electricity markets. Historical arguments for subsidy removal have been the welfare of the poor, but through transitional policies such as in-kind transfers directed at solar incentives potential welfare damage can be averted. Given the intimate relationship government has with the energy sector in most resource-rich countries, it is essential to detangle this bond with more decentralized solutions and ensure that the oversight role taken by governments does not increase the bottlenecks in the process.

References

[1] W. Steffen, W. Broadgate, L. Deutsch, O. Gaffney, C. Ludwig, The trajectory of the anthropocene: the great acceleration, Anthr. Rev. 2 (1) (2015) 81–98.
[2] W. Steffen, et al., Trajectories of the earth system in the anthropocene, Proc. Natl. Acad. Sci. 115 (33) (2018) 8252–8259.
[3] W. Steffen, L. Hughes, A. Pearce, Climate Change 2015: Growing Risks, Critical Choices. Climate Council of Australia, Sydney, 2015. 76 p.
[4] IPCC, 2014: Climate Change 2014: Mitigation of Climate Change. Contribution of Working Group III to the Fifth Assessment Report of the Intergovernmental Panel on Climate Change [Edenhofer, O., R. Pichs-Madruga, Y. Sokona, E. Farahani, S. Kadner, K. Seyboth, A. Adler, I. Baum, S. Brunner, P. Eickemeier, B. Kriemann, J. Savolainen, S. Schlömer, C. von Stechow, T. Zwickel and J.C. Minx (eds.)]. Cambridge University Press, Cambridge, United Kingdom and New York, NY, USA.
[5] H.J. Schellnhuber, S. Rahmstorf, R. Winkelmann, Why the right climate target was agreed in Paris, Nat. Clim. Change 6 (7) (2016) 649.
[6] R. Muceka, T. Kukeera, Y. Alokore, K. Noara, S. Groh, Integrating a solar PV system with a household based backup, in: Presented at the Africa-EU Renewable Energy Research and Innovation Symposium 2018 (RERIS 2018), January 23–26, 2018, National University of Lesotho on Occasion of NULISTICE 2018, 2018, p. 43.
[7] V.R. Barros, C.B. Field, Climate Change 2014: Impacts, Adaptation, and Vulnerability. Part B: Regional Aspects, Cambridge University Press, Cambridge, 2014.
[8] J.A. Mathews, S. Kidney, Financing climate-friendly energy development through bonds, Dev. South. Afr. 29 (2) (2012) 337–349.
[9] Y. Mulugetta, F. Urban, Deliberating on low carbon development, Energy Policy 38 (12) (2010) 7546–7549.

[10] H. Winkler, A. Marquand, Changing development paths: from an energy-intensive to low-carbon economy in South Africa, Clim. Dev. 1 (1) (2009) 47–65.
[11] A.E. Camacho, Adapting governance to climate change: managing uncertainty through a learning infrastructure, Emory LJ. 59 (2009) 1.
[12] I. Ackah, R. Kizys, Green growth in oil producing African countries: a panel data analysis of renewable energy demand, Renew. Sustain. Energy Rev. 50 (2015) 1157–1166.
[13] IRENA, Global Energy Transformation: A Roadmap to 2050, 2019 ed., International Renewable Energy Agency,Abu Dhabi, 2019.
[14] P.A. Trotter, S. Abdullah, Re-focusing foreign involvement in sub-Saharan Africa's power sector on sustainable development, Energy Sustain. Dev. 44 (2018) 139–146.
[15] J. Friedrichs, Global energy crunch: how different parts of the world would react to a peak oil scenario, Energy Policy 38 (8) (2010) 4562–4569.
[16] A.S. Oyewo, A. Aghahosseini, D. Bogdanov, C. Breyer, Pathways to a fully sustainable electricity supply for Nigeria in the mid-term future, Energy Convers. Manage. 178 (2018) 44–64. Available from: https://doi.org/10.1016/j.enconman.2018.10.036.
[17] G.C. Wu, et al., Strategic siting and regional grid interconnections key to low-carbon futures in African countries, Proc. Natl. Acad. Sci. (2017). Available from: https://doi.org/10.1073/pnas.1611845114. 201611845.
[18] ESMAP, Tracking SDG7: The Energy Progress Report 2019, 2019.
[19] EIA, International Energy Statistics. <https://www.eia.gov/beta/international/data/browser/#/?pa = 0000000000000000000008000000000000002&c = ruvvvvvfvtvnvv1urvvvvfvvvvvvfvvvo-u20evvvvvvvvvnvvuvo&ct = 0&tl_id = 5-A&vs = INTL.57-6-AFG-BB.A&cy = 2018&vo = 0&v = H&s = INTL.57-6-NGA-BB.A~ ~INTL.57-6-DZA-BB.A~ ~INTL.57-6-AGO-BB.A~ ~INTL.57-6-LBY-BB.A~ ~INTL.57-1-DZA-TBPD.A~ ~INTL.57-1-AGO-TBPD.A~ ~INTL.57-1-NGA-TBPD.A~ ~INTL.57-1-LBY-TBPD.A>, 2019 (accessed 14.11.19).
[20] W. Adzawla, M. Sawaneh, A.M. Yusuf, Greenhouse gasses emission and economic growth nexus of sub-Saharan Africa, Sci. Afr. 3 (2019) e00065.
[21] O. Serdeczny, et al., Climate change impacts in Sub-Saharan Africa: from physical changes to their social repercussions, Reg. Environ. Change 17 (6) (2017) 1585–1600.
[22] ICTSD, Africa's Energy Transition: Opportunities and Challenges for Decent Work. <http://www.ictsd.org/bridges-news/bridges-africa/news/africa%E2%80%99s-energy-transition-opportunities-and-challenges-for-decent>, 2018 (accessed 21.11.19).
[23] I. Ackah, O. Alabi, A. Lartey, Analysing the efficiency of renewable energy consumption among oil-producing African countries, OPEC Energy Rev. 40 (3) (2016) 316–334.
[24] A.K. Aliyu, B. Modu, C.W. Tan, A review of renewable energy development in Africa: a focus in South Africa, Egypt and Nigeria, Renew. Sustain. Energy Rev. 81 (2018) 2502–2518.
[25] G. Schwerhoff, M. Sy, Developing Africa's energy mix, Clim. Policy 19 (1) (2019) 108–124.
[26] P.A. Trotter, M.C. McManus, R. Maconachie, Electricity planning and implementation in sub-Saharan Africa: a systematic review, Renew. Sustain. Energy Rev. 74 (2017) 1189–1209.
[27] B. Fattouh, R. Poudineh, R. West, The rise of renewables and energy transition: what adaptation strategy for oil companies and oil-exporting countries? Oxford Institute for Energy Studies, 2018.
[28] IEA, Africa Energy Outlook 2019, International Energy Agency, November 2019.
[29] Africa Progress Panel, Power People Planet: Seizing Africa's Energy and Climate Opportunities: Africa Progress Report 2015, 2015.

[30] IEA, CO_2 Emissions From Fuel Combustion, 2018 ed., OECD/IEA, Paris, 2018.
[31] Z. Abada, M. Bouharkat, Study of management strategy of energy resources in Algeria, Energy Rep. 4 (2018) 1–7.
[32] S. Haddoum, H. Bennour, T.A. Zaïd, Algerian energy policy: perspectives, barriers, and missed opportunities, Glob. Chall. 2 (8) (2018) 1700134.
[33] A. Bouraiou, et al., Status of renewable energy potential and utilization in Algeria, J. Clean. Prod. (2019) 119011.
[34] IMF, Algeria 2016 Article IV Consultation—Press Release and Staff Report, Washington, DC, 2016.
[35] Kenya National Bureau of Statistics, Economic Survey 2018, 2018.
[36] World Bank, IEA, IRENA, UNSD, and WHO, Tracking SDG 7: The Energy Progress Report, 2019.
[37] Energy Regulatory Commission, Updated Least Cost Development Plan Study Period: 2017–2037, 2018.
[38] J.K. Kiplagat, R.Z. Wang, T.X. Li, Renewable energy in Kenya: resource potential and status of exploitation, Renew. Sustain. Energy Rev. 15 (6) (2011) 2960–2973.
[39] K. Mokveld, S. von Eije, Final Energy Report, 2018.
[40] World Bank, Global Tracking Framework, World Bank, Washington, DC, 2015.
[41] A. Jerome, S. Adjibolosoo, D. Busari, Addressing oil related corruption in Africa: is the push for transparency enough? Rev. Hum. Factor. Stud. 11 (1) (2005).
[42] N. Shaxson, Oil, corruption and the resource curse, Int. Aff. 83 (6) (2007) 1123–1140.
[43] S. Usman, The opacity and conduit of corruption in the Nigeria oil sector: beyond the rhetoric of the anti-corruption crusade, J. Sustain. Dev. Afr. 13 (2) (2011) 294.
[44] M. Vagliasindi, Implementing Energy Subsidy Reforms: Evidence From Developing Countries, The World Bank, 2012.
[45] IMF, Nigeria: 2019 article IV consultation-press release; staff report; and statement by the executive director for Nigeria, IMF. <https://www.imf.org/en/Publications/CR/Issues/2019/04/01/Nigeria-2019-Article-IV-Consultation-Press-Release-Staff-Report-and-Statement-by-the-46726>, 2019 (accessed 25.11.19).
[46] N.L. Lam, E. Wallach, C.-W. Hsu, A. Jacobson , P. Alstone, P. Purohit, Z. Klimont, The Dirty Footprint of the Broken Grid: The Impacts of Fossil Fuel Back-up Generators in Developing Countries. International Finance Corporation, Washington D.C., USA, 2019.
[47] M.O. Oseni, Get rid of it: to what extent might improved reliability reduce self-generation in Nigeria? Energy Policy 93 (2016) 246–254.
[48] UNEP, ATLAS of Africa Energy Resources, United Nations Environment Programme, 2017.
[49] M.O. Oseni, Improving households' access to electricity and energy consumption pattern in Nigeria: renewable energy alternative, Renew. Sustain. Energy Rev. 16 (6) (2012) 3967–3974.
[50] ODI, Accelerating Access to Electricity in Africa With Off-Grid Solar Off-Grid Solar Country Briefing: Nigeria, Overseas Development Institute (ODI), United Kingdom, 2016.
[51] S.K. Jensen, Angola's Infrastructure Ambitions Through Booms and Busts, *Chatham House*, 2018, p. 40.
[52] AFREC, Africa Energy Database, Algiers, 2015.
[53] SE4ALL, Rapid Assessment and Gap Analysis—Angola, 2015.
[54] International Finance Corporation, Creating Markets in Angola: Opportunities for Development Through the Private Sector, World Bank, 2019.

[55] Government of Angola, Presidential Decree No. 141/18 of 7 June 2018, 2018.
[56] Paris Climate Agreement targets https://unfccc.int/files/essential_background/convention/application/pdf/english_paris_agreement.pdf)
[57] C. David, S. Baoping, W.H.P. Ian, S. Louis, How Large Are Global Energy Subsidies? International Monetary Fund, May 2015.
[58] IEA, Fossil-fuel subsidies. <https://www.iea.org/weo/energysubsidies/>, 2019 (accessed 01.11.19).
[59] M.B.J. Clements, D. Coady, M.S. Fabrizio, M.S. Gupta, M.T.S.C. Alleyne, M.C.A. Sdralevich, Energy Subsidy Reform: Lessons and Implications, International Monetary Fund, 2013.
[60] F.J.A. del Granado, D. Coady, R. Gillingham, The unequal benefits of fuel subsidies: a review of evidence for developing countries, World Dev. 40 (11) (2012) 2234–2248.
[61] C.R. Laderchi, Transitional policies to assist the poor while phasing out inefficient fossil fuel subsidies that encourage wasteful consumption, in: Contribution World Bank G20 Finance Ministry Central Bank of Government, 2014.
[62] T. Laan, C. Beaton, Strategies for reforming fossil-fuel subsidies: practical lessons from Ghana, France and Senegal, Fr. Senegal, April 26, 2010.
[63] C. Ramstein, et al., State and Trends of Carbon Pricing 2019, The World Bank, 2019.
[64] IRENA, Solar PV in Africa: Costs and Markets, 2016.
[65] G.R. Timilsina, L. Kurdgelashvili, P.A. Narbel, Solar energy: markets, economics and policies, Renew. Sustain. Energy Rev. 16 (1) (2012) 449–465. Available from: https://doi.org/10.1016/j.rser.2011.08.009.
[66] M. Amaza, Scaling mini-grids: what African countries can learn from each other, Power for all. <https://www.powerforall.org/news-media/articles/scaling-mini-grids-what-african-countries-can-learn-each-other>, 2019 (accessed 23.01.19).
[67] B.J. Cook, Arenas of power in climate change policymaking, Policy Stud. J. 38 (3) (2010) 465–486.
[68] G. Kavlak, J. McNerney, J.E. Trancik, Evaluating the causes of cost reduction in photovoltaic modules, Energy Policy 123 (2018) 700–710.
[69] S. Hallegatte, J. Rentschler, J. Rozenberg, Lifelines, World Bank, Washington, DC, 2019.
[70] S.G. Banerjee, F.A. Moreno, J. Sinton, T. Primiani, J. Seong, Regulatory Indicators for Sustainable Energy, The World Bank, 2017.
[71] J. Urpelainen, RISE to the occasion? A critique of the World Bank's regulatory indicators for sustainable energy, Energy Res. Soc. Sci. 39 (2018) 69–73. Available from: https://doi.org/10.1016/j.erss.2017.10.034.
[72] D. Klenert, et al., Making carbon pricing work for citizens, Nat. Clim. Change 8 (8) (2018) 669–677. Available from: https://doi.org/10.1038/s41558-018-0201-2.
[73] A. Sarzynski, J. Larrieu, G. Shrimali, The impact of state financial incentives on market deployment of solar technology, Energy Policy 46 (2012) 550–557. Available from: https://doi.org/10.1016/j.enpol.2012.04.032.
[74] Energy and Petroleum Regulatory Authority, Renewable—renewable energy and energy efficiency regulations (public notice). <https://renewableenergy.go.ke/index.php/news/3>, 2012 (accessed 31.10.19).
[75] V. Vigolo, R. Sallaku, F. Testa, Drivers and barriers to clean cooking: a systematic literature review from a consumer behavior perspective, Sustainability 10 (11) (2018) 4322.

[76] M. Jürisoo, F. Lambe, M. Osborne, Beyond buying: the application of service design methodology to understand adoption of clean cookstoves in Kenya and Zambia, Energy Res. Soc. Sci. 39 (2018) 164–176.
[77] P.A. Trotter, R. Maconachie, M.C. McManus, Solar energy's potential to mitigate political risks: the case of an optimised Africa-wide network, Energy Policy 117 (2018) 108–126.
[78] Algeria, Complementary Finance Act (LFC), 2009.

Section VI

Geothermal Power Technology

An up-to-date perspective of geothermal power technology

7

Abel F. Hernández Ochoa, Alfonso Aragón Aguilar and José M. Franco Nava
National Institute of Electricity and Clean Energies, Cuernavaca, Mexico

7.1 Introduction

Geothermal energy may become one of the main energy resources in the future as it is a clean and renewable energy source in the long term if it is managed in a sustainable way. Geothermal power technology makes use of the heat available within the Earth's crust and the impact of its utilization on the environment is considerably low as compared to fossil fuel technologies. Geothermal energy is generally available worldwide, although its exploitation for electric power generation has traditionally taken place near tectonic or volcanic regions where hydrothermal systems generally exist within the first few kilometers below the terrestrial surface [1].

Because traditional hydrothermal developments for electricity generation have been constructed worldwide, this type of geothermal setting is emphasized. However, within the second half of the past century, the majority of geothermal fields of this type have already been identified and are currently being exploited. This is why new geothermal developments make use of medium or low enthalpy/temperature resources, which are more globally distributed.

Yet high enthalpy resources hosted in low permeability rocks are now being developed as engineered geothermal systems (EGS)—which are artificially created by generating fracture networks within this kind of geothermal target rock—aimed at electricity generation or the direct use of heat.

There is no one type of geothermal system; each system has unique characteristics, properties, and settings. Geothermal plays are defined as geological settings that are capable of holding potentially recoverable heat and can be classified in terms of their tectonic environment, heat source origin, and predominant process of heat transfer. According to Harvey et al. [2], geothermal plays may be cataloged as shown in Table 7.1. The most common type of geothermal system, globally speaking, corresponds to a convection heat transfer mechanism, as the conduction mechanism is only found in Europe, mainly in Germany. Nevertheless, more than one geothermal play type may exist in real geothermal systems.

In this chapter, the focus will mainly be placed on the technical issues of geothermal power technology. Yet geothermal exploration policies associated with geothermal projects are important issues that may help to assess the decision of getting involved in a geothermal project. The latter issues are beyond the scope of this

Sustainable Fuel Technologies Handbook. DOI: https://doi.org/10.1016/B978-0-12-822989-7.00008-1

Table 7.1 Classification of geothermal systems (as defined in [2]).

Geothermal play type	Geological controls	Geological setting
Convection-dominated		
Magmatic	Intrusion of different ages, hydrothermal	Active to extinct volcanic fields (convergent, divergent, transform faults, hot spots, plumes)
Extensional domain	Active faults, amagmatic, high porosity, high permeability	Active rifts, metamorphic core complexes, back-arc basins, segmented strike-slip faults
Conduction-dominated		
	Faults, fractures, lithofacies, diagenesis	Sedimentary basins, basement provinces, orogenic belts

chapter, but the reader may refer to Speer et al. [3], or bibliographical sources therein, to gain knowledge on these issues.

This chapter is conceptually divided into the broad aspects of geothermal power technology. First, the main aspects of the preliminary survey and exploratory stages for promising areas of geothermal interest, including the methods and technologies that are usually employed, are presented. Then, the phases of prefeasibility study and exploratory drilling are introduced in order to assess and reduce, respectively, the risk associated with a geothermal project.

Next, the phase of drilling production wells, which would allow a better assessment of the geothermal play potential and a better knowledge-based risk management, is explained. Concepts on project planning and field development are presented once the feasibility stage has had a positive outcome. Once a plant has been built up, monitoring of the reservoir during start-up should be carried out before proceeding to the operation of the geothermal technology plant.

Finally, the energy conversion technologies that are most widely used to obtain the best of geothermal resources, which include the most common designs of geothermal power plants, are presented and illustrated. A final section of conclusions end this chapter.

7.2 Preliminary survey stage

Any geothermal power development project must start with a preliminary survey of the regional area of interest. The geographical scale is generally regional, encompassing areas greater than 1000 km^2 [4]. The purpose of the preliminary survey is to find geological settings capable of containing geothermal systems that can be economically exploited. The preliminary study consists mainly of desktop work, including searching and analyzing the available data from geology, hydrology, thermal manifestations, drilling of water, or any other type of well, satellite imagery, mineral explorations as well as information collected from communities.

The technical desk work must be complemented with field work in order to verify important aspects of regional and structural geology as well as of thermal manifestations. Field work should include the investigation of possible environmental or social obstacles that may impede the development of the possible project.

The preliminary survey must be in accordance with the relevant regulations on land use, indigenous rights, the environment, and access to the communities of the country or province in which the area of interest is located. Excluded areas such as national parks, archaeological sites, and environment-protected, urban, wetland, or volcanic-risk areas, among others, are not suitable for geothermal development.

Regions where there are conflicts over land use can be considered, but it is advisable to resolve or avoid any possible difficulties from the start of the project before beginning exploration work. Nowadays, it is essential to have communication with communities and local authorities from the beginning of a project in order to let them know the positive and negative impacts of a possible geothermal development and listen to their ideas, interests, and opinions.

It is important to encourage the participation of communities, as far as possible, from the beginning of the plan, and to maintain continuous feedback with communities throughout the development of the project through meetings and surveys in addition to complying with the regulations of different levels of government. It is necessary that the developer foresees possible obstacles of a social or environmental nature in order to solve them as soon as possible.

The availability of land roads, water resources, and electric power lines should be investigated from the beginning of a project in order to include the associated costs in the absence of some of these.

Some countries have geothermal rights legislation, others lack specific regulations, while others consider the laws of mineral resources or water rights. The time required for obtaining exploration and exploitation permits depends on the type of legislation in the country in which the site is located. The exploration and exploitation rights of geothermal resources may be privately or publicly owned, in accordance with the laws that apply to the area under survey. This implies the payment of a rent or negotiations with the owner to have permission to exploit the resource.

In Table 7.2, the most important information that should be obtained during a preliminary survey is summarized [2].

The period of time required to carry out the preliminary survey can be anywhere from a few months to one or more years, depending on the size of the area, its geological/geothermal complexity, the technical information available, the number of communities in the area as well the ease/complexity of obtaining permits and financing.

The preliminary study must allow the concessionaire to determine the most likely geothermal play of the area of study as well as the geological, environmental, social, and legal aspects in which the project must operate. In the next stage of exploration, the objective is to minimize the critical gaps in geological information that the preliminary study has located. This will allow the explorer to develop a power plant legally and socially if the existence of a profitable geothermal resource is confirmed.

Table 7.2 Information to be collected in a geothermal preliminary survey.

Type of information	Information to be collected
Technical	• Bibliographic/purchasable topographic, geological, geophysical, regional heat flow, and other maps within the license area. • Bibliographic/purchasable data on geology, geophysics, geochemistry, isotopy, surface and subsurface temperatures from existing wells, and seismicity records. • Technical reports of geological, geophysical, hydrological, tectonic, seismic, and thermal manifestation studies as well as of well-logging of water, geophysical, or of any other type. • Purchasable data of satellite imagery, aerial photogrammetry, DEM, and LIDAR. • Map of land use restrictions.
Social/ environmental	• Possible needs of direct energy use in communities (i.e., district or greenhouse heating/cooling, etc.). • Environmental and social conditions.
Infrastructure	• Availability/lack of infrastructure (roads, water, telecommunications, and transmission lines).
Legal/regulatory	• Legal framework in the ownership of resources. • National and local laws and regulations.
Financial/political	• Power markets and power purchase agreements (PPA) or feed-in tariff. • Political and financial stability.

Note: *DEM*, Digital elevation model; *EC*, electrical conductivity; *GIS*, geographic information system; *LIDAR*, laser imaging detection and ranging.

7.3 Exploration stage

The most common geothermal resources can be categorized by type into (1) convective geothermal systems in volcanic and or tectonically active environments, (2) thermal fluids in deep sedimentary aquifers, and (3) low-permeability hot dry rock.

In the first two types, thermal fluids can transport energy to the surface; these are traditionally the most exploited systems. While the third type is the most widely available on a global scale, it lacks sufficient permeability and fluid to be exploited without a suitable technical design that allows for heat extraction. This is achieved by developing improved or EGS through the generation of interconnected fracture networks and an artificially or naturally injected working fluid that transports heat to a geothermal plant through one or more production wells [5].

Each geothermal system is unique and has internal characteristics and processes that make it different from others. Therefore there are no recipes for the exploration and exploitation of geothermal systems. Geothermal exploration should be based on understanding these characteristics and processes through field information.

As previously mentioned, most geothermal technology in exploration and exploitation that has been developed to date has been designed for the first two types of resources, and they are the focus of this chapter. However, many future advances in the development and innovation of technological tools and methodologies for prospecting and exploiting EGS resources can be expected [5,6].

The purpose of the exploration stage is to reduce the uncertainty related to the estimation of the main parameters of a geothermal system (temperature, depth, extension, permeability, etc.,), before going on to the exploratory drilling phase, through the collection of new geoscientific data in a profitable way [2,4,7,8].

Exploration generally begins at the regional level and progresses to smaller areas as the data indicate the most attractive spots. A common practice is to sample existing surface manifestations and wells, if any. Then, the surface and deeper layers of the area are studied using geological, geochemical, and geophysical methods. At this stage, the environmental studies that are carried out will constitute baseline information that will form part of the environmental impact studies required by the law in some countries.

Due to the high cost associated with drilling exploratory wells, a key goal of this stage is to maximize knowledge of the drilling target to the extent that the project budget allows. In Fig. 7.1, some of the most used techniques in the exploration

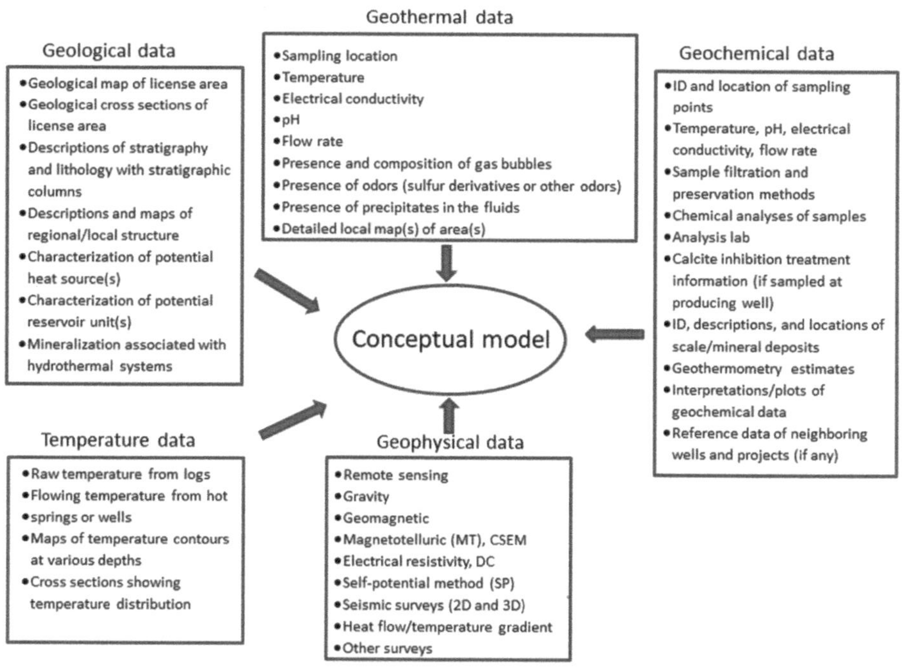

Figure 7.1 Applicable geoscientific techniques and potentially collectible datasets in the geothermal exploration stage.
Source: Modified from C. Harvey, G. Beardsmore, I. Moeck, H. Ruter, S. Bauer, Best Practices Guide for Geothermal Exploration (English), World Bank Group, Washington, DC, 2014. <http://documents.worldbank.org/curated/en/190071480069890732/Best-practices-guide-for-geothermal-exploration/> (accessed 27.04.20).

stage are presented. In the end, the collected field data must be analyzed, interpreted, and integrated to obtain a solid conceptual model of the potential geothermal system.

7.3.1 Geology

To understand the operation of a geothermal system, it is necessary to know the geology of the project area to well within the regional geological and tectonic framework. This allows for both the flow of the geothermal fluid, which moves through faults and permeable rocks, and the thermal anomalies indicated by the alteration of minerals to be estimated. Geological exploration can begin on a regional scale and then focus on favorable areas to identify the structural conduits through which geothermal fluids ascend.

It is important to infer the possible source of heat from a geothermal system, which could be related to active magmatic sources or to a large regional heat flux. A felsic composition of magma indicates possible shallow heat sources, while a mafic composition is related to deeper magma chambers [2]. The younger the igneous rocks (Pliocene or newer), the more likely they are to be associated with significant heat intrusion.

Geothermal manifestations are direct signs of subsurface hot water; whereas areas with no thermal manifestations, but which show evidence of their previous presence, are also of interest. Indicators of past geothermal activity include hot spring deposits, hydrothermally bleached or altered areas, and silica cementing in shallow deposits. Furthermore, mineral deposits and clay alteration zones are significant indicators of a geothermal system.

Using surface geology studies and well drilling data in the region of interest, the lithostratigraphic sequence in the area can be defined. Permeability is a key property of geothermal systems; thus, some lithologies have a greater potential to become the host rock of the reservoir, since they have a high primary and/or secondary permeability. The cap rocks, with low permeability, constitute the upper physical boundary of a geothermal reservoir and, thus, need to be mapped out.

Deep regional faults are important because they can serve as pathways or barriers for fluid flow as well as to establish or improve secondary permeability [2,4]. The local geological structure is also of key importance; geothermal systems are generally associated with structural maxima, and dipping units can transport deep geothermal fluids over significant distances.

As exploration activities begin to provide subsurface information, 2D cross sections can be drawn; this helps to visualize the stratigraphy and geological structures of the geothermal play. If available, data from existing wells can help to obtain the characteristics of geological layers. And as drilling data become available, a 3D geological model can be integrated using all available information (Fig. 7.2). These types of models, implemented in visualization platforms (e.g., [9]), can be useful in the planning of the drilling of wells. Nevertheless, it is important to build a 3D model in the early stages of a project.

Figure 7.2 Lithological surfaces and deep faults are shown in a 3D geological model.

At the end, the geological analysis must define the regional and local geology, stratigraphy, and structures of the area. The analysis should specify the potential units that might host a reservoir. Further data needs must also be pointed out for later studies.

7.3.2 Geochemistry

In the initial exploration stage, the interpretation of geothermometry is helpful in assessing the temperatures and size of the potential geothermal system. This gives an early signal as to whether the explored resource can sustain the construction of a power plant. Later, geochemistry focuses on determining the sources and pathways of geothermal fluids and evaluating operational problems that would accompany the development of the project (e.g., wellbore scaling, corrosion, and the accumulation of noncondensable gases).

Geochemical studies involve field work on geothermal features, which may include hot/mineral springs and seeps, geothermal/groundwater wells, geysers, mud pots, fumaroles, solfataras, gas seeps, and steaming ground [2,4]. Geothermal manifestations must be characterized by sampling and analyzing fluids, steam, and/or gases. Fluid samples must be preserved and analyzed for silica, cations, anions, and isotopes in water and sulfate.

The main geothermometers are grouped as silica, cation, and isotopic geothermometers, each depending on different premises. In general, hot springs with higher flow rates and temperatures provide the most useful and reliable results. Yet it is advisable to calculate and compare a set of silica, cation, and isotope geothermometers, so that an experienced geochemist can suggest the most appropriate geothermometer temperatures.

A novel advancement in regional geochemical exploration is carbon dioxide (CO_2) soil gas studies. Taking measurements of CO_2 fluxes and concentrations with a portable meter is quite practical and may corroborate the results of geological

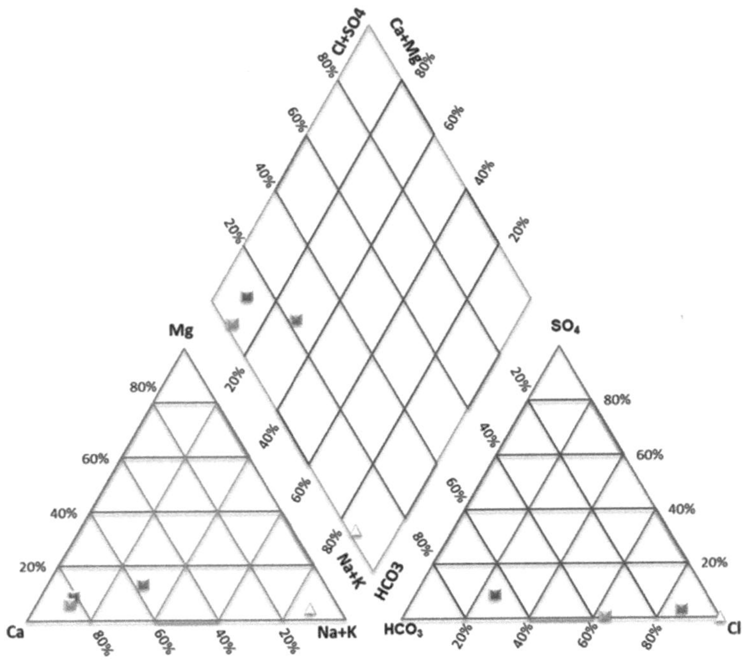

Figure 7.3 Plot of spring water samples on a Piper diagram.

surveys. Active faults can also be located through radon detection methods and radon isotope studies.

Geochemical data are usually presented on diagrams, graphs, drawings, maps, and tables (Fig. 7.3). To evaluate the characteristics of geothermal fluids, chemical parameters can be plotted against each other. Various types of plots can help to create mixture models, identify end members, and, thus, infer the evolution of geothermal fluid composition. Such charts can shed light on whether mixing, boiling, and/or dilution is occurring in the system. On the other hand, annex reports must interpret the data and propose conclusions. Table 7.3 suggests the content that an annex geochemical report should include [2].

Expected results of geochemistry studies would include the temperature distribution of the reservoir, a range for the maximum temperature in the system, a notion on the mixing of fluids, and a list of information that should be covered in the ensuing phases of exploration.

7.3.3 Geophysics

Geophysical exploration is an essential part of a geothermal project. Its studies allow for important properties of rocks and fluids as well as the existence and shape of reservoirs and permeable pathways to be deduced, with reasonable certainty, before drilling. Choosing the most suitable and profitable geophysical techniques

Table 7.3 Suggested content of a geochemical data report.

Maps	Sample locations with local/assigned names for the geothermal features Contours showing sample points and their soil survey data values as well as tables with locations, values, and characteristics of the sample points
Tables	Summary of fluid geochemistry of the sampled locations, including field parameters (location, temperature, EC, pH, flow rate, gas bubbles, odors, precipitates) Summary of gas geochemistry of the sampled geothermal locations, including: • Field parameters, namely location, temperature, flow rate, odors • Geochemical analyses of NH_3, H_2S, CO_2, CH_4, H_2, N_2, Ar, He, SO_2, HCl, HF, O_2. • Also, $^3He/^4He$, $^{40}Ar/^{36}Ar$, noble gas concentrations and ratios, and stable isotopes in steam condensate; should include standard deviation of each sample and/or other evidence of quality control on analyses • Total flux and makeup of noncondensable gases for any production well Geothermometry results, including silica (quartz, chalcedony, and amorphous glass), cation (Na-K-Ca, Na-K-Ca-Mg, Na/K, K-Mg), and sulfate water isotope (^{18}O)
Graphs	Piper diagrams K (mg/L) versus Na (mg/L) δD versus δ18O Ternary plot of the major anions (SO4-HCO3-Cl) Ternary plot of Na-K-Mg, including scales for the Na/K and K-Mg geothermometers Na-K-Ca geothermometer temperature versus Cl concentration Temperature of Na/K/Ca (°C) versus temperature of K/Mg (°C) Discharge temperature (°C) versus Cl (mg/L) A ternary plot of N2, CO2/100, and 100*Ar Giggenbach gas ratio grids, including H2/Ar versus CO2/Ar, H2/Ar versus T, CH4/CO2 versus CO/CO2, CO/CO2 versus H2/Ar
Inferences and conclusions	Estimated reservoir temperature at depth Origin of geothermal fluids Locations of aquifers and reservoirs (in 2D or 3D) Mixing between aquifers Sources of recharge to the geothermal system Pathways of discharge from the geothermal system Potential for corrosion and/or scaling by the geothermal fluids

Source: Modified from C. Harvey, G. Beardsmore, I. Moeck, H. Ruter, S. Bauer. Best Practices Guide for Geothermal Exploration (English). Washington, DC: World Bank Group, 2014. <http://documents.worldbank.org/curated/en/190071480069890732/Best-practices-guide-for-geothermal-exploration/> (accessed 27.04.20).

requires the knowledge of experienced geothermal scientists and depends primarily on the type of geothermal play.

Many types of geophysical surveys are available to choose from. These include gravity and magnetic studies; electrical and electromagnetic resistivity surveys, particularly magnetotelluric or controlled source electromagnetic methods; and active/passive seismic techniques, among others. Heat flow surveys include shallow temperature gradient drilling.

Methods of geophysical exploration are presented in Table 7.4, which lists the physical principles, the main equipment, the characteristics, and the expected results for each method. The content of Table 7.4 is based on the work of Harvey et al. [2] as well as on documents by Bracke et al. [8], Coolbaugh et al. [10], and OLADE/BID [4].

7.3.4 Lab measurements on rock samples

When samples of the rock units in the area (from drilled bores or outcrops) can be analyzed in a lab to determine their properties, all the geological, geochemical, and geophysical surveys can be better interpreted. Several properties, among others, could be measured for each sample in the lab, namely density, porosity, permeability, seismic wave velocities, thermal conductivity, rock strength, electrical resistivity, magnetic susceptibility, remnant magnetism, mineralogy, structural fabric, radiogenic element (K, U, Th) concentrations, oxygen isotope composition, geochemical assays, and fluid inclusion studies.

Rock sample analyses are a low-cost way to greatly enhance geophysical surveys and a drilling program. Care should be taken to obtain fresh samples, whenever possible, for each rock unit. Weathered or mineralogically altered samples should be used for determining rock property values that will be used to constrain survey studies. Or else, if altered rocks are measured, one should keep in mind the implicit uncertainty of their estimated measurements.

7.3.5 Mechanical stress field

Principal mechanical stresses are the Cartesian, orthogonal components of the stress vector **S**: $S_1 > S_2 > S_3$. They affect the alignment and distribution of permeable pathways in geothermal reservoirs and govern the evolution of engineered systems in the course of hydraulic shearing. One of the components is vertical and the other two are horizontal. Their relative magnitudes determine the stress regime of the rock, including normal faulting, strike-slip faulting, reverse faulting, or a hybrid transitional state. An assessment of the magnitude/orientation of S_1, S_2, and S_3 can help anticipate zones of maximum fracture permeability. This should be estimated as soon as possible in the course of a project.

When measuring the mechanical stresses in the field, one should be aware of affecting factors such as neighboring faults, mechanical anisotropy in rocks, borehole location/orientation, and the employed technical measurement method. Moeck

Table 7.4 Geophysical methods and their characteristics [2,4,8,10].

Geophysical method	Physical basis	Equipment, processing, and characteristics	Expected results
Gravimetric	Small variations in the Earth's gravitational force due mainly to bulk density changes of the rock sequence beneath the survey site.	Spring-based or high-precision superconducting gravimeters. Data are collected at specific surface sites or along a path during airborne surveys. A differential GPS is used to record the location and precise elevation. The gravimetric method is relatively inexpensive.	Reliable Bouguer map showing gravitational field strength variations that can be related to geological features such as geometry and depth of rock units or faults.
Magnetic	Spatial changes in the magnetic properties of near-surface rocks cause local variations in the strength/direction of the Earth's magnetic field. Geomagnetic surveys map these changes and highlight anomalies. Most variations in the magnetic field can be attributed to different concentrations of the minerals magnetite (Fe_3O_4) and maghemite (Fe_2O_3) in the near surface.	A magnetometer measures the orientation and strength of the magnetic field using proton-rich fluids surrounded by an electric coil. Surveys are often performed along paths during airborne surveys. Magnetic records from a static base station are used to correct the survey data for the drift of the magnetic field strength due to atmospheric effects. Airborne data must also be corrected for heading and level. This method is relatively inexpensive and simple to implement.	Reliable magnetic anomaly map revealing variations in magnetic properties that can be associated to geological features such as geometry and depth of intrusions, dikes, etc. Magnetic surveys are most appropriate for obtaining a general view of the underground.
Magnetotelluric (MT)	The method responds to the Earth's electrical resistivity structure by recording time series of natural, low frequency, electric/magnetic fields at the Earth's surface, then interpreting the data in the frequency domain. Variations in the Earth's magnetic field induce electric, telluric currents in the	The horizontal electric/magnetic fields at the Earth's surface are measured using electrodes and magnetometers buried in the ground. The magnetometers are induction coils for frequencies >0.01 Hz and of fluxgate type for lower frequencies. The shallow resistivity structure is estimated using	3D apparent resistivity block model derived from stacked 1D or 2D inversions (incorporating static shift corrections from TDEM data), or a full 3D inversion. The inversion model should indicate regions of contrasting electrical resistivity consistent with the

(*Continued*)

Table 7.4 (Continued)

Geophysical method	Physical basis	Equipment, processing, and characteristics	Expected results
	subsurface. Recording data over a wide frequency spectrum gives information on a great thickness of the ground; lower frequency records demand longer collection times.	higher frequency data, then the deeper resistivity structure is progressively calculated from the longer wavelength bulk resistivity data and the shallow resistivity estimates. MT interpretations become less precise at greater depths.	conceptual model of the geothermal system.
Controlled source electromagnetic (CSEM)	CSEM is the general term for electromagnetic (EM) sounding methods that use an active transmitter and antenna to generate EM source fields. The primary EM fields are generated by current waveforms passed through a loop or grounded dipole. These fields generate eddy currents in the ground, which induce secondary EM fields that can be measured at the surface using a receiver coil or receiver electrodes. The induced field depends on the electrical resistivity of the subsurface rocks.	For frequency domain methods, the transmitter produces a sinusoidal wave at a fixed frequency selected for the chosen depth of exploration lower (higher) frequencies for greater (shallower) depths. Some methods work in the time domain by recording the decline of a secondary EM field induced by a finite pulse. In most instruments, the current is transmitted as a periodic square wave. It is advisable to record multiple receiver measurements relative to the same transmitter station to enhance the signal-to-noise ratio. Upon processing, data are portrayed as apparent resistivity.	Interpreted anomaly maps and profiles and 2D/3D images of the interpreted structures. Modeling and inversion can be carried out in 1D, 2D, or 3D. The main advantage of CSEM methods over the MT method is that CSEM data acquisition is faster. But the depth of imaging for CSEM is generally shallower than that of MT.
Electrical resistivity (DC)	The resistivity of soils and rocks depends mainly on pore water content and its resistivity. In this method, the electrical field produced in the ground around a current electrode is a function of the electrical resistivity of the surrounding soils/rocks. An electrical current is	Electrical resistivity direct-current (DC) soundings also image the vertical apparent resistivity distribution of the underground. Surveys can be designed to generate 1D soundings or 2D profiles. Many different electrode configurations can be applied, including the	Interpreted anomaly maps and profiles and 2D/3D images of interpreted structures. Modeling and inversion can be carried out in 1D, 2D, or 3D.

(Continued)

Table 7.4 (Continued)

Geophysical method	Physical basis	Equipment, processing, and characteristics	Expected results
	generated between two electrodes in the ground and a voltage difference is measured between two other electrodes not connected to an active current.	dipole–dipole array, Schlumberger array, Wenner array, pole–pole array, and others. The exact array parameters need to be accurately known before a reliable inversion can be performed.	
Self-potential (SP)	Potentials occur naturally about different materials, near distinct concentrations of electrolytic solutions, and due to flowing fluids. Spontaneous potentials can be caused by mineralization gradients, electrochemical action, geothermal activity, and bioelectric generation of vegetation.	In SP surveys, electrodes are separated by a constant distance (5–10 m), and they are moved to a new location for each measurement. Alternately, one electrode stays in a fixed place and the other electrode is moved around so that voltages can be measured across the survey area This is a relatively low-cost method, but its results may be difficult to understand.	Interpreted anomaly maps and profiles of interpreted structures such as major boundaries or tracing faults. The SP method is used frequently in preliminary surveys, on mapping major boundaries, or in tracing faults.
Seismic	The seismic method detects differences in the acoustic impedance (the product of the density and seismic velocity of rocks). The concept of the seismic (reflection) method is to generate elastic waves from the source into the subsurface, where each layer surface reflects a part of the wave and permits the rest to refract through. The reflected waves are detected at the surface by a number of seismic devices (i.e., geophones) that record the motion of the ground.	Seismic reflection methods utilize induced vibrations or single sources (e.g., an explosive source) at the surface. Reflection seismology is one of the more costly geophysical methods and demands significant permitting works, broad fieldwork coordination, and sophisticated data processing. 3D data surveys can require crews of hundreds of workers depending on the survey size.	Interpreted cross sections or slices and cubes. 2D/3D images (detailed pictures) of interpreted structures such as faults and rock discontinuities, including their location and orientation. Interpreted sections usually depict the most important seismic reflectors (i.e., contacts between rock units) and faults.

(*Continued*)

Table 7.4 (Continued)

Geophysical method	Physical basis	Equipment, processing, and characteristics	Expected results
Passive seismic	The passive seismic method depends on natural tremors or earthquakes and volcanic events or other tectonic activity as sources within the geothermal field or from outside it. Seismometers are located on the ground surface or in shallow boreholes and the seismic signals are continuously recorded. When seismic waves travel through a layered/fractured medium, the shear component can split into two components. Shear wave splitting may provide additional value from passive seismic surveys.	The measured seismic components are usually processed using tomographic techniques in which source locations and velocity distribution are inverted jointly. The outcome is a 3D picture of the subsurface velocity distribution. Passive seismic surveys must be performed for a significantly longer time than reflection seismic studies to collect a similar amount of data. Yet the method is comparatively inexpensive, despite the fact that it requires instrument deployment over long enough periods to record a sufficient number of events for useful analysis.	3D block of the velocity distribution. Mapping the hypocenters of seismic events in seismically active areas has proved useful for identifying active faults. Processing of shear wave splitting may indicate potential routes of fluid flow within a geothermal reservoir.
CSEM	Shallow ground temperatures can be used to quantify the conductive heat flow. Atypical values of conductive heat flow may suggest a shallow temperature anomaly. Temperatures within the top 30 m of the ground are regularly influenced by daily and seasonal atmospheric temperature variations. The magnitude of the effect depends directly on the extent of the temperature cycle and decays exponentially with depth.	Mapping the relative ground temperature at a 2 m-depth can depict relatively hot areas identified with subsurface geothermal anomalies. If temperature mapping is surveyed jointly with a heat flow gradient well method, both techniques can decrease the number of temperature-gradient wells drilled in order to outline thermally anomalous zones. Ground temperature can be estimated with a sensor that is deployed at a 2 m depth using handheld hammer drills.	Maps or profiles of ground temperature. Higher levels of conductive heat flow may indicate a shallow subsurface anomaly or delineate the subsurface extent of an outflow zone. It may be used jointly with a heat flow gradient method to reduce the number of drilled boreholes or augment temperature data.

(*Continued*)

Table 7.4 (Continued)

Geophysical method	Physical basis	Equipment, processing, and characteristics	Expected results
Heat flow gradient	Heat flow drilling provides the most direct knowledge into the temperature distribution within a geothermal system. Heat flow wells are usually drilled down to less than 500 m deep, having a relatively slim diameter (up to 6–15 cm) compared to production wells. Differences in the shallow temperature field can lead to the determination of the deeper thermal structure. A significant difference between temperature surveys and gravity/magnetic surveys is that boreholes are needed to measure the temperature field.	The first goal of this method is to get temperature gradient and rock property information to enhance the certainty of temperature and depth predictions. The second goal is to validate geophysical survey data or to acquire additional geochemical data. Heat flow surveys are not invariably recommended or feasible. Its viability depends on the characterization of the geothermal system, the accessibility of a fitting drill rig and experienced drillers, and the rock properties underground. Slim holes may be drilled with small drill rigs, as commonly used for mineral or groundwater drilling.	Joint temperature mapping and heat flow surveys may provide approximated horizontal/vertical temperature distributions, an outline of convective zones, clues of fluid migration pathways, and an estimate of conductive heat loss from the survey area. This method, used in conjunction with geothermometry, can assess both the temperature and the depth limit of a reservoir.
Temperature contour maps and cross sections	Same as in the methods of temperature mapping and heat flow gradient.	Same as in the methods of temperature mapping and heat flow gradient.	Temperature data from different depths and locations can be gridded and presented as temperature-contour maps. Otherwise isotherms can be plotted on geological cross sections. Gridded maps of surface conductive heat flow may be helpful for conductive geothermal plays.

[11] suggests performing certain steps to assess the principal stress components in the exploration stage (before the drilling phase), namely:

1. Look for the regional maximum horizontal stress (S_H) orientation in the World Stress Map [12]. S_H could be either S_1 or S_2. Be aware that some regions lack information.
2. Build a geological structural model for your site. This should put limits on the possible stress regimes.
3. Estimate the depth of the reservoir and the rock density of overburden. This should help estimate the vertical stress component.
4. Estimate the density of reservoir fluid. This should allow for the assessment of the formation pressure and the effective stresses.
5. Use the concept of limited stress ratios (frictional constrains). This should help to estimate both horizontal stresses.

The confidence level for the estimations in points 1 through 5 goes in ascending order from lowest (point 1) to low (point 2), then up to medium confidence level for points 3 through 5. The method of Moeck [11] leads one to obtain the orientations of the principal stress components and then to define the stress regime.

Alternatively, the local stress regime might be attained from focal mechanism solutions for local earthquakes. Fault data from geological or seismic surveys may also point out the present stress regime in areas of active faulting. If existing boreholes are present near the area, then drilling-induced borehole breakouts and tensile fractures might indicate the directions of the horizontal stress components.

7.3.6 Conceptual model

As shown in Fig. 7.1, a conceptual model of a geothermal system contains the interpretation of the results of the preliminary survey and exploration stages of the project area. If the drilling phase is later undertaken, the conceptual model would be enriched iteratively every time new data become available from the important drilling phase as well as from subsequent exploration phases. Thus the conceptual model is always improvable, incorporating new details and confidence to the entire model.

The conceptual model should be supported by a solid integrated geological model; if possible, on a GIS database that consolidates and allows for the visualization of each type of data in layers. In this way, maps and sections can be generated more readily.

A satisfactory conceptual model should include the geoscientific system, surface geothermal manifestations, reservoir properties, fluid flow directions, spatial distribution of isotherms, heat source location, and a sketch of the reservoir, including its depth, temperature, and geometry. The conceptual model should honor all available data and information with enough certainty to defend and locate wells for the test drilling phase, targeting deep, full-diameter wells in the direction of lithological units and/or structures that are deemed most suitable to produce geothermal fluids at profitable rates and temperatures.

7.3.7 Numerical model

Having a reasonable knowledge of the geothermal system under study, one may construct a numerical model using the measured/inferred properties of the main

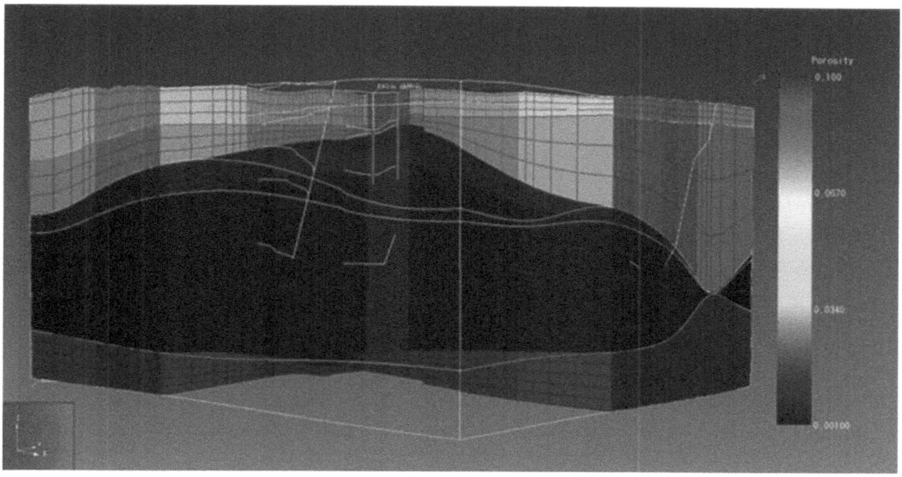

Figure 7.4 A numerical model with geological layers showing different porosity values.

rock units, primary and secondary permeabilities, migration fluid pathways, geothermal fluid and gas properties, and inferred heat source temperature/flux (Fig. 7.4). The numerical model should be constrained to reproduce measured temperatures, pressures, and outflow discharges [13].

As rock properties, pressure, and temperature data from the drilling phase become available, both the conceptual and numerical models will most likely undergo modifications. Early numerical simulations on the future performance of the reservoir may be carried out by considering different exploitation scenarios. This should help forecast the impact that different proposed geothermal exploitation schemes would have on the resource. If it is decided that the geothermal project is to continue, a sustainable production plan should be designed according to the available knowledge of the geothermal system.

7.4 Prefeasibility stage

The conceptual model of a geothermal system is the primary guide during a test drilling phase to aim deep full-diameter wells to lithological units and/or structures that are most likely to produce commercial rates of geothermal fluid. The conceptual model and nontechnical studies are assimilated into a prefeasibility report that incorporates a financial model (including an estimated power capacity of the planned power plant) to predict investment returns and justify the next phase, that is, the significantly higher level of funding required for the test drilling phase and well testing.

This decision should be made when the characterization of the reservoir is still uncertain and proceeding to test drilling means significant financial commitments for the project. Thus the prefeasibility study should recommend going for or against the project, taking into account all the relevant elements.

The objectives of the prefeasibility study are (1) to show good possibilities that the geothermal system can support economic power production, (2) to reduce the financial risks related to the test drilling phase, and (3) to establish a business case to seek financial support from public, private, or institutional organizations to carry out the project. The prefeasibility report should include a risk analysis that addresses the uncertainties of the project.

The prefeasibility report must be aimed at potential financiers of the project and should justify proceeding to the test drilling phase. The requirements for the prefeasibility study include the technical and nontechnical data and information that are available up to this stage, including:

- Preliminary information (power market, infrastructure, environmental, and social issues, etc.).
- Environmental impact and resource protection (environmental and social impact statement).
- Collection of baseline data (baseline environmental and social data).
- Literature review (academic publications, technical reports, maps, data, etc., from government agencies, previous leaseholders, internet searches, etc.).
- Active geothermal features (location, names, characteristics of geothermal features on maps, tables, etc.).
- Geology (geological maps, structural maps, stratigraphic columns, cross sections, reports, literature reviews, etc.).
- Geochemistry (fluid and gas geochemical data, maps, tables, drawings, and plots, with interpreted and estimated temperatures from geothermometry, etc.).
- Geophysics (collected and interpreted geophysical data on maps, cross sections, 3D models, uncertainties, data gaps, etc.).
- Drilling data (locations, temperature and pressure logs, temperature contours on geological maps, etc.).
- Conceptual model (maps and 3D models; uncertainties about depths, temperatures, faults, lithologies, porosity/permeability, etc., that remain after the exploration should be clearly disclosed to potential financiers).
- The prefeasibility study and financial justification to proceed.

For a detailed example of a suggested prefeasibility report, the reader is referred to Appendix A1 of Harvey et al. [2].

7.5 Exploratory drilling

The final process of the exploration consists of the evaluation of all the data, including the technical, environmental, and social data, before committing to the decision of a given location to start the drilling of a deep well. This is an important decision because it involves large financial commitments for the project.

A geothermal project presents risks to success in its early stages because geochemical studies, geological exploration, and geophysics are developed from the surface. Even though these studies are of high precision there is always the uncertainty of their validation. Corroboration of the latter can be achieved by drilling a well, which allows to know the type and properties of the formation rock, as well as the thermodynamic state of the reservoir. Knowledge of the above can confirm

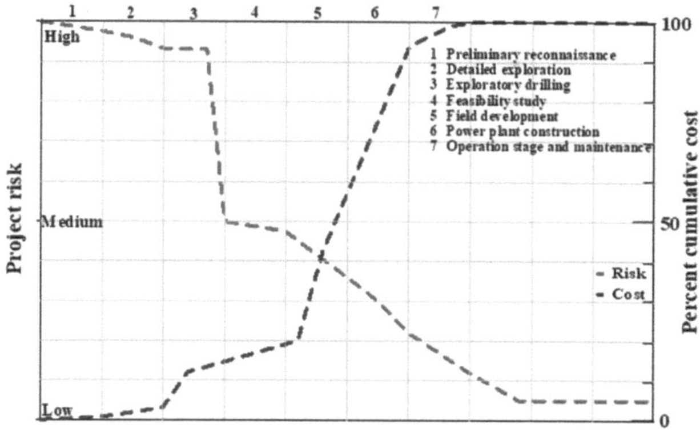

Figure 7.5 Typical profiles of risks and costs in a geothermal project taking into account each stage of its development [15,16].

the veracity of the aforementioned studies. Conversely, the first stages do not involve large costs until the drilling begins. However, drilling leads to a decrease in project risks.

The main barrier to exploiting geothermal energy is the high cost of drilling. Geothermal electricity plants are traditionally built on the edges of tectonic plates where high temperature geothermal resources are available near the surface. Besides, it takes an average of 5–7 years to develop geothermal power projects to the point of commercial deployment, which is a long return on investment [14].

Drilling typically accounts for 30%–50% of the total development cost for electricity generation. Drilling two boreholes to a depth of 3000 m can cost up to US$14 million, while piping costs vary from US$200 to US$6000 by meter in urban areas.

However, in defense of exploratory drilling wells, it is appropriate to emphasize that these provide a direct view of the rock formation and help to define, with different measurements, the feasibility of the studied zone. During the drilling stage, different measurements are logged, focused on obtaining the major information quantity, which allows for formation evaluation.

Fig. 7.5 shows a graph of the behavior of risks in a geothermal project [15,16], and its relationship with the investment costs according to the stages of the same. Barnett and Quinlivan [17] developed an estimate of installed capacity costs for plants from 20–50 MWe in New Zealand. A research focused on obtaining the costs of drilling wells around the world was carried out and the results, as published by different authors, are shown in Fig. 7.6.

7.6 Well drilling

The drilling of the first few wells in any project represents the period of the greatest risk. In the exploratory stage, it is usually recommended to drill three or four deep wells to validate the commercial feasibility of the exploitation of the resource.

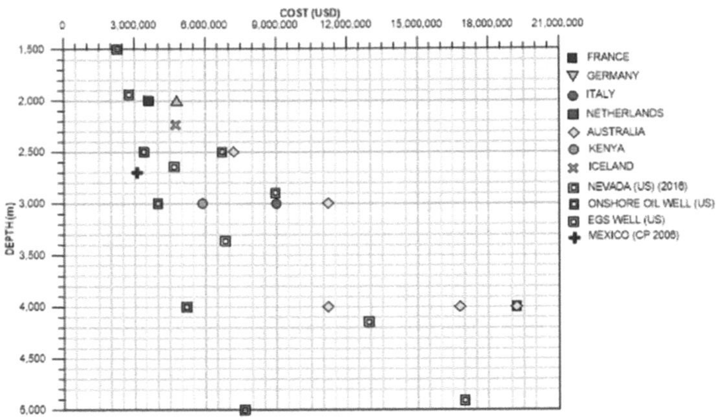

Figure 7.6 Drilling costs of wells worldwide, taking into account their respective depths and locations.
Source: Modified from J.M. Vaca-Serrano, Modelos de costeo de pozos geotérmicos aplicado para el caso del campo geotérmico de Cerro Prieto BC, Geotermia 21 (1) (2008) 51–58 [18]; L. Shevenell, The estimated costs as a function of depth of geothermal development Wells drilled in Nevada, GRC Trans. 36 (2012) 121–128 [19]; S. Thorhallsson, B.M. Sveinbjornsson, Geothermal drilling cost and drilling effectiveness, United Nations University, Geothermal Training Programme, LaGeo, Santa Tecla, El Salvador, 2012, 10 p. [20]; S. Gul, V. Aslanoglu. Drilling and well completion cost analysis of geothermal wells in Turkey, in: Proceedings Forty-Third Workshop on Geothermal Reservoir Engineering, SGP-TR-213, Stanford University, Stanford, California, February 12–14, 2018, 20 p. <https://pangea.stanford.edu/ERE/pdf/IGAstandard/SGW/2018/Gul.pdf/> (accessed 27.04.20) [21].

Likewise, the drilling of at least one injection well must be considered jointly. It is important to consider that drilling wells, their logs, and their tests significantly improve the understanding of the available resource.

Due to the uncertainty in geothermal systems, it is necessary to drill four or five wells before making a decision on whether to retire from the exploration zone. In this stage, drilled wells represent a reliable basis for a preliminary estimation of the energy reserve, the identification of the production thicknesses, the petrophysical characteristics of the formation, and an approximate evaluation of the system area. The basic information that must be carefully collected at this stage includes:

- Drilling speeds.
- The inlet and outlet temperatures of the drilling fluids.
- Chemical sampling of fluids and gases that come out with the circulation of the drilling fluid.
- Sampling (meter by meter) of the perforation cuttings for further analysis.
- Mineralogical control.
- Lithology control of the well (each meter), for the construction of its column.
- Record of the circulation losses of drilling fluids.
- Temperature and pressure profiles, logged approximately every 400 m, using resting time series of 0, 6, 12, 18, 24, and 30 h. The objective is to determine the profiles in

undisturbed state and the static pressures and temperatures to be used in the numerical models of the reservoir.
- Injection tests when the target depth is reached in the well to analyze them by means of pressure transient techniques and determine the petrophysical characteristics of the reservoir area.
- If the well presents thermodynamic characteristics of production, its surface facilities, valves, pipes, silencer, measuring instruments (manometers, thermometers, and sampling holes) should be enabled to make a discharge test to different openings and develop the initial characteristic curve of the well production.
- During the production test, samples of the fluid produced at each orifice change should be taken for their chemical characterization.

The series of pressure—temperature logs, together with all the parameters that have been determined, allow for the characteristics of the well to be identified in order to obtain an appropriate definition of the presence of the appropriate range of exploitation. After the well drilling reaches the intervals of interest, transient pressure tests are carried out.

Some of the logged parameters during the drilling stages of the well are summarized in a representative graph, similar to that shown in Fig. 7.7. From this graph, the technical staff can see correlations between thermodynamic properties, lithology, rock properties, and mechanical well completion, which is useful for making decisions on its behavior.

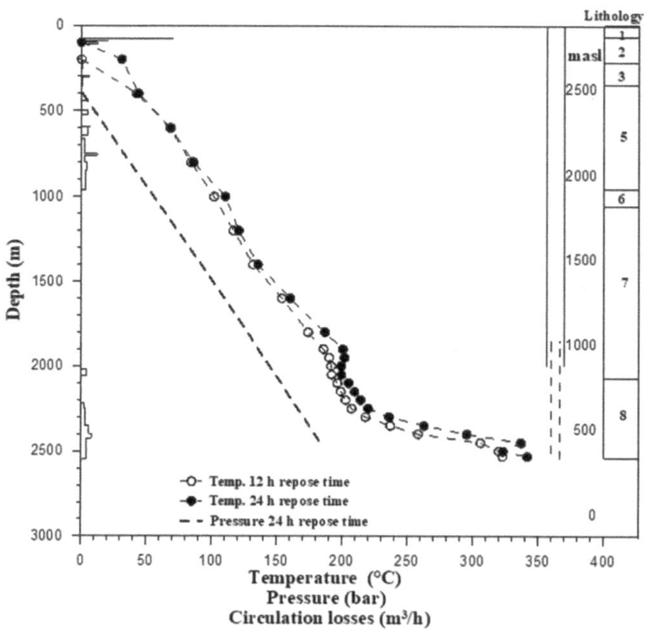

Figure 7.7 Correlation of different well information, recovered during its drilling and completion stages.

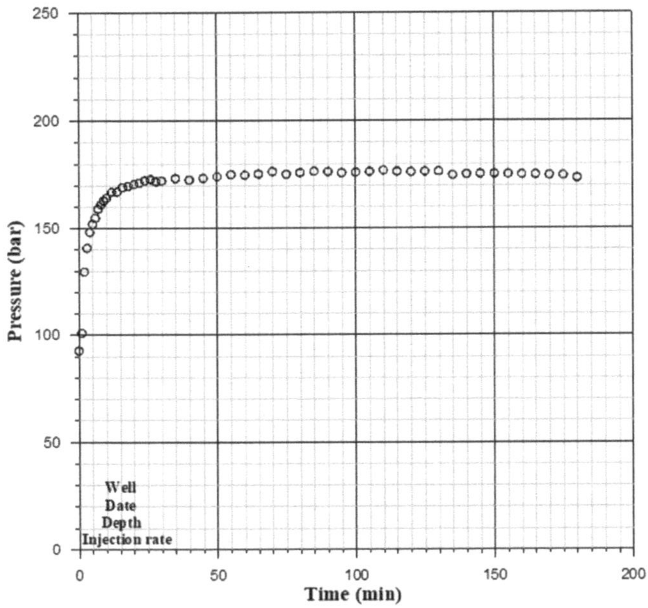

Figure 7.8 Typical graph showing behavior of a transient pressure test by water injection at a constant flow rate.

Detailed reservoir information is essential to the reservoir engineer in order for them to analyze the present behavior and future performance of the reservoir. Pressure transient testing is designed to provide the engineer with a quantitative analysis of the reservoir properties. A transient test is essentially conducted by creating a pressure disturbance in the reservoir and recording the pressure response at the wellbore, that is, bottom-hole flowing pressure (p_{wf}) as a function of time.

The graph in Fig. 7.8 shows the typical behavior of a transient pressure test, by water injection at a constant flow rate. From Figure 7.8, both well storage and transient and pseudosteady state periods can be identified. Analyses of the transient test are useful for determining petrophysical properties such as permeability (k), porosity (ϕ), static pressure (p_s), drainage radius (r_e), and skin factor (s), among others.

If both permeable zones and thermodynamic conditions suitable for exploitation are found, the decision of the completion of the well is made. During well completion, the design of production pipes that are going to be introduced into the well is carried out. The next step is to proceed to conditioning the facilities on the surface of the well; giving a period of heating for the pipes and even measuring their elongation on the surface.

Performing daily measurements on the pressure and temperature at the well head, it is possible to determine if the heating has been uniform in the pipes in order to determine the appropriate moment to carry out the initial discharge test and obtain its production characteristic curve. The parameters of the characteristic curve (pressure–flow–enthalpy) are the basis for the well characterization.

Taking all previous considerations, the results of the first wells are aimed at:

- Obtaining a better approach to the heat source.
- Determining the average productivity as well as the probability that the same trend will continue in future drilling.
- Selecting new sites, target depths, and the construction of road infrastructure.
- Mechanically designing the subsequent exploitation and reinjection wells.
- Developing a preliminary design of the power plant and the steam transport system.

7.7 Project planning

The first wells drilled in an area allow for the existence of a viable project to be located and confirmed and, from the information recovered, their risks are substantially reduced. With all the information obtained, it is feasible to prepare a report. At this point, the developer is able to figure out the general plan of the project. The incorporation of the risk analysis in the economic models of the project provides support for the request of the financing that is necessary.

The feasibility report is designed to provide confidence in the viability of the project and facilitate its financing. Even though the well does not produce continuously, from this stage and in the subsequent ones, a constant monitoring of the behavior parameters of the wells must be maintained such as:

- The production behavior (pressure, flow, enthalpy, production hole diameter, etc.).
- Fluid and gas sampling (chemical and isotopic analysis).
- Production tests (according to the possibilities in each well).

7.8 Field development

As quickly as the exploratory wells give successful results, feasibility studies are carried out in order to define the new actions of the project, which are mainly focused on locating wells to be drilled. With the technical support of exploratory wells, these wells are drilled with the clear objective of finding the thickness and production depth.

Information related to drilling equipment, work crews, and supervisors include:

- Detailed specifications of the drilling equipment for the subsequent wells.
- Analysis of the experience, in geothermal wells, of the drilling company.
- Design of drilling programs for the subsequent wells to be drilled (depths, thicknesses, casing pipes, production liners, surface installations, etc.).
- List and classification of possible specialized service providers.
- Knowledge and classification of the work teams for the project and the technical support of their advisors.

The drilled wells in this project stage are for field development and during their drilling must recover necessary information, including, as mentioned in exploratory wells drilling:

- Drilling speeds.
- The inlet and outlet temperatures of the drilling fluids.
- Chemical sampling of fluids and gases that come out with the circulation of the drilling fluid.
- Sampling (meter by meter) of the perforation cuttings for further analysis.
- Mineralogical control.
- Lithology control of the wells (each meter), for the construction of their columns.
- Record of the circulation losses of drilling fluids.
- Temperature and pressure profiles, logged approximately every 400 m, using resting time series of 0, 6, 12, 18, 24, and 30 h. The objective is to determine the profiles in undisturbed state and the static pressures and temperatures to be used in the numerical models of the reservoir.
- Injection tests when the target depth is reached in the well to analyze them by means of pressure transient techniques and to determine the petrophysical characteristics of the reservoir area.
- If the well presents thermodynamic characteristics of production, its surface facilities, valves, pipes, silencer, measuring instruments (manometers, thermometers, and sampling holes) should be enabled to make a discharge test to different openings and develop the initial characteristic curve of the well production.
- During the production test, samples of the fluid produced at each orifice change should be taken for their chemical characterization.

Determined parameter values from wells drilled at this stage of field development are useful for designing mechanical well completions, which are related to production thickness and capacities of surface production equipment. Due to thermodynamic bottom conditions prevailing in geothermal systems, only thermodynamic parameters (temperature and pressure) are logged with mechanical tools. However, these measurements are used for determining interval thickness, depth, static level, and the thermodynamic state of the fluid wells, among others. The graphs shown in Figs. 7.9 and 7.10 show the correlation of the temperature and pressure profiles of different wells, used for defining the characteristics of the conceptual model.

With information on each well, the conceptual model of the field can be developed in order to carry out the simulation of different exploitation scenarios in order to go forward with the power generation plant capacity design. With known well production parameters, their characteristic production curves are defined in order to be used for continuous exploitation. The next step is to develop the feasibility report, which should include these elements:

- Civil works necessary (access roads, leveling, possible locations of the generation plant, etc.).
- Design of field backup wells.
- Design of the electric power generation plant.
- Construction of the main access roads.
- Budgeting.
- Terms of agreement for the sale of electric power.
- Projections of the income–expenditure budget.

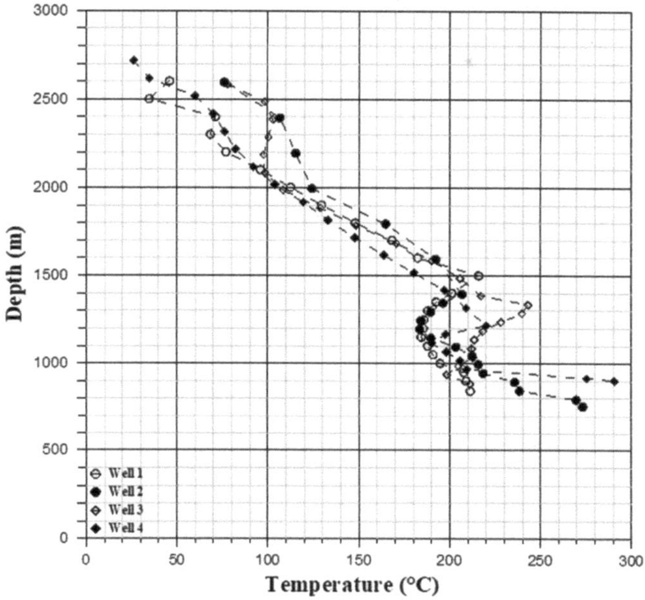

Figure 7.9 Temperature profiles of drilled wells for field development, which can be used for defining the thickness and depth of the production zone [22].

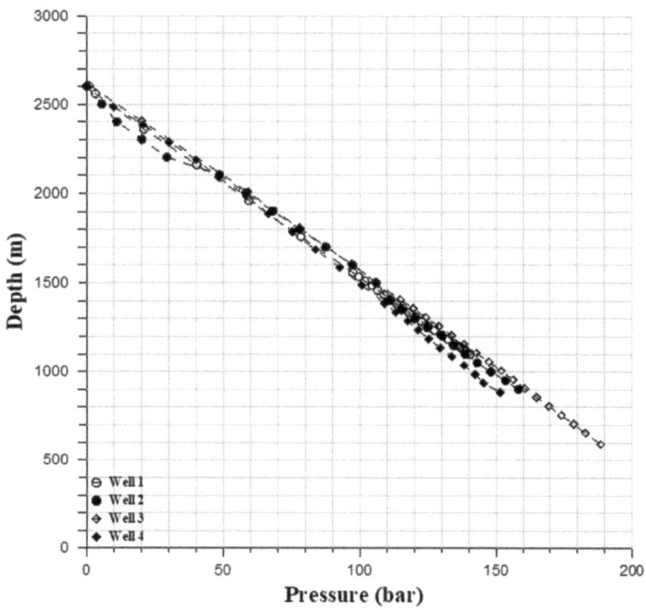

Figure 7.10 Pressure profiles of drilled wells for field development, which can be used for defining the depth of the static level at the initial conditions of the reservoir [22].

This stage is characterized by the growth of the field in production wells and the increase of the production flow until the supply requirements of the generation plant are satisfied. Also, an environment-friendly eviction of the brine produced must be sought through reinjection wells.

The data and useful information that project partners must analyze may include: Information on the commercial aspects of the project.

- Plan for the sale of electric power and its wholesale or retail sale.
- Description of the project structure.
- Analysis of financial projections, including expenses for unforeseen contingencies during drilling or costs of maintenance operations and/or stimulation.
- Description and value of the goods and equipment to be insured.

Information of the company responsible for the project.

- Basic information of the company (type, organization, owners, financial movements, related companies, etc.).
- Experience of managers in the management of drilling projects and the development of natural resource projects.
- Experience of the work team.
- Number and type of support staff.

During this stage, the development of transient pressure tests for the wells that are finishing drilling is continued as far as possible. The results of these tests are useful for characterizing the reservoir in its undisturbed state.

7.9 Reservoir monitoring at start-up and operation

After finishing the construction of the plant, the start-up and operation phases of the plant begin. Having secured the supply of the resource (with the drilling of wells) for the start of the operation of the plant, the main objective is aimed at optimizing production. This is achieved through appropriate designs, applying criteria for reservoir engineering, and well production.

Likewise, it is necessary to emphasize the drilling of injection wells for the clean evacuation of the brine, preserving the environment, and seeking reservoir recharge. During the operation, it must seek to minimize costs and ensure the reliable delivery of the resource to the generation plant.

A timeline of the likely total duration of a given project and each of its stages is shown in Table 7.5 [4,16,23]. It should be emphasized that the quality of the information during the first two stages is the basis for obtaining a location for the first well to be drilled, and, if necessary, ensuring its success and, therefore, the continuity of the project. However, in the case that the first well drilled is not a producer, it is recommended not to discard the project, but rather to try to drill three or four exploratory wells; always applying the best technical criteria to select their locations.

After the exploitation stage comes the operation and maintenance stages. This stage involves the operation and maintenance activities of the electricity generation

Table 7.5 Activities and the likely duration of the various stages of a business plan for a geothermal project [4,16,23].

	Exploration			**Exploitation**		
Phases	Year 1	Year 2	Year 3	Year 4	Year 5	Year 6
	Recognition	Exploration	Exploratory drilling	Feasibility	Field development	Construction
Generic activities	Licenses acquisition	Geology detailed	Drilling and evaluation of exploration wells	Financing analysis	Well drilling	Construction of the plant
	Regional surface geological exploration	Geochemistry	Parameter evaluation	Conceptual design	Exploitation designs	Drilling and evaluation of wells
	Geochemical exploration	Geophysical studies	Static reservoir characterization	Reservoir evaluation	Plant design	Construction of vapor ducts and surface equipment
		Location of sites to drill	Delimitation of the exploration area	Feasibility report	Design of steam transport and equipment	Admission tests and turbine treading
Main objectives	Surface recognition	Confirmation of geothermal conditions	Geothermal reservoir confirmation	Technoeconomic feasibility	Technoeconomic infrastructure	Commercial operation startup

plant as well as the maintenance of the wells, the transport systems, and the accessory equipment for the operation of the plant.

7.10 Energy conversion

At the end 2018, the installed capacity worldwide was around 13.329 GW [24] and the production prices for electricity varied from 3 to 25 ct/kWh or even higher depending on the resource [25].

The energy conversion technology used to obtain the best of geothermal resources depends on the reservoir properties (e.g., geological, geophysical, geochemical, thermodynamic, and physicochemical, among others). Worldwide, from a commercial and successful point of view, three types of geothermal power plant technologies have been used to convert hydrothermal fluids into electricity. The options available to developers are (1) dry steam plants that make use of a direct flow geothermal steam, (2) flash steam power plants that use a mixture of liquid water and steam (single, double, and triple), (3) binary cycle geothermal plants, and (4) flash/binary combined cycle plants. The power unit size usually ranges from 25 to 117 MWe, and may utilize a simple or multiple flash system, flashing the fluid in a series of vessels at successively lower pressures to maximize the extraction of energy from the geothermal fluid. A brief overview of these technologies is given here.

7.10.1 Dry steam technology

Dry steam plants make use of direct flow geothermal steam. The steam travels directly to a turbine that converts thermal energy into mechanical energy, which drives an electric generator that produces electricity that is then distributed to the grid (see Fig. 7.11). It is important to mention that primary steam eliminates the

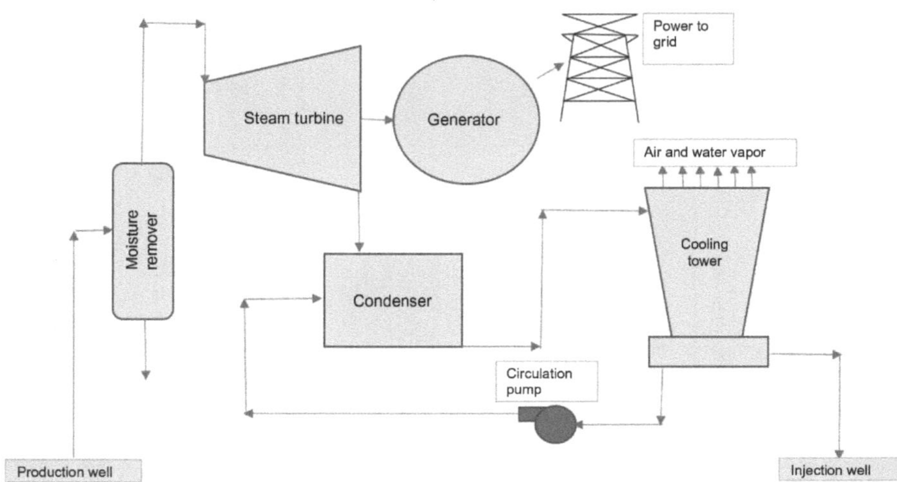

Figure 7.11 Basic dry steam technology.

need to burn fossil fuels to run a turbine. These plants emit only excess steam and minor amounts of gases.

Dry steam power plant systems were the first type of geothermal power generation plants built (they were first used at Larderello in the Tuscany region of Italy in 1904). Steam technology is still effective today and is currently in use at The Geysers in northern California, the world's largest single source of geothermal power [26].

Dry steam energy conversion technology contributed 22% of the worldwide installed capacity by the end 2014 [27].

7.10.2 Flash steam power plants

Today, flash steam plants are the most common type of geothermal power generation plants in operation. Fluid at temperatures greater than 360°F (182°C) is pumped under high pressure into a tank at the surface held at a reduced pressure, causing some of the fluid to rapidly vaporize. This results in an increase in the steam fraction of the fluid, providing a greater proportion of steam available. Then the vapor drives a turbine, which drives a generator. The liquid phase of the fluid that remains in the tank may then be reinjected into the reinjection well or can be flashed again in a second tank to extract even more energy prior to reinjection.

When the mixture temperature is over 210°C, a single flash setup is generally used, as shown in Fig. 7.12. In this type of configuration, the geothermal fluid is extracted from the production well and sent to a cyclonic separator where the liquid and vapor phases of the mixture are separated due to a difference in densities [28,29]. The electricity generation process starts when the primary vapor passes from the separator to a steam turbine and finally to a generator to complete the process. The remaining liquid phase (also known as brine) that remains in the separator is sent to a reinjection well, which in turn receives cooling water from a

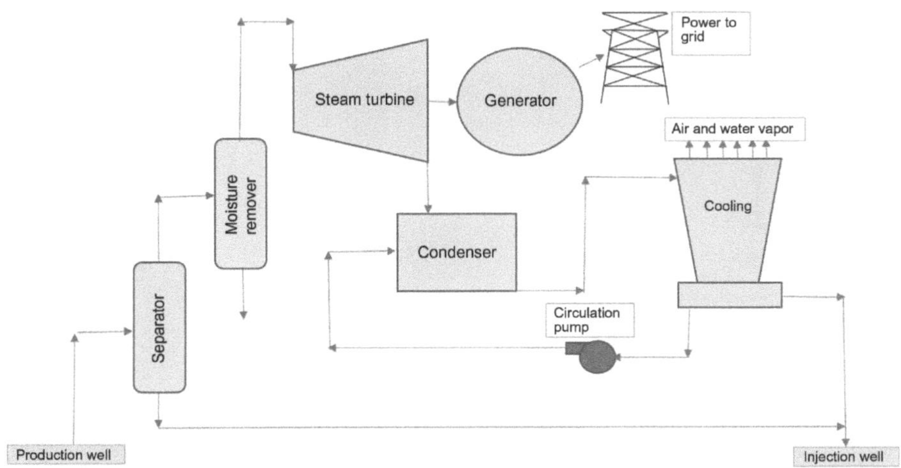

Figure 7.12 Single flash steam power plant.

condensation process that is designed to cause a change in the steam state at the turbine outlet and to obtain the condensed steam in the form of pure water.

The use of single flash conversion technology contributed to 41% of the world's geothermal power installed capacity at the end of 2014 [27].

As mentioned previously, in order to improve the efficiency of the process, the brine from the first separator (high-pressure separator) is sent to a low-pressure separator, where the steam is fed to the turbine. The turbine is designed in such a way that the pressure difference over the first stages is the same as the pressure difference between the high- and low-pressure separators. The brine from the low-pressure separator is sent to a reinjection well, as the return condition of the geothermal fluid. The steam from the high-pressure separator enters the turbine. The low-pressure steam enters the turbine a few stages later. In this case, a dual-admission turbine is used (see Fig. 7.13); however, two separate tandem compound turbines may be used. Thus the steam is then expanded through the turbine (or turbines) down to the condenser.

Double flash power plants increase both the efficiency of the process by approximately 35% [30] and the power production by 15% in relation to the single flash setup [31].

The use of double flash conversion technology contributed to 21% of the world's geothermal power installed capacity in 2014 [27].

Triple flash system power plants (see Fig. 7.14), utilize a series of vessels for flashing the work fluid at successively lower pressures to maximize the extraction of energy from it. This process is implemented in order to obtain additional use of

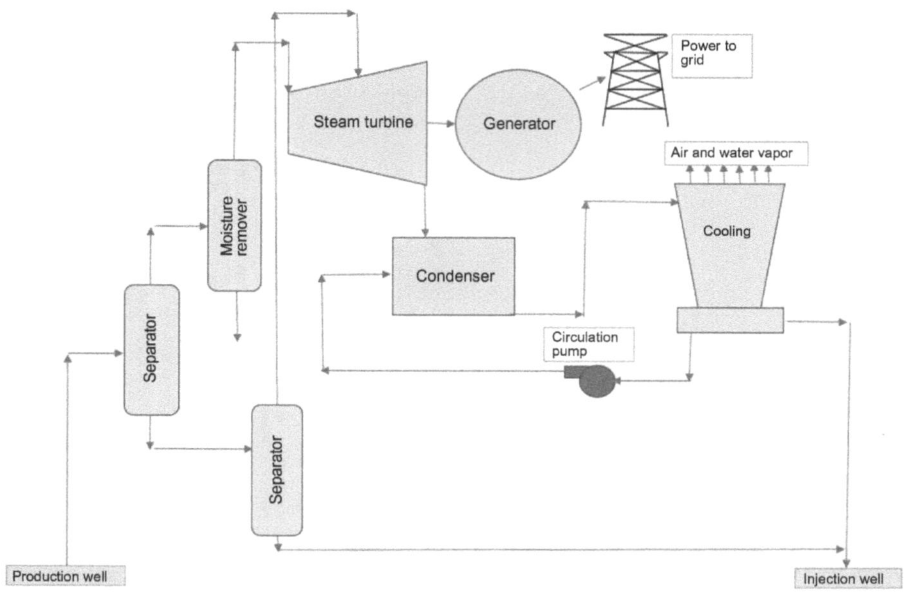

Figure 7.13 Double flash steam power plant.

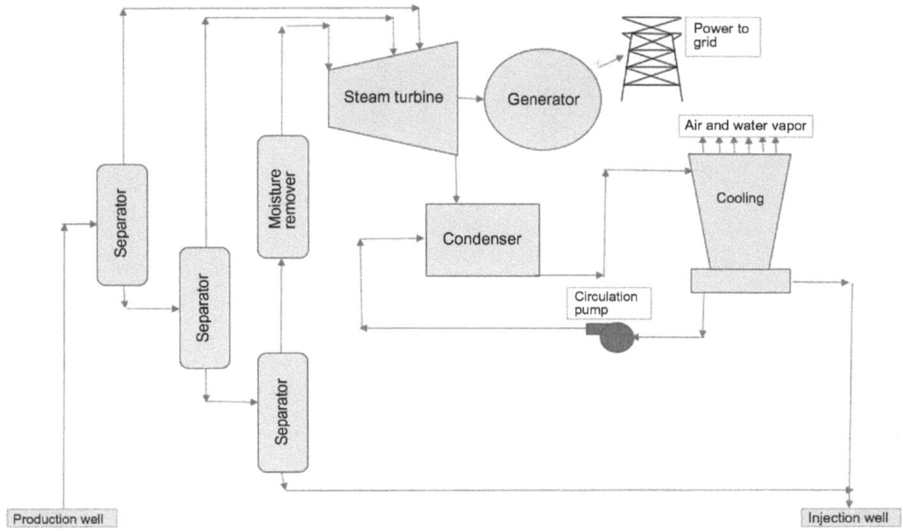

Figure 7.14 Triple flash steam power plant.

the energy available in the remaining liquid phase coming from the double flash cycle. The use of triple flash conversion technology contributed to 1% of the world's geothermal power installed capacity in 2014 [27].

According to Moon and Zarrouk [32], during the flash process, the pressure drop results in an increase in the silica concentration of the brine. This silica can build up a layer of solid deposit on the internal surfaces of pipelines, flash plants, and turbines, thus, impeding the flow of the fluid and leading to a drop in conversion efficiency and high maintenance costs. Other factors affecting the conversion efficiency are noncondensable gas (NCG) content, heat loss from equipment, turbine and generator efficiency, and power plant auxiliary equipment (e.g., fans, pumps, and gas extraction system). Among other problems associated with elevated levels of NCG in geothermal steam power systems, the gases reduce the heat transfer efficiency of the power plant condensers, resulting in a reduction of the turbine power output [33] and a decrease in the specific expansion work in the turbine [34].

7.10.3 Binary cycle power plants

Binary cycle power plants operate on lower-temperature waters, that is, 74°–177°C. These plants (see Fig. 7.15) use the heat of the hot water to boil a "working fluid," usually an organic compound with a low boiling point such as isobutene (e.g., isopentane, propane, freon, or ammonia). This working fluid is then vaporized in a heat exchanger and used to turn a turbine. The geothermal water and the working fluid are confined to separate closed loops. This is done through the use of a heat exchanger used to transfer the heat energy from the geothermal water to the "working fluid" in a conventional Rankine cycle, or alternatively, a Kalina

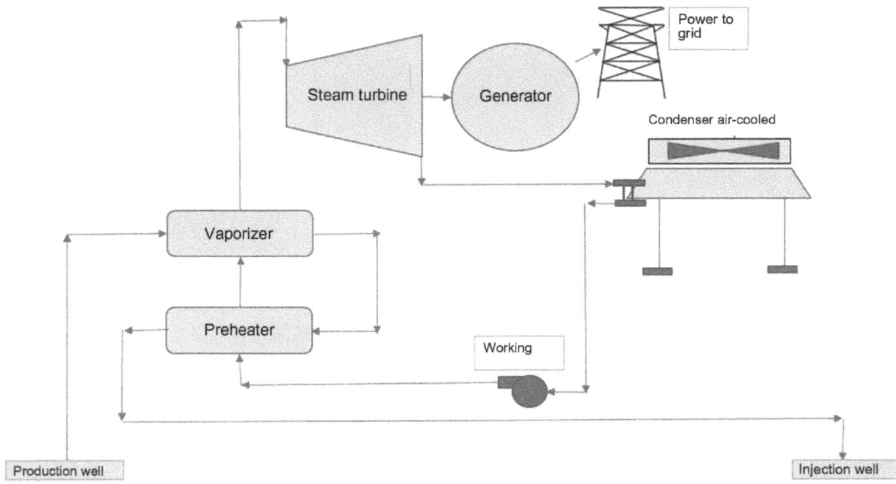

Figure 7.15 Binary cycle power plants.

cycle. The secondary fluid vaporizes into gaseous vapor and (like steam) the force of the expanding vapor turns turbines that power generators. If the power plant uses air cooling, the geothermal fluids never make contact with the atmosphere before they are pumped back into the underground geothermal reservoir, effectively making the plant emission free. The fluid in a binary plant is recycled back to the heat exchanger and forms a closed loop [35–37].

The use of binary conversion technology contributed to 12% of the world's geothermal power installed capacity in 2014 [27].

7.10.4 Flash/binary combined cycle

Geothermal power plants often start their operating life cycle with a flash unit or units configuration; however, it is well known that depending upon of field characteristics and performance, it is possible to extent the power of the initial unit or units by adding a binary unit that uses a heat exchanger to transfer the heat energy from the geothermal water (waste geothermal fluid) to the so called "working fluid" or secondary fluid, as used in a binary cycle.

In this case, an attractive alternative is generated to use the available heat and increase the output power of the facilities. In this combined or hybrid flash-binary plant [26], initially a single flash plant is operated by itself and the waste liquid or "working-fluid" from the cyclone separators is sent directly to injection wells. A binary cycle is inserted to tap into the reinjection pipeline where it extracts some heat and thereby lowers the temperature of the waste brine prior to injection. The additional power generated by the binary cycle is gained from the same amount of geothermal fluid being extracted from the reservoir. A simplified schematic of such an arrangement is given in Fig. 7.16.

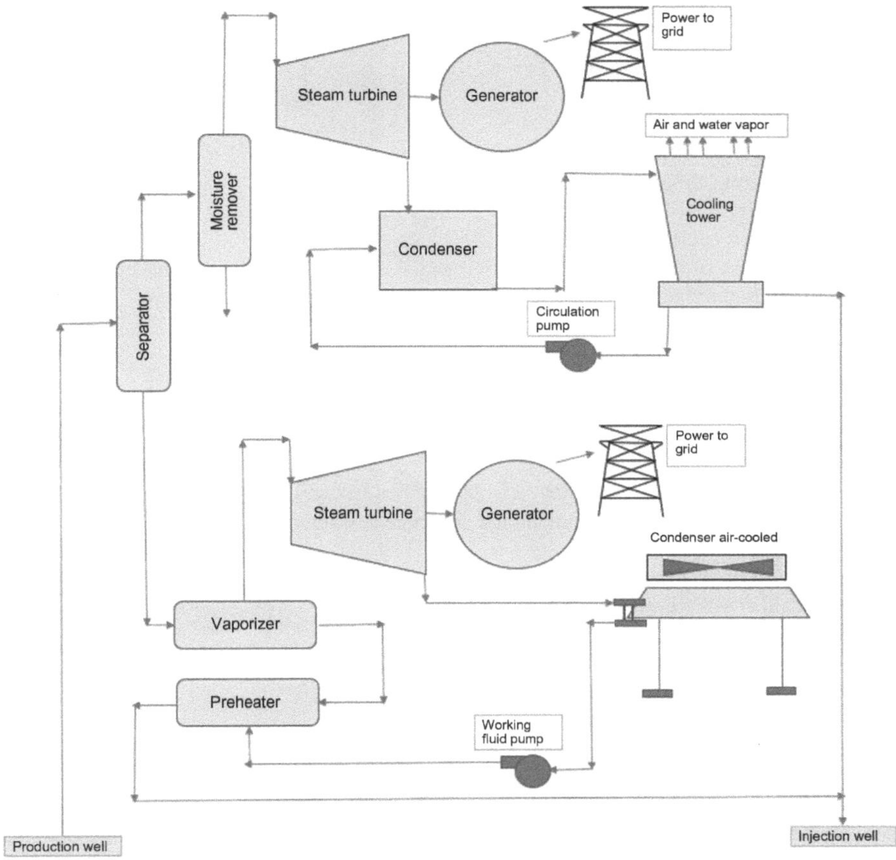

Figure 7.16 Flash/binary combined cycle.

7.10.5 Development of a geothermal power plant

There are three main items relating to the construction phase of the development of a geothermal power plant, including:

- Project description.
- Design philosophy.
- Particular technical requirements.

The information on the company responsible for the project may include:

Project description.

- Features of the installation.
- Terminal points for the water.
- Drain.
- Sewage.
- Environmental impact and water supply permits.

Design philosophy.
- Efficiency.
- Performance requirements.
  - Operating regime.
  - Starting and take of load.
    - Number of starts and load variations.
    - Startup time and load taking curve.
- Availability.
- Control and operation requirements.
- Maintenance of the installations.
- General requirements.
  - Equipment life.
  - Performance criteria.
  - Protection against the environment.
- Rules, regulations, laws, codes, and applicable specifications.
  - Laws, regulations, criteria, norms, and local codes.
  - International standards.

Particular technical requirements.
- Requirements of main equipment and associated systems.
  - Proposed equipment.
  - Steam turbo generators.
  - Main cooling system.
- Electrical requirements.
  - Frequency and voltage limit.
  - Substation of the power plant.
  - Electric generators.
  - Excitation system.
  - Main transformers and medium voltage systems.
  - Supervision and control of the power plant.
  - Electrical protection.
  - Single line diagram.
- Civil engineering requirements.
- Environmental requirements.
- Environmental management system.

7.11 Conclusion

Geothermal energy is a natural resource, the characteristics of which make it renewable and sustainable. It is generally available worldwide, although its exploitation for electric power generation has traditionally taken place near tectonic or volcanic regions.

Potential developments of EGS aimed at electricity generation or at the direct use of heat are more geographically distributed.

New geothermal developments make use of medium or low enthalpy/temperature resources, which are more globally distributed.

Any geothermal power development project must start with a reconnaissance of the regional area of interest, followed by a local exploration stage in which geological, geophysical, and/or geochemical studies are performed.

A conceptual model of the area is constructed before proceeding to a prefeasibility stage, in which data analysis and risk assessment are carried out prior to recommending whether a drilling phase should proceed or not.

The exploration, geological, geochemical, and geophysical stages allow for the regional area to be delimited to a focused area where the first exploratory wells to be drilled can be located.

The quality of the information obtained, its analysis, and interpretation support the technical bases of the stages of the recognition, prefeasibility, and feasibility of a geothermal project.

The risks in the exploration stage are high due to the uncertainties of not yet having direct information from the subsurface; however, the costs are low, which may be 15% or 20% of the total cost of the project.

The costs of the project increase during the well drilling stage for the development of the field as well as in the construction of the plant, however, the risks decrease noticeably.

The characterization of the productivity of wells allows for the appropriate designs of their exploitation to be established and also for their efficiency and useful life to be increased.

As of the startup of the plant, costs are minimized because the working fuel is a natural resource; only operating and maintenance costs are considered for wells and facilities. It is necessary to maintain a balance between the extracted mass and the recharge input to increase the efficiency of the operation.

In a geothermal plant process, energy can be fully utilized by sequentially using fluid discharges from different process stages to generate electricity under conditions of lower plant acceptance in a second and third flash.

The energy conversion technology used to obtain the best of geothermal resources is always under research, development, and innovation; in this chapter, only the main options available to developers are presented.

Other alternatives of integrated flash-binary configurations for particular ranges of temperature resources are available. Furthermore, total-flow systems are under research.

Acknowledgments

The authors wish to express their gratitude to the authorities of the Instituto Nacional de Electricidad y Energías Limpias (INEEL) for their support to publish this work. Many thanks go as well to Siomara López Blanco, Adriana Paredes Soberanes, and Emmanuel Hiram Hernández Arias for the kind elaboration of some figures.

References

[1] IGA, What Is Geothermal, International Geothermal Association webpage, 2020. <https://www.geothermal-energy.org/explore/what-is-geothermal/> (accessed 27.04.20).

[2] C. Harvey, G. Beardsmore, I. Moeck, H. Ruter, S. Bauer, Best Practices Guide for Geothermal Exploration (English), World Bank Group, Washington, DC, 2014. <http://documents.

worldbank.org/curated/en/190071480069890732/Best-practices-guide-for-geothermal-exploration/> (accessed 27.04.20).
[3] B. Speer, R. Economy, T. Lowder, P. Schwabe, S. Regenthal, Geothermal exploration policy mechanisms: lessons for the United States from international applications. United States, 2014. <https://doi.org/10.2172/1134132>.
[4] OLADE/BID, Guía para estudios de reconocimiento y prefactibilidad geotérmicos. Quito, Ecuador, 1994a.
[5] S.-M. Lu, A global review of enhanced geothermal system (EGS), Renew. Sustain. Energy Rev. 81 (part 2) (2018) 2902–2921. Available from: https://doi.org/10.1016/j.rser.2015.08.026.
[6] E. Jolie, et al., GEMex–A Mexican–European research cooperation on development of superhot and engineered geothermal systems, in: Proceedings, Forty Third Workshop on Geothermal Reservoir Engineering, February 12–14, 2018, Stanford University, Stanford, CA, SGP-TR-213, 2018. <https://pangea.stanford.edu/ERE/pdf/IGAstandard/SGW/2018/Jolie2.pdf/> (accessed 27.04.20).
[7] OLADE/BID, Guía para la Evaluación del Potencial Energético en zonas Geotérmicas durante las etapas previas a la factibilidad. Quito, Ecuador, 1994b.
[8] Geothermal exploration best practices – geology, exploration drilling, geochemistry, geophysics, in: R. Bracke, C. Harvey, H. Rueter (Eds.), IGA ACADEMY Report 01/2013, IGA Service GmbH, Bochum, Germany, 2013.
[9] J. Ahrens, B. Geveci, C. Law, ParaView: An End-User Tool for Large Data Visualization, Visualization Handbook, Elsevier, 2005.
[10] M.F. Coolbaugh, C. Sladek, J.E. Faulds, R.E. Zehner, G.L. Oppliger, Use of rapid temperature measurements at a 2-meter depth to augment deeper temperature gradient drilling, in: Proceedings of the Thirty-Second Workshop on Geothermal Reservoir Engineering, January, 22–24, 2007. Stanford University. Stanford, CA. SGP-TR-183, 2007, 8 p. <https://pangea.stanford.edu/ERE/pdf/IGAstandard/SGW/2007/coolbaug.pdf/> (accessed 27.04.20).
[11] I. Moeck, Iterative development of stress models for deep geothermal reservoirs, in: Proceedings of the 74th EAGE Conference and Exhibition, June 4–7, 2012, Copenhagen, Denmark, 2012, 5 pp.
[12] O. Heidbach, M. Rajabi, K. Reiter, M. Ziegler, World Stress Map 2016, GFZ Data Services, 2016. Available from: https://doi.org/10.5880/WSM.2016.002.
[13] J. Newson, W. Mannington, F. Sepulveda, R. Lane, R. Pascoe, E. Clearwater, et al., The application of 3D modelling and visualisation software to reservoir simulation: leapfrog and TOUGH2, in: Proceedings, Thirty-Seventh Workshop on Geothermal Reservoir Engineering, January 30–February 1, Stanford University, Stanford, CA, 2012, SGP-TR-194, 2012. <https://pangea.stanford.edu/ERE/pdf/IGAstandard/SGW/2012/Newson.pdf/> (accessed 27.04.20).
[14] European Commission, Geothermal energy, SETIS information for decision making. <http://www.renewables-made-in-germany.com/en/renewables-made-im-germany-start/geothermal-energy/market-development.html/>, 2015 (accessed 27.04.20).
[15] IPCC, Special report on the intergovernmental panel on climate change, Managing the Risks of the Extreme Events and Disasters to Advance Climate Change Adaptation, Cambridge University Press, New York, 2012. 582 pp.
[16] M. Gehringer, V. Loksha, Geothermal Handbook: Planning and Financing Power Generation. ESMAP Technical Report 002/12, World Bank, Washington, DC, 2012. © World Bank. License: CC BY 3.0 IGO. <https://openknowledge.worldbank.org/handle/10986/23712/> (accessed 27.04.20.).

[17] P. Barnett, P. Quinlivan, Assessment of Current Costs of Geothermal Power Generation in New Zealand, SKM Report for New Zealand Geothermal Association, 2009.
[18] J.M. Vaca-Serrano, Modelos de costeo de pozos geotérmicos aplicado para el caso del campo geotérmico de Cerro Prieto BC, Geotermia 21 (1) (2008) 51–58.
[19] L. Shevenell, The estimated costs as a function of depth of geothermal development Wells drilled in Nevada, GRC Trans. 36 (2012) 121–128.
[20] S. Thorhallsson, B.M. Sveinbjornsson, Geothermal drilling cost and drilling effectiveness, United Nations University, Geothermal Training Programme, LaGeo, Santa Tecla, El Salvador, 2012, 10 p.
[21] S. Gul, V. Aslanoglu, Drilling and well completion cost analysis of geothermal wells in Turkey, in: Proceedings Forty-Third Workshop on Geothermal Reservoir Engineering, February 12–14, 2018, SGP-TR-213, Stanford University, Stanford, CA, 2018, 20 p. <https://pangea.stanford.edu/ERE/pdf/IGAstandard/SGW/2018/Gul.pdf/> (accessed 27.04.20).
[22] A. Aragón-Aguilar, G. Izquierdo-Montalvo, S. López-Blanco, V. Arellano-Gómez, Reservoir engineering analysis applied to low permeability formations and high temperature, in: Proceedings World Geothermal Congress, Melbourne, Australia, April 2015, 11 pp.
[23] IGA Service GmbH, Geothermal exploration best practices: a guide to resource data collection, analysis, and presentation for geothermal projects, Bochum University of Applied Sciences (Hochschule Bochum), 2013, 74 p.
[24] IRENA, Renewable Capacity Statistics 2019, 2019, p. 40. <https://www.irena.org/publications/2019/Mar/Capacity-Statistics-2019-IRENA_RE_Capacity_Statistics_2019.pdf/> (accessed 27.04.20).
[25] L. Wissing, Introduction and overview, in: IEA Geothermal, International Workshop on Geothermal Energy, 2019. <http://iea-gia.org/workshop-presentations/2019-gran-canaria-geothermal-energy-workshop//2019%2004%2008%20Lothar%20Wissing.pdf/> (accessed 27.04.20).
[26] R. DiPippo, Combined and hybrid geothermal power systems, Geotherm. Power Gener. (2016). Available from: https://doi.org/10.1016/B978-0-08-100337-4.00014-0.
[27] R. Bertani, Geothermal power generation in the world 2010–2014 update report, Geothermics 60 (2016) 31–43. <https://www.geothermal-energy.org/pdf/IGAstandard/WGC/2015/01001.pdf/> (accessed 27.04.20).
[28] F. Rivas, et al., Diseño y evaluación de separadores y secadores tipo Webre: Estado del arte, in: Asociación Geotérmica Mexicana – Memorias del XXII Congreso Anual. Marzo 2015. <https://www.researchgate.net/publication/281376243_Diseno_y_evaluacion_de_separadores_y_secadores_tipo_Webre_Estado_del_arte/link/55e488ef08aecb1a7ccb8137/download/> (accessed 27.04.20).
[29] S.J. Zarrouk, M.H. Purnanto, Geothermal steam-water separators: design overview, Geothermics 53 (2014) 236–254. <https://www.researchgate.net/publication/265128616_Geothermal_steam-water_separators_Design_overview/> (accessed 27.04.20).
[30] C. Tomasini-Montenegro, E. Santoyo-Castelazo, H. Gujba, R.J. Romero, E. Santoyo, Life cycle assessment of geothermal power generation technologies: an updated review, Appl. Therm. Eng. 114 (2017) 1119–1136. Available from: https://doi.org/10.1016/j.applthermaleng.2016.10.074.
[31] S. Jalilinasrabady, R. Itoi, Flash Cycle and Binary Geothermal Power Plant Optimization, GRC Transactions, 2012, p. 36.<http://pubs.geothermal-library.org/lib/grc/1030365.pdf/>. (accessed 27.04.20).

[32] H. Moon, S.J. Zarrouk, Efficiency of geothermal power plants: a worldwide review, in: New Zealand Geothermal Workshop 2012 Proceedings, November 19–21, 2012, Auckland, New Zealand, 2012. <https://www.geothermal-energy.org/pdf/IGAstandard/NZGW/2012/46654final00097.pdf/> (accessed 27.04.20).
[33] M. Vorum, E. Fitzler, Comparative Analysis of Alternative Means for Removing Noncondensable Gases From Flashed-Steam Geothermal Power Plants, National Renewable Energy Lab, Golden, CO, 2000. <https://www.nrel.gov/docs/fy00osti/28329.pdf/> (accessed 27.04.20).
[34] H.E. Khalifa, E. Michaelides, Effect of non condensable gases on the performance of geothermal steam power systems, Department of Engineering, Brown University, Providence, RI, 1978. <https://digital.library.unt.edu/ark:/67531/metadc1110882/m2/1/high_res_d/6081081.pdf/> (accessed 27.04.20).
[35] B. Goldstein, G. Hiriart, R. Bertani, C. Bromley, L. Gutiérrez-Negrín, E. Huenges, et al., Geothermal energy, in: O. Edenhofer, R. Pichs-Madruga, Y. Sokona, K. Seyboth, P. Matschoss, S. Kadner, T. Zwickel, P. Eickemeier, G. Hansen, S. Schlömer, C. von Stechow (Eds.), IPCC Special Report on Renewable Energy Sources and Climate Change Mitigation, Cambridge University Press, Cambridge and New York, 2011. <https://www.ipcc.ch/report/renewable-energy-sources-and-climate-change-mitigation/> (accessed 27.04.20.).
[36] A. Kagel, The State of Geothermal Technology. Part 2. Surface Technology, Publication by the Geothermal Energy Association for the U.S. Department of Energy, 2008. <https://geothermalcommunities.eu/assets/elearning/7.34.Geothermal%20Technology%20-%20Part%20II%20(Surface).pdf/> (accessed 27.04.20).
[37] P. Valdimarsson, Geothermal power plant cycles and main components, in: Presented at "Short Course on Geothermal Drilling, Resource Development and Power Plants", Organized by UNU-GTP and LaGeo, January 16–22, 2011, Santa Tecla, El Salvador, 2011. <https://orkustofnun.is/gogn/unu-gtp-sc/UNU-GTP-SC-12-35.pdf/> (accessed 27.04.20).

Section VII

Marine Power Technology

Marine power technology—wave energy

8

Abdus Samad and R. Suchithra
Wave Energy and Fluids Engineering Laboratory, Department of Ocean Engineering, Indian Institute of Technology Madras, Chennai, India

8.1 Introduction

For thousands of years, humankind has continuously utilized fossil fuels to generate energy. These conventional energy sources are not only being depleted, but are causing massive damage to the environment. Challenges faced by conventional energy sources are inadequate for the growing energy demand as their consumption is rapidly increasing which further contributes to climate change such as global warming. Studies have reported that the world's energy demand would nearly double the present demand by 2050. Carbon dioxide emissions from conventional sources are projected to peak before 2030, and over the 21st century, the cumulative CO_2 emissions will result in global warming by an increase in more than 2°C compared to the temperature during the pre-industrial time. Due to these major challenges, much research effort has been made in the search for renewable and sustainable forms of energy that have less impact on the environment. The economic, environmental, and geo-political policies emphasize the reduction of global conventional energy resources, resulting in revolutionizing energy harvesting from renewable sources.

Naturally occurring energy sources such as wind, hydropower, solar, biomass, geothermal, and oceans are present-day renewable energy resources. Over the past few decades there has been tremendous work in the wind and solar energy sectors. The World Energy Council report estimated that the rapid growth projection of wind and solar power would contribute about one-third of the world's energy supply by 2050. However, there is still research needed on applying more scientific and advanced technological routines to enhance the performance and efficiency of such energy harvesting devices. Khaligh and Onar [1] described several renewable energy sources such as wind, solar, ocean tidal, ocean thermal, and ocean wave energy harvesting, along with various topologies and different kinds of power electronic systems used for the grid connection of renewable energy devices.

Ocean energy is one of the least explored renewable energy sources. The ocean covers about two-thirds of the Earth and contains a vast amount of untapped energy. Energy extraction from the ocean can be achieved by various phenomena such as temperature or salinity gradient, waves, currents, and tidal variations. Among all these possibilities, ocean waves are the most promising form of energy.

Sustainable Fuel Technologies Handbook. DOI: https://doi.org/10.1016/B978-0-12-822989-7.00009-3

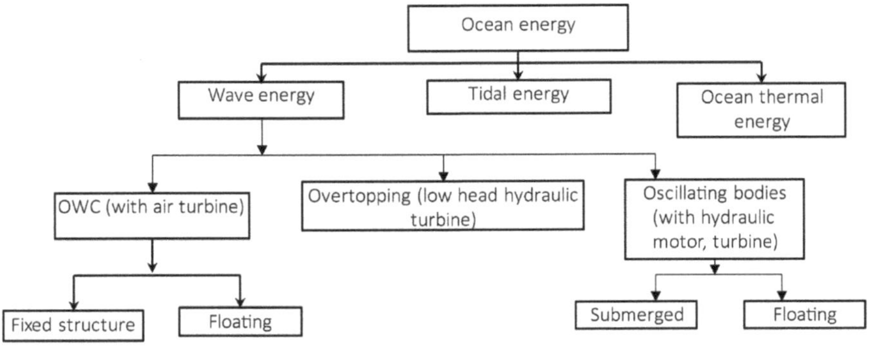

Figure 8.1 Classification of ocean energy.

There are three basic ways to extract energy from the ocean. Fig. 8.1 represents the classification of ocean energy. This chapter focuses on the investigation into wave energy converters.

8.2 Wave energy

Ocean waves are created by the action of wind over the ocean surface caused by atmospheric disturbance due to solar irradiation. Waves are an enormous source of energy and the entire ocean contains a potential of about 8000–80,000 TW/year, with each wave crest containing 10–50 kW/m [2]. Waves have the highest energy density compared to other available renewable energy sources [3]. Wave power offers a good correlation between resource and demand. Unlike wind and solar power, ocean waves produce power continuously. They offer multiple locations for energy harvesting from the shoreline to deep water. It is estimated that approximately 1-2 TW of electricity production could be produced from the ocean waves [4]. A major solution to the worldwide energy crisis could be met by harnessing the immense power of ocean waves.

The average annual wave power density distribution for all the global points is shown in Fig. 8.2, where the units of the values are in kW/m, which corresponds to power per width of the wave front. The arrows on the plot graph show the mean-best direction. The figure demonstrates the energy regions of global oceans from the mid to high latitude temperature storm belts of both hemispheres.

8.2.1 Wave climate at the Indian coast

The Indian coastline is about 7500 km and can generate wave power of around 40 GW. Harnessing at least 10%–20% of this resource would help the growing energy demand in India. Worldwide there was a quest for harvesting energy renewable energy resources, the attention towards wave energy was initially during the

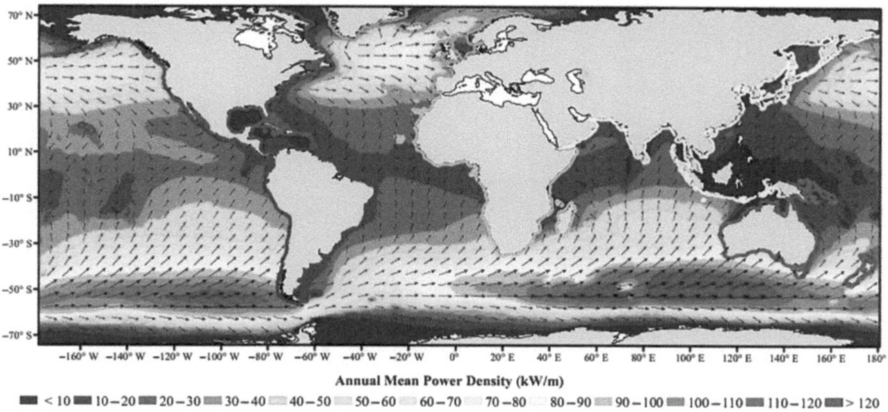

Figure 8.2 Global annual wave power density distribution [5].

cold war period. In India, initial work on wave energy started at the Ocean Engineering Center of IIT Madras during 1982. An early study on wave-energy potential along the Indian coast, with seasonal variability, was reported by Rao and Sundar [6]. Their work found that the highest estimate of wave-energy potential along the west coast of India was about 47 kW/m for the region between 15–25°N and 70°E and 34 kW/m for the east coast (the region between 15–25°N and −85°E). Thus with the help of these wave-energy potential estimates, the development of the first wave energy converter (WEC) project was commissioned in 1983 [7]. Vizhinijam, Kerala, was chosen for deployment of the pilot project. The significant wave height and mean wave period at this site are 1.5–1.8 m and 6–8 s, respectively. A nearshore oscillating water column (OWC) based WEC was built with a Wells turbine (diameter of 2 m) as the power take-off (PTO) system. The plant could produce electrical power of 150 kW [8] on a continual basis (Fig. 8.3). In 2004 the National Institute of Ocean Technology (NIOT) proposed a project to use the OWC system for the desalination of ocean water using the power produced from the wave energy converter system. But due to several technical and maintenance issues the plant was decommissioned in 2011 and handed over to the Harbor Department. Even after facing several difficulties, a tremendous amount of work has been carried out over the past decade by IIT Madras and NIOT toward the development of wave energy converters.

8.3 Types of wave energy converters

The development of WEC devices generally depends on the different concepts and design that can be used to extract energy from the wave-energy resources. These designs and ideas are classified on several bases. One of the bases of classification is the device size and operating principles. They can be categorized as attenuators,

Figure 8.3 Indian wave energy plant [8].

point absorbers, and terminators. Attenuators are devices that have a large horizontal structure perpendicular to the wave-propagation direction, whereas point absorbers are usually small devices relative to the wavelength of the prevailing waves. Terminators are generally long devices placed parallel to the incident-wave field. WECs can also be classified based on the location at which they are installed, such as offshore, nearshore, and shore mounted. Typically offshore devices are floating structures, nearshore devices are situated where the incident-wave field is affected by the water depth, thus are usually bottom mounted. The shore-mounted devices are usually connected firmly to the shore. A large number of wave energy converter prototypes are tested throughout the world. Many studies have addressed different converter topologies [9–11]. Many devices are being investigated, but are in the research and development stages, with only a small number being deployed in the ocean and tested. There are more than 1000 types of wave energy-conversion techniques that have been patented in several countries. Different designs address the categories of wave energy devices based upon location, wave climate, size, and working principal.

8.3.1 Wave energy converters based on location

Based on their location, WECs are classified as nearshore and offshore devices. Nearshore devices are generally designed in the shallow water region and often attached to the seabed. Converters are located at the shoreline and can be placed above the sea in shallow water, integrated into a breakwater, like in a dam, or fixed to a cliff. Thus this type of device generally has a stationary base with an oscillating body above it. These devices have several advantages like easy construction, maintenance, and reduced cable length used for power transmission. The main disadvantage of this type of device is the reduced wave power in the shallow water region, which limits the harnessing potential. On the other hand, offshore devices are deployed in the deep ocean where the depth of the ocean is relatively large compared to wave height. The main advantage of this type of device is that it has access to a vast amount of wave energy because of the higher content in deep-water waves [12]. However, the drawbacks include it being difficult to construct, maintain, should be designed to withstand extreme sea conditions, and the need for longer power cable length for transmission.

8.3.2 Wave energy converters based on the working principle

Based on their working principle, WECs are categorized as pressure differential devices, floating structures, overtopping devices, and impact devices. The pressure differential devices convert the hydraulic energy of waves into pneumatic energy. A major prototype developed in this range is OWC and Archimedes wave swing (AWS). Among all ocean wave extraction methods, OWC is most exploited and reputed.

8.3.2.1 Archimedes wave swing

The idea of AWS was developed around 1993 by Gardner and Breugel. This device behaves as an oscillating mass-spring-damper system. AWS is a nearshore device operating in intermediate depth and consists of a hollow cylinder and a lid called the floater. The cylinder is filled with air and is fixed on the sea bottom. The direction of the motion of the floater is vertical. When a wave crest is above the floater, it sinks due to the increased pressure acting upon it and, as a result, the air pressure inside the cylinder also increases. When the wave trough is above the AWS, the movement of the floater is in an upward direction due to the decrease in pressure of the air inside the cylinder. The main advantage of this type of device is better because it can withstand severe weather condition, as it is a submerged device. Its design is simple, so its risk of failure is also less. A schematic of the ASW system is shown in Fig. 8.4A. The AWS wave energy conversion concept, its full-scale prototype design (Fig. 8.4B) process, construction, and installation are given in Prado et al. [15].

8.3.2.2 Wave dragon

One model of overtopping devices is the wave dragon. It is an overtopping type, floating, wave-energy converter generally constructed offshore. A schematic of this overtopping device is shown in Fig. 8.5A. It extracts energy by process of overtopping of waves up a ramp into a reservoir. The waves are focused on a ramp by using two reflectors. These overtopping waves fill a floating reservoir which are

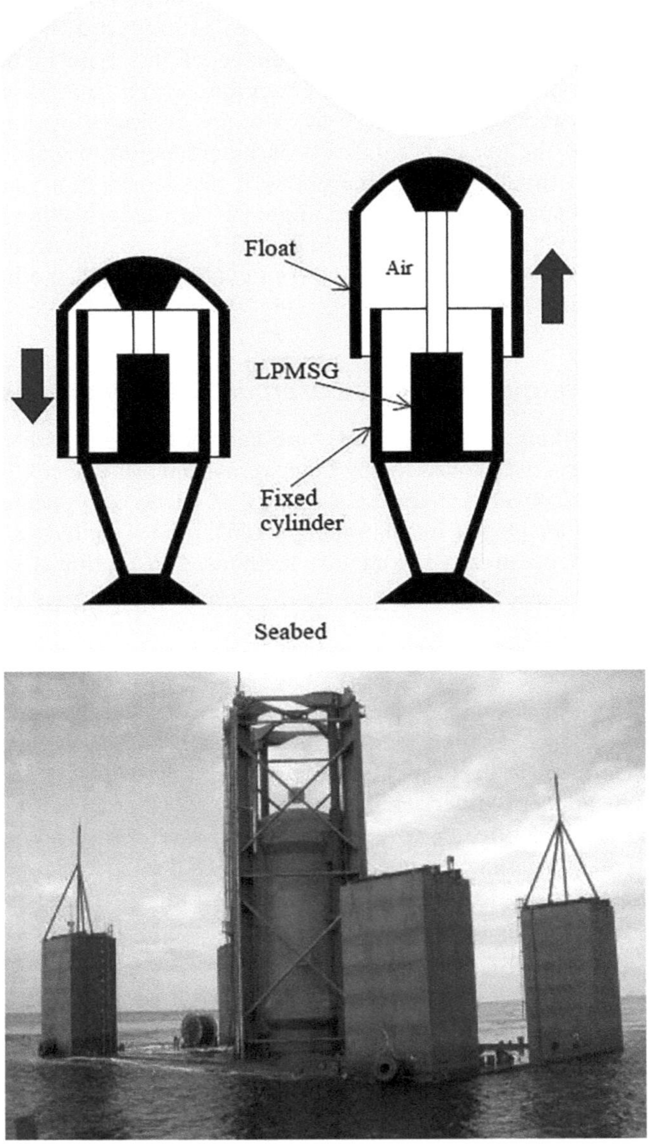

Figure 8.4 AWS (A) Schematic [13]; (B) prototype of AWS wave energy system before submersion [14].

higher than mean-sea level. This rise in water level is used to run a specially designed hydroturbine and is thus used for power production. A fully automated prototype (Fig. 8.5B) of a wave dragon is deployed in northern Denmark [17].

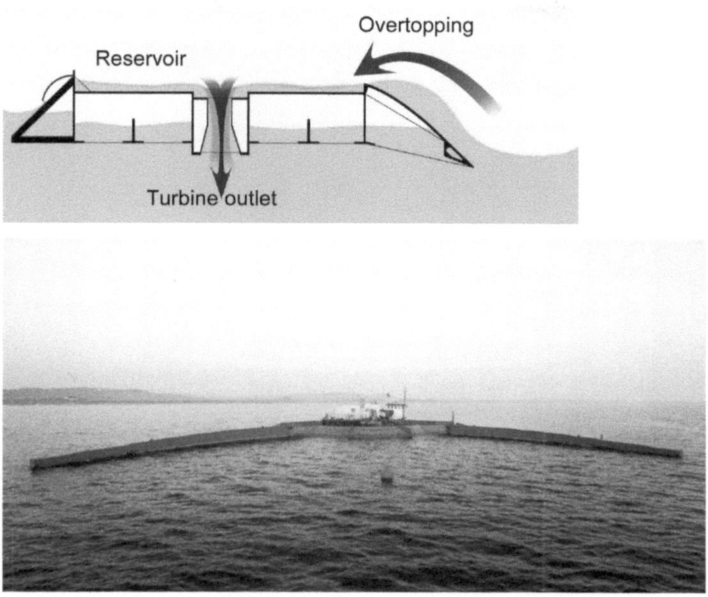

Figure 8.5 Wave dragon: (A) Schematic; (B) Wave dragon prototype [16].

8.3.2.3 Point absorbers

Point absorbers are a class of wave-energy devices that trap the heaving motion of the wave energy and convert it into electrical energy (Fig. 8.6A). These are usually wave activated heaving bodies. These devices have dimensions comparatively smaller than the incident-wave field. A few types of point absorbers are submerged devices that absorb energy using the pressure gradient. These buoys move with the heave motion of waves, thus by absorbing these motions, electricity is produced. This is realized through a hydraulic or mechanical PTO stage in-between the wave and a rotatory electrical generator, or the heaving device can be directly coupled with a linear generator (no intermediate stages).

Examples of these devices are Ocean Power Technologies' (OPT) (Fig. 8.6B) power buoy, these devices operate based on the principle of point absorbers. The OPT power buoy uses a reference plate to provide the orientation for the PTO. OPT has used different PTO methods such as oil hydraulics. OPT functions on a variety of deployment projects and has accompanied sea trials for both a 40 and 150 kW version of their technology.

8.3.2.4 Attenuator type wave energy converter

The attenuator type of wave energy converters are based on the principle of the wave activated bodies. These devices comprises of large field bodies deployed in parallel to the direction of the waves. An example of such WECs are the Pelamis

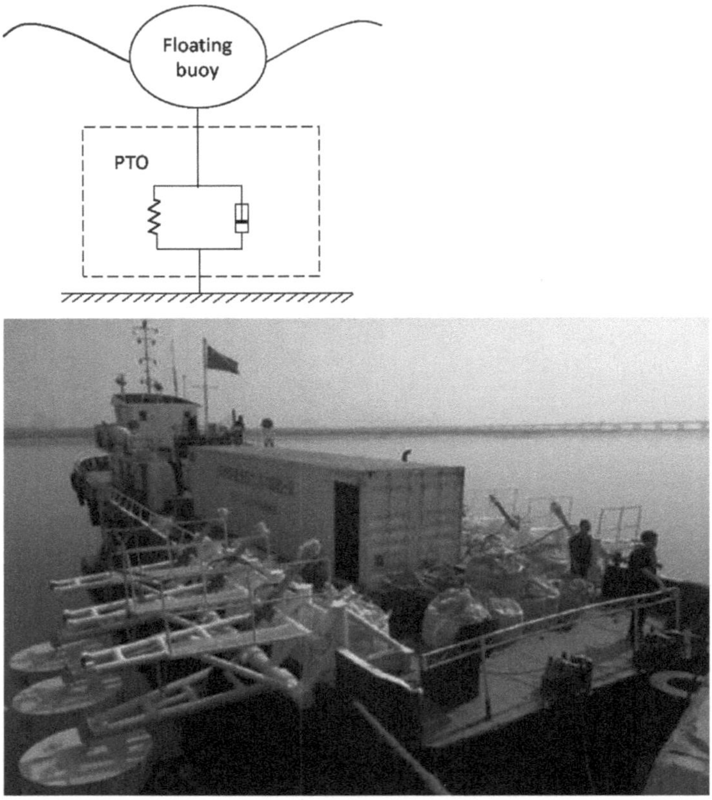

Figure 8.6 Point absorber: (A) Schematic; (B) National Ocean Technology Centre (NOTC), Tianjin, 10 kW WEC [18].

Figure 8.7 Attenuator device—Pelamis WEC [19].

WECs (Fig. 8.7) which are long, thin, semisubmerged floating devices with its long dimension aligned parallel to the direction of the waves. Usually they are an articulated, snake-like assembly with several partitions that bend (possibly twist) as the wave incident on them. The subsections are linked with universal joints that allow them to move according to the wave action. These devices absorb energy from the waves as they pass laterally along them, producing a progressive attenuation. There is current research toward the design of flexible attenuator devices that change their shape or volume, and compressing fluid as part of the PTO system.

8.3.2.5 Oscillating water column–wave energy converter

Among the numerous wave energy-extraction systems available, OWC–WEC is the most popular with a large number of prototypes all over the world [20–24]. Masuda and Mccormick [25] developed the first modern OWC device on a floating navigational buoy equipped with an air turbine. Subsequently countries like Norway, Japan, Sweden, the UK, Portugal, Denmark, India, Korea, Spain, and China developed and tested floating and fixed OWC prototypes.

An OWC–WEC plant by Voith Hydro Wavegen's land installed marine energy transformer (LIMPET) [23] was installed on the Scottish island of Islay on a breakwater system. Another prototype of an OWC device is the Mutriku plant [26] in Spain. It is an OWC consisting of 16 air turbines incorporated in breakwater.

The OWC–WEC (Fig. 8.8A and B) converts the wave motion into a reciprocating airflow that drives an air turbine. For any device to absorb wave energy, a mechanical moving interface is required to convert the oscillating nature of wave motion into a usable form of energy. OWC-based wave-energy conversion technology uses air turbines to convert the kinetic energy of air into useful mechanical energy.

An OWC contains two parts, namely the air chamber and PTO system. The air chamber absorbs the power from the incident waves and transmits it to the air inside it. It can also be designed with a hinged deflector, positioned perpendicular to the direction of the wave, which can be used to oscillate the air column utilizing the horizontal particle velocity of the wave. The chamber has an opening below the mean-sea level. When the wave is incident on the OWC system, water is forced into the chamber, which in turn forces the air to escape through the turbine [9]. The PTO is typically an air turbine coupled with a generator that converts airflow into energy. The main attraction of OWC is its simplicity as there are less mechanical moving parts in the system and no moving parts inside the water. Its concept is flexible and can be implemented on a wide range of collectors along the coastline, in nearshore devices, or in offshore structures. The concept of this device is advantageous as it uses not only wave power but also the power from the tidal motions, with the necessary mechanical isolation within the water surface and air chamber in order to get better efficiency. The wave motion generates a low frequency in the order of 0.1 Hz which must be matched with a typical electrical frequency in the order of 50 Hz. OWC–WEC is a class of device that avoids this type of problem with the help of high-speed air turbines [28]. The high-speed rotation can be directly connected to an electrical generator. Optimal design and construction of the device are the most critical issues in this technology. The geometry of the oscillating chamber can be improved by understanding the process of energy conversion at each stage. The major available OWC devices were divided into fixed and floating structures. It can be seen from the operation of the OWC–WEC

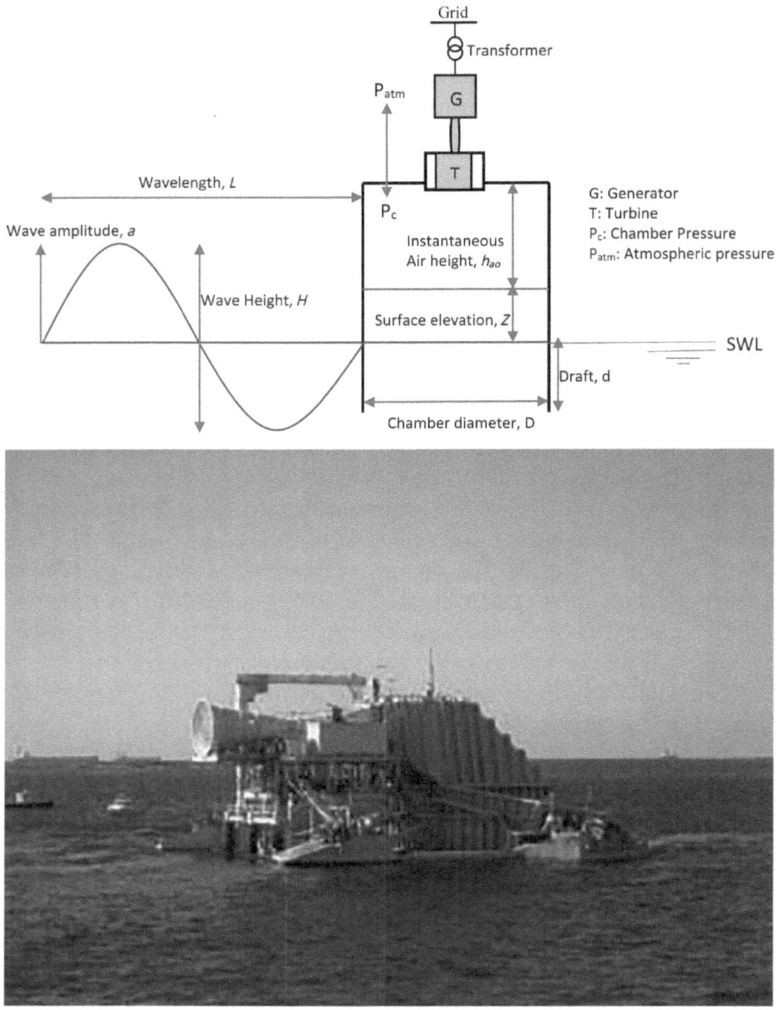

Figure 8.8 OWC–WEC: (A) Schematic; (B) Oceanlinx wave energy conversion unit at Port Kembla in Australia [27].

that the self-rectifying air turbine plays a vital role in the energy conversion process, thus it is important to understand the working of these turbines.

8.4 Power take-off systems for wave energy converters

The mechanism that converts the absorbed wave energy into usable forms of energy is known as the PTO system. Usually the WECs have a primary and secondary

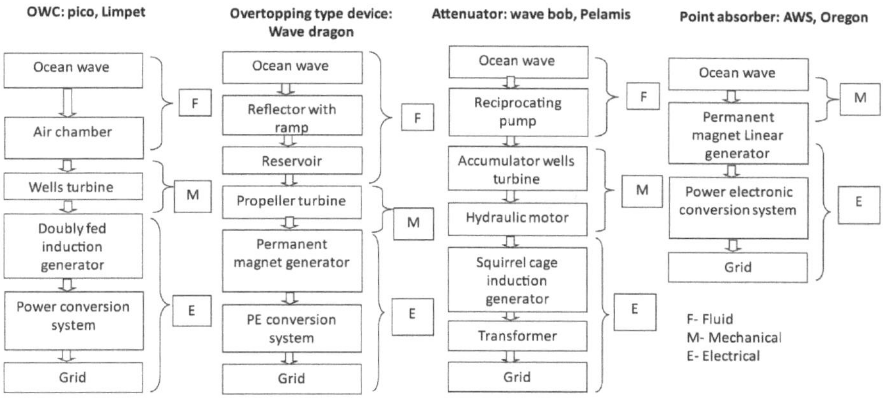

Figure 8.9 Conversion stages of wave-energy converters.

converter. For example in case of OWC−WECs the primary converter is the closed chamber and the turbines are the secondary converter that convert the air oscillation inside the chamber into mechanical energy. The operation of PTO greatly affects the performance of the overall WEC. As it affects the efficiency of energy absorption from the waves, its contribution toward the mass and size of the overall system and also affect the dynamics of the WEC. WECs are usually deployed in challenging environments, and thus the PTOs should be robust, reliable, require less maintenance, and withstand harsh environmental conditions. The types of PTOs also vary depending on the design of WECs. Some of the PTOs used are mechanical, hydraulic, and direct-drive PTOs. The direct drive PTOs can be direct electrical drives or direct mechanical drives.

WECs with different PTO systems are conceptually designed, however few of the WECs are developed as prototypes are tested and deployed in real seas, such as Mighty whale, Oceanlinx, wave roller, Oyster, Pelamis, Sakata (Japan), Tapchan, Pico (Portugal), Limpet (Scotland), Mutriku [29], Indian (Vizhinjam wave power plant, near Trivandrum, Kerala state) [30]. Fig. 8.9 shows the conversion stages of these wave energy converters. OWC is a standout among the other converters as it is the most industriously utilized strategy to harness energy from ocean waves.

8.4.1 Power take-off systems for oscillating water column−wave energy converters

The PTO for most OWC devices is the air turbine. In the early days, unidirectional turbines with a valve rectification system were used for wave-energy conversion. A common type of self-rectifying air turbine used for OWC devices is the axial flow turbine. Familiar axial turbines for OWC−WEC are the Wells turbine [31−34] and the bidirectional impulse turbine with guide vanes (GVs) [35]. Radial flow machines were tested experimentally by McCormick et al. [36], where the radial configuration of the impulse turbine with GVs was proposed as an

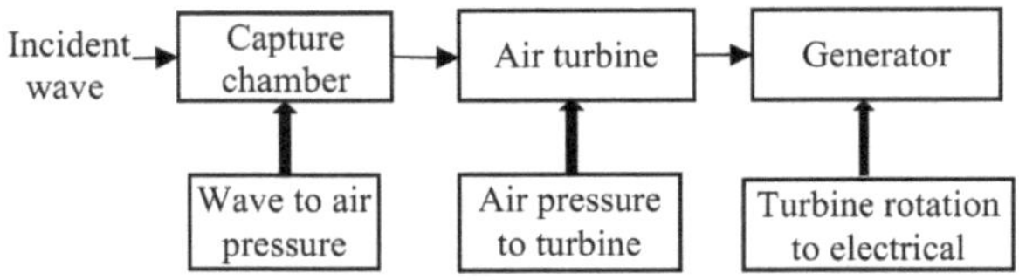

Figure 8.10 Conversion stages in an OWC.

alternative for the axial machines. However, in later years, the unidirectional system was replaced with a self-rectifying turbine. The various stages of an OWC (Fig. 8.10) are:

1. Wave to air pressure inside the capture chamber
2. Air pressure to mechanical rotation in turbine
3. Mechanical rotation to electrical power in the generator

8.4.1.1 *Wells turbine*

Research on air turbines for wave energy converters began in the 1970s by Dr. Wells, a former professor at the Queen's University of Belfast. He recommended a self-rectifying axial 90 degrees air turbine as a suitable device for OWCs. This simple form of air turbine involves a number of symmetrical airfoil blades located around a hub. On account of simplicity and efficient operation, Wells turbines are extensively employed for ocean wave energy absorption [31,32,34]. The Wells turbine is shown in Fig. 8.11A and B. The efficient design of a mechanical system to capture wave energy poses an engineering challenge. The device needs to be adequate for drawing suitable energy from a comparatively calm sea state with wave heights of only 1–5 m. It must also withstand severe wave conditions where wave heights can exceed 15 m. Under the hostile conditions of a corrosive environment, simplicity and reliability become paramount design constraints. Studies by Setoguchi et al. [38], showed the analysis of the Wells turbine for the exhalation phase separately and reported the difference in energy transfer during acceleration and deceleration. The numerical analysis of Ghisu et al. [39] showed the difference in energy transfer during acceleration and deceleration for inhalation and exhalation. This effect is caused by the compressibility effects of the OWC chamber. The experimental studies for the bidirectional airflow conditions were performed on a Wells turbine and flow characteristics were observed by Paderi and Puddu [40]. Various works reported on the design modification using optimization methods to enhance the performance of the Wells turbine from a wave energy and fluids engineering laboratory [41–47]. Their work mainly involved increasing the operating range, improving the peak efficiency, and to improve the torque coefficient.

8.4.1.1.1 Principle of operation of Wells turbine

The Wells turbine is an axial flow air turbine made up of a symmetric aerofoil blades fixed around a central hub. Blades are set at 90 degrees-stagger angles. The turbine works under the general principle of aerodynamics [31,48]. The relative velocity, V_w, acts at angle, α, known as the angle of attack, to the blade. The

Figure 8.11 Wells turbine: (A) Schematic [37]; (B) The turbine used in the OWC Plant, Vizinjham, India.

relative velocity causes a lift force, L, perpendicular to the direction of V_w and a drag force, D, in the direction of V_w. These lift and drag forces are resolved as:

Tangential force,

$$F_T = L\sin a - D\cos a \tag{8.1}$$

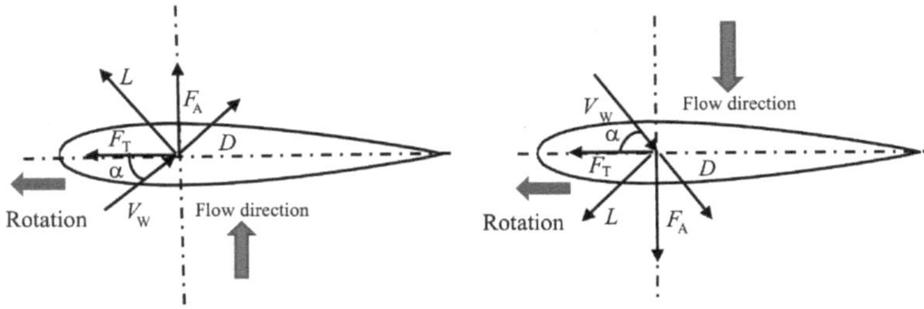

Figure 8.12 Forces acting on Wells turbine blade.

Axial force,

$$F_A = L\cos a + D\sin a \tag{8.2}$$

The tangential force direction does not change with a change in airflow direction (Fig. 8.12). The forces increase with the increase of angle of attack [49]. After a certain angle of attack, the lift force decreases and drag force increases on the aerofoil blade.

8.4.1.2 Bidirectional impulse turbine

Another common choice of a self-rectifying air turbine is the bidirectional impulse turbine. This turbine has two rows of symmetrical GVs and one rotor blade row with the same blade inlet and outlet angles. The flow leaving the downstream GV is always axial. Impulse turbines have been employed in bottom standing OWCs, rated 2×250 kW, which is functioning at Jeju Island, South Korea. A 1 m diameter impulse turbine with a self-pitch controlled GV was designed by the NIOT in India [50]. The turbine was tested in real-sea conditions on the west coast of India, and the conclusion was that even though the self-pitch–controlled GV gave better performance, it suffered a major drawback because of its complex mechanical design and high-maintenance cost. Thus the self-pitch controlled GV was replaced by a fixed GV setup. Maeda et al. [51], performed an experimental analysis on the fixed GV impulse turbine and proposed an enhancement in operation by changing the GV setting angle. Jayashankar et al. [52], studied a twin unidirectional impulse turbine having the advantage of operating two unidirectional turbines for inflow and outflow. The unidirectional turbine operation provides higher efficiency in the operating region of the exhalation, and in other operating ranges, the efficiency drop is lower than the mean efficiency of bidirectional turbines [53]. Thus a third of the flow is wasted in the inhalation phase of operation because of the high incidence losses and the overall torque production is affected due to the coupling of two turbines in the same shaft. Liu et al. [54], reported experimental steady-state performance of the bidirectional impulse turbine using a constant flow test setup and showed scatter efficiency diagrams for steady-state operation and reported the peak performance characteristics.

The impulse turbine is shown in Fig. 8.13A and B. An impulse turbine superior over the Wells turbine regarding operating range and starting characteristics. Various works

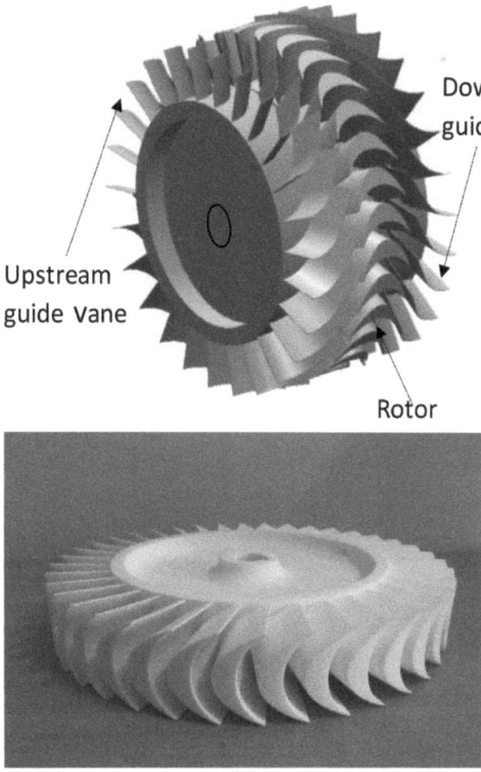

Figure 8.13 Bidirectional impulse turbine: (A) Schematic; (B) The turbine used in experimental testing.

reported on design modification using optimization techniques to alter the performance of an impulse turbine with GVs from wave energy and fluids engineering laboratory [55–61]. Their work was mainly focused on increasing the operating range, improving the peak efficiency, and improving the torque coefficient.

In order to design such a turbine, a detailed full order model of the turbine must be developed and the airflow around the turbine must be analyses to predict its performance. For such a study, high power computational methods like computational fluid dynamics or experimental testing in the OWC test rig is required.

8.4.1.2.1 Principle of operation of impulse turbine

To calculate the power output, the velocity triangle approach was followed (Fig. 8.14). The axial and tangential velocities at the inlet of the rotor are calculated as:

Absolute velocity,

$$c_2 = \frac{v}{\sin(\alpha_2)} \tag{8.3}$$

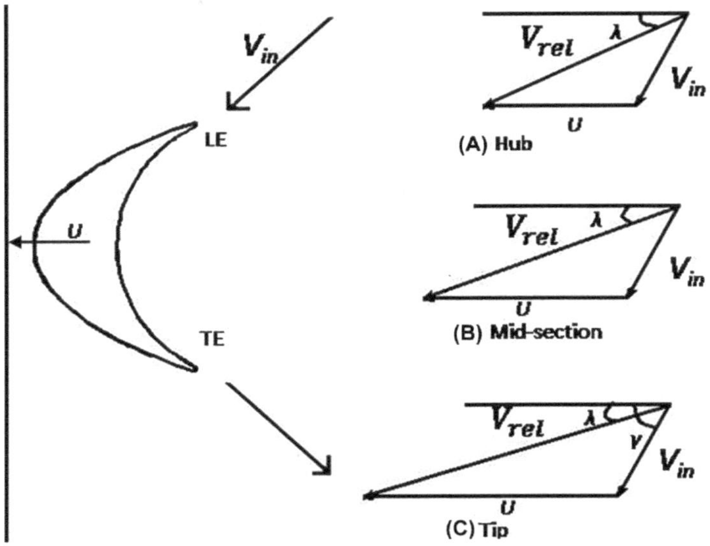

Figure 8.14 Velocity triangle for the calculation of rotor blade work or power output.

The tangential component of absolute velocity,

$$c_{2u} = \sqrt{c_2^2 - v^2} \tag{8.4}$$

At the exit of the rotor or inlet of second GV the relative velocity w_3, the tangential component of relative velocity, the tangential component of absolute velocity is calculated. The formulae are:

$$w_3 = \frac{v}{\sin(\beta_{3b})} \tag{8.5}$$

$$w_{3u} = \sqrt{w_3^2 - v^2} \tag{8.6}$$

$$c_{3u} = w_{3u} - u \tag{8.7}$$

Once the velocities are calculated, the blade work and power output of the turbine is calculated. The formulae are:

$$W_{bl} = u(c_{2u} + c_{3u}) \tag{8.8}$$

$$P = \dot{m} W_{bl} \tag{8.9}$$

From the graph of efficiency and flow coefficient, a polynomial equation has been generated and the efficiency values were calculated for different flow coefficients and substituted to estimate C_t and C_a, as given by:

$$\frac{c_t}{c_a} = \eta\phi \tag{8.10}$$

$$\text{Torque}, T = \frac{P}{\omega} \tag{8.11}$$

$$C_t = \frac{2T}{\rho(v^2 + u^2)lbzr_m} \tag{8.12}$$

$$\text{Pressuredrop}, \Delta P = \frac{c_a\rho\left(v^2 + u^2\right)lbzv}{2Q} \tag{8.13}$$

Fig. 8.11 shows the velocity triangle for the calculation of rotor blade work or power output.

8.4.2 Hydraulic power take-offs

Hydraulic PTOs are usually employed in systems where the energy conversion is based on the motion of the body actuated by the action of waves (Fig. 8.15). These PTOs are effective when the electrical devices are directly incompatible with the WEC. In hydraulic PTOs the energy capture is facilitated with the help of a hydraulic motor. These hydraulic motors act as an interface between the WEC and the electrical motors. Hydraulic PTOs are mainly adapted for point absorbers and attenuator-type WECs.

The main concern while using a hydraulic PTO for WECs are that, they are extremely sensitive in terms of performance. As these devices contain the fluid which has to address the environmental impacts in the times of leakage. However, biodegradable oils can be used to address these environmental issues. Another main concern regarding hydraulic PTOs is hydraulic storage, usually for the gas accumulators, where these storage units become heavy and thereby affect the dynamics of the system. As these hydraulic systems are comprised of many moving parts, the proper sealing of the piston is a major concern, which increases the maintenance cost of the overall system.

8.4.3 Direct mechanical drive systems

A direct mechanical drive system converts the motion of the wave into electrical energy with the help of mechanical translational systems. The mechanical system acts as the only interface between the waves and the rotary electrical generator.

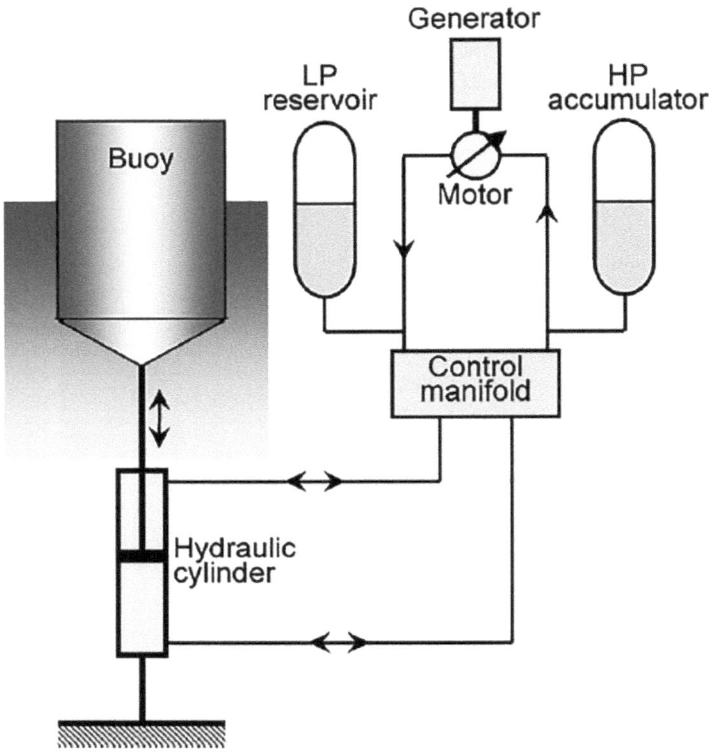

Figure 8.15 Schematic of a hydraulic PTO [10].

Examples of the direct drive mechanical system is the gear and pully system, and a flywheel system integrated with a rotatory-based system.

8.4.4 Direct electrical drive systems

The direct electrical drive is a systems that contain the moving part of the electrical generator directly activated by the wave motion. These are wave-activated electric generators that operate on the same principles as alternating current generators. Permanent magnets are employed for such PTOs. Direct electrical drive PTOs are usually linear generators (Fig. 8.16). When considering a WEC actuated by the heave motion of a wave, there is relative motion between the moving part and the stator. During this relative motion the coils in the moving part cuts the flux developed by a permanent magnet, inducing electrical current in the coil. When the wire loop undergoes relative motion between the magnets, the magnets move due to the wave-induced motion on the heaving buoy. In such systems the linear generators are the primary converters. These devices exhibit a 2 degree of freedom mechanical system generating electrical power by electromagnetic induction. As there are lesser

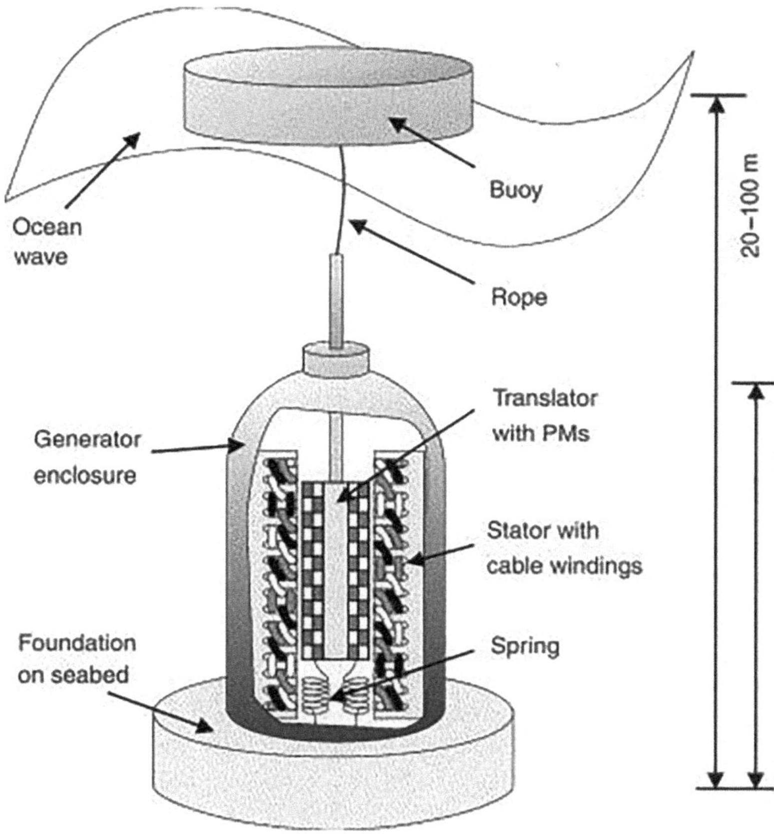

Figure 8.16 Schematic of a direct electrical PTO [62].

conversion stages they operate with minimum loss and reduced complexity. Fig. 8.14 shows a schematic of a direct electrical drive PTO attached to a buoy. The stator contains the windings mounted on a relatively stationary assembly. These PTOs are effective as they operate with a low force, high-speed input, and thus in these WECs the generator usually operates with relatively lower speeds compared to a rotary-type electrical generator.

8.4.5 Rotary electrical generators

The selection of a generator is an important factor when considering the overall efficiency of the wave energy plant. O'sullivan and Lewis [63] investigated the suitability of different generator topologies for wave-energy converters and made a comparative analysis. Traditional fossil fuel power plants use online synchronous generators that operate at a fixed speed. These generators can operate effectively for few MW and are easily controllable as the input is constant mechanical power.

On the contrary, the characteristics of renewable energy systems are completely different from the conventional energy systems such as thermal power plants. Renewable energy power plants have to cope with variable input. The input may be in the form of wind, waves, or tides; these energies are random, irregular, and unpredictable. If we consider a fixed-speed generator for renewable energy applications, it will extract only a small portion of available energy. On the other hand, if there is a heavy gust of wind or swell of waves, then the synchronous generator shaft will experience a shock load. Thus researchers in renewable energy are mainly concentrating on variable-speed generators. As these generators can operate at different speeds, maximum energy extraction from renewable energy sources is possible.

A list of different rotary generators used in wave energy converting devices include:

1. Fixed- and variable-speed electric generators [63]
   - Squirrel cage induction generator (SCIG)
2. Variable-speed electric generators
   - Doubly fed induction generator (DFIG)
   - Permanent magnet synchronous generators (PMSGs)

The selection of generators also depends on the type of wave energy device under study.

A WEC uses the turbine-generated torque or force to drive an electrical generator. However, the limits on rotational speed of the mechanical PTO also depend on the type of generator used. Thus a variable-speed generator is usually preferred for WECs [63], such that the turbine rotation can be tuned to the wave climate. Rotating electrical generators are broadly classified as fixed-speed drives and variable-speed drives. Fixed-speed electrical machines are usually used in thermal and hydroelectric power plants where the input flow conditions are kept constant. However, for renewable power sources, fixed-speed machines would not be effective to extract a large portion of the available source. Thus variable-speed drives are usually preferred for the renewable energy sector. The benefits of using variable-speed drives in the wind-energy industry are reported in the literature [64–67]; however, the similar deduction is not as clear in the ocean wave-energy sector.

8.4.5.1 Fixed-speed generators

The most common type of fixed-speed electrical drive is the SCIG. The SCIG needs soft starters to reduce the initial inrush current through the stator. These starters limit the speed during the initial period, and the speed can be increased gradually providing a ramp-up. This system needs a power-factor correction unit such as a capacitor bank to compensate for the reactive power component. As these fixed-speed drives are directly connected to the grid, they are cost-effective by eliminating the power electronic components in the circuit without adding any harmonics into the distribution network. However, these drives are not flexible and are not fully controllable on the generation side. Thus PTO systems with these drives can be controlled only in the transmission stage and cannot be controlled in the pneumatic or mechanical stage of energy transfer.

In the wave energy industry, the Pelamis WEC [68] is coupled with a fixed-speed generator (basically SCIG) that is connected to a variable displacement hydraulic motor. The performance at fixed speed is then tuned with the help of the motor by changing its displacement to control its torque. However, on the generator side, only a soft starter is used to ramp-up the speed and to correct the power factor of the output power.

Synchronous generators are an important part of the power generation system using fossil fuels. As these generators run on an effectively constant speed, they are synchronized to electric grids. The output of synchronous generators is effectively optimized and are currently in use. The main problem comes with renewable energy resources. The efficiency of the fixed-speed generators is less in the renewable energy sector due to the variable speed nature of renewable energy sources. The control strategies should be well optimized for getting maximum efficiency from the system. Thus under variable sea conditions it is difficult for synchronous generators to produce output efficiently. The fixed-speed generators will be affected by severe shock loads on the generator shaft. Considering these mechanical aspects, the solution led to the increased use of asynchronous generators with variable-speed operations. The features of renewable power sources, which are extremely variable in nature, recommends that variable-speed generators are beneficial over fixed-speed generators to absorb a larger fraction of the existing renewable power source.

8.4.5.2 Variable-speed generators

For variable-speed drives, different configurations have been suggested by various WEC developers [69]: A DFIG for an OWC converter [23,70], a PMSG is used for the wave energy device known as Lifesaver [71] and for the Wave dragon converter [17], and a variable-speed induction generator by Wave star [72].

The selection of the generator also depends on its robustness to withstand the unpleasant atmospheric conditions around the sea since the salinity present in the air affects metal and reduces the efficiency of the movable parts in the machine. Therefore if movable parts are less, then the machine will require less maintenance and will be more cost-efficient. However, the recent development of the power electronics industry has led to the increased use of switching devices for fully controllable variable-speed drives with the options of using a DFIG or PMSG for generating electricity in wave-energy devices.

The reason for choosing these two generators is that, the DFIG is cost-efficient and delivers required power with a smaller size of generators, so it requires less space. The power electronic circuits required for DFIG are rated at 30%–50% of the system-rated power. So its cost, size, and power efficiency are optimum for a variable-speed system. In the case of PMSG, the cost is higher because of the high cost of large permanent magnets and the power electronics circuits are rated at 100% of the generator rated power. But the reason for using PMSG is that it is more reliable and requires low maintenance and as it is brushless, the turbine is directly coupled to the generator shaft, which reduces the gearbox mechanism and its maintenance cost.

8.5 Future of wave energy

The economic, environmental, and geo-political constraints on the use of conventional energy resources have led to a revolution in energy harvesting from renewable sources. Therefore more scientific and advanced methods have evolved to enhance the efficiency of devices that harvest energy from various ecological-friendly resources and convert them into electrical energy.

In the future ocean wave energy is a promising renewable energy source to meet the global energy demand. Several WEC devices have been investigated but only a few of them have proceeded to the second phase and even fewer are near to the final stage to undergo commercial deployment. WECs have higher installation costs and are high maintenance, making the development of more robust and reliable designs necessary. Nearshore devices are expected to grow as they have easy accessibility and ease of maintenance. The past few decades have seen tremendous research work on designing efficient WECs and application of this technology to enhance operation. Thus the future of WECs will undergo a first phase in which a large number of scaled down models of prototype will be tested in laboratories and with field testing of these devices. The next stage is designing full-scale models and testing in real sea. This phase is possible only with more funding from government. Finally, after improving the performance of the WECs the energy produced can be supplied to the grid, followed by increasing the number of devices in allocation, making it a wave energy farms which will guarantee the continuous energy production to supply the grid.

8.6 Summary

This chapter focuses on marine power technology by discussing wave-energy converters. The potential and kinetic energy confined in ocean wave oscillations are a powerful driving force that can be utilized to generate usable forms of energy. The different types of wave energy concepts are explained along with the power take-off systems for such converters. In conclusion, the chapter considers the promising future of the WEC applications.

References

[1] A. Khaligh, O.C. Onar, Energy Harvesting: Solar, Wind, and Ocean Energy Conversion Systems, Taylor and Francis Group, LLC, 2010, pp. 223–303.

[2] S.A. Sannasiraj, V. Sundar, Assessment of wave energy potential and its harvesting approach along the Indian coast, Renew. Energy 99 (2016) 398–409. Available from: https://doi.org/10.1016/j.renene.2016.07.017.

[3] A. Clément, et al., Wave energy in Europe : current status and perspectives, Renew. Sustain. Energy Rev. 6 (5) (2002) 405–431. Available from: https://doi.org/10.1016/S1364-0321(02)00009-6.

[4] J. Falnes, A review of wave-energy extraction, Mar. Struct. 20 (4) (2007) 185–201. Available from: https://doi.org/10.1016/j.marstruc.2007.09.001.

[5] K. Gunn, C. Stock-Williams, Quantifying the global wave power resource, Renew. Energy 44 (2012) 296–304. Available from: https://doi.org/10.1016/j.renene.2012.01.101. Elsevier Ltd.

[6] T.V.S.N. Rao, V. Sundar, Estimation of wave power ptential along the Indian coastline, Energy 10 (1982) 839–845.

[7] V. Raju, M. Ravindran, Wave energy: potential and programme in India, Renew. Energy 10 (2) (1997) 339–345.

[8] N. Sharmila, et al., Wave powered desalination system, Energy 29 (11) (2004) 1659–1672. Available from: https://doi.org/10.1016/j.energy.2004.03.099.

[9] B. Drew, A.R. Plummer, M.N. Sahinkaya, A review of wave energy converter technology, Proc. Inst. Mech. Eng. Part A: J. Power Energy 223 (8) (2009) 887–902. Available from: https://doi.org/10.1243/09576509JPE782.

[10] A.F.D.O. Falcão, Wave energy utilization: a review of the technologies, Renew. Sustain. Energy Rev. 14 (3) (2010) 899–918. Available from: https://doi.org/10.1016/j.rser.2009.11.003.

[11] I. López, et al., Review of wave energy technologies and the necessary power-equipment, Renew. Sustain. Energy Rev. 27 (2013) 413–434. Available from: https://doi.org/10.1016/j.rser.2013.07.009.

[12] S.A. Sebo, Ocean power: wave and tidal energy review, Marit. Stud. Manag. 2 (4) (2004) 202–214. Available from: https://doi.org/10.1080/03088837500000027.

[13] M.I. Marei, M. Mokhtar, A.A. El-Sattar, MPPT strategy based on speed control for AWS-based wave energy conversion system, Renew. Energy 83 (2015) 305–317. Available from: https://doi.org/10.1016/j.renene.2015.04.039.

[14] H. Polinder, M. Scuotto, Wave energy converters and their impact on power systems, Future Power Systems, 2005... (2005) 1–9. Available at:. Available from: http://ieeexplore.ieee.org/xpls/abs_all.jsp?arnumber = 1600483.

[15] M. Prado, H. Polinder, Direct drive in wave energy conversion—AWS full scale prototype case study, in: 2011 IEEE Power and Energy Society General Meeting, IEEE (2011) 1–7. doi:10.1109/PES.2011.6039720.

[16] J.P. Kofoed, et al., Prototype testing of the wave energy converter wave dragon, Renew. Energy 31 (2) (2006) 181–189. Available from: https://doi.org/10.1016/j.renene.2005.09.005.

[17] J. Tedd, et al., Wave dragon, prototype wave power production, In: A.A.M. Sayigh (Ed.), Proceedings of the 9th World Renewable Energy Congress: WREC IX, Florence, Italy, Pergamon Press (August 2006).

[18] A. Kolios, et al., Reliability assessment of point-absorber wave energy converters, Ocean Eng. 163 (2018) 40–50. Available from: https://doi.org/10.1016/j.oceaneng.2018.05.048 (May).

[19] R. Yemm, et al., Pelamis: experience from concept to connection, Philos. Trans. R. Soc. A: Math. Phys. Eng. Sci. 370 (1959) (2012) 365–380. Available from: https://doi.org/10.1098/rsta.2011.0312.

[20] M. Folley, R. Curran, T. Whittaker, Comparison of LIMPET contra-rotating wells turbine with theoretical and model test predictions, Ocean Eng. 33 (2006) 1056–1069. Available from: https://doi.org/10.1016/j.oceaneng.2005.08.001 (August 2001).
[21] T.V. Heath, A review of oscillating water columns, Philos. Trans. R. Soc. A: Math. Phys. Eng. Sci. 370 (1959) (2012) 235–245. Available from: https://doi.org/10.1098/rsta.2011.0164.
[22] M. Ravindran, et al., The Indian wave energy program – an overview, Teri. Inf. Dig. Energy 7 (3) (1997) 173–188. Available at:. Available from: https://inis.iaea.org/search/search.aspx?orig_q = RN:29004501 (accessed 9.02.18).
[23] T.J.T. Whittaker, Islay LIMPET wave power plant. Technical report. Queen's University of Belfast, 2002.
[24] T.J.T. Whittaker, et al., The Limpet wave power project – the first years of operation, Renew. Energy 1997 (2004) 1–8.
[25] Y. Masuda, M.E. Mccormick, Experiences in pneumatic wave energy conversion in Japan, in: M.E. McCormick, Y.C. Kim (Eds.), Utilization of Ocean Waves—Wave to Energy Conversion, American Society of Civil Engineers, 1986, pp. 1–33.
[26] Y. Torre-Enciso, et al., Mutriku wave power plant: from the thinking out to the reality, in: 8th European Wave and Tidal Energy Conference (EWTEC 2009), pp. 319–328. Available at: <http://www.mendeley.com/catalog/mutriku-wave-power-plant-thinking-reality/%5Cnhttp://tethys.pnnl.gov/sites/default/files/publications/Torre-Enciso_et_al_2009.pdf> (2009).
[27] A.F.O. Falcão, J.C.C. Henriques, Oscillating-water-column wave energy converters and air turbines: a review, Renew. Energy 85 (2015) 1–34. Available from: https://doi.org/10.1016/j.renene.2015.07.086.
[28] D.V. Evans, R. Porter, Hydrodynamic characteristics of an oscillating water column device, Appl. Ocean Res. 17 (3) (1995) 155–164. Available from: https://doi.org/10.1016/0141-1187(95)00008-9.
[29] I. Heras-Saizarbitoria, I. Zamanillo, I. Laskurain, Social acceptance of ocean wave energy: a case study of an OWC shoreline plant, Renew. Sustain. Energy Rev. 27 (2013) 515–524. Available from: https://doi.org/10.1016/j.rser.2013.07.032.
[30] S. Anand, et al., Performance estimation of bi-directional turbines in wave energy plants', J. Therm. Sci. 16 (4) (2007) 346–352. Available from: https://doi.org/10.1007/s11630-007-0346-1.
[31] L.M.C. Gato, A.F.D.O. Falcão, Aerodynamics of the Wells turbine, Int. J. Mech. Sci. 30 (6) (1988) 383–395. Available from: https://doi.org/10.1115/1.3239616.
[32] S. Raghunathan, O.O. Ombaka, Effect of frequency of air flow on the performance of the Wells turbine, Int. J. Heat. Fluid Flow. 6 (2) (1985) 127–132.
[33] S. Raghunathan, C.P. Tan, Performance of the Wells turbine at starting, J. Energy 6 (6) (1981) 430–431. Available from: https://doi.org/10.2514/2.579.
[34] S. Raghunathan, C.P. Tan, Effect of blade profile on the performance of the Wells self-rectifying air turbine, Int. J. Heat. Fluid Flow. (1985) 1–6.
[35] K. Kaneko, T. Setoguchi, S. Raghunathan, Self-rectifying turbines for wave energy conversion, Int. J. Offshore Polar Eng. 2 (3) (1992) 1988–1990.
[36] M.E. McCormick, J.G. Rehak, B.D. Williams, An experimental study of a bidirectional radial turbine for pneumatic wave energy conversion, Oceans 1992—Proc. Master. Ocean Technol. 2 (1992) 866–870. Available from: https://doi.org/10.1109/OCEANS.1992.607698.
[37] T.K. Das, P. Halder, A. Samad, Optimal design of air turbines for oscillating water column wave energy systems: a review, Int. J. Ocean Clim. Syst. 8 (1) (2017) 37–49. Available from: https://doi.org/10.1177/1759313117693639.

[38] T. Setoguchi, et al., Air-turbine with self-pitch-controlled blades for wave energy conversion (Estimation of performances in periodically oscillating flow), Int. J. Rotating Mach. 3 (4) (1997) 233–238. Available from: https://doi.org/10.1155/S1023621X97000213.

[39] T. Ghisu, P. Puddu, F. Cambuli, Physical explanation of the hysteresis in Wells turbines: a critical reconsideration, J. Fluids Eng. 138 (11) (2016) 111105. Available from: https://doi.org/10.1115/1.4033320.

[40] M. Paderi, P. Puddu, Experimental investigation in a Wells turbine under bi- directional flow, Renew. Energy 57 (2013) 570–576. Available from: https://doi.org/10.1016/j.renene.2013.02.016.

[41] P. Halder, et al., High performance ocean energy harvesting turbine design: a new casing treatment scheme, Energy 86 (June) (2015) 219–231. Available from: https://doi.org/10.1016/j.energy.2015.03.131.

[42] P. Halder, M.H. Mohamed, A. Samad, Wave energy conversion: design and shape optimization, Ocean Eng. 150 (2018) 337–351. Available from: https://doi.org/10.1016/j.oceaneng.2017.12.072 (December 2017).

[43] P. Halder, A. Samad, Tip leakage effect on Wells turbine, Renew. Energy Offshore (2015) 555–560.

[44] P.M. Kumar, et al., Performance enhancement of Wells turbine: combined radiused edge blade tip, static extended trailing edge, and variable thickness modifications, Ocean Eng. 185 (2019) 47–58. Available from: https://doi.org/10.1016/j.oceaneng.2019.05.041 (May).

[45] P.M. Kumar, A. Samad, Introducing Gurney flap to Wells turbine blade and performance analysis with OpenFOAM, Ocean Eng. 187 (2019) 106212. Available from: https://doi.org/10.1016/j.oceaneng.2019.106212 (May).

[46] R. Suchithra, K. Ezhilsabareesh, A. Samad, Development of a reduced order wave to wire model of an OWC wave energy converter for control system analysis, Ocean Eng. 172 (2019) 614–628. Available from: https://doi.org/10.1016/j.oceaneng.2018.12.013 (July 2018).

[47] T.K. Das, A. Samad, The effect of midplane guide vanes in a biplane wells turbine, J. Fluids Eng., Trans. ASME 141 (5) (2019) 1–13. Available from: https://doi.org/10.1115/1.4041600.

[48] S. Raghunathan, C.P. Tant, Aerodynamic performance of a Wells air turbine, J. Energy 7 (1983) 226–230 (January).

[49] R.E. Sheldahl, P.C. Klimas, Aerodynamic characteristics of seven symmetrical airfoil sections through 180-degree angle of attack for use in aerodynamic analysis of vertical axis wind turbines. Sandia National Laboratories Energy Report. The United States of America, 1981.

[50] S. Santhakumar et al., Performance of an impulse turbine based wave energy plant, in: Proceedings of Eighth International Offshore and Polar Engineering Conference. Montreal, Canada (1998) 75–80.

[51] H. Maeda, et al., Performance of an impulse turbine with fixed guide vanes for wave power conversion, Renew. Energy 17 (4) (1999) 533–547. Available from: https://doi.org/10.1016/S0960-1481(98)00771-X.

[52] V. Jayashankar, et al., A twin unidirectional impulse turbine topology for OWC based wave energy plants, Renew. Energy 34 (3) (2009) 692–698. Available from: https://doi.org/10.1016/j.renene.2008.05.028.

[53] B. Pereiras, P. Valdez, F. Castro, Numerical analysis of a unidirectional axial turbine for twin turbine configuration, Appl. Ocean Res. 47 (2014) 1–8. Available from: https://doi.org/10.1016/j.apor.2014.03.003.

[54] Z. Liu, et al., Steady state performance of an axial impulse turbine for oscillating water column wave energy converters, Energy 141 (2017) 1–10. Available from: https://doi.org/10.1016/j.energy.2017.09.068.
[55] K. Ezhilsabareesh, S.H. Rhee, A. Samad, Shape optimization of a bidirectional impulse turbine via surrogate models, Eng. Appl. Comput. Fluid Mech. 12 (1) (2018) 1–12. Available from: https://doi.org/10.1080/19942060.2017.1330709.
[56] K. Ezhilsabareesh, R. Suchithra, A. Samad, Performance enhancement of an impulse turbine for OWC using grouped grey wolf optimizer based controller, Ocean Eng. 190 (August) (2019) 106425. Available from: https://doi.org/10.1016/j.oceaneng.2019.106425.
[57] A. George, et al., Experimental analysis of turbine-chamber coupling for wave energy conversion, Int. J. Energy Res. 42 (15) (2018) 4770–4782. Available from: https://doi.org/10.1002/er.4230.
[58] R. Badhurshah, et al., High efficiency design of an impulse turbine used in oscillating water column to harvest wave energy, Renew. Energy 121 (2018) 344–354. Available from: https://doi.org/10.1016/j.renene.2018.01.028.
[59] R. Badhurshah, A. Samad, Multiple surrogate based optimization of a bidirectional impulse turbine for wave energy conversion, Renew. Energy 74 (2015) 749–760. Available from: https://doi.org/10.1016/j.renene.2014.09.001.
[60] B. Ranjith, P. Halder, A. Samad, High-performance ocean energy harvesting turbine design—detailed flow analysis with blade leaning strategy, Proc. Inst. Mech. Eng. Part A: J. Power Energy 0 (0) (2018). Available from: https://doi.org/10.1177/0957650918787692. p. 095765091878769.
[61] R. Suchithra, K. Ezhilsabareesh, A. Samad, Optimization based higher order sliding mode controller for efficiency improvement of a wave energy converter, Energy 187 (2019) 116111. Available from: https://doi.org/10.1016/j.energy.2019.116111.
[62] K. Thorburn, M. Leijon, Farm size comparison with analytical model of linear generator wave energy converters, Ocean Eng. 34 (5–6) (2007) 908–916. Available from: https://doi.org/10.1016/j.oceaneng.2006.05.017.
[63] D.L. O'sullivan, A.W. Lewis, Generator selection and comparative performance in offshore oscillating water column ocean wave energy converters, IEEE Trans. Energy Convers. 26 (2) (2011) 603–614. Available from: https://doi.org/10.1109/TEC.2010.2093527.
[64] S. Mathew, G.S. Philip, *Advances in Wind Energy Conversion Technology, Environmental Science and Engineering*, Springer-Verlag, 2011. Available from: http://dx.doi.org/10.1007/978-3-540-88258-9.
[65] J. Milewski, et al., Wind Energy Conversion Systems, Fuel Cells, Springer, 2012. Available from: http://dx.doi.org/10.1007/978-1-4471-2201-2.
[66] W. Bin, et al., Power Conversion and Control of Wind Energy Systems, John Wiley & Sons, Inc., Hoboken, NJ, 2011. Available from: http://dx.doi.org/10.1017/CBO9781107415324.004.
[67] D. Rekioua, Wind Power Electric Systems, Springer, 2014. Available from: http://dx.doi.org/10.1007/978-1-4471-6425-8.
[68] R. Henderson, Design, simulation, and testing of a novel hydraulic power take-off system for the Pelamis wave energy converter, Renew. Energy 31 (2006) 271–283. Available from: https://doi.org/10.1016/j.renene.2005.08.021.
[69] K. Chau, W. Li, C. Lee, Challenges and opportunities of electric machines for renewable energy, Prog. Electromagn. 42 (2012) 45–74. Available at: <http://search.ebscohost.com/login.aspx?direct = true&profile = ehost&scope = site&authtype = crawler&jrnl = 19376472&AN = 79351976&h = 8oV3MTk51ZHdngJWXapuDNBjMCV073y%

2BsQJR1NeDyCjEz8k7e0tpbsc%2BJu82BtGaccDO32rds9VMKyEwJfJh1Q%3D%3D&crl = c>.
[70] F.-X. Faÿ, et al., Sea trial results of a predictive algorithm at the Mutriku wave power plant and controllers assessment based on a detailed plant model, Renew. Energy 146 (2019) 1725–1745. Available from: https://doi.org/10.1016/j.renene.2019.07.129.
[71] J. Sjolte, G. Tjensvoll, M. Molinas, Power collection from wave energy farms, Appl. Sci. 3 (2) (2013) 420–436. Available from: https://doi.org/10.3390/app3020420.
[72] M.M. Kramer, et al., Performance evaluation of the Wavestar prototype, in: 9th European Wave and Tidal Energy Conference (2011).

Section VIII

Hydrogen Fuel

Renewable hydrogen production by water electrolysis

9

Ernesto Amores, Mónica Sánchez, Nuria Rojas and Margarita Sánchez-Molina
Centro Nacional del Hidrógeno (CNH2), Puertollano, Ciudad Real (Spain)

9.1 Introduction

The increase of pollutant emissions generated by the use of fossil fuels, together with the growth of the population worldwide and the increase of energy demand in emerging economies, bring to light the need of looking for an alternative energy system not based on finite and exhaustible resources such as coal, petroleum, or natural gas. To these reasons energy safety and stability should be added as most deposits and fields are located only in a few areas of the planet.

In this context, renewable energies serve as essential tools to solve these problems and change the current energy model to a more sustainable, safe, and environment-friendly one, especially in those countries where energy dependence is high. Nevertheless, renewable energies have a great variability in their operation as they depend on the weather conditions existing at any given moment. Due to this, it is necessary to establish some mechanisms that allow these fluctuations to be mitigated. In this sense, energy storage becomes a vital element to achieve greater flexibility in these power generation systems, especially in more decentralized or distributed power grids; this would allow for generation to be adjusted to demand.

Nowadays, there are different storage systems capable of storing energy for anywhere between several seconds and some hours such as capacitors, flywheels, or batteries (electrochemical storage) to compressed air energy storage systems and hydroelectric pumping stations that allow for energy to be stored for much longer periods of time. However, it is necessary to use energy storage systems of a greater capacity than those currently available to ensure a higher penetration of renewable energy sources. In this context, hydrogen may play an important role in the storage of large renewable surpluses.

Unlike fossil fuels, hydrogen is not a natural source of energy, but an energy carrier; and due to this, it has to be produced from other sources. Nowadays, hydrogen is mainly produced from steam methane reforming (SMR) because of its low cost, in a process based on the reaction of methane with steam at a high temperature [1]. Nevertheless, hydrogen production without the consumption of fossil fuels is the key to achieving a more sustainable energy model.

Hydrogen production from renewable energies, thus, becomes one of the greatest challenges for the future. It will not only reduce the volume of pollutant emissions,

Sustainable Fuel Technologies Handbook. DOI: https://doi.org/10.1016/B978-0-12-822989-7.00010-X

but will also decrease fossil fuel dependence; besides, it may be used as a way of storing renewable energies, thus, offering the possibility of transporting and using that energy at another time by converting it into electricity though a fuel cell for different applications. It would also be possible to use it in mobility through fuel cell electric vehicles and even through its transformation into new energy carriers (such as synthetic methane or ammonia, by processes called *power-to-X*).

Water electrolysis, thermochemical cycles, and photocatalytic processes are some of the most important processes that are not based on fossil fuels used to produce hydrogen. Nowadays, the efficiencies of the latter two are low and they are uncompetitive in economic terms [1,2]. However, the electrolytic rupture of water molecules is a technology more and more promising for hydrogen production on a large scale, whenever the energy produced comes from renewable sources.

Nevertheless, only 4% of the hydrogen generated worldwide currently comes from renewable sources (mainly electrolysis); meanwhile, almost 50% of it is obtained from SMR [3]. Fig. 9.1 summarizes the main energy sources from which hydrogen can be extracted as well as the most important processes to obtain it.

9.2 Principles of water electrolysis

Water electrolysis is a process known for more than 200 years [2]. It consists of the decomposition of water molecules into oxygen and hydrogen by applying a determined amount of energy (electric current and heat), causing a global reaction of oxidation-reduction (Eq. 9.1):

$$H_2O \rightarrow H_2 + \frac{1}{2}O_2 \tag{9.1}$$

A direct current is applied by a pair of electrodes connected to an electric power supply. The positive electrode (+) is known as the anode (*oxygen evolution reaction*, OER) and the negative one (−) is the cathode (*hydrogen evolution reaction*, HER). In the electrodes, there is a transfer between electrons and ions. Negative

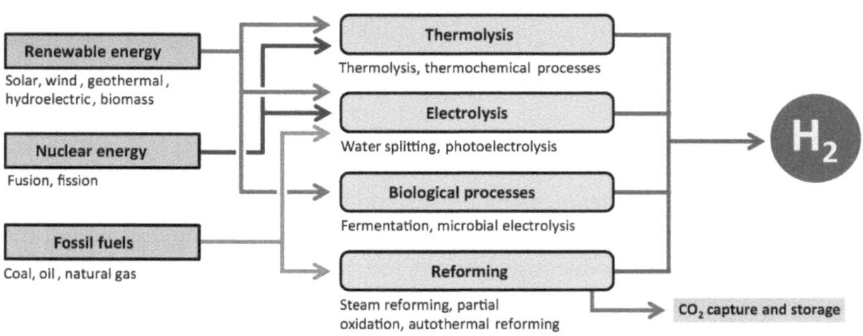

Figure 9.1 Main processes for hydrogen production.

ions or anions give electrons to the anode (+) and positive ions or cations take electrons from the cathode (−).

9.2.1 Thermodynamics of water electrolysis

9.2.1.1 Equilibrium conditions

The electrolysis reaction is an endothermic process that, thus, requires an external energy supply in order to be produced ($\Delta G > 0$). This energy can be supplied as electricity or heat. Assuming the reaction is reversible, the energy necessary for this process can be quantified using the laws of thermodynamics (Eq. 9.2):

$$\Delta H = \Delta G + T \times \Delta S \tag{9.2}$$

where ΔH shows the total energy (enthalpy) that must be supplied to divide the water molecule. This energy can have a thermal origin (entropy, $T \times \Delta S$), which represents the system irreversibilities, or an electric one (Gibbs free energy, ΔG). In standard conditions (25°C, 1 bar), these parameters take the values of $\Delta h^0 = 285.88$ kJ/mol, $\Delta g^0 = 237.23$ kJ/mol, and $T \times \Delta s^0 = 48.65$ kJ/mol [2].

Fig. 9.2A shows the variations of the different terms in Eq. (9.2) regarding the temperature. It is observed that the electricity demand (ΔG) is reduced as the temperature (T) at which electrolysis is verified increases. This reduction is much more significant when the process takes place with steam instead of liquid water. The major change noticed in ΔH and $T \times \Delta S$ at 100°C is due to water vaporization. Concerning the pressure, its influence is not relevant in the total energy (ΔH) necessary for the process [4].

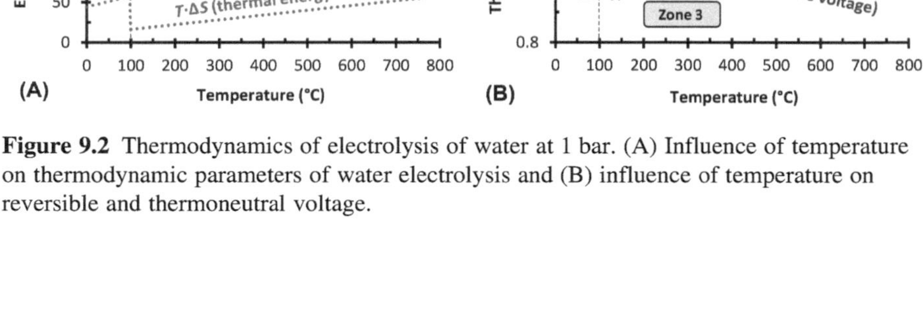

Figure 9.2 Thermodynamics of electrolysis of water at 1 bar. (A) Influence of temperature on thermodynamic parameters of water electrolysis and (B) influence of temperature on reversible and thermoneutral voltage.

9.2.1.2 *Minimum cell voltage*

9.2.1.2.1 Reversible voltage

Reversible or equilibrium cell voltage (U_{rev}) is defined as the minimum voltage necessary to be applied between the anode and cathode for electrolysis to happen. From a thermodynamic point of view, the work necessary to be supplied in order to carry out this process at a constant temperature and pressure is equal to the increase of Gibbs free energy (ΔG). If the electrolyzer used can only exchange electricity and does not add heat, this equation is verified:

$$U_{rev} = \frac{\Delta g}{z \times F} \tag{9.3}$$

where U_{rev} is the reversible voltage (in V), z is the number of electrons transferred in water electrolysis (2e$^-$), and F is the Faraday constant (96,500 C/mol).

When working in equilibrium conditions (25°C, 1 bar), the reversible potential can be defined by the Gibbs free energy in standard conditions (U^0_{rev}). Thus the minimum thermodynamic tension to divide water into hydrogen and oxygen at 25°C and 1 bar is 1.229 V, whenever conditions are ideal and there are no resistances. The value of the reversible voltage in other conditions is described by the Nernst equation (Eq. 9.4):

$$U_{rev} = U^0_{rev} + \frac{\mathrm{R} \times T}{z \times F} \times \ln\left(\frac{p_{O_2}^{1/2} \times p_{H_2}}{a_{H_2O}}\right) \tag{9.4}$$

where p_{H_2} and p_{O_2} are the partial pressure of the reaction products, a_{H_2O} is the water activity and R is the absolute gas constant (8.314 J/K/mol).

9.2.1.2.2 Thermoneutral voltage

When the electrolytic process is carried out under adiabatic conditions, the enthalpy of the total reaction (ΔH) must be supplied only by electric current, which is required to maintain the reaction without generating or absorbing heat. Thus as it happens with the reversible voltage, the enthalpy variation allows the thermoneutral or enthalpic cell voltage (U_{tn}) to be defined by Eq. 9.5:

$$U_{tn} = \frac{\Delta h}{z \times F} \tag{9.5}$$

In equilibrium conditions (25°C, 1 bar), the thermoneutral voltage (U^0_{tn}) is 1.48 V. The difference between reversible and thermoneutral voltage is known as thermal or entropic voltage and represents the minimum overvoltage regarding the reversible voltage that must be applied to the electrolytic cell to start the water dissociation. In standard conditions, this potential is 0.25 V [5].

9.2.1.3 Theoretical operating ranges in water electrolysis

As mentioned previously, $U^0_{rev} = 1.23$ V and $U^0_{tn} = 1.48$ V in standard operation conditions. Fig. 9.2B shows how these potentials vary with temperature. As is observed, the U_{tn} remains almost constant (with a determined value below or above 100°C) due to the fact that the variation of the reversible voltage (associated with ΔG) in the face of temperature changes is compensated by the variation of thermal energy ($T \times \Delta S$), in accordance with Eq. (9.2) and Fig. 9.2A.

Fig. 9.2B also allows for the different zones to be defined where an electrolysis cell can operate if the tension applied between the cell electrodes (U) is fixed and the operating temperature (T) is left variable to any intensity value (I) circulating through the electrolytic cell [6]:

- *Zone 1* is produced when, at a particular temperature, a voltage greater than the thermoneutral voltage for that pressure and temperature is applied. Electrolysis is produced in these conditions, and the energy excess is degraded in heat ($\dot{Q}_1$), which has to be removed through the refrigeration system of the electrolyzer:

$$\dot{Q}_1 = (U - U_{tn}) \times I \tag{9.6}$$

- *Zone 2* occurs when a potential between the thermoneutral and the reversible voltage is applied ($U_{tn} > U > U_{rev}$). In these conditions, electrolysis is produced as long as it provides thermal energy ($\dot{Q}_2$) of the same value as the result of the difference between the thermoneutral and the applied voltage:

$$\dot{Q}_2 = (U_{tn} - U) \times I \tag{9.7}$$

- *Zone 3* is the situation caused by applying a voltage below the reversible voltage ($U < U_{rev}$); the consequence of this is the impossibility of electrolysis being carried out.

9.2.2 Characterization of the water electrolysis process

9.2.2.1 Kinetic losses and overpotentials

For practical purposes, when an electrolysis cell is working, the potential is always higher than the reversible voltage (U_{rev}) and even higher than the thermoneutral voltage (U_{tn}) due to the irreversibilities of the real electrolysis process, which may be expressed as overpotentials ($\eta^{\vee}$). These factors are shown in efficiency losses or a raise in the required voltage compared to the theoretical one. These overpotentials can be classified as [2,5,7]:

- *Activation overpotential* ($\eta^{\vee}_{act}$): This is due to the activation energy of the electrochemical reactions that happen in the electrodes. It is a loss caused by the limitation of the charge transfer speed. The choice of a suitable catalyst and increasing the operating temperature can reduce this overvoltage value. The equation of Butler–Volmer is used to calculate the activation overpotentials of both electrodes; it describes the dependence of the current density regarding the electrode potential.

- *Ohmic overpotential* ($\eta^{\vee}_{ohm}$): This is due to ionic resistance in the electrolyte, to the electric resistance in the inner parts of the electrolytic cell, and to the contact resistance among these parts. They are proportional to the current density and depend on the kind of material used, the design, and the temperature. This overpotential can be calculated as the addition of the different electric resistances of the process according to Ohm's law.
- *Concentration or diffusion overpotential* ($\eta^{\vee}_{conc}$): It takes into consideration the limitations in the mass transport happening, especially at high current densities. The greater influence comes from limitations in the transport of hydrogen (H_2) and oxygen (O_2) products because if they are not eliminated as fast as they are produced, their concentration in the reaction sites increases and, thus, decreases the reaction kinetics. Losses due to concentration can be disregarded at low currents. Overvoltages due to diffusion can be estimated from the Nernst equation and the species concentration.

Therefore the total potential of an electrolysis cell (U) will be given by the addition of the reversible voltage and all the overpotentials mentioned as they are not isolated in practice. However, some of these overvoltages are normally insignificant compared to others depending on the current, therefore, the kinetics of the global reaction will be dominated by the highest overpotential [7]:

$$U = U_{rev} + \left(\eta^{\vee}_{act} + \eta^{\vee}_{ohm} + \eta^{\vee}_{conc}\right) \quad (9.8)$$

The polarization curve of an electrolyzer models the kinetics of water electrolysis and exposes the effect of the different irreversibilities (Fig. 9.3). Besides, it allows for both the different operation points where the electrolysis cell works and the process efficiency to be determined.

In Fig. 9.3, it is observed that the behavior of the polarization curve for intermediate current densities is linear due to the predominance of ohmic overpotentials. However, a much more pronounced slope can be observed for low current densities

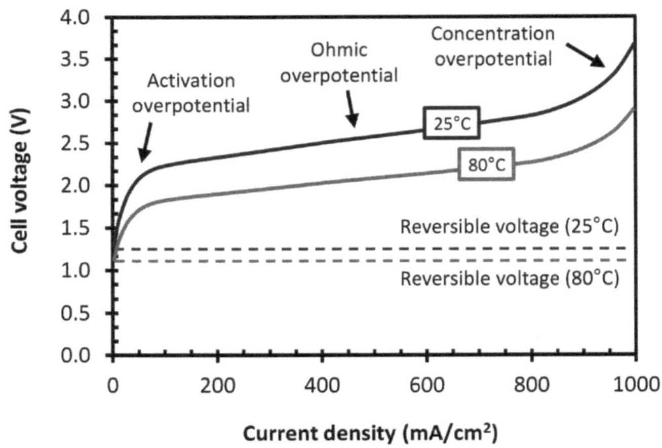

Figure 9.3 Polarization curve (i-U) of an electrolysis cell at different temperatures (<100°C).

(<100 mA/cm^2) due to the activation overpotentials. Likewise, commercial electrolyzers usually work at current densities that avoid reaching the zone of concentration overpotentials as the potential would raise too much and the system efficiency would be significantly reduced.

On the other hand, it can be observed that the curves corresponding to a higher temperature are under those corresponding a lower temperature. This means that it is convenient to work at the highest temperatures possible with regard to reducing the voltage necessary to carry out the electrolytic process, whenever damage to the materials or the parts of the electrolyzer is not produced.

9.2.2.2 *Faraday law of electrolysis: theoretical hydrogen production*

The current density used during water electrolysis is a critical operation parameter as it directly determines the amount of H_2 obtained. The relationship between the amount of H_2 produced ($\dot{m}_{prod,th}$) and the current associated to the electrochemical process (I) is given by Faraday law (Eq. 9.9). Thus an increase in the electrical current implies a raise in the H_2 generation:

$$\dot{m}_{prod,th} = \left(\frac{M}{F \times z}\right) I \tag{9.9}$$

where z is the number of electrons involved in water electrolysis ($2e^-$), F is the Faraday constant (96,500 C/mol), M is the molecular weight of the substance (g/mol), and I is the electrical current (A). However, given that one electrolysis cell produces little H_2, these cells join each other to form an electrolysis *stack*, whose main configurations and components will be discussed.

9.2.2.3 *Efficiency and electrochemical performance*

9.2.2.3.1 Voltage efficiency

Voltage efficiency (η_U) is defined as the relationship between the thermoneutral voltage (U_{tn}) and the real voltage (U) applied to the electrolysis cell:

$$\eta_U = \frac{U_{tn}}{U} \tag{9.10}$$

This efficiency quantifies the suitability with which electrolysis is made as the irreversibilities appearing in this process are shown as overpotentials, as indicated in Eq. (9.8). Thus the voltage efficiency is a design parameter useful for optimizing the electrolysis equipment.

From a practical point of view, it is possible to obtain voltage efficiencies higher than the unit if the electrolyzer operates with a voltage lower than the thermoneutral one, whenever thermal energy is supplied to the system. Fig. 9.4A shows the variation in the voltage efficiency according to the current density and the temperature.

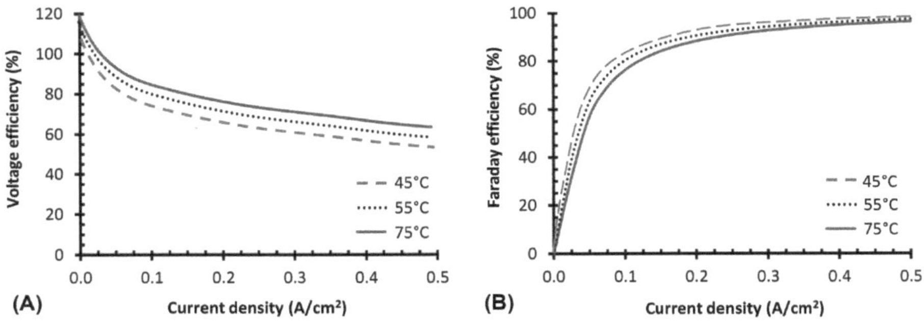

Figure 9.4 Efficiency of an electrolyzer versus current density (<100°C). (A) Voltage efficiency and (B) Faraday efficiency.

9.2.2.3.2 Faraday efficiency

The current or Faraday efficiency (η_F) relates the theoretical electrical current (I_{th}) needed to carry out the electrolysis reaction with the real intensity circulating through the cell (I). Thus it is a parameter that provides information about the current fraction circulating through the cell without being involved in the electrolysis. By means of the Faraday law described in Eq. (9.9), this efficiency can be expressed as the relationship between the real H_2 volume produced ($\dot{m}_{prod,real}$) and the theoretical H_2 volume ($\dot{m}_{prod,th}$):

$$\eta_F = \frac{I_{th}}{I} \Rightarrow \frac{\dot{m}_{prod,real}}{\dot{m}_{prod,th}} \tag{9.11}$$

Fig. 9.4B shows the relationship between the current density and Faraday efficiency for different temperatures. As is observed, the lower the process temperature, the higher the Faraday efficiency achievable.

9.2.2.3.3 Overall efficiency

The overall efficiency (η_{net}) is defined as the existing relationship between the variation of the process enthalpy and the total energy provided to the electrolysis cell per H_2 mole produced. If the only energy supply comes from the electricity provided in cell terminals, the thermal energy externally applied to the cell will be nonexistent [all the heat would come from the one generated as shown in Eq. (9.6)]. Therefore the efficiency would be equivalent to:

$$\eta_{net} = \eta_F \times \eta_V \tag{9.12}$$

9.2.3 Main types of water electrolysis

There are two fundamental parameters that define the kind of electrolysis being carried out, namely temperature and pH [2]. As stated previously, the temperature dependence has a strong influence on the potential necessary for electrolysis since the reversible voltage tends to decrease as the thermal contribution ($T \times \Delta S$) increases, following Gibbs free energy (ΔG) dependence on the temperature (see Fig. 9.2). This way, electrolysis can be carried out in two different manners regarding the process temperature:

- At a *low temperature*: In these processes, the operating temperature is usually 60°C–80°C so the power consumption is high, on the order of the energy contained in the H_2 produced. Therefore this procedure is only justified to produce H_2 in situ in small amounts and in the absence of other kinds of supplies, or to produce H_2 on a large scale by its integration with renewable energies when a way of storing the exceeding renewable potential is found.

  Concerning pH, when water electrolysis happens under neutral conditions (pH = 7), the theoretical potential difference between the oxygen evolution semireaction (OER) and the hydrogen evolution semireaction (HER) is 1.23 V, as shown in Fig. 9.5. However, the anode tends to acidify due to H^+ generation while electrolysis takes place; at the same time, the cathode alkalizes because OH^- is released. As a consequence, the pH surrounding the anode and the cathode varies, thus, increasing the thermodynamic voltage of the process. Due to this, concentrated solutions of H^+ (acid pH) or OH^- (basic pH) are required to avoid concentration overpotentials since these are the most mobile ionic species [2]. As shown in Fig. 9.5, the potential difference between the anode and cathode is always 1.23 V [8], even though the characteristic potentials of each semireaction vary both in a basic medium (+0.401 V OER, −0.829 HER) and in an acid medium (+1.229 V OER, 0 HER). This allows for two ways of carrying electrolysis out at a low temperature to be defined:

  Electrolysis in an acid medium: In this case, H^+ ions are responsible for the ionic transport; thus, a H_2SO_4 or HCI solution can be used as an electrolyte. However, electrolyzers using liquid acid electrolytes rarely exist due to severe corrosion problems.

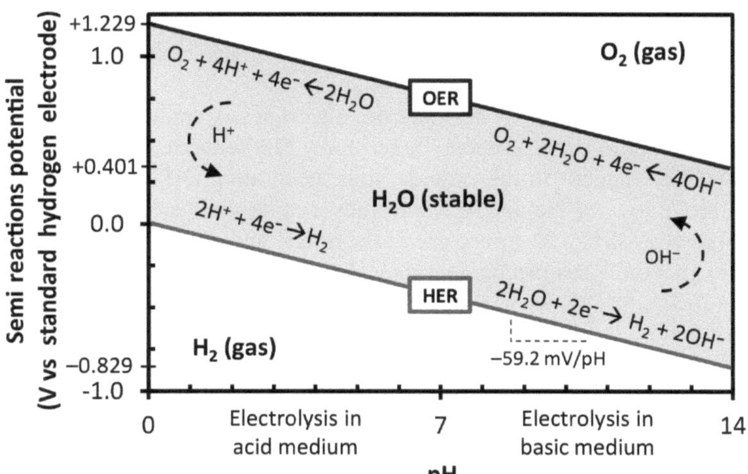

Figure 9.5 Potential variation in the HER and OER semireactions with respect to pH (standard conditions).

The development of a solid polymer electrolyte (SPE) where H^+ ions remain contained, guaranteeing both ionic transport and gas separation, has allowed this technology to be developed using *proton exchange membrane (PEM) electrolyzers.*
Electrolysis in a basic medium: In these processes, OH^- ions are responsible for ionic transport. The electrolyte used is generally a concentrated solution of potassium hydroxide (KOH) or sodium hydroxide (NaOH). The anode and cathode are separated by a porous medium (diaphragm) that prevents the gases generated in each electrode from being mixed. This is the fundamental principle of *alkaline electrolyzers*, which was the first technology to be developed historically as many metals passivate in highly basic mediums and then corrosion is prevented. On the other hand, the development of anion exchange membranes (AEM) has allowed for the use of SPE systems in alkaline electrolysis similar to PEM electrolyzers, thus, combining the advantages of both technologies.

- At a *high temperature*: In these processes, power consumption decreases significantly, although steam and a thermal source of high temperature are needed to carry out electrolysis. Technology also has to be modified, becoming solid oxide in *solid oxide electrolysis cell (SOEC) electrolyzers.*

Table 9.1 shows a comparison of the main technologies for water electrolysis (alkaline, PEM, and solid oxide).

9.3 Alkaline water electrolysis

9.3.1 Technology and operating principle

An alkaline electrolytic cell (Fig. 9.6) is basically made of two electrodes immersed in a liquid electrolyte and a porous diaphragm acting as a separator between the cathode and the anode chambers, which allows for ionic transport (OH^-), but is gas impermeable. The liquid electrolyte is an alkaline aqueous solution generally based on KOH or NaOH.

When the required potential between the electrodes is applied, the semireactions in anode and cathode simultaneously take place. The reduction reaction of water to hydrogen (HER) happens in the cathode, thus, generating OH^- ions. The OH^- ions generated move toward the anode where they donate their electrons and oxygen is produced (OER) [14,15]. The pressure in the anode and cathode chambers has to be the same to guarantee the mechanical integrity of the diaphragm:

Anode (OER):

$$2OH^- \rightarrow \frac{1}{2}O_2 + H_2O + 2e^- \tag{9.13}$$

Cathode (HER):

$$2H_2O + 2e^- \rightarrow H_2 + 2OH^- \tag{9.14}$$

Table 9.1 Comparison between the main technologies for water electrolysis [2,5,9–14].

	Alkaline electrolysis	**PEM electrolysis**	**Solid oxide electrolysis**
Semireactions	• Anode: $2OH^- \rightarrow \frac{1}{2}O_2 + H_2O + 2e^-$ • Cathode: $2H_2O + 2e^- \rightarrow H_2 + 2OH^-$	• Anode: $2H_2O \rightarrow O_2 + 4H^+ + 4e^-$ • Cathode: $2H^+ + 2e^- \rightarrow H_2$	• Anode: $2O^{2-} \rightarrow O_2 + 4e^-$ • Cathode: $H_2O + 2e^- \rightarrow O^{2-} + H_2$
Electrodes and catalysts	• Anode: Ni, Fe, metal oxides • Cathode: Ni/Ni-Co, metal oxides	• Anode: Ti + Ti/RuO_2, IrO_2 • Cathode: Graphite, Ti + Pt	Anode: Ceramic (Mn, La, Cr), Ni Cathode: Zr + Ni/CeOx
Separator	Diaphragm (NiO, PAM, ZrO_2 in PPS, usually Zirfon)	Membrane (PEM)	Membrane (ceramic)
Electrolyte	KOH (about 32 wt.%), NaOH (about 18 wt.%)	Solid polymer electrolyte (PFSA, usually Nation)	Y_2O_3-ZrO_2, Sc_2O_3-ZrO_2, MgO-ZrO_2, CaO-ZrO_2
Ion transport	OH^-	H^+	O^{2-}
Temperature (°C)	60–85	50–80	800–1000
Pressure (bar)	<30	<35	1–5
Current density (mA/cm^2)	200–600	≤2000	≤2000
Energy consumption (kWh/Nm^3H_2)	4.3–5.0	4.2–4.7	3.2–3.7
Costs (/kW)	1000–1500	1500–2000	Prototypes
Global efficiency (%)	60%–70%	70%–80%	85%–95%
Lifetime (hours)	>95,000	50,000–75,000	In research
Advantages	Technology well tested and commercially developed. It has greater durability and lower cost	Solid electrolyte. High pressure of the gases produced. Operation at high current densities	Solid electrolyte. Low electricity consumption. Lower demand on water quality

(Continued)

Table 9.1 (Continued)

	Alkaline electrolysis	**PEM electrolysis**	**Solid oxide electrolysis**
Disadvantages	Corrosive liquid electrolyte. Low pressure of the generated gases. It requires a purification stage for the hydrogen produced	High costs of catalysts and membranes. Great requirement in inlet water	Limited lifetime due to thermal cycles. Limitations on applications due to operating temperatures
Cell scheme	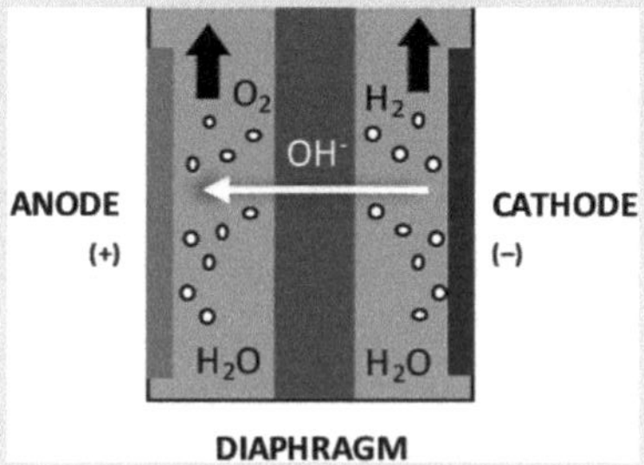	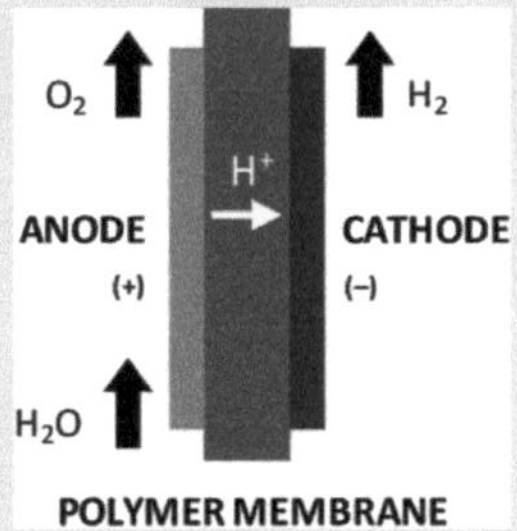	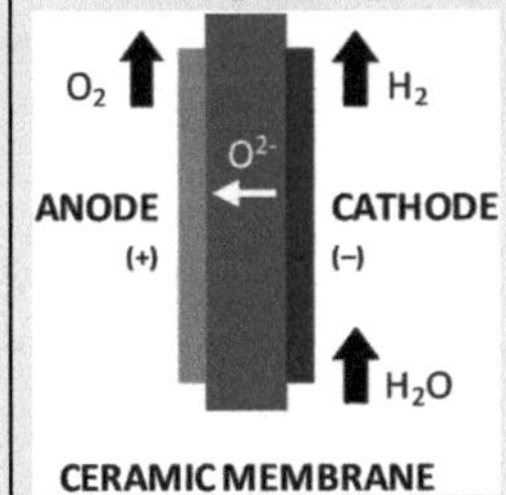

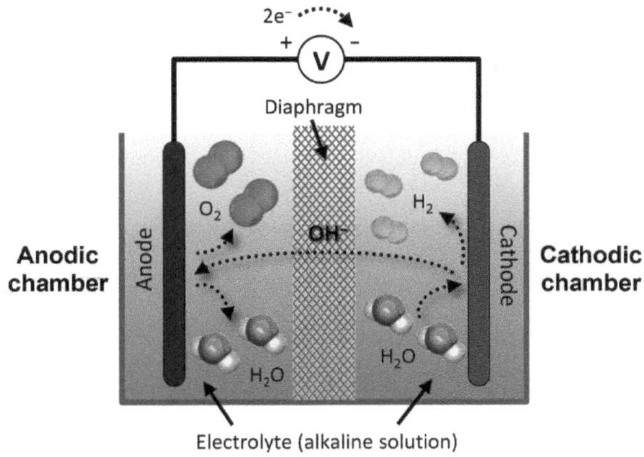

Figure 9.6 General scheme and operation of an alkaline electrolysis cell.

9.3.2 Components of an alkaline electrolysis cell

9.3.2.1 Electrodes and catalysts

Electrodes are responsible for conducting the electric current to the anode and cathode chambers. The positive electrode is called the anode; the OER is produced in its interphase with the electrolyte. The HER is produced in the negative electrode or cathode. These electrodes must have a high resistance to corrosion, good conductivity, suitable catalytic properties, and good structural integrity [9,10].

In alkaline electrolysis, electrodes usually based on nickel (Ni), Co, Ru, or Mo are used [9,10,16,17] as it is not necessary to use materials based on noble metals when electrolysis takes place at a high pH. Despite the electrode acting as a catalyst itself, if the electrode is covered or doped with a more stable or active layer, the charge transfer is facilitated and then the activation energy of the reactions is reduced.

The cathode is usually made of Ni or Raney Ni, which shows a high initial electrocatalytic activity for the HER and an excellent stability in alkaline mediums [18]. However, its overpotential increases after some time of cathode polarization in alkaline electrolysis; this is attributed to the generation of a nonconductive Ni hydride [19]. Due to this, Ni is mainly used in alloys with metals (Mo, Zn, Fe, Co, Cr, W) and/or metallic oxides (MoO_2, MoO_3, CeO_2, IrO_2, RuO_2) [20–23]. This kind of electrocatalyzer is usually plated over Ni by conventional electroplating techniques (Fig. 9.7A).

The anode is also commonly made of Ni, but like the cathode, it is necessary to counteract its deactivation mechanisms. The addition of iron improves the efficiency of the O_2 evolution. In this case, the use of electrocatalyzers also contribute to the reduction of the overpotentials. The most commonly used electrocatalyzers are those based on Ni and Co such as Ni and spinel type Co_3O_4, Ni and La-doped Co_3O_4 [24], MnOx-modified Au [25], 10% Li-doped Co_3O_4 [26], $La_{0.5}Sr_{0.5}CoO_3$ and $Ni_{0.2}Co_{0.8}LaO_3$ [27].

Figure 9.7 Scanning electron microscope (SEM) images of the surface morphology of electrodes: (A) Ni-Co cathode and (B) Ni foam.

On the other hand, the electrode morphology affects the ohmic resistance of the electrolysis cell. Thus different kinds of electrodes such as screens, meshes, metalized particles, porous foams, etc., have been developed, reaching the greatest electrocatalytic activity when the electrode surface is larger [28]. Of these, Ni foams (Fig. 9.7B) have the lowest initial ohmic resistance [29].

The mentioned materials generally offer a good resistance to corrosion in alkaline solutions and a good electrochemical activity at a reasonable cost. This makes the cost of alkaline electrolyzers relatively low compared to other electrolysis technologies. For this reason, it has been the most widely used technology for H_2 production on a large scale in the past few decades [10].

9.3.2.2 Separators: diaphragms and membranes

The function of the diaphragm or membrane is to separate the electrolytic cell into two chambers (Fig. 9.6); one in contact with the anode (known as the anode chamber) and the other in contact with the cathode (the cathode chamber). Electrolyte circulation takes place inside both chambers, where the generated gas ascends. The diaphragm or membrane has to be stable in the usual working conditions of alkaline electrolyzers (alkaline solution, highly oxidizing conditions in the anode, and reducing conditions in the cathode) and also has to only allow for the transfer of OH^- ions between both chambers.

9.3.2.2.1 Diaphragms

Diaphragms are the most widely used elements as separators. In the past, asbestos diaphragms were used due to their excellent durability. However, this material is toxic and dangerous to health (being cancerous) so it has been replaced by other materials. Most asbestos substitutes are compound materials based on microporous polymers or ceramic materials. Nowadays, the most widely used materials are based on polysulfone and zirconium oxide as an inorganic charge known as Zirfon PERL from the company Agfa [9,30]. Other diaphragms also used are, namely reinforced polyethersulfone [31], glass fiber–reinforced polyphenylene sulfide (Ryton) [32], and Ni oxides [33].

9.3.2.2.2 Anionic exchange membrane

It is being attempted to replace the diaphragm with an AEM, eliminating the use of a liquid and highly corrosive electrolyte such as NaOH or KOH solutions [11].

The use of AEMs allows for the electrodes to be manufactured directly over the membrane surface, thus, making the development of new configurations for alkaline electrolytic cells possible. These new designs aim at reducing the ohmic overpotentials associated with the separators that are currently being used in alkaline electrolysis. In this way, higher current densities are achieved, thus, improving electrolysis efficiency [34].

AEMs are made of a main polymeric skeleton to which functional groups that carry out the anion exchange (OH^-) are connected. The most widely used polymers are polysulfone and crosslinked polystyrene with divinylbenzene; while the most common groups of ion exchange are $-NH_3^+$, $-RNH_2^+$, $-RN^+$, $=R_2N^+$, $-R_3P^+$, $-R_2S^+$ or quaternary ammonium salts [35]. Nowadays, there are different commercial AEM such as A-201 (Tokuyama Corporation), Fumasep (Fumatech BWT), Selemion, Sustainion, Xion (XergyInc), and AlkaMem (AFC Energy), among others.

Nevertheless, there is still no alkaline polymer electrolyte that has satisfying electrochemical properties in alkaline electrolysis conditions. To this, problems of mechanical integrity or the negative effect of temperature on the stability of the membrane itself are added, thus, limiting its use in temperatures lower than 50°C–60°C [36,37].

9.3.2.3 Electrolyte

The electrolyte provides the ions responsible for the electrical conduction inside the electrolysis cell (OH^-). In this case, the electrolyte also supplies the reagent of the reaction. Water solutions of KOH around 32% (w/w) or NaOH at 18% (w/w) are commonly used to minimize losses due to electrolyte resistance since concentration values above or below these reduce the electrolyte conductivity [38,39]. As shown in Fig. 9.8, the maximum conductivity is obtained at KOH concentrations of 30%–35% by weight in water, improving significantly with temperature increases.

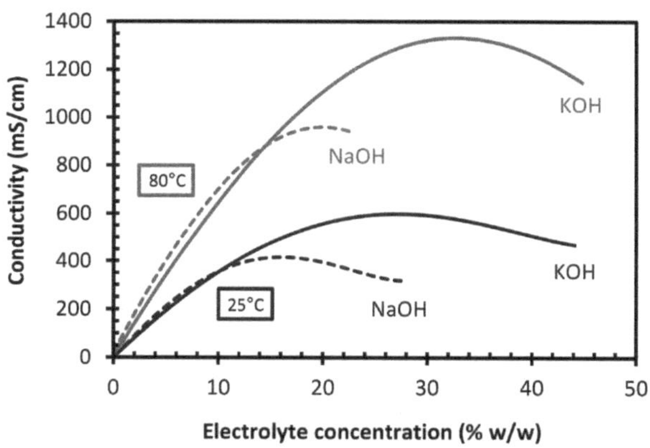

Figure 9.8 Conductivity of NaOH and KOH solutions at different temperatures.

It must also be considered that the electrolyte concentration increases as electrolysis takes place since only the water and not the KOH is electrolyzed, so deionized water has to be continuously supplied to the process.

9.3.3 Configuration of an alkaline electrolyzer

9.3.3.1 Main cell designs

In general terms, there are two main configurations for an alkaline electrolysis cell, as shown in Fig. 9.9. These configurations are generally found in a bipolar configuration, following a filter-press assembly where the diaphragm is the intermediate element [16].

The first configuration, called *gap cell*, reflects the case of a traditional cell in which two flat electrodes are immersed in a liquid electrolyte (Fig. 9.9A). There is a separating medium (diaphragm) between the electrodes so that gases (H_2 and O_2) are developed in different chambers without getting mixed. The distance between the electrodes and the membrane is the cell gap and it can range from some microns to a few millimeters [9,40]. This separation determines the ohmic drop associated to the electrolyte in a cell.

A proper selection of gap cell is vital for an electrolysis cell to work correctly. When high current densities are applied, the efficiency increases significantly when

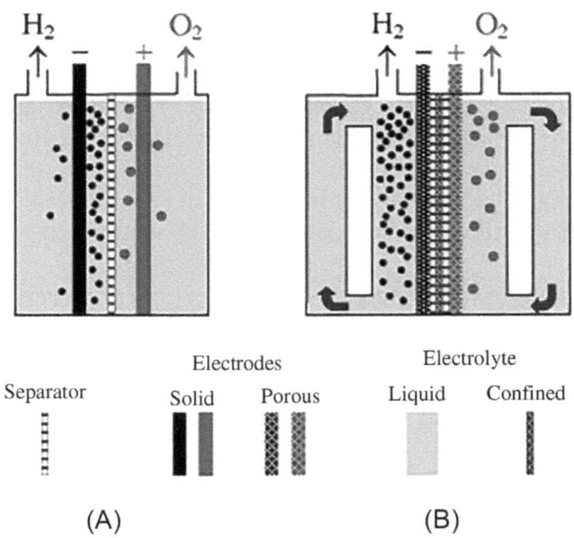

Figure 9.9 Main designs of alkaline water electrolysis cells. (A) Gap cell or conventional design and (B) zero-gap cell configuration.
Source: Reproduced from S. Marini, P. Salvi, P. Nelli, R. Pesenti, M. Villa, M. Berrettoni, et al. Advanced alkaline water electrolysis. Electrochim. Acta 82 (2012) 384–391. https://doi.org/10.1016/j.electacta.2012.05.011 from ELSEVIER.

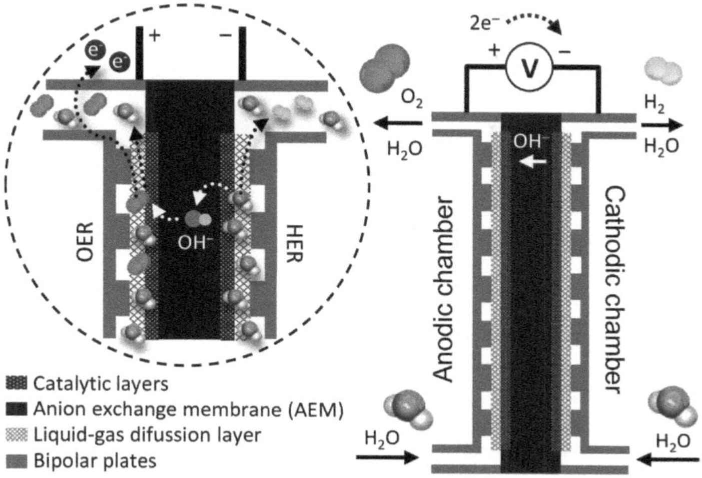

Figure 9.10 Scheme of an alkaline exchange membrane electrolysis cell.

the space between electrodes is reduced, up to a limit distance, under which the potential is increased. Therefore there is an optimum distance between the electrodes of the configuration in Fig. 9.9A that depends on the current density and other operating parameters [41]. This fact can be explained considering that a distance reduction implies a lower resistance to ionic transport in the electrolyte, but a distance too small increases the bubble fraction significantly since the volume of the electrolytic chamber is too low.

The zero-gap cell configuration came about in order to solve some of the problems associated with the previous design and reduce the ohmic overpotentials of the electrolyte [9]. This design, as shown in Fig. 9.9B, "ideally" allows for bubbles between the electrodes to be eliminated as the cathode and anode are directly placed over the separating medium so that the electrolyte remains contained in the diaphragm [40]. For this to happen, the electrodes used have to be porous (screens, meshes, foams). Thanks to these designs, higher current densities can be reached without an excessive potential increase. Most current commercial electrolyzers use this design [9].

In more current configurations (Fig. 9.10), as stated previously, the diaphragm is replaced by the use of an AEM. This results in a configuration known as SPE, which is structurally similar to a PEM electrolyzer. In this way, the use of a corrosive liquid electrolyte (KOH or NaOH solutions) is avoided [16,40].

9.3.3.2 Stack configuration

One single electrolysis cell produces a small amount of H_2, so different cells are usually grouped to form a stack. Depending on the electric connection of these cells, there are two possible configurations [6,10]:

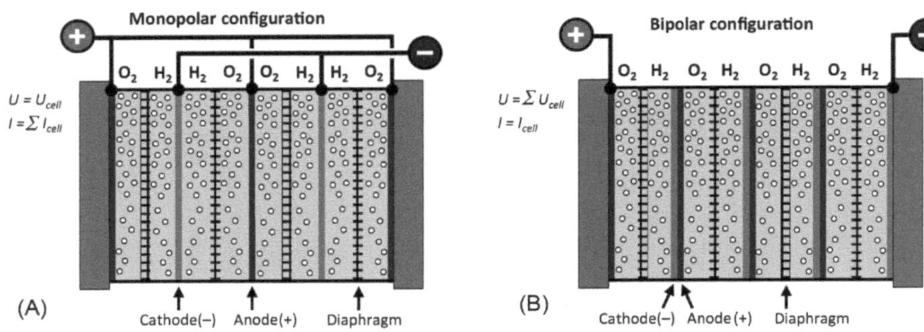

Figure 9.11 Configuration of an electrolysis stack. (A) Monopolar and (B) bipolar.

- In the *monopolar configuration* (Fig. 9.11A), the cell layout is in parallel so that the same electrochemical reaction is produced on both sides of each electrode. They have a simple and strong design, although they require a lot of space compared to bipolar configurations. Their advantage is that cells can easily be separated and replaced in case of failure, without having to stop the other cells from working.
- In the *bipolar configuration* (Fig. 9.11B), cells are serially arranged, that is, the anode of a cell is connected to the cathode of the next one. In this case, anode and cathode are assembled together, thus, allowing for more compact designs and a wider range of working temperatures and pressures. This is the design used in commercial electrolyzers.

9.3.4 Developments, limitations, and future perspectives

Alkaline electrolysis is a mature technology currently used on an industrial scale. There is commercially available equipment that offers high capacities of producing H_2 in the range of MW [13].

Compared to other technologies, the key advantages of this one are its proven durability, its maturity, availability, and low specific costs (these are systems not based on components containing noble metals). Regarding the disadvantages, these are mainly the low current densities and the use of a highly corrosive liquid electrolyte [42].

In general, commercial systems usually work at temperatures between 65°C and 90°C and typical pressures around 10–30 bar. The purity of the produced H_2 is about 99.5%–99.9% (vol.) because the gases generated in the anode and cathode can easily undergo diffusion through the porous diaphragm, which is used as a separator. Therefore it is necessary to add a purification stage based on de-oxo reactors capable of eliminating O_2 traces and increasing the purity up to >99.999% in case H_2 of a greater purity is needed. The overall efficiencies of these systems are in the range of 51% to 60% in relation to the low heating value of the H_2 produced, as long as the operation takes place in nominal conditions between 0.25 and 0.45 A/cm^2. In practice, H_2 production is limited to an operating range of 20% to 100% of the nominal power in order to prevent the formation of inflammable mixes due to gas diffusion through the diaphragm, which takes place mostly at low current densities (<0.1 A/cm^2) [13,42].

Concerning the durability of these systems, commercial equipment currently offers around 95,000 operating hours with a typical degrading ratio of 1−3 μV/h, which properly satisfies industrial requirements (25 working years). However, these systems are clearly influenced by working with variable charges. Operating in the presence of dynamic charges as well as frequently getting switched on and off reduce their durability, thus, reducing the quality of the gas obtained and reaching lower efficiencies [9,43].

Nowadays, the efforts of research and development in this technology are mainly aimed at solving the greatest disadvantages of this kind of electrolysis, which are the low current density and the problems derived from working in dynamic conditions. Due to this, the future challenges are [2,9,11,44−47]:

- Developing *advanced materials to improve the catalytic activity* in the electrodes, thus, reducing the overvoltages both for the cathode reaction and the anode reaction.
- Optimizing the *space between the electrodes* to reduce the ohmic losses and make it possible to work at high current densities. Nowadays, the distance between the electrodes is usually below 1 mm. In addition, a zero-gap cell design is used in a lot of configurations where the space is minimized since the electrodes are in direct contact with the diaphragm. However, these configurations can limit the release of the generated gases as the space between the electrodes is small, therefore, new designs are needed that allow for the flow distribution to be improved.
- Developing AEMs as separators in electrolysis cells, thus, eliminating the use of a liquid electrolyte. This kind of membrane will allow for the ohmic overpotentials associated with the separating medium currently used (diaphragms) to be reduced, reach higher current densities, and reduce diffusion (crossover) between the anode chamber and the cathode chamber.
- Developing *high temperature alkaline electrolyzers* with operating temperatures up to 150°C, which would increase the electrolyte conductivity and so the electrochemical reaction rate. The main drawback is the corrosion of the cell materials at these temperatures.

9.4 Proton exchange membrane water electrolysis

9.4.1 Technology and operating principle

This kind of electrolyzer is based on the use of a PEM immersed in deionized water, which separates the cathode and anode semi-cells (Fig. 9.12). This membrane acts as an SPE, allowing for the transport of protons (H^+) generated in the anode and separating the produced gases. Thanks to the use of these membranes, the operation response is faster and electrolyzers are more compact and of a lower volume [2,12].

As shown in Fig. 9.12, water is supplied through the anode channels in a PEM electrolysis cell. In the anode, the OER takes place to release O_2 and H^+, as shown in Eq. (9.16). Then, the resulting H^+ ions go through the ionic exchange membrane, while the electrons circulate through the external circuit up to the cathode; this closes the electric circuit that supplied the power necessary to carry out the process. In the cathode, the protons that come through the membrane are

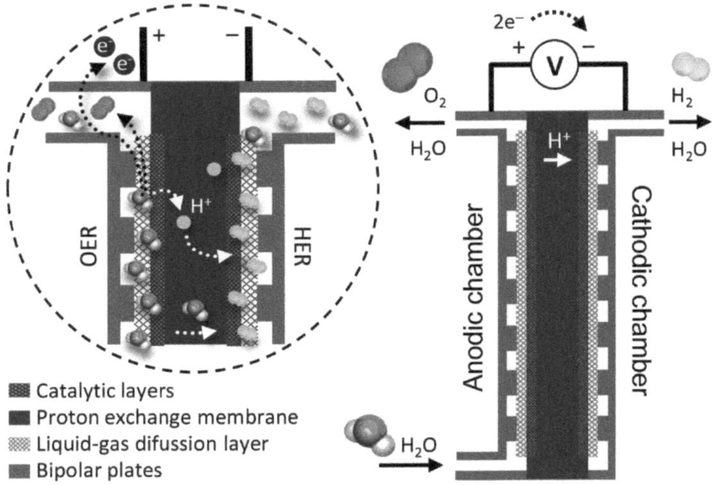

Figure 9.12 General scheme and operation of a PEM electrolysis cell.

recombined with the electrons coming through the external circuit, thus, producing H_2 as per Eq. (9.15) [13]:

Cathode (HER):

$$2H^+ + 2e^- \rightarrow H_2 \tag{9.15}$$

Anode (OER):

$$2H_2O \rightarrow O_2 + 4H^+ + 4e^- \tag{9.16}$$

9.4.2 Components of a proton exchange membrane electrolysis cell

9.4.2.1 Membrane-electrode assembly

The membrane-electrode assembly (MEA) consists of the assembly of three components: membrane and electrodes (anode and cathode), being the membrane (PEM) the central part of this assembly. This assembly can be made in two ways, according to the site where the catalytic layer is deposited, namely over the membrane (Fig. 9.13A) or over the liquid–gas diffusion layers (Fig. 9.13B).

9.4.2.1.1 Electrodes and catalysts

The main function of the electrodes is to contribute to the contact between reagents and reaction sites. In order to do this, it is necessary to facilitate the electron and proton conduction, the evacuation of the gaseous products generated in the reaction, and the water coming to the catalytic sites. The electrocatalytic function will be

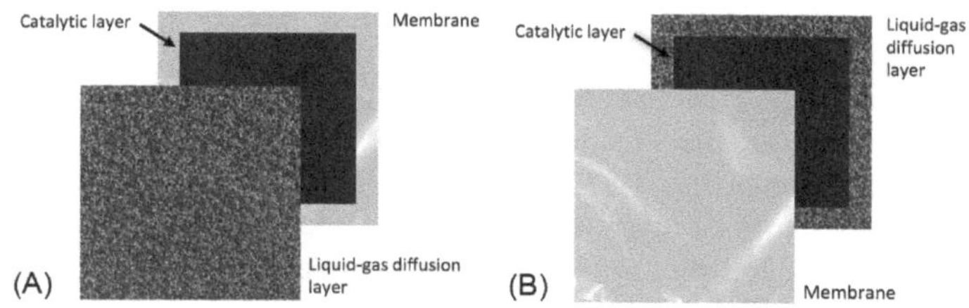

Figure 9.13 Different membrane-electrode assemblies. (A) MEA with catalytic layer deposited on membrane and (B) MEA with catalytic layer deposited on liquid–gas diffusion layer.

determined by the material used, generally a metallic oxide deposited over a support with a large superficial area (Fig. 9.13). The anode semireaction promotes the OER, while the cathode semireaction promotes the HER.

OER has slower kinetics so it is the limiting reaction of the process. Due to this, stable electrocatalyzers with high catalytic activities and selectivity are necessary to favor O_2 evolution. Thus RuO_2 and IrO_2 are the most widely used catalysts in the OER. RuO_2 shows better performances, but is less stable than IrO_2 when facing the highly corrosive environment of the anode side of a PEM electrolysis cell. This acid medium is mainly caused by the presence of H^+, fluorides coming from the perfluorosulfonic membrane, from the presence of O_2, and from the overpotential that the electrode is subject to. Thus mixes of both oxides are still the most widely used solutions as anode catalysts in PEM electrolysis [48]. The addition of mixed transition metal oxides such as Nb, Ta, or Ti to those already used like Ru and Ir is another of the most widely studied methods for the anode [49–51], aiming at reducing the amount of noble metals needed.

Regarding the cathode, platinum (Pt) has traditionally been used as a catalyst due to its proven activity to promote the HER and its high chemical stability [12]. Decreasing the catalyst loading through the improvement of Pt nanoparticle dispersion is fundamental in PEM electrolysis cost reduction [12,52]. Nowadays, there are reports of projects in which Pt loadings of around 0.2 mg_{Pt}/cm^2 have been reached in the cathode [53,54]. Other alternatives to Pt such as Pd are being evaluated by the scientific community for use as a catalyst in the HER [55].

The use of noble metals as catalysts is one of the reasons that PEM electrolysis implies high H_2 production costs. The electrodes are around 10% of the cell cost, from which 8% belongs to the electrocatalyzers [56]. Therefore the development of this technology can achieve a decrease in the catalyst loading by means of improving the efficiency of the depositing methods or even by the total replacement of noble metals in the electrodes by other alternatives [55,57].

9.4.2.1.2 Proton exchange membrane: electrolyte

The membrane is one of the most important components of a PEM cell, playing a key role in the purity of the produced gases and in the durability of the system. The most widely used are polymer membranes of perfluorosulfonic acid (PFSA) such as

Nafion, Fumapem, Flemion, Aquivion, and Aciplex. These membranes have properties that make them suitable for use as solid electrolytes with excellent thermal and chemical stability, good mechanical resistance, and high proton conductivity. Nowadays, Nafion membranes (Nafion 115, 117, and 212) are the most widely used in PEM electrolysis [12].

The main disadvantages of these membranes include their high cost, the presence of fluoride in the polymer structure (which can migrate and create pitting corrosion in the metallic elements of the electrolyzer), their thickness (which increases the ohmic resistance and reduces the performance at high current densities), and their mechanical behavior at high temperatures [5]. For these reasons, Nafion membranes have been modified with different oxides to improve their thermal properties. Other alternative membranes to Nafion that are being developed include polybenzimidazoles, polyether ether ketones, and sulfonated polyphenyl quinoxalines [5].

9.4.2.2 Liquid–gas diffusion layers

Liquid–gas diffusion layers (LGDL) are porous metallic screens that do not take part in electrochemical reactions. Their function is to make the electric current pass on, transporting it from the bipolar plates to the MEA and guaranteeing the water access to the electrodes homogeneously as well as the removal of the gas produced. This process takes place in acid mediums ($pH < 4$) with high overpotentials (~ 2 V) and in the presence of gas, which implies severe conditions for the LGDL, especially in the anode side where the medium is also highly oxidizing due to the presence of O_2.

The porosity of this material is a key factor in its operation; if the pores are too small, they can cause problems in the transport of water and gas, whereas if the porosity increases, gas evacuation is facilitated, but significant ohmic losses are introduced in the contact sites between the LGDL and the MEA as well as between the LGDL and the bipolar plate. Sintered Ti at a high temperature from powders is the most commonly used material since the existing pores allow gas and water to flow through it while it offers a large contact area with the electrodes; this gives it a higher conductivity and also a higher resistance in the acid medium of the cell. Titanium in the form of a mesh or foam is also commonly used in this application, although with lower conductivities that the sintered Ti [12].

9.4.2.3 Bipolar plates

Bipolar plates are 30% of the size and 50% of cost of a cell, so a good design for this component is essential [56]. They are located in the ends of each of the cells and they have some mechanized channels that allow water to go toward the LGDL and MEA as well as for the removal of the generated gases. Thus these flow distribution channels are responsible for the thermofluid dynamic behavior inside an electrolysis cell, which directly affects the overall performance of the process. The channel size is determined by the power requirements, while the flow geometry is the result of the position of the water intake manifolds and those for gas removal,

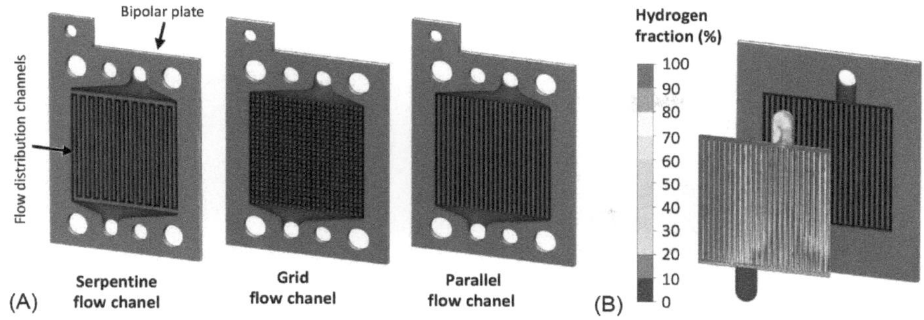

Figure 9.14 Bipolar plates and flow distribution channels in a PEM cell. (A) Main types of flow field channel designs (serpentine, grid, parallel) and (B) flow distribution and H_2 fraction determined by computational fluid dynamics (CFD) when straight parallel channels are used.

the selected topology, the heat management, and the limitations of the manufacturing process. All these functions reveal the importance of a proper design of the bipolar plates and their distribution flow channels (Fig. 9.14). In addition, bipolar plates are essential for the stack configuration as they connect the cells to each other.

On the other hand, since materials with high corrosion resistance and high electrical conductivity are needed, metals with specific coatings are commonly used. The most common option is still the use of Ti with a Pt coating on the surface to prevent TiO_2 from being generated. This makes the cost of this component almost half of the total cost of the cell [56]. Alternatives to the use of Ti/Pt are being investigated, although it is not easy to find a material showing the same performance in terms of conductivity and durability as Ti/Pt at a lower cost. In the past few years, different authors [58–61] have evaluated alternatives to Ti, from metals like Mo, Ta, or W to stainless steel coated with nitride multilayers and metallic carbides by means of techniques such as the deposition of thin layers on a vapor phase (PVD). Using stainless steel with protecting and conducting layers allows for the price of the manufacturing process to be reduced substantially; this becomes an important reduction in the H_2 productions costs.

9.4.3 Configuration of a proton exchange membrane electrolyzer

As discussed previously, the catalytic layers (electrodes) can be deposited about the membrane or on the LGDL; these electrodes are commonly used together with the membrane to form the MEA. On both sides of the MEA and their current and flow distributors (LGDL), the bipolar plates are located. In an electrolysis cell, all these components are maintained structurally joined at a suitable compaction pressure, which guarantees the right electric contact among all the components. The pressure

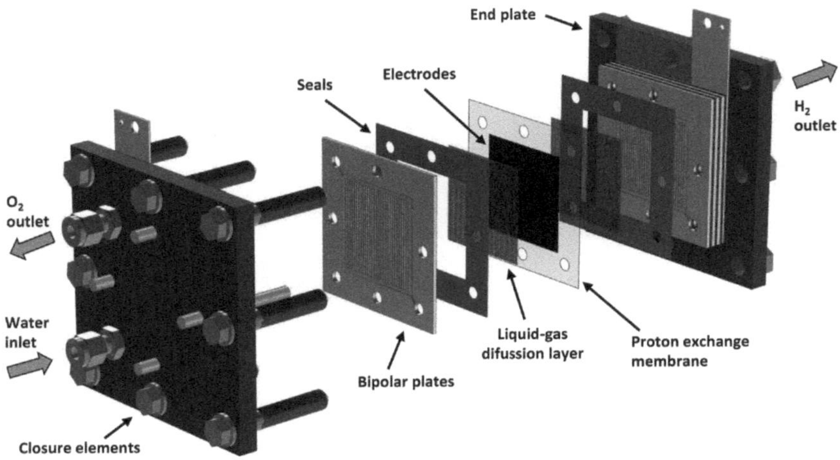

Figure 9.15 Configuration of a PEM electrolysis stack.

necessary to close a cell is applied through the end plates and the closure elements needed (screws, nuts, and bolts).

These cells can be stacked to increase the H_2 production capacity, thus, forming a stack in the same conditions as described for the alkaline electrolyzer. Bipolar configurations are the more common in PEM electrolysis. In these designs, cells are connected in series in a stack by the bipolar plates, which were called this because one of their faces is in the cell anode and the other face is in the cathode of the next one (Fig. 9.15).

These stacks are structurally compact, but they need sealing elements to maintain the gases generated in the two separate chambers, thus, avoiding explosive mixtures and maintaining the anode and cathode sides electrically isolated, but allowing for electron flow through the bipolar plates (Fig. 9.15). A correct pressure that holds it together also allows for the electric conductivity to be improved among all the components [62].

9.4.4 Developments, limitations, and future perspectives

PEM electrolysis is a modular technology that is also easily scalable. It could be designed depending on the final application, according to the amount of H_2 required. These systems work at a high current density (around 2 A/cm^2) at a temperature between 50°C and 80°C.

One of the main advantages of PEM electrolysis systems is the possibility to produce high-purity ($>$99.99%) and high-pressure hydrogen in the same facility. PEM electrolysis technology shows a fast start-up in the face of variable power sources (such as renewable energy), making it possible to work in all the nominal operating ranges (0%–100%). The use of a solid electrolyte also implies not having to handle a liquid electrolyte, as is the case with the alkaline technology.

Disadvantages of these systems include their cost due to the use of precious metals as catalysts (like Ru, Ir, Pt) and the exposure of materials to highly acidic media. Therefore it is necessary to obtain the maximum efficiency of the stack at minimum costs in order to develop PEM electrolysis. Since 2010, PEM electrolysis systems have followed a decreasing trend in terms of the power consumption (kWh/kg), that is, the efficiency in H_2 generation has progressively improved. Estimations for 2030 [43] show that it will be possible to reach a power consumption around 45 kWh/kg H_2, when working at 1000 mA/cm^2.

Nevertheless, the efficiency of a PEM electrolyzer is also affected by voltage degradation, that is, the overpotential to be applied to an electrolysis cell so that it maintains a constant H_2 production. According to data from different PEM electrolyzer manufacturers, the durability reported is around 50,000–75,000 operation hours with a degradation that can reach 4–8 μV/h of operation [13,43].

9.5 Solid oxide water electrolysis

9.5.1 Technology and operating principle

SOEC are devices that replace part of the electric energy required to divide water for thermal energy. A SOEC is structurally similar to a solid oxide fuel cell (SOFC) working in reverse, where a solid oxide electrolyte acts as a ceramic membrane, allowing O^{2-} ions to pass. As shown in Fig. 9.16, the cathode is fed in these cells with steam that is reduced, generating H_2 and O^{2-} in the cathode–electrolyte interphase. Oxide ions then migrate through the electrolyte to the anode, where they oxidize to produce O_2 [5,63]:

Anode (OER):

$$2O^{2-} \rightarrow O_2 + 4e^- \tag{9.17}$$

Cathode (HER):

$$2H_2O(\text{steam}) + 4e^- \rightarrow 2H_2 + 2O^{2-} \tag{9.18}$$

The use of higher temperatures (500°C–1000°C) increases the efficiency of the electrolyzer by reducing the overpotentials of the process. This allows the cell to work even below the thermoneutral voltage [13]. To do so, it is necessary for the components to be resistant to the high temperatures of these processes, therefore, the materials used in SOEC are usually ceramic [9,64]. However, it is necessary to have steam and a thermal source of a high temperature to carry out the process; this may sometimes limit the applications of this technology.

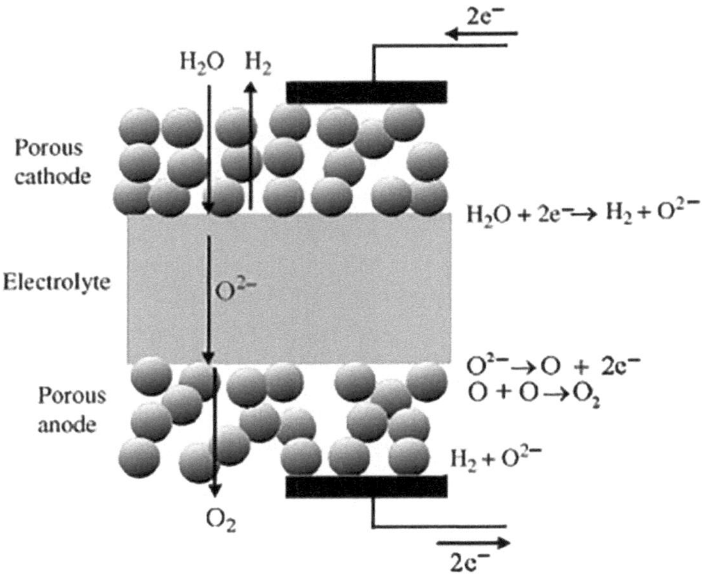

Figure 9.16 General scheme and operation of a solid oxide electrolysis cell.
Source: Reproduced from M. Ni, M.K.H. Leung, D.Y.C. Leung, Technological development of hydrogen production by solid oxide electrolyzer cell (SOEC). Int. J. Hydrog. Energy 33 (9) (2008) 2337–2354. https://doi.org/10.1016/j.ijhydene.2008.02.048 from ELSEVIER.

9.5.2 Components of a solid oxide electrolysis cell

9.5.2.1 Electrolyte

The solid electrolyte has to be chemically stable and have good ionic conductivity and low electric conductivity. In addition, it has to be as thin as possible to minimize ohmic overpotentials [63]. It generally consists of a mixed oxide based on zirconium oxide stabilized with yttrium oxide (ZrO_2/Y_2O_3), which is known as YSZ (*yttria-stabilized zirconia*) and acts as a ceramic membrane. Oxide ion transport happens by diffusion through the crystalline structure of the membrane (Fig. 9.16). The electrodes are located on both sides of this ceramic membrane [2].

9.5.2.2 Electrodes

The electrodes must be made of ceramic materials with a suitable porosity and pore size, making them capable of conducting electrons and facilitating mass transport (H_2 and O_2) as well as having the ionic conductivity to allow for the migration of the O^{2-} species. The rate of thermal expansion of both electrodes (cathode and anode) must be similar to the electrolyte to avoid any failure of the latter due to a high mechanical tension. Electrochemical reactions take place in the electrode/electrolyte/gas interphases of both chambers, known as the *three-phase boundary*. In addition, the need of electrocatalysts is less significant when compared to the requirements at a low temperature due to the high temperatures of these processes; for this reason, it is possible to operate without using precious metals as catalysts [2].

The cathode allows for steam and H_2 diffusion and provides active sites for steam reduction (HER). Ni has a high electrochemical reactivity and so it is widely used in SOEC. Aiming at increasing the reaction rates, particles of Ni can be mixed with ion conducting particles that are generally made of the same material as the electrolyte such as YSZ. This kind of electrode is called *cermet* (ceramic–metal). Thus Ni–YSZ is the most widely used material as a cathode in this kind of electrolysis cell [65].

For the anode (OER), materials are mainly based on conducting oxides with a structure similar to perovskite (ABO_3). In particular, one of the best materials is lanthanum manganite ($LaMnO_3$) doped with alkaline earth metals and rare earth elements. This compound is a semiconducting oxide with a structure similar to ABO_3 and has a p-type electronic conductivity due to the valence changes in manganese. $LaMnO_3$ doped with strontium (LSM) is considered to be the reference compound since it is particularly suitable for operations over 800ºC [9]. However, as it is a purely electric conductor, it is generally mixed with YSZ (the material used as the electrolyte) to create an LSM–YSZ compound that shows good conductivity, both ionic and electric, and also improves its stability and integration with the electrolyte mentioned.

9.5.2.3 Interconnectors

The cell also has interconnectors that are used for transporting the electric current between the cells, guaranteeing gas distribution inside and acting as a separator between the anode of a cell and the cathode of the contiguous one (similar to the bipolar plates of a PEM electrolysis cell). Therefore it has to be made of a material with excellent electric conductivity and great corrosion resistance. Another important feature that interconnectors must have is related to their mechanical properties. As these elements link different cells and have to prevent the rupture of the same while operating, the materials used as interconnectors must have thermal expansion rates similar to ceramic elements; while at the same time, they must be malleable enough to be stamped, cut, or mechanized. For these reasons, SOEC interconnectors can be made of metallic or ceramic materials. However, the manufacturing cost of ceramic interconnectors is high. Thus metallic alloys have attracted more and more attention due to their low cost, high tenacity, and excellent capacity of manufacturing. In particular, some kinds of stainless steels with certain modifications or coatings (since some elements of these steels can migrate and deactivate the electrodes) offer the necessary features for SOEC [9]. The most widely used steels are ferritic such as Crofer 22 APU [66].

It is important to remark that, besides all the specific requirements of each of the materials, the whole set has to be chemically inert to each other to avoid interdiffusion and chemical reaction phenomena, which may form isolating phases that could reduce the cell performance. On the other hand, thermomechanical stability among the materials is fundamental, mainly during thermal cycles (switching on and off).

9.5.3 Solid oxide electrolysis cell and stack configuration

Cells are the smallest units of an electrolysis stack. Aiming at guaranteeing that each cell has a good mechanical resistance and the electrodes have a porosity suitable for gas diffusion, cells are designed with the cathode, anode, or electrolyte being thicker with the purpose of one of them able to mechanically withstand the other two. Therefore the main designs are electrolyte supported cells, electrode supported (anode and cathode) cells, and metal supported cells.

In terms of stack design, most developments have been focused on two kinds of configurations [67,68]:

- *Planar configuration* (Fig. 9.17A): This consists of a stack of monocells of a flat geometry in a structure with separators and interconnectors that make gas circulation possible and electrically connect the terminals of each unit with the rest of the series. It has the advantage of an internal serial connection of cells, which avoids charge losses as well as allowing for more compact designs, high power density, and lower manufacturing costs. As the main disadvantage, it can be highlighted that they may have seal problems between the layers that form part of the cell.
- *Tubular configuration* (Fig. 9.17B): This consists of two hollow cylinders with walls made of the cathode and anode layers. The electrolyte is placed between these two and is grouped by internal connections. Steam is fed through the inside of the tube and is reduced to H_2 and O^{2-}. Subsequently, the OER takes place in the anode, where O_2 is extracted from the external layer of the tubular cell. This configuration has attracted increasing attention in the past few years due to its inherent advantages such as a fast-starting capacity and a high resistance to heat and thermal cycles. However, it shows the disadvantage of an external serial connection of cells, which causes bigger ohmic losses than in the planar configuration.

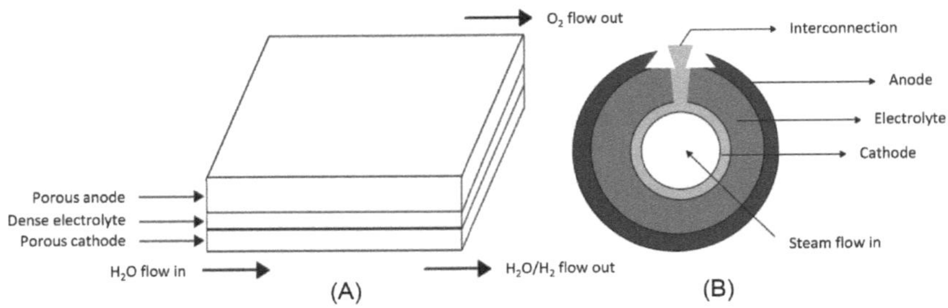

Figure 9.17 SOEC configurations. (A) Planar SOEC and (B) tubular SOEC (end view). *Source*: Reproduced from M. Ni, M.K.H. Leung, D.Y.C. Leung, Technological development of hydrogen production by solid oxide electrolyzer cell (SOEC). Int. J. Hydrog. Energy 33 (9) (2008) 2337–2354. https://doi.org/10.1016/j.ijhydene.2008.02.048 from ELSEVIER.

9.5.4 Developments, limitations, and future perspectives

SOEC electrolyzers are the most electrically efficient electrolyzers because they operate at high temperatures, although they are the less developed electrolysis technology. As previously mentioned, they use a solid electrolyte and do not show so many problems related to liquid/gas distribution (as electrolysis at a low temperature) due to the fact that they work with temperatures way above the boiling point [1,5]. In addition, they do not require using precious metals as catalysts due to the operation conditions in which SOECs work [2].

The current densities that these systems can reach are in the range of PEM electrolysis systems, but they are usually kept below 1 A/cm^2 due to the high degradation at those temperatures. The corresponding cell voltages are about 1.3 V, which constitute a power consumption much lower than in the other technologies (around 3.2 kWh/Nm^3 H_2). Efficiencies can become higher than 90% by taking into account both the electrical component and the thermal use.

However, the use of high operating temperatures requires using costly materials and more complex manufacturing methods [1]. The components of a cell can show different thermal expansion rates so the start/stop cycles become damaging for the interfacial contact between the electrodes and the electrolyte; this leads to the delamination of the different layers and an increase of the internal cell resistance. Other problems associated with component degradation include the deactivation of the electrodes on the side of the cathode due to the Ni hardening and the gradual agglomeration and reoxidation during electrolysis [65]. In addition, the different components have to be extremely thin to increase the conductivity properties, thus, making the cell assembly highly complex and fragile [5]. For these reasons, numerous investigations shall be still devoted to optimize the materials, components, and systems used before this technology can reach commercial deployment.

Nevertheless, this technology has great potential for H_2 production taking advantage of the energy generated from different thermal sources such as solar concentration power plants, geothermal energy, or nuclear reactors [5]. Besides, the possibility of using reversible devices that can work as an electrolyzer or as a fuel cell is open since a solid oxide cell is essentially a solid oxide fuel cell working in reverse.

Finally, it should be pointed out that the overall efficiency of SOEC technology can be improved by co-electrolysis processes, using steam and carbon dioxide (CO_2) to produce synthesis gas ($H_2 + CO$), which is interesting from an industrial point of view in connection with the production of different synthetic fuels [2]. This process has attracted great attention due to its high conversion and energy efficiency, which provides some opportunities to reduce CO_2 emissions, mitigate global warming, and store renewable energy. Nevertheless, the high operating temperature and the difficulty to activate the double bond $C=O$ of CO_2 requires the use of more specific cathodes for SOEC, so efforts are being made to develop suitable materials with a high catalytic activity and an excellent long-term stability to electro-reduction CO_2/H_2O [69].

Regarding the durability of SOEC systems, degradation ratios are in the range of 0.4 to 6 %/1000 h at temperatures between 650°C and 850°C and current densities between 0.26 and 1 A/cm^2 [13].

9.6 Electrolysis system technology

9.6.1 System lay-out and balance of plant

Although the stack is the main element of an electrolysis system, it is necessary to have a series of systems and equipment that allow for the correct operation of the electrochemical cells as well as for obtaining gases in the required conditions. This set of auxiliary systems is what constitutes the balance of plant (BoP) [39].

In general terms, the BoP of a low-temperature electrolysis system (alkaline and PEM) is formed by several subsystems, including (Fig. 9.18):

- *Gas production subsystem*: This is where the water electrolysis process actually takes place. It is made up of the *stack*, where O_2 and H_2 bubbles are generated and pulled together with the liquid to the top part of the *stack*, from where they flow into the separation subsystem.
- *Gas–liquid separation subsystem*: The gases generated (bubbles) leave the *stack* immersed in a liquid, which can be water in the case of PEM electrolysis or an aqueous electrolyte (KOH or NaOH solution) in the case of alkaline electrolysis. These two-phase mixtures are carried to deposit-separators, where they are divided into two streams with gas at the top (H_2 or O_2, depending on the deposit) and water or electrolyte at the bottom. This liquid is then recirculated to the *stack*. This recirculation can be done by natural convection or using recirculation pumps.
- *Deionized water subsystem*: This is necessary to restock the water consumed during electrolysis. The water supplied must be deionized with a particular purity and conductivity to avoid a possible contamination of the catalyst and the membrane or diaphragm.
- *Pressure control subsystem*: The way in which pressure is created in an electrolyzer consists of retaining the O_2 and H_2 being produced in the system so that the pressure increases as the gas is accumulated up to the set point of the pressure control valves. In the case of alkaline electrolysis systems, this process is especially important due to the design of the cells themselves, which use porous diaphragms as separators; this produces a diffusion of the gases through the diaphragm micropores, even at only 5 mbar difference in the pressure between the cathode and anode chambers. To minimize this pressure difference, the deposit-separators are usually connected by communicating vessels, thus, allowing for the level and pressure of the systems to be controlled more accurately. In the case of PEM electrolyzers, this effect is minimized due to the use of a polymeric membrane that is practically gas impermeable; this allows for different pressures to be used in the cathode and anode chambers.
- *Thermal management subsystem*: An important part of the electric energy supplied to the stack is transformed into heat due to the irreversibilities of the process. In order to maintain the temperature, a refrigeration system is necessary. This control is made by cooling the water or the electrolyte, which is recirculated to the cells.
- *Drying subsystem*: The gaseous currents that leave the deposit-separators go through a conditioning system made up of condensate traps, coalescence filters, cooling systems,

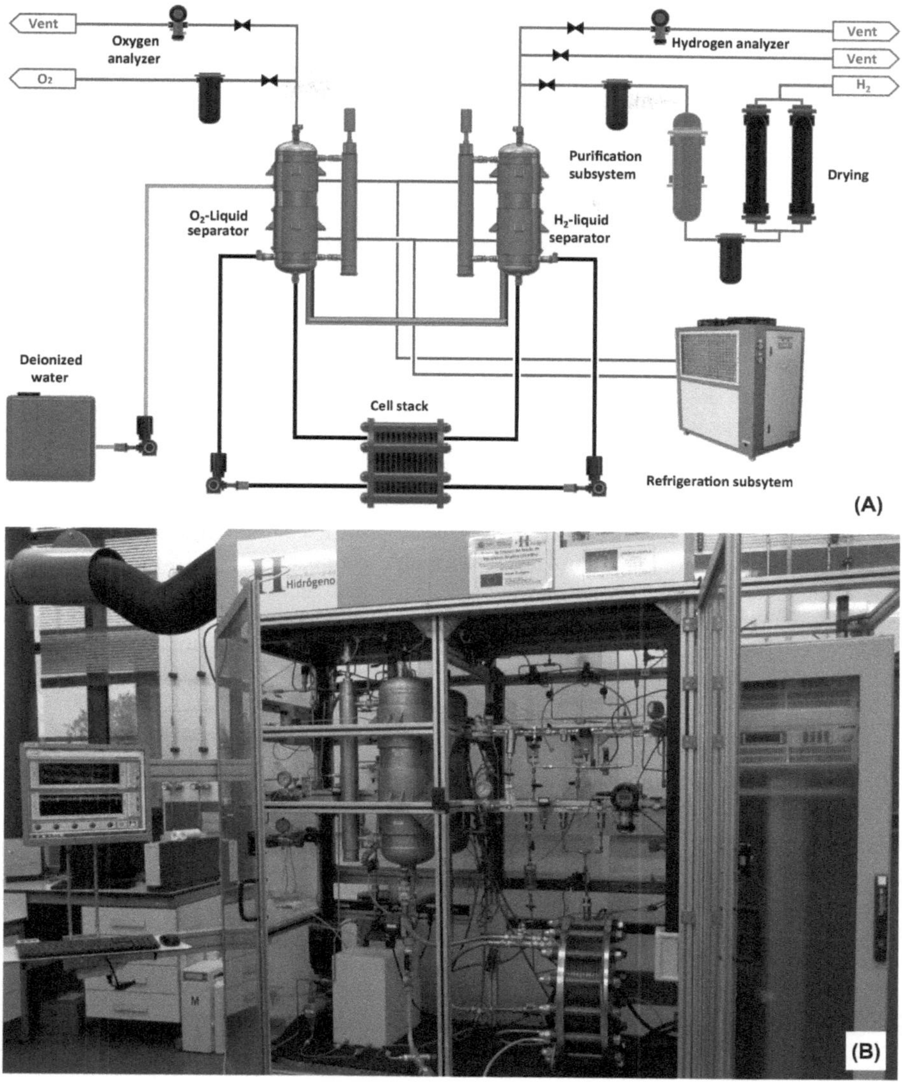

Figure 9.18 Balance of plant of a low-temperature electrolyzer. (A) Simplified process flow diagram and (B) 15 kW alkaline water electrolysis test-bench developed by *Centro Nacional del Hidrógeno* (CNH2, Spain).

and desiccant filters. These allow for the maximum amount of liquid possible to be removed, which is dragged by the gaseous stream, thus, obtaining H_2 as dry as possible.

- *Purification subsystem*: In cases where H_2 of a higher purity is needed, it is necessary to install a de-oxo reactor to remove the O_2 pollutants that have been dragged. These systems are based on a catalyst, usually made of palladium, where the residual O_2 reacts with H_2 by an exothermic reaction and then produces water vapor that is condensed later in order to be removed. This system is especially necessary in alkaline electrolysis, where the purity generated is lower than that in PEM electrolysis systems.

- *Subsystem of analysis of the gases produced*: A small amount of the H_2 and O_2 produced is sent to analyzers in order to quantify the purity of the gases produced.
- *Power supply subsystem*: The electric power that is supplied to the *stack* has to be continuous so it is necessary to incorporate into the system power electronics suitable for integration with the power supplies available.
- *Monitoring and control subsystem*: This aims at gathering all the information coming from sensors and takes the necessary actions to allow for the proper and safe operation of the system.

In the case of high-temperature electrolyzers (SOEC), it is also necessary to consider subsystems such as:

- *Subsystem of stack heating*: Due to the high operating temperatures of solid oxide cells, it is necessary for the stack to be integrated into a furnace capable of heating up and maintaining the temperature of the fluids coming into the stack as well as the elements of the electrolytic cells themselves.
- *Steam production subsystem*: Unlike low-temperature electrolyzers, these electrolysis processes need the water to be supplied as steam at the operating temperatures required.
- *Subsystem of air supply*: A preheated air flow is normally introduced through the anode. This air stream is used to control the temperature and guarantee the correct temperature distribution inside the stack, thus, avoiding a degradation increase.
- *H_2/steam supply subsystem:* The steam previously generated is mixed with H_2 and then introduced into the cathode chamber. This mix must contain a minimum H_2 percentage of 5% in order to keep the Ni electrode reduced and prevent its oxidation.
- *Nitrogen supply subsystem:* This is an essential part for the starting of the system, allowing for the heating and conditioning of the electrolytic cells at the operating temperature.

The final configuration of each of these subsystems will depend on the electrolysis system, the technology considered, the capacity of H_2 production, and the main operating parameters. In addition, the different electrolysis technologies have the advantage of being modular, therefore, the sizes of electrolysis systems can go from some watts to a MW scale joining several systems or stacks. This makes it a highly flexible technology regarding its final application.

9.6.2 Manufacturers and review of commercial electrolyzers

The fast development that H_2 technologies have experienced in the past few years has caused the appearance of a wide range of electrolyzer manufacturers in the past decade [70–72]. Nevertheless, the number of manufacturers is different for electrolyzers based on alkaline technology, PEM, or solid oxide due to the different stages of technology development. PEM electrolysis has taken off in the past few years showing a great development in terms of MW, while alkaline electrolysis is a mature technology since electrolysis plants of a great capacity have existed for decades [9]. Concerning SOEC systems, only some SOFC manufacturers have shown precommercial developments for this technology to date [73]; there is only equipment in the range of a few kilowatts, mainly because of the problems arising from component degradation. In Table 9.2, some of the main electrolyzer manufacturers have been summarized.

Table 9.2 Manufacturers and review of commercial electrolyzers (not exhaustive list).

Technology	Company	Location	Nominal power [kW][a]	Hydrogen rate [Nm^3/h][b]	Energy consumption [kWh/Nm^3][c]	Maximum pressure [bar][d]	References[e]
Alkaline electrolysis	ELB Elektrolyse Technik	Germany	1–7000	200–1400	4.30	30	http://elektrolyse.de/wordpress/
	Ener Blue	Switzerland	5–1000	1–200	4.30	60	http://www.ener-blue.com
	ErreDue	Italy	3–850	0.66–170	5.30	30	https://www.erreduegas.it
	GreenHydrogen	Denmark	390	1–90	5.50	35	http://sreenhydrosen.dk
	HT-Hydrotechnik	Germany	200–700	40–140	5.00	15	http://www.ht-hydrotechnik.de
	Hydrogen Pro	Norway	50–1000	10–210	5.35	30	http://www.hydroeen-pro.com/
	Hydrogenics	Canada	20–300	4–60	4.90	10	https://www.hydroeenics.com
	Idroenergy	Italy	2–400	0.40–80	5.60	6	https://idroenergy.it
	IHT Industrie Haute Technology	Switzerland	4000	760	4.45	32	http://www.iht.ch/
	McPhy	France	2–4000	0.40–800	4.50	30	https://mcphy.com/
	NEL Hydrogen	Norway	2200	485	4.40	30	https://nelhydrogen.com/
	PERIC	China	2800	600	4.60	15	http://www.peric.ac.cn/en/index.asp
	Sagim	France	2.50–25	0.50–5	5.25	8	https://saeim-eip.com/en/products/
	Suzhou Jingli	China	10–4700	2–1000	4.80	15	https://www.iineli-hydroeenplant.com
	Teledyne Energy Systems	USA	300–400	56–80	5.00	10	http://www.teledvnees.com
	Tianjin Mainland	China	0.50–5000	0.10–1000	5.00	30	http://www.cnthe.com
	Enapter (former ACTA)[f]	Italy	2.50	0.50	4.80	30	https://www.enapter.com/aem-electrolyser/
PEM electrolysis	Angstrom Advanced	USA	5–60	1–10	4.00	4	http://www.angstrom-advanced.com
	AREVA H2Gen	France	25–1000	5–200	4.70	35	https://www.arevah2gen.com/
	GINER ELX	USA	150–1000	30–200	4.80	40	https://www.ffinerelx.com/electrolyzer-systems

(*Continued*)

Table 9.2 (Continued)

Technology	Company	Location	Nominal power [kW][a]	Hydrogen rate [Nm^3/h][b]	Energy consumption [kWh/Nm^3][c]	Maximum pressure [bar][d]	References[e]
	GreenHydrogen	Denmark	4.95–390	1–90	5.00	35	http://ereenhvdroeen.dk
	H2B2	Spain	3.20–1000	0.50–200	4.70	40	http://h2b2.es/hydrogen/
	Hydrogenics	Canada	5–1250	1–250	5.20	30	https://www.hydrogenics.com/
	H-TEC	Germany	225–1000	47–210	4.90	30	https://www.h-tec-systems.com
	ITM Power	UK	12–1100	2.40–220	4.80	30	https://www.itm-power.com/
	Proton OnSite (acquired by NEL)	USA	2000	400	4.53	30	https://www.Drotononsite.com/
	Siemens	Germany	1125	225	4.76	35	https://new.siemens.com/global/en/products/enerev/renewable-enerev/hvdroeen-solutions.html
Solid oxide electrolysis	Sunfire	Germany	150	40	3.70	10	https://www.sunfire.de/

[a]Calculated for 5 kW of power for 1 Nm^3 H_2 (when data is not available);
[b]Nominal production ranges for the different product offered;
[c]Specific energy consumption defined by the manufacturer (5 kWh/Nm3 when not available);
[d]Maximum pressure without external compression system;
[e]Some manufacturers supply electrolysis systems consisting of several stacks (modular design);
[f]*AEM*, Alkaline electrolysis membrane.

9.6.3 Perspective of costs

Million tons of H_2 are produced worldwide every year, mainly from fossil fuels (>96%), by SMR (49%) or fuel partial oxidation (29%); only 4% is produced using renewable energies. From this H_2 production, only 1% is generated for energy purposes [71].

However, H_2 is becoming a key part in the energy transition due to its great potential as an energy carrier and its use as a clean fuel. The fact that high purity H_2 can be produced from electricity without any CO_2 emission due to water electrolysis will allow a large number of renewable energies to be channeled to different sectors [15]. Based on these premises, it is estimated that H_2 may come to cover up to 25% of the total energy demand in the European Union by 2050 [74]. Increases in the efficiencies of electrolysis technologies and decreases in their production costs will be key factors for their development.

According to the possible scenarios available currently, different capital expenditures (CAPEX) can be established for each of the electrolysis technologies; although it is necessary to take into consideration that the greater the level of development and implementation of a technology is, the more reliable these data are. Thus the costs of alkaline and PEM electrolysis systems are supported by suppliers and manufacturers that have hundreds of kW installed so far.

In this sense, the cost of alkaline electrolysis systems is currently in the range of 2500–5000 €/kW when the power is under 100 kW, while the cost decreases to 800–1500 €/kW at the MW level. Nevertheless, the alkaline technology may reach a cost around 580 €/kW by 2030, according to the existing cost reduction perspectives (Fig. 9.19). For their part, the operational expenditures (OPEX) of these systems are usually around 2%–3% of the CAPEX [13,43].

On the other hand, PEM electrolysis technology shows a CAPEX around 1500–2000 €/kW for power over 500 kW. The operational expenditures as well as the CAPEX are above those of the alkaline technology, being around 3%–5% of the CAPEX per year [13]. However, the PEM technology shows a greater capability of reducing costs according to the existing estimations, going from 1500 €/kW in

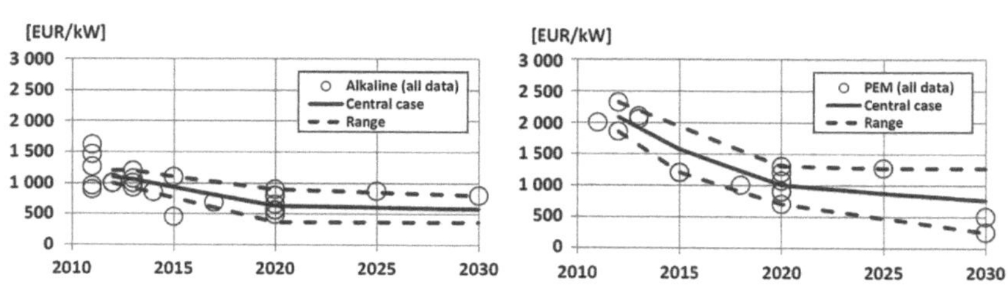

Figure 9.19 Capital cost perspectives for alkaline and PEM electrolysis systems. *Source*: Reproduced from L. Bertuccioli, A. Chan, D. Hart, F. Lehner, B. Madden, E. Standen, Development of Water Electrolysis in the European Union. FCH JU Funded Studies: Element Energy, E4tech Sarl. https://www.fch.europa.eu/sites/default/files/FCHJUElectrolysisStudy_FullReport%20(ID%20199214).pdf, 2014 (accessed 27.12.19). from FCH JU Funded Studies.

2020 to 760 €/kW in 2030 (Fig. 9.19). Despite that, the alkaline technology still shows the lowest cost prospects [9].

Regarding the SOEC technology, there is great uncertainty about its capital expenditures due to its precommercial status with the current CAPEX of these systems being around 2000–3000 €/kW [13]. However, there is also a great prospect of reducing the costs to 1400 €/kW by 2030, when it reaches a production scale of hundreds of MW per year [13,73].

9.7 Integration with renewable energies

Water electrolysis allows for high purity H_2 to be obtained and its production cost is determined by the cost of the electricity used to a great extent. Due to this, producing H_2 efficiently from renewable energies may be the necessary nexus for a proper energy transition. As shown in Fig. 9.20, H_2 as an energy carrier will allow for the integration of renewable energies by managing and storing the excesses generated; on the other hand, it will also allow for this energy to be channeled to different sectors of society such as transport, building, and industry, carrying out a gradual decarbonization of these [72,74,75]:

- *Re-electrification*: The H_2 produced can be reused again to produce power on a large scale through turbines, internal combustion engines, or fuel cells, thus, turning this power back to the grid with the purpose of adjusting demand and generation (*power-to-power* processes).

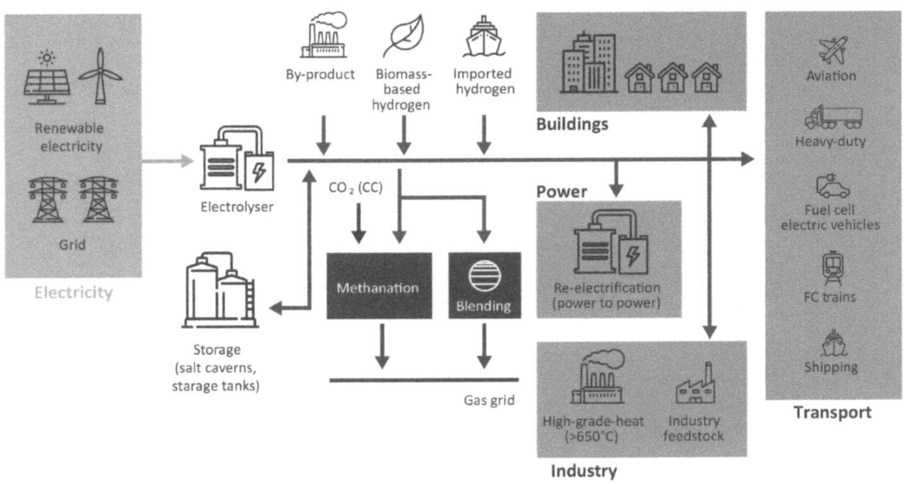

Figure 9.20 Integration of renewable energy sources into end uses by means of hydrogen. *Source*: Reproduced from E. Taibi, R. Miranda, W. Vanhoudt, T. Winkel, J.-C. Lanoix, F. Barth, Hydrogen from Renewable Power: Technology Putlook for the Energy Transition, International Renewable Energy Agency (IRENA), Abu Dhabi. https://www.irena.org/-/media/Files/IRENA/Agency/Publication/2018/Sep/IRENA_Hydrogen_from_renewable_power_2018.pdf, 2018 (accessed 30.12.19) from IRENA.

- *Transport*: H_2 can be used as an alternative fuel in FCVEs, offering a transport option low in carbon and a performance comparable to those of conventional vehicles. It is especially interesting in the sector of heavy-duty vehicles (ships, trains, buses, etc.,), where H_2 is the most promising option.
- *Residential*: Current systems would be able to keep working as usual without implementing any modifications to natural gas systems, but reducing emissions by means of injecting pure H_2 or synthetic natural gas previously produced by methanizing H_2 (*power-to-gas* processes) into the natural gas grid. This way, a gradual decarbonization of the natural gas infrastructure would take place. This also opens up the possibility to integrate cogeneration systems based on fuel cells.
- *Industry*: H_2 is currently used as a raw material in main different industry sectors on a large scale, therefore, the decarbonization of the process in which it is applied is achieved if it is produced from low-carbon sources.

Up to now, numerous demonstrative projects have been carried out in which the electrolytic production of H_2 has been integrated with renewable energies, especially with wind and PV energies. Most of these projects have been based on alkaline electrolysis as it is the most industrially developed technology (there are currently systems of up to 6 MW available). In fact, large installations based on alkaline electrolysis (> 100 MW) have been developed through the 20th century in remote areas with great electric excesses, especially in hydroelectric plants. Some examples are Aswan (Egypt, 1960) with 40,000 Nm^3/h through 288 electrolyzers, Glomfjord (Norway, 1949) with 27,100 Nm^3/h, and Kwekwe (Zimbabwe, 1974) with 27,100 Nm^3/h [13].

On the other hand, the development of *power-to-gas* systems to generate synthetic methane from the CO_2 and H_2 produced via electrolysis in the past few years has caused an important leap in the scale of electrolysis technologies, especially in PEM. This technology has currently come to be commercialized in systems greater than 1 MW since these electrolyzers were used before, mainly for producing H_2 on a small scale [9,56]. In the case of SOEC, the systems available are still in the scale of a few kilowatts [13], many of them in the investigation stage.

Nevertheless, commercial electrolyzers are designed to produce H_2 constantly under nominal operating conditions. However, some difficulties associated with the variability and intermittence of renewable energy sources are caused by their strong dependence on weather conditions. These fluctuations may cause different problems in electrolysis systems such as corrosion, explosive mixture generation, pressure drops, or changes in the internal temperature.

To analyze these effects, different studies have been carried out to improve the integration between renewable energies and electrolyzers, in terms of operation and durability. Some of these solutions pass through the optimization of the designs and materials that constitute an electrolyzer. Other solutions point to an improvement of the BoP control to face the fast fluctuations that can happen when the electric power comes from a renewable source. The use of batteries and other accumulation elements is also seen as an excellent strategy that, together with suitable power electronics, ensures a supply to the electrolyzer as stable and gradual as possible.

References

[1] W.-Y. Chen, J. Seiner, T. Suzuki, M. Lackner (Eds.), Handbook of Climate Change Mitigation, Springer, New York, 2012. Available from: https://doi.org/10.1007/978-1-4419-7991-9.

[2] L.M. Gandía, G. Arzamendi, P.M. Diéguez, Renewable Hydrogen Technologies: Production, Purification, Storage, Applications and Safety, Elsevier, Amsterdam, 2013. Available from: https://doi.org/10.1016/C2011-0-05152-9.

[3] J.I. Linares Hurtado, B.Y. Moratilla Soria, El hidrógeno y la energía, Colección Avances de Ingeniería, Asociación Nacional de Ingenieros del ICAI;, Madrid, 2007. ISBN 978-84-932772-9-1.

[4] A. Goñi-Urtiaga, D. Presvytes, K. Scott, Solid acids as electrolyte materials for proton exchange membrane (PEM) electrolysis: review, Int. J. Hydrog. Energy 37 (4) (2012) 3358–3372. Available from: https://doi.org/10.1016/j.ijhydene.2011.09.152.

[5] C. Coutanceau, S. Baranton, T. Audichon, Hydrogen Energy and Fuel Cells Primers - Hydrogen Electrochemical Production, Academic Press, London, 2017 [chapter 3]. Available from: https://doi.org/10.1016/C2016-0-01050-1.

[6] M.F. Rosa Iglesias, Estudio teórico y experimental sobre la producción de H2 electrolítico a partir de energía solar fotovoltaica [dissertation], Universidad de Sevilla, Sevilla (Spain), 2003.

[7] E. Amores, J. Rodríguez, C. Carreras, Influence of operation parameters in the modeling of alkaline water electrolyzers for hydrogen production, Int. J. Hydrog. Energy 39 (25) (2014) 13063–13078. Available from: https://doi.org/10.1016/j.ijhydene.2014.07.001.

[8] E. McCafferty, Introduction to Corrosion Science, Springer-Verlag, New York, 2010 [chapter 6]. Available from: https://doi.org/10.1007/978-1-4419-0455-3.

[9] A. Godula-Jopek (Ed.), Hydrogen Production: by Electrolysis, Wiley-VCH, Weinheim (Germany), 2015. Available from: https://doi.org/10.1002/9783527676507.

[10] K. Zeng, D. Zhang, Recent progress in alkaline water electrolysis for hydrogen production and applications, Prog. Energy Combust. Sci. 36 (3) (2010) 307–326. Available from: https://doi.org/10.1016/j.pecs.2009.11.002.

[11] D. Pletcher, X. Li, Prospects for alkaline zero gap water electrolysers for hydrogen production, Int. J. Hydrog. Energy 36 (23) (2011) 15089–15104. Available from: https://doi.org/10.1016/j.ijhydene.2011.08.080.

[12] M. Carmo, D.L. Fritz, J. Mergel, D. Stolten, A comprehensive review on PEM water electrolysis, Int. J. Hydrog. Energy 38 (12) (2013) 4901–4934. Available from: https://doi.org/10.1016/j.ijhydene.2013.01.151.

[13] A. Buttler, H. Spliethoff, Current status of water electrolysis for energy storage, grid balancing and sector coupling via power-to-gas and power-to-liquids: a review, Renew. Sustain. Energy Rev. 82 (3) (2018) 2440–2454. Available from: https://doi.org/10.1016/j.rser.2017.09.003.

[14] A. Ursúa, L.M. Gandía, P. Sanchis, Hydrogen production from water electrolysis: current status and future trends, Proc. IEEE 100 (2) (2012) 410–426. Available from: https://doi.org/10.1109/JPROC.2011.2156750.

[15] J.O. Jensen, V. Bandur, N.J. Bjerrum, S.H. Jensen, S. Ebbesen, M. Mogensen, et al. Pre-investigation of water electrolysis (PSO-F&U 2006-1-6287). KI DTU, Risø DTU, DONG Energy, 2008.

[16] P.K. Shen, C.-Y. Wang, S.P. Jiang, X. Sun, J. Zhang (Eds.), Electrochemical Energy: Advanced Materials and Technologies, CRC Press, Boca Raton, 2016. Available from: https://doi.org/10.1201/9781351228756.

[17] D. Hall, Alkaline water electrolysis anode materials, J. Electrochem. Soc. 132 (2) (1985) 41–48. Available from: https://doi.org/10.1149/1.2113856.

[18] M. Ďurovič, J. Hnát, C.I. Bernäcker, T. Rauscher, L. Röntzsch, M. Paidar, et al., Nanocrystalline $Fe_{60}Co_{20}Si_{10}B_{10}$ as a cathode catalyst for alkaline water electrolysis: impact of surface activation, Electrochim. Acta 306 (20) (2019) 688–697. Available from: https://doi.org/10.1016/j.electacta.2019.03.107.

[19] D.M. Soares, O. Teschke, I. Torriani, Hydride effect on the kinetics of the hydrogen evolution reaction on nickel cathodes in alkaline media, J. Electrochem. Soc. 139 (1) (1992) 98–105. Available from: https://doi.org/10.1149/1.2069207.

[20] I.A. Raj, Nickel-based, binary-composite electrocatalysts for the cathodes in the energy-efficient industrial production of hydrogen from alkaline-water electrolytic cells, J. Mater. Sci. 28 (1993) 4375–4382. Available from: https://doi.org/10.1007/BF01154945.

[21] N.V. Krstajić, L.J. Gajić-Krstajić, U. Lačnjevac, B.M. Jović, S. Mora, V.D. Jović, Non-noble metal composite cathodes for hydrogen evolution. Part I: the Ni-MoOx coatings electrodeposited from Watt's type bath containing MoO_3 powder particles, Int. J. Hydrog. Energy 36 (11) (2011) 6441–6449. Available from: https://doi.org/10.1016/j.ijhydene.2011.02.105.

[22] Z. Zheng, N. Li, C.Q. Wang, D.Y. Li, Y.M. Zhu, G. Wu, Ni-CeO_2 composite cathode material for hydrogen evolution reaction in alkaline electrolyte, Int. J. Hydrog. Energy 37 (19) (2019) 13921–13932. Available from: https://doi.org/10.1016/j.ijhydene.2012.07.102.

[23] L. Vázquez-Gómez, S. Cattarin, P. Guerriero, M. Musiani, Influence of deposition current density on the composition and properties of electrodeposited Ni + RuO_2 and Ni + IrO_2 composites, J. Electroanal. Chem. 634 (1) (2009) 42–48. Available from: https://doi.org/10.1016/j.jelechem.2009.07.009.

[24] R.N. Singh, D. Mishra, Anindita, A.S.K. Sinha, A. Singh, Novel electrocatalysts for generating oxygen from alkaline water electrolysis, Electrochem. Commun. 9 (6) (2007) 1369–1373. Available from: https://doi.org/10.1016/j.elecom.2007.01.044.

[25] M.S. El-Deab, M.I. Awad, A.M. Mohammad, T. Ohsaka, Enhanced water electrolysis: electrocatalytic generation of oxygen gas at manganese oxide nanorods modified electrodes, Electrochem. Commun. 9 (8) (2007) 2082–2087. Available from: https://doi.org/10.1016/j.elecom.2007.06.011.

[26] M. Hamdani, M.I.S. Pereira, J. Douch, A.A. Addi, Y. Berghoute, M.H. Mendonca, Physicochemical and electrocatalytic properties of Li-Co_3O_4 anodes prepared by chemical spray pyrolysis for application in alkaline water electrolysis, Electrochim. Acta 49 (9-10) (2004) 1555–1563. Available from: https://doi.org/10.1016/j.electacta.2003.11.016.

[27] H. Wendt, H. Hofmann, V. Plzak, Materials research and development of electrocatalysts for alkaline water electrolysis, Mater. Chem. Phys. 22 (1-2) (1989) 27–49. Available from: https://doi.org/10.1016/0254-0584(89)90030-8.

[28] Q. Radiguè, G. Thunis, J. Proost, On the use of 3-D electrodes and pulsed voltage for the process intensification of alkaline water electrolysis, Int. J. Hydrog. Energy 44 (56) (2019) 29432–29440. Available from: https://doi.org/10.1016/j.ijhydene.2019.06.156.

[29] Z. Xie, P. He, L. Du, F. Dong, K. Dai, T. Zhang, Comparison of four nickel-based electrodes for hydrogen evolution reaction, Electrochim. Acta 88 (2013) 390–394. Available from: https://doi.org/10.1016/j.electacta.2012.10.057.

[30] Ph Vermeiren, J.P. Moreels, A. Claes, H. Beckers, Electrode diaphragm electrode assembly for alkaline water electrolysers, Int. J. Hydrog. Energy 34 (23) (2009) 9305–9315. Available from: https://doi.org/10.1016/j.ijhydene.2009.09.023.

[31] J. Otero, J. Sese, I. Michaus, M. Santa María, E. Guelbenzu, S. Irusta, et al., Sulphonated polyether ether ketone diaphragms used in commercial scale alkaline water electrolysis, J. Power Sources 247 (2014) 967–974. Available from: https://doi.org/10.1016/j.jpowsour.2013.09.062.

[32] K. Scott, R. Hughes (Eds.), Industrial Membrane Separation Technology, Springer, Dordrecht (Netherlands), 1996 [chapter 7]. Available from: https://doi.org/10.1007/978-94-011-0627-6.

[33] T. Ohmori, K. Tachikawa, K. Tsuji, K. Anzai, Nickel oxide water electrolysis diaphragm fabricated by a novel method, Int. J. Hydrog. Energy 32 (18) (2007) 5094–5097. Available from: https://doi.org/10.1016/j.ijhydene.2007.07.055.

[34] L. Xiao, S. Zhang, J. Pan, C. Yang, M. He, L. Zhuang, et al., First implementation of alkaline polymer electrolyte water electrolysis working only with pure water, Energy Environ. Sci. 5 (2012) 7869–7871. Available from: https://doi.org/10.1039/C2EE22146B.

[35] I. Vincent, D. Bessarabov, Low cost hydrogen production by anion exchange membrane electrolysis: a review, Renew. Sustain. Energy Rev. 81 (2) (2018) 1690–1704. Available from: https://doi.org/10.1016/j.rser.2017.05.258.

[36] M. Paidar, V. Fateev, K. Bouzek, Membrane electrolysis—history, current status and perspective, Electrochim. Acta 209 (10) (2016) 737–756. Available from: https://doi.org/10.1016/j.electacta.2016.05.209.

[37] J. Hnát, M. Paidar, J. Schauer, J. Žitka, K. Bouzek, Polymer anion selective membranes for electrolytic splitting of water. Part I: stability of ion-exchange groups and impact of the polymer binder, J. Appl. Electrochem. 41 (2011) 1043–1052. Available from: https://doi.org/10.1007/s10800-011-0309-9.

[38] R. Gilliam, J.W. Graydon, D.W. Kirk, S.J. Thorpe, A review of specific conductivities of potassium hydroxide solutions for various concentrations and temperatures, Int. J. Hydrog. Energy 32 (3) (2007) 359–364. Available from: https://doi.org/10.1016/j.ijhydene.2006.10.062.

[39] M. Sánchez, E. Amores, L. Rodríguez, C. Clemente-Jul, Semi-empirical model and experimental validation for the performance evaluation of a 15 kW alkaline water electrolyzer, Int. J. Hydrog. Energy 43 (45) (2018) 20332–20345. Available from: https://doi.org/10.1016/j.ijhydene.2018.09.029.

[40] S. Marini, P. Salvi, P. Nelli, R. Pesenti, M. Villa, M. Berrettoni, et al., Advanced alkaline water electrolysis, Electrochim. Acta 82 (2012) 384–391. Available from: https://doi.org/10.1016/j.electacta.2012.05.011.

[41] N. Nagai, M. Takeuchi, M. Nakao, Influences of bubbles between electrodes onto efficiency of alkaline water electrolysis, in: Proceedings of PSFVIP-4, Chamonix (France), 2003.

[42] M. Lehner, R. Tichler, H. Steinmüller, M. Koppe, Power-to-gas: Technology and Business Models, Springer, Cham (Switzerland), 2014. Available from: https://doi.org/10.1007/978-3-319-03995-4.

[43] L. Bertuccioli, A. Chan, D. Hart, F. Lehner, B. Madden, E. Standen, Development of Water Electrolysis in the European Union, FCH JU Funded Studies: Element Energy, E4tech Sarl, 2014. <https://www.fch.europa.eu/sites/default/files/FCHJUElectrolysisStudy_FullReport%20(ID%20199214).pdf> (accessed 27.12.19).

[44] X. Li, F.C. Walsh, D. Pletcher, Nickel based electrocatalysts for oxygen evolution in high current density, alkaline water electrolysers, Phys. Chem. Chem. Phys. 13 (2011) 1162–1167. Available from: https://doi.org/10.1039/C0CP00993H.

[45] J. Fischer, H. Hofmann, G. Luft, H. Wendt, Fundamental investigations and electrochemical engineering aspects concerning an advance concept for alkaline water electrolysis, AIChe J. 26 (5) (1980) 794–802. Available from: https://doi.org/10.1002/aic.690260513.

[46] Y. Leng, G. Chen, A.J. Mendoza, T.B. Tighe, M.A. Hickner, C.-Y. Wang, Solid-state water electrolysis with an alkaline membrane, J. Am. Chem. Soc. 134 (22) (2012) 9054–9057. Available from: https://doi.org/10.1021/ja302439z.

[47] J.C. Ganley, High temperature and pressure alkaline electrolysis, Int. J. Hydrog. Energy 34 (9) (2009) 3604–3611. Available from: https://doi.org/10.1016/j.ijhydene.2009.02.083.

[48] S. Song, H. Zhang, X. Ma, Z. Shao, R.T. Baker, B. Yi, Electrochemical investigation of electrocatalysts for the oxygen evolution reaction in PEM water electrolyzers, Int. J. Hydrog. Energy 33 (19) (2008) 4955–4961. Available from: https://doi.org/10.1016/j.ijhydene.2008.06.039.

[49] A.T. Marshall, S. Sunde, M. Tsypkin, R. Tunold, Performance of a PEM water electrolysis cell using $Ir_xRu_yTa_zO_2$ electrocatalysts for the oxygen evolution electrode, Int. J. Hydrog. Energy 32 (13) (2007) 2320–2324. Available from: https://doi.org/10.1016/j.ijhydene.2007.02.013.

[50] C. Hao, H. Lv, C. Mi, Y. Song, J. Ma, Investigation of mesoporous niobium-doped TiO_2 as an oxygen evolution catalyst support in an SPE water electrolyzer, ACS Sustain. Chem. Eng. 4 (3) (2015) 746–756. Available from: https://doi.org/10.1021/acssuschemeng.5b00531.

[51] M. Morimitsu, R. Otogawa, M. Matsunaga, Effects of cathodizing on the morphology and composition of $IrO_2Ta_2O_5$/Ti anodes, Electrochim. Acta 46 (2-3) (2000) 401–406. Available from: https://doi.org/10.1016/S0013-4686(00)00598-3.

[52] A. Brouzgou, S.Q. Song, P. Tsiakaras, Low and non-platinum electrocatalysts for PEMFCs: current status, challenges and prospects, Appl. Catal. B: Environ. 127 (2012) 371–388. Available from: https://doi.org/10.1016/j.apcatb.2012.08.031.

[53] T. Audichon, E. Mayousse, S. Morisset, C. Morais, C. Comminges, T.W. Napporn, et al., Electroactivity of RuO_2–IrO_2 mixed nanocatalysts toward the oxygen evolution reaction in a water electrolyzer supplied by a solar profile, Int. J. Hydrog. Energy 39 (30) (2013) 16785–16796. Available from: https://doi.org/10.1016/j.ijhydene.2014.07.170.

[54] Y. Shao, Y. Cheng, W. Duan, W. Wang, B. Li, Y. Lin, et al., Nanostructured electrocatalysts for PEM fuel cells and redox flow batteries: a selected review, ACS Catal. 5 (12) (2015) 7288–7298. Available from: https://doi.org/10.1021/acscatal.5b01737.

[55] S.U.B. Ramakrishna, D.S. Reddy, S.S. Kumar, V. Himabindu, Nitrogen doped CNTs supported Palladium electrocatalyst for hydrogen evolution reaction in PEM water electrolyser, Int. J. Hydrog. Energy 41 (45) (2016) 20447–20454. Available from: https://doi.org/10.1016/j.ijhydene.2016.08.195.

[56] D. Bessarabov, H. Wang, H. Li, N. Zhao (Eds.), PEM Electrolysis for Hydrogen Production: Principles and Applications, CRC Press, Boca Raton, 2016. Available from: https://doi.org/10.1201/b19096.

[57] S.S. Kumar, S.U.B. Ramakrishna, D. Bhagawan, V. Himabindu, Preparation of $Ru_xPd_{1-x}O_2$ electrocatalysts for the oxygen evolution reaction (OER) in PEM water electrolysis, Ionics 24 (2018) 2411–2419. Available from: https://doi.org/10.1007/s11581-017-2359-4.

[58] P. Lettenmeier, R. Wang, R. Abouatallah, B. Saruhan, O. Freitag, P. Gazdzicki, et al., Low-cost and durable bipolar plates for proton exchange membrane electrolyzers, Nat. Sci. Rep. 7 (2017) 44035. Available from: https://doi.org/10.1038/srep44035.
[59] L. Mendizabal, A. Oedegaard, O.E. Kongstein, S. Lædre, J. Walmsley, J. Barriga, et al., TaN_X coatings deposited by HPPMS on SS316L bipolar plates for polymer electrolyte membrane fuel cells: correlation between corrosion current, contact resistance and barrier oxide film formation, Int. J. Hydrog. Energy 42 (5) (2017) 3259–3270. Available from: https://doi.org/10.1016/j.ijhydene.2017.01.070.
[60] S. Lædre, O.E. Kongstein, A. Oedegaard, H. Karoliussen, F. Seland, Materials for proton exchange membrane water electrolyzer bipolar plates, Int. J. Hydrog. Energy 42 (5) (2017) 2713–2723. Available from: https://doi.org/10.1016/j.ijhydene.2016.11.106.
[61] A.S. Gago, S.A. Ansar, B. Saruhan, U. Schulz, P. Lettenmeier, N.A. Cañas, et al., Protective coatings on stainless steel bipolar plates for proton exchange membrane (PEM) electrolysers, J. Power Sources 307 (2016) 815–825. Available from: https://doi.org/10.1016/j.jpowsour.2015.12.071.
[62] E. Borgardt, L. Giesenberg, M. Reska, M. Müller, K. Wippermann, M. Langemann, et al., Impact of clamping pressure and stress relaxation on the performance of different polymer electrolyte membrane water electrolysis cell designs, Int. J. Hydrog. Energy 44 (42) (2019) 23556–23567. Available from: https://doi.org/10.1016/j.ijhydene.2019.07.075.
[63] M. Ni, M.K.H. Leung, D.Y.C. Leung, Technological development of hydrogen production by solid oxide electrolyzer cell (SOEC), Int. J. Hydrog. Energy 33 (9) (2008) 2337–2354. Available from: https://doi.org/10.1016/j.ijhydene.2008.02.048.
[64] P.T. Moseley, J. Garche (Eds.), Electrochemical Energy Storage for Renewable Sources and Grid Balancing, Elsevier, Amsterdam, 2015. Available from: https://doi.org/10.1016/C2012-0-01253-7.
[65] Ch Neofytidis, E. Ioannidou, L. Sygellou, M. Kollia, D.K. Niakolas, Affecting the H_2O electrolysis process in SOECs through modification of NiO/GDC: experimental case of Au-Mo-Ni synergy, J. Catal. 373 (2019) 260–275. Available from: https://doi.org/10.1016/j.jcat.2019.04.002.
[66] J. Yoo, S.-K. Woo, J.H. Yu, S. Lee, G.W. Parl, $La_{0.8}Sr_{0.2}MnO_3$ and $(Mn_{1.5}Co_{1.5})O_4$ double layer coated by electrophoretic deposition on Crofer22 APU for SOEC interconnect applications, Int. J. Hydrog. Energy 34 (3) (2009) 1542–1547. Available from: https://doi.org/10.1016/j.ijhydene.2008.12.005.
[67] S. Fogel, H. Kryk, U. Hampel, Simulation of the transient behavior of tubular solid oxide electrolyzer cells under fast load variations, Int. J. Hydrog. Energy 44 (18) (2019) 9188–9202. Available from: https://doi.org/10.1016/j.ijhydene.2019.02.063.
[68] T. Ishihara, Z. Tan, CO_2 electrolysis of tubular type solid oxide cells using $LaGaO_3$ electrolyte thin film prepared by dip-coating method, ISIJ Int. 59 (4) (2019) 619–622. Available from: https://doi.org/10.2355/isijinternational.ISIJINT-2018-395.
[69] X. Zhang, Y. Song, G. Wang, X. Bao, Co-electrolysis of CO_2 and H_2O in high-temperature solid oxide electrolysis cells: recent advance in cathodes, J. Energy Chem. 26 (5) (2017) 839–853. Available from: https://doi.org/10.1016/j.jechem.2017.07.003.
[70] M. David, C. Ocampo-Martínez, R. Sánchez-Peña, Advances in alkaline water electrolyzers: a review, J. Energy Storage 23 (2019) 392–403. Available from: https://doi.org/10.1016/j.est.2019.03.001.
[71] J. Gummer, J. King, N. Chater, R. Heaton, P. Johnson, C. Le Quéré, et al., Members of the Committee on Climate Change. Hydrogen in a Low-carbon Economy, Committee

on Climate Change (CCC), London, 2018. <https://www.theccc.org.uk/wp-content/uploads/2018/11/Hydrogen-in-a-low-carbon-economy.pdf> (accessed 23.12.19).

[72] E. Taibi, R. Miranda, W. Vanhoudt, T. Winkel, J.-C. Lanoix, F. Barth, Hydrogen from Renewable Power: Technology Outlook for the Energy Transition, International Renewable Energy Agency (IRENA), Abu Dhabi, 2018. Available from: https://www.irena.org/-/media/Files/IRENA/Agency/Publication/2018/Sep/IRENA_Hydrogen_from_renewable_power_2018.pdf (accessed 30.12.19).

[73] R. Anghilante, D. Colomar, A. Brisse, M. Marrony, Bottom-up cost evaluation of SOEC systems in the range of 10–100 MW, Int. J. Hydrog. Energy 43 (45) (2018) 20309–20322. Available from: https://doi.org/10.1016/j.ijhydene.2018.08.161.

[74] Fuel Cells and Hydrogen Joint Undertaking (FCH JU), Hydrogen Roadmap Europe: A Sustainable Pathway for the European Energy Transition, FCH JU;, Luxembourg, 2019. Available from: https://www.fch.europa.eu/sites/default/files/Hydrogen%20Roadmap%20Europe_Report.pdf (accessed 30.12.19).

[75] Hydrogen Council, How Hydrogen Empowers the Energy Transition, Hydrogen Europe;, Brussels, 2017. Available from: https://hydrogencouncil.com/wp-content/uploads/2017/06/Hydrogen-Council-Vision-Document.pdf (accessed 30.12.19).

Microbial production of hydrogen 10

Ioanna Ntaikou
Foundation for Research and Technology, Institute of Chemical Engineering Sciences, Patras, Greece

10.1 Molecular basis of microbial hydrogen production

During microbial metabolism, either in the case of heterotrophs or autotrophs, oxidation of organic compounds is carried out for the generation of the necessary chemical compounds and energy to support microbial growth. The oxidation process results in the generation of electrons that must be deposited on other chemical compounds ensuring thus electrical neutrality. Under aerobic conditions, the final electrons molecular oxygen acceptor is reduced toward water (aerobic respiration). However, in environments where anaerobic or anoxic conditions prevail, other elements or compounds need to act as electron acceptors such as nitrate and nitrite ions that lead to the generation of molecular nitrogen or sulfate ions that lead to the generation of hydrogen sulfide. Protons are also a good electron acceptor, leading to molecular hydrogen production during their reduction. However, not all microorganisms obtain the ability to reduce the same electron acceptors, since the presence of specific enzymatic systems is required. In case of hydrogen-producing microorganisms there are two recognized groups of enzymes, namely nitrogenases and hydrogenases, that may or may not coexist.

10.1.1 Nitrogenases

Nitrogenases are metalloenzymes, and are divided into three types based on the metal cofactor that occupies their active center: (1) nitrogenases-FeMo, (2) nitrogenases-FeV, and (3) nitrogenases-FeFe. The latter are found in phototrophic bacteria and cyanobacteria and are responsible for the conversion of atmospheric nitrogen to ammonia as nitrogen source for the microbial growth. This reaction requires the presence of reducing agent and ATP consumption, and the concomitant reduction of protons to molecular hydrogen (Eq. 10.1):

$$N_2 + 8H^+ + 8e^- + 16ATP \rightarrow 2NH_3 + \mathbf{H_2} + 16ADP + 16Pi \quad (10.1)$$

However, even in the case of absence of molecular nitrogen, nitrogenases can also catalyze the production of molecular hydrogen via the reduction of protons, using ferredoxin as electron donor. It has been reported that molecular hydrogen

Sustainable Fuel Technologies Handbook. DOI: https://doi.org/10.1016/B978-0-12-822989-7.00011-1

can be generated from organic acids with simultaneous production of carbon dioxide [1]. Such reactions are thermodynamically unfavorable though ($\Delta G^o > 0$) as shown in Eq. (10.2), according to which molecular hydrogen is generated via the bioconversion of acetic acid:

$$CH_3COOH + 2H_2O + \text{``}\mathbf{hv}\text{''} \rightarrow 2CO_2 + 4H_2 \tag{10.2}$$

$$\Delta G^o = +75.2 \ \text{kJ/mol}$$

In this context, it should be mentioned that since nitrogenases are quite vulnerable to oxygen and decompose quickly, in case of cyanobacteria, which are autotrophic microorganisms and generate molecular oxygen during photosynthesis, nitrogenases are located in specially modified cells called heterocystes where almost anoxic conditions prevail. In particular, the maintenance of low-oxygen levels in their interior is achieved by their extremely thick cell walls, thus acting as a barrier which eliminates reducing diffusion, as well as by their high respiratory rate leading to carbon dioxide production.

10.1.2 Hydrogenases

Hydrogenases are metalloenzymes, like nitrogenases, and play a central role in microbial metabolism, catalyzing the reversible oxidation of molecular hydrogen by reactions (10.3) and (10.4):

$$H_2 + A_{ox} \rightarrow 2H^+ + A_{red} \tag{10.3}$$

$$2H^+ + D_{red} \rightarrow H_2 + D_{ox} \tag{10.4}$$

Proton uptake (molecular hydrogen oxidation) occurs during the reduction of electron acceptors such as oxygen in aerobic respiration, nitrate and sulfate ions in anaerobic respiration, and oxygen and hydrogen sulfide (for prokaryotes only) in photosynthesis. Molecular hydrogen production occurs during the fermentation of pyruvate, in which the protons are receptors of the electron surplus. In these reactions coenzymes such as NAD and NADP, ferredoxins, and cytochromes c3 and c6 act as donor (D) and acceptor (A) electron. Molecular hydrogen production through NAD and ferredoxin is carried out according to the reactions (10.5) and (10.6) [2]:

$$NADH + H^+ \rightarrow NAD^+ + H_2 \tag{10.5}$$

$$\Delta G^o = +18.1 \ \text{kJ/mol}$$

$$2Fd_{(red)} + 2H^+ \rightarrow 2Fd_{(ox)} + H_2 \tag{10.6}$$

$$\Delta G^o = +3.1 \ \text{kJ/mol}$$

ΔG^o is positive in both reactions, meaning that those reactions can only be realized with the consumption of energy (in the form of ATP). The main hydrogenases are summarized based on their biochemical action in Table 10.1.

Hydrogenases are mainly found in archaebacteria and bacteria, but occur in eukaryotic microorganisms as well. Depending on the type of metal ion present in their active center, they are subdivided into Ni-Fe-hydrogenases and single-Fe-hydrogenases, to which the subclass of free Fe-S hydrogenases also belongs. Fe-hydrogenases are typically occurred in hydrogen-producing microorganisms, such as bacteria and eukaryotic microorganisms, while Ni-Fe-hydrogenases are typically found in hydrogen-consuming microorganisms, such as archaebacteria and bacteria. Like most redox metalloenzymes, hydrogenases are extremely sensitive to oxygen, high temperatures, carbon monoxide, cyanide anion, and several other environmental factors.

10.2 Hydrogen-producing microorganisms

Hydrogen-production capacity has been observed in many microbial species that vary taxonomically, physiologically and biochemically and which may function singly or as a consortium of similar types or mixed cultures. Gray and Gest [3] first classified hydrogen-producing microorganisms into four main groups:

- Strictly anaerobic heterotrophs without a cytochrome system (clostridia, micrococci, methylobacteria, etc.)
- Strictly anaerobic heterotrophs with a cytochrome system (*Desulfovibrio desulfuricans*)
- Facultative anaerobic heterotrophs with cytochromes that produce hydrogen through the degradation of formic acid (enterobacteriacea, etc.).
- Photosynthetic bacteria and algae that produce hydrogen through NADH harnessing solar energy.

Nandi and Sengupta [4], following similar criteria, classified hydrogen-producing organisms according to their tolerance to oxygen and the ability to exploit sunlight as a source of energy. Following this classification, some representative groups and genera of the hydrogen-producing microorganisms studied to date are further analyzed.

10.2.1 Obligatory anaerobic hydrogen producing microorganisms

In this class only prokaryotic microorganisms are included, with the order Clostridia being the most widely studied. Among them, *Clostridium* sp. include spore forming anaerobic bacilli that lack a cytochrome system for oxidative phosphorylation and the ATP produced during fermentation through substrate-level phosphorylation. Species of the genus *Clostridium* that have been isolated and exhibit high hydrogen yields include *Clostridium butyricum, Clostridium welchii,*

Table 10.1 Biochemical classification of hydrogenases.

Enzyme	Reaction	Encoding[a]
Hydrogen dehydrogenase (H_2:NAD^+ oxidoreductase) $NADP^+$	$H_2 + NAD^+ \leftrightarrow H^+ + NADH$	EC 1.12.1.2
Hydrogen dehydrogenase (H_2:$NADP^+$ oxidoreductase)	$H_2 + NADP^+ \leftrightarrow H^+ + NADPH$	EC 1.12.1.3
Cytochrome-c3 hydrogenase	H_2 + 2ferricytochrome $c_3 \leftrightarrow 2\ H^+$ + 2 ferrocytochrome c_3	EC 1.12.2.1
Ferredoxin hydrogenase	$H_2 + 2Fd_{oxidised} \leftrightarrow 2\ H^+ + 2Fd_{reduced}$	EC 1.12.7.2
Coenzyme F_{420} hydrogenase	H_2 + coenzyme $F_{420(oxidized)} \leftrightarrow$ coenzyme $F_{420(reduced)}$	EC 1.12.98.1
Hydrogenase (acceptor)	$H_2 + A \leftrightarrow AH_2$	EC 1.12.99.6
Hydrogen:quinone oxidoreductase	H_2 + menaquinone $\leftrightarrow$ menaquinol	EC 1.12.5.1
5,10-methenyltetrahydromethanopterin hydrogenase	H_2 + 5,10-methenyltetrahydromethanopterin $\leftrightarrow H^+$ + 5,10-methylenetetrahydromethanopterin	EC 1.12.98.2

[a]https://enzyme.expasy.org

Clostridium pasteurianum, Clostridium beijerinckii, Clostridium felsineum. Hydrogen yields up to 2.81 mmol/mmol-glucose was reported from *C. beijerinckii* [5]. Clostridia produce hydrogen via the metabolism of different C5 and C6 monosaccharides and disaccharides [6] as well as complex carbohydrates since different species possess fibrolytic [7] and/or amylolytic [8] activities. Clostridia are also able to metabolize other carbon sources such as glycerol [9] yielding to lower hydrogen production. Other genus of clostridia with fibrolytic properties that is reported to produce hydrogen in high yields from carbohydrate based substrates is *Ruminococcus* sp. [10,11]; yields up to 2.58 mil H_2/mol hexose have been reported from *Ruminococcus albus* using different monosaccharides, lignocellulosic biomass and wastes as carbon sources [12,13].

Methyltrophic bacteria belong to the same category. Egorov et al. [14] firstly isolated NAD-dependent formic dehydrogenase from a methylotrophic bacterium, thus demonstrating the ability to produce hydrogen from organic fuels through methylotrophic fmicroorganisms. Subsequent studies have indicated the possibility of producing hydrogen from methane, methanol, formaldehyde, and formic acid through the methylotropes *Methylomonas albus BG8* and *Methyloshinus trichosporium OB3b*, reaching yields up to 2.45 mol H_2/mol substrate [15].

The ability of archaebacteria to produce molecular hydrogen has also been extensively studied. The hyperthermophilic archaebacterium *Pyrococcus furiosus* produces hydrogen from carbohydrates and peptides. The microorganism also appears to be capable of oxidizing pyruvate [16], aldehyde [17], formaldehyde [18], and 2-ketoglutaric acid [19]. These oxidations are carried out by ferredoxin-linked redox reductants, and the reduced ferredoxin is then recycled by producing either molecular hydrogen or hydrogen sulfide via hydrogenases or sulfur hydrogens.

10.2.2 Facultative aerobic hydrogen producing microorganisms

These microorganisms can grow either in the presence or absence of oxygen, adjusting their metabolism. The most widely studied species of this class are *Enterobacter aerogenes*, *Escherichia coli*, and *E. aerogenes* in terms of their hydrogen production efficiency and their metabolism. Early studies have shown that *E. coli* produces hydrogen and carbon dioxide from formic acid in the absence of oxygen [20,21] via the enzymatic complex hydrogen–hydrogen lyase [3]. The molecular yield of hydrogen from *E. coli* cultures on glucose was found to range from 0.9 to 1.2 with the highest observed for immobilized cell cultures [22,23]. On the other hand, the genus *Enterobacter*, having high growth rates, can metabolize a variety of substrates, while the hydrogen yield does not seem to be inhibited by high hydrogen partial pressure [24,25], as recently reported by Boshagh et al. [26] who reported yields up to 2.2 mol/mol glucose in immobilized cultures of *E. aerogenes*.

10.2.3 Photosynthetic hydrogen producing microorganisms

Hydrogen production through photosynthetic processes is carried out by eukaryotic (i.e., green algae) and prokaryotic (i.e., cyanobacteria and other photosynthetic bacteria) via different mechanisms. Specifically, the phototropic green algae and cyanobacteria, using sunlight and carbon dioxide as sources of energy and carbon, respectively, are able to produce sugars through the process of photosynthesis via the reduction of carbon dioxide and water according to the reaction:

$$6CO_2 + 6H_2O + \text{``}\mathbf{hv}\text{''} \rightarrow C_6H_{12}O_6 + 6O_2 \quad (10.7)$$

$$\Delta G^{\circ} = +\,2844.41\ \text{kJ/mol}$$

However, under anaerobic conditions they divert water directly to hydrogen and oxygen using solar energy as shown in (10.8):

$$H_2O + \text{``}\mathbf{hv}\text{''} \rightarrow H_2 + 1/2O_2 \quad (10.8)$$

$$\Delta G^{\circ} = +\,374.5\ \text{kJ/mol}$$

The latter requires only water and sunlight, and therefore the prospect of using them is of great interest, both economically and environmentally. About a third of the cyanobacteria also produce hydrogen through the reaction of atmospheric nitrogen fixation according to the reaction of Eq. (10.1), thus exhibiting higher rates of hydrogen production. In particular, it has been reported that nitrogen-fixing species of the genus *Anabaena* could produce 0.91–4.2 μmol H_2/mg chl a/h [27], quite higher than the respective values obtained in case of nonnitrogen-fixing species like *Synechococcus* (0.02–0.66 μmol H_2/mg chl a/h), *Microcystis* (0.16 μmol H_2/mg chl a/h), *Synecocystis* (0.07–0.13 μmol H_2/mg chl a/h), and *Gloebacter* (1.38 μmol H_2/mg chl a/h) [28].

On the other hand, photosynthetic bacteria such as purple bacteria cannot produce hydrogen directly from water and require an organic carbon source for their growth. In those species hydrogen is produced by the consumption of organic acids, such as acetic and lactic acid [29], according to reactions (10.2) and (10.9), respectively.

$$C_3H_6O_3 + H_2O + \text{``}\mathbf{hv}\text{''} \rightarrow 6H_2 + 3CO_2 \quad (10.9)$$

$$\Delta G^{\circ} = +\,52\ \text{kJ/mol}$$

ΔG^o of these reactions is much less than the energy required for the evolution of molecular hydrogen via the reaction (10.8). In addition, there is a wide range of organic substrates that can be exploited, whereas their bioconversion is complete. Apart from acetate and lactate, other fatty acids and carbohydrates such as sugars, starch, and cellulose can be exploited. *Rhodobacter sphaeroides* is maybe the most-studied microorganism that can generate molecular hydrogen metabolizing organic substrates. Initial studies with *R. sphaeroides* reported the hydrogen-production efficiency of the bacterium against different organic acids such as acetate, lactate, butyrate, and malate [30–34], while many studies reported the exploitation of complex organic feedstocks such as sugar, refinery wastewater [35], brewery wastewaters [36], and hydrolyzed straw biomass [37].

10.3 Processes of microbial hydrogen production

10.3.1 Biophotolysis

Biophotolysis-based hydrogen production can be carried out via direct or indirect means as identified by whether light is irradiated during hydrogen evolution [38]. A brief description of the principles, systems, and main bottlenecks for the practical application of the two processes are given next.

10.3.1.1 Direct biophotolysis

Direct biophotolysis is an attractive process since solar energy is used to convert water, a readily available substrate, into oxygen and hydrogen (Eq. 10.10):

$$H_2O + \text{``}\mathbf{hv}\text{''} \rightarrow H_2 + 1/2O_2 \tag{10.10}$$

$$\Delta G^o = +\,374.5\ \text{kJ/mol}$$

The latter was first observed by Benemann et al. [39] in a chloroplast-ferrodoxin-hydrogenase system, although the existence of such a reaction in green algae had been suggested much earlier [40]. Certain species of green algae possess a reversible hydrogenase (Fe-hydrogenase) and could be used in a direct biophotolysis process. Among them *Chlamydomanas reinhardtii* was reported to exhibit a 22% conversion efficiency of solar energy to molecular hydrogen during in vivo biophotolysis under low-light intensity and low-partial oxygen pressure since the activity of Fe-hydrogenase is extremely sensitive to oxygen. Although not systematically studied, it appears that the oxygen partial pressure must be below 0.1%, corresponding to <1 μmole of oxygen in the liquid phase for the simultaneous production of oxygenand molecular hydrogen. However, maintaining such a partial oxygen pressure is almost impossible in practice due to the large amounts of carrier gas and energy required to remove the generated oxygen [41].

Different strategies have been proposed for overcoming the limitations of the process including the engineer oxygen-sensitive hydrogenases with enhanced oxygen tolerance [42], controlling the level of oxygen in the range of hydrogenase/nitrogenase by inhibiting oxygen evolution [43], and introducing oxygen binding proteins [44].

10.3.1.2 Indirect biophotolysis

Indirect biophotolysis consists of two stages in series, that is, the photosynthetic production of carbohydrates and the subsequent fermentation of the later for hydrogen production, as shown by the reactions (10.11) and (10.12).

$$6H_2O + 6CO_2 + \text{``}hv\text{''} \rightarrow C_6H_{12}O_6 + 6O_2 \tag{10.11}$$

$$6H_2O + C_6H_{12}O_6 + \text{``}hv\text{''} \rightarrow 12H_2 + 6CO_2 \tag{10.12}$$

These reactions take place in different parts of the microbial cell, making feasible the temporal and/or spatial separation of oxygen and hydrogen evolution [38]. This separation is actually the natural solution to the incompatibility of oxygen and hydrogen evolution, which is among the greatest limitations of direct biophotolysis, whereas at the same time facilitating subsequent purification of the produced hydrogen [45].

The ability of indirect biophotolysis has been identified in algae and more attractively in cyanobacteria, such as *Anabaena*, *Spirulina*, *Calothrix Synechococcus*, and *Geobacter*. Among them, research is focused on the species of *Anabaena variabilis*. Specifically, due to its high hydrogen evolution performance [27], emphasis has been given to increase the activity of hydrogen producing enzymes and to develop genetically engineered strains of *Anabaena* sp.

10.3.2 Microbial water–gas shift

The water–gas shift reaction results in the formation of hydrogen and carbon dioxide (Eq. 10.13).

$$CO + H_2O \rightarrow CO_2 + H_2 \tag{10.13}$$

$$\Delta G^{o} = -20 \text{ kJ/mol}$$

Two different metalloenzymes, a monofunctional [NiFe]-carbon monoxide dehydrogenase ([NiFe]-CODH) and an energy-conserving [NiFe]-hydrogenase, are implicated in the process [46]. Carbon monoxide dehydrogenase is responsible for the oxidation of carbon monoxide, whereas the second enzyme, the energy-conserving hydrogenase, catalyzes the reduction of two protons toward molecular hydrogen by the released electrons.

Thermophilic and mesophilic species are included in the hydrogenogenic carboxydotrophs, with thermophilic species being more abundant than the mesophilic ones, suggesting that high temperature could facilitate the process, possibly by increasing gas diffusion rates [47]. The archaea *Thermococcus onnurineus*, *Calderihabitans maritimus*, and *Thermincola ferriacetica* and the bacteria *Thermincola carboxydiphila*, *Thermoanaerobacter thermohydrosulfuricus*, *Caldanaerobacter subterraneus pacificus*, and *Carboxydocella pertinax* are some of the thermophilic carbon monoxide-converting microorganisms that have been studied for their hydrogen-evolving capacity [48].

The biological water–gas shift seems appealing compared to conventional chemically catalyzed processes, but as in the analyzed photosynthetic bioprocesses for hydrogen evolution, certain hurdles must be addressed in order to be applicable. First, since the feedstock is gaseous its availability to the bacteria is rather poor due to gas-to-liquid mass transfer limitations. Toward the optimization of the concentration of carbon monoxide in the feedstock, the operation of the bioreactors system under high pressures (1 MPa) has been investigated, increasing mass transfer rates and, thus, the rate of carbon monoxide conversion. Another limiting factor is the toxicity of carbon monoxide to the biocatalysts. Indeed, carbon monoxide is reported as potentially inhibitory for both anabolic and catabolic processes of the microorganisms due to its high affinity to metalloenzymes. As such, despite the latest advances noted in the field, the application of microbial water gas shift remains at laboratory scale and improvements in system productivity and hydrogen yield are required [48].

10.3.3 Photofermentative hydrogen production

During photofermentative hydrogen production, different organic substrates are oxidized under anaerobic conditions and illumination, generating hydrogen and carbon dioxide [49]. The key enzyme for the photofermentative process is nitrogenase, which under nitrogen-deficient conditions can catalyze the evolution of molecular hydrogen using light energy and reduced compounds (such as organic acids) as the electron donors and where ferredoxin acts as the electron carrier [50]. Organic acids that have been investigated as electron donors during photofermentation are malic, lactic, succinic, acetic, propionic, and butyric, and have been used either as carbon sources in defined media or as parts of actual waste streams [51].

The theoretical hydrogen yield obtained by photofermentation varies according to the carbon source, thus ranging from 4 mol H_2/mol substrate to 12 mol H_2/mol substrate when is based on organic acids, as show in the reactions (10.14–10.18) [52].

$$\text{(glucose) } C_6H_{12}O_6 + 12H_2O + hv \rightarrow 12H_2 + 2CO_2 \tag{10.14}$$

$$\text{(acetic) } CH_3COOH + 2H_2O + hv \rightarrow 4H_2 + 2CO_2 \tag{10.15}$$

$$\text{(malic)}COOHCH_2CHOHCOOH + 3H_2O + hv \rightarrow 6H_2 + 4CO_2 \tag{10.16}$$

$$\text{(propionic)}CH_3CH_2COOH + 4H_2O + hv \rightarrow 7H_2 + 3CO_2 \tag{10.17}$$

$$\text{(butyric)}CH_3CH_2CH_2COOH + 6H_2O + hv \rightarrow 10H_2 + 4CO_2 \tag{10.18}$$

Different strategies of photofermentation are reported using monocultures of photosynthetic microorganisms or cocultures of photosynthetic with fermentative ones. In the first case, substrates including organic acids, sugars, industrial wastewaters, and acidic effluents derived from dark fermentative hydrogen-production processes are directly bioconverted by photosynthetic microorganisms. However, there are certain limitations in this process, including low-hydrogen production rates, substrate inhibition, and high-illumination requirements. In the second case, a coculture of dark and photohydrogen producers is cultivated in the same reactor where organic acids are generated by dark-hydrogen producers and are simutlaneously consumed by photosynthetic hydrogen producers. Following that strategy, substrate inhibition may be prevented and high hydrogen yields can be obtained. Since, however, growth rates of the different types of microorganisms vary, the fermentation conditions have to be carefully selected and optimized in order to maximize the substrate conversion and hydrogen yields [53].

Both prokaryotic and eukaryotic phototrophic microorganisms are reported as capable of performing photofermentation. Types of photosynthetic bacteria include green sulfur bacteria, purple sulfur bacteria, and nonsulfur bacteria (PNSB) [54], whereas photofermentation was also identified in green algal species [55].

The efficiency of photofermentative hydrogen production can be evaluated by: (1) the conversion efficiency of light energy to hydrogen, called photochemical efficiency (PE); (2) the hydrogen production yield; and (3) the yield coefficient of the produced hydrogen in relation to the consumed carbon source [1]. The latter could be successfully fulfilled in case of waste-derived, low-cost carbon source usage. Therefore it can be assumed that the efficient utilization of light energy is of outmost importance for the construction of a feasible energy production system [56], while the PE is calculated by the Eq. (10.19) [57]:

$$PE = \frac{[\text{combustion enthalpy of } H_2]}{[\text{light energy at surface of reactor}]} \times 100\% \tag{10.19}$$

Factors affecting PE include the wavelength and intensity of light, the cell density of the biocatalysts, the volume-to-ratio surface of the culture (reactor geometry) and the ratio of light penetration. It is widely accepted that optimal light utilization is indispensable for maximal hydrogen production [51].

10.3.4 Dark fermentation

Fermentative hydrogen production or dark fermentation is maybe the most widely studied microbial hydrogen-production process. It is realized under anaerobic conditions via either strict anaerobic or facultative anaerobic bacteria. Various organic compounds including proteins, lipids, and carbohydrates can be used as fermentative substrates; however, the fermentation of carbohydrates lead to higher yields and productivities. The maximum theoretical hydrogen yield that is used for the assessment of the efficiency of hydrogen production process is estimated on the basis of hexoses (glucose) fermentation according to two main fermentative pathways: (1) butyric acid fermentation (clostridial-type fermentation, Fig. 10.1); and (2) mixed-acid fermentation (enterobacterial-type fermentation by facultative anaerobes, Fig. 10.2).

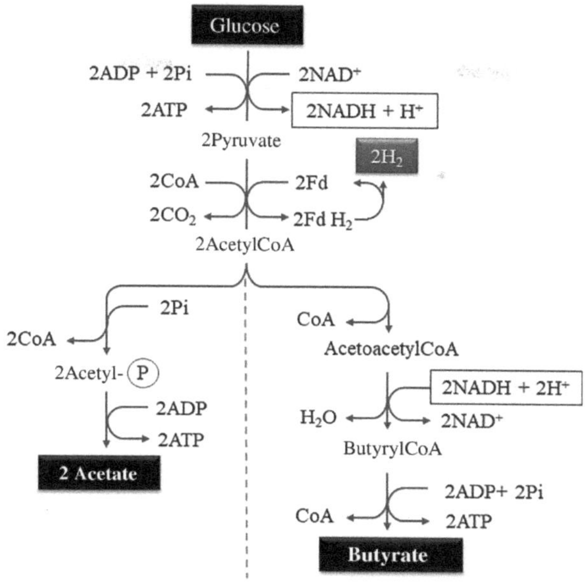

Figure 10.1 Biohydrogen production from clostridial-type fermentation.

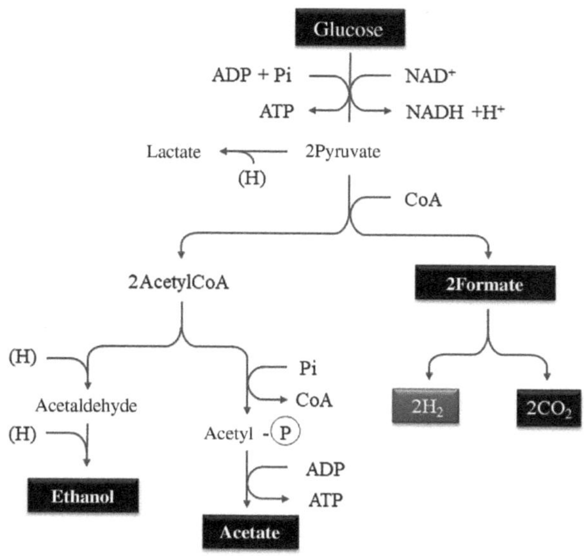

Figure 10.2 Biohydrogen production from enterobacterial-type fermentation.

During the clostridial-type pathway the theoretical hydrogen yield is 4 mol of hydrogen per mol of consumed glucose when the sole organic metabolic product is acetate, as shown in Eq. (10.20) [58]. However, in the case where butyrate is the sole organic metabolic product the theoretical hydrogen yield is 2 mol of hydrogen per mol of consumed glucose as shown in Eq. (10.21) [58]:

$$C_6H_{12}O_6 + 4H_2O \rightarrow 2CH_3COO^- + 2HCO_3^- + 4H^+ + 4H_2 \quad (10.20)$$

$$\Delta G^o = -206.3 \ \text{kJ/mol}$$

$$C_6H_{12}O_6 + 2H_2O \rightarrow CH_3CH_2CH_2COO^- + 2HCO_3^- + 3H^+ + 2H_2 \quad (10.21)$$

$$\Delta G^o = -254.8 \ \text{kJ/mol}$$

The maximum theoretical hydrogen yield of 4 moles/mol glucose could also be estimated via the bioconversion of sugars to acetate and formate, that is, the enterobacterial type pathway, according to Eqs. (10.22) and (10.23) [20,59]:

$$C_6H_{12}O_6 + 2H_2O \rightarrow 2CH_3COO^- + 2HCOO^- + 4H^+ + 2H_2 \quad (10.22)$$

$$\Delta G^o = -209.1 \ \text{kJ/mol}$$

$$2HCOOH + 2NAD^+ \rightarrow 2CO_2 + 2NADH + 2H^+ \quad (10.23)$$

$$\Delta G^o = -6 \ \text{kJ/mol}$$

The reaction described by Eq. (10.11) is achieved via the enzyme formate-hydrogen lyase, consisting from a formate dehydrogenase and a hydrogenase [3].

In all these cases pyruvate is the intermediate metabolite, the production of which is conducted via the EMP glycolytic pathway according Eq. (10.24).

$$C_6H_{12}O_6 \rightarrow 2CH_3COCOO^- + 4H^+ + 2NADH \quad (10.24)$$

$$\Delta G^o = -112.1 \ \text{kJ/mol}$$

During the EMP pathway, 2 moles of NADH are generated from which 2 extra moles of molecular hydrogen can further be produced per mole of hexose ($NADH + H^+ \rightarrow NAD^+ + H_2$). It should be noted that acetyl coenzyme A (acetyl-CoA) is considered the key metabolite during hydrogen production via dark fermentation. The fate of acetyl-CoA will determine whether the maximum hydrogen production will be 4 or 2 mol/mol of consumed hexoses, and whether the final yield reaches the maximum. The archived hydrogen yield during dark fermentation is also controlled by the flux of carbon toward other metabolic products. The fermentation of sugars, apart from acetic formic and butyric acid, can lead to the generation of alcohols such as butanol and isopropanol via acetyl-CoA and also directly via pyruvate to the production of lactic acid (homolactic fermentation) and alcohols such as ethanol (alcoholic fermentation) and 2,3-butanediol. Lactate and ethanol can also be generated through heterolactic fermentation. The evolution of such metabolites is not accompanied by molecular hydrogen production, thus leading to the reduction of the final hydrogen yield.

The final distribution of metabolic products depend highly on the culturing condition, either when using monocultures [13] or mixed cultures [60,61] as biocatalysts. In the case of *E. coli*, it has been reported that when pH drops below 7, the equilibrium shifts to lactic acid production [62,63] resulting in reduced hydrogen yield. In such cases appropriate conditions are chosen to favor the metabolism of the microorganism to produce hydrogen, as well as the predominance of the hydrogen-producing microorganisms in the case of using mixed cultures.

Other factors that could affect the metabolic shift and, consequently, hydrogen yields, include temperature, nutrients limitation, substrate concentration, and hydraulic retention time (HRT) in the case of continuous cultures, as well as the concentration of oxygen, in the case of facultative anaerobes. When hydrogen production is based on mixed microbial in which hydrogen-producing and consuming microorganisms may coexist, the regulation pH [64], HRT, and maintenance of low-partial hydrogen pressure are considered as key factors that may favor the dominance of hydrogen producers, thus leading to high hydrogen yields [65].

10.3.5 Microbial electrolysis cells

Microbial electrolysis cell (MEC) is a technology for hydrogen production closely related to microbial fuel cell (MFC) via which an electric current is generated during the degradation of organic compounds by electroactive microorganisms. An MFC consists of an anode and a cathode, typically separated by an ion exchange membrane. The anode is colonized by microorganisms that catalyze the oxidation of organic substrates toward carbon dioxide, electrons, and protons using the electrode as direct electron acceptor. The collected electrons flow from the anode to the cathode through an external circuit. The electrons that reach the cathode combine with the protons that diffused from the anode through the membrane and oxygen provided (passively) from air, resulting in the generation of water as the final product, whereas the electrical potential generated by the electron flux is up to 0.3 V. In MECs this process is partially reversed by applying an electric current from an outside power source as shown in Fig. 10.3. The combined voltage is enough to reduce protons, resulting to the evolution of molecular hydrogen in the cathode. As such, while the anode process of MECs is identical to that of MFCs, the cathode process is similar to that of a water electrolyzer [66].

The efficiency of hydrogen production is highly affected by the type of substrate used as the electron donor in the anode. When acetate is used as a substrate, the process can be described in electrochemical reactions (10.25–10.27) [67,68]:

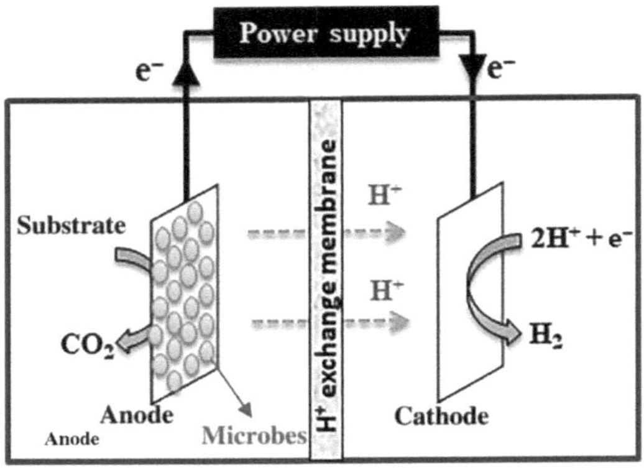

Figure 10.3 Schematic illustration of a typical dual-chamber MEC.

Anode:

$$CH_3COO^- + 4H_2O \rightarrow 2HCO_3^- + 9H^+ + 8e^- \quad (10.25)$$

Cathode:

$$9H^+ + 8e^{--} \rightarrow 4H_2 \quad (10.26)$$

Overall reaction:

$$CH_3COO^- + 4H_2O \rightarrow 2HCO_3^- + H^+ + 4H_2 \quad (10.27)$$

$$\Delta G^o = 104.6 \text{ kJ/mol}$$

It should be mentioned that the final reduction of protons to molecular hydrogen is a thermodynamically, nonspontaneous reaction and this is the reason why the current process requires external energy input to be realized.

MEC is considered a promising sustainable technology for wastewater treatment with simultaneous energy recovery in the form of biofuels. In order to assess the overall performance of an MEC, different parameters must be taken into account and optimized. These include: (1) the coulombic efficiency of the system (CE), (i.e., the number of electrons or molecular hydrogen recovered in the circuit over the number of electrons or H_2 theoretically available from the substrate); (2) the hydrogen recovery (i.e., the ratio of the hydrogen recovered, and the maximum theoretical hydrogen produced based on substrate utilization); (3) the hydrogen yield (i.e., the theoretical number of moles H_2 produced based on substrate utilization); and (4) the substrate utilization expressed as chemical oxygen demand (COD) removal, which is described by the reaction:

$$\Delta COD = \frac{COD_f - COD_{in}}{COD_{in}} \times 100\% \quad (10.28)$$

where, COD_{in} is the initial COD concentration of the electrolyte and COD_f is the final COD concentration of the effluent after each batch cycle. It is reported that lactic and acetic acid result to 82% efficiency, while the respective values for cellulose or glucose are close to 63%, whereas the efficiency of normal water electrolysis is 60%–70%.

Different types of microorganisms with electroactive properties have been identified as adequate for applications in MECs, including archaea, the cyanobacterium *Cyanothece* 51142 [69], dechlorinating bacteria such as *Dehalococcoides* spp. and *Desulfitobacterium* spp., as well as methanogenic and homoacetogenic microorganisms [70].

10.4 Bioreactors

10.4.1 Photobioreactors

A photobioreactor is a culture system of microorganisms in which illuminance has to be assessable to the cultivated cells that carry out a light-dependent process. A successful photobioreactor concept should have a simple design, low material and production costs, and should also ensure a high degree of light bioconversion and satisfactory gas exchange [71]. The specific requirements of each bioreactor systems depend on the type of microorganisms that will be used as biocatalyst and the process that will be followed for photobiological hydrogen production. According to Adessi and Philippis [71] the photobioreactors used for cultivation of PNSB must be closed systems, adequate for maintaining anaerobic conditions and prevent dispersion of molecular hydrogen.

The most important issues to be considered in designing a photobioreactor are related to light intensity, light quality and sources, and light distribution into the photobioreactors. Different geometries have been tested for hydrogen production with PNSB with tubular and flat panel bioreactors being the most widely studied. Tubular photobioreactors are horizontal [72], inclined (10–30 degrees) [73], or vertical [74] transparent tubes. The hydrogen production rates per illuminated surface of tubular reactors is reported to be efficient, but the main limitation for their large-scale application remains the ground area occupancy, which highly affects the cost of the process.

Flat panel reactors are rectangular transparent boxes that can be either vertically placed or inclined. Their design (limited thickness) results to short mixing-induced light/dark cycles, which can decrease hydrogen productivity. In general, flat panel photobioreactors are difficult to scale-up over a few liters of volume, but they offer the possibility to arrange a set of reactors one behind the other at an appropriate distance to increase the ground-area productivity [73]. Moreover, their construction and modification are rather easy, making them ideal for base-studies on light distribution and mixing.

The operational mode of photobioreactors can be batch, fed-batch, and continuous [75], whereas the mixing strategy should be selected based on the limitations of the biocatalysts. As such, the selection of airlift or bubble column bioreactors can be applied for cyanobacteria or algae cultivation but is not adequate for anoxygenic bacteria [76]. In terms of system illuminance, either artificial light sources (i.e., florescent lamps, halogen lamps, optical fibers, neon tubes, light emitting diodes, and photosynthetically active radiations), or direct sunlight can be applied. Sunlight has the advantage of being a free and abundant light source, which can be used for direct irradiance of the bioreactor or amplified by the use of solar-energy-excited optical fibers [77]. However, a drawback of sunlight could be the periodicity of the light source, which is not limiting for certain biocatalysts [65].

10.4.2 Fermentors

The selection of the proper type of bioreactor is widely acknowledged as crucial for the overall efficiency of the process, since it influences the reactor microenvironment, microbial population, hydrodynamic behavior, substrate-consortia contact, and so forth. [78]. The operational mode that is selected for fermentative hydrogen production can be either batch, continuous, or semicontinuous. Batch fermentations are usually applied at small-scale for the investigation of key parameters affecting the process, but the sustainability of scale-up applications can only be achieved via continuous fermentations.

The continuous stirred tank reactor (CSTR) is the most widely applied and studied continuous reactor system due to its simple construction, ease of operation, and effective homogenous mixing [65]. In a conventional CSTR, biomass is continuously suspended and the retention time of the microbial cells, that is, the solids retention time (SRT) is the same with the HRT of the system lacking the danger of possible microbial washout when low HRT as applied. Physical retention of microbial biomass can be achieved by methodologies such as flocs or granules formation, microbial immobilization on inert materials, formation of microbial biofilms, or use of retentive membranes [79]. In these systems, the SRT is independent of HRT, allowing high-cell concentrations and high-hydrogen volumetric production rates with relatively small reactor volumes. It should be noted that the relatively high SRT might allow the possible establishment of slow-growing methanogenic populations, leading to consumption of produced hydrogen [65].

10.4.3 Microbial electrolysis cells configurations

The simplest MEC configuration includes two chamber systems, where the anode and cathode separated by a semipermeable membrane, thus allowing the transfer of protons (Fig. 10.3). Single-chambered MEC systems have further been developed aiming for the reduction of costs and simplification of reactor design [80]. Both types may vary in terms of geometry and rectangular, tubular, cylindrical, and bottle-shaped reactors have been developed, aspiring to achieve higher-hydrogen production rates and current densities. As in fermentative hydrogen-production systems, the operational mode of MECs can be batch, semicontinuous (fed-batch), or continuous. The batch mode is mostly used during initial adaptation of electrogenic microorganisms in mixed cultures, and also as a concept of proof to evaluate the catalytic activity of cathode materials [81] or for comparatively testing the efficiency of different organic compounds as substrates [82].

One of the challenges in scaling-up MECs is the cost of the cathode and the cathode catalysts, since most MECs use platinum applied on carbon cloth [83,84] or carbon paper [85]. In order to improve the performance and economic feasibility of MECs, platinum needs to be replaced with alternative low-cost cathode materials such as stainless steel [82] or microbial biocathodes [86]. Though promising, this is still an experimental method for hydrogen production that has not evolved beyond the laboratory since certain microbiological, technological, and economic challenges need to be resolved before large-scale implementation.

10.5 Storage and purification

While direct and indirect photolysis systems produce pure hydrogen, dark-fermentation and photofermentation processes are able to produce a mixed biogas containing primarily hydrogen and carbon dioxide, as well as lesser amounts of methane, carbon monoxide, hydrogen sulfide, and/or ammonia. Moreover, the hydrogen content in the gas phase is almost lower than 50%. PEMFCs require hydrogen of high purity ($>99\%$) and cannot tolerate carbon monoxide at concentrations higher than 10 ppm. In order to remove carbon dioxide, methane, and/or contaminating gases (e.g., carbon monoxide), purification of the biogas is essential. Up to now, membrane technologies based on palladium have been proposed as hydrogen purifiers in industrial scale applications [87].

10.6 Conclusion

The microbial processes for biohydrogen production include direct and indirect biophotolysis, photofermentation, dark-fermentation, and MEC. These technologies refer to processes conducted via bacteria or algae under mild conditions and do not rely on fossil fuels or electricity. The fact that substrates that are exploited for microbial hydrogen production can be water or low/zero cost, organic substrates such as waste and residual biomass make these processes appealing on the economic and environment-friendly points of view.

Although several studies have been devoted on understanding and optimizing these technologies for a number of decades, microbial hydrogen production is far from a mature technology and there is still a long way to go before their industrial application is feasible. This is mainly due to the fact that several engineering issues still need to be addressed, including the identification of effective and cost efficient separation/purification methodologies, appropriate designs of bioreactors, and maintenance of robust biocatalysts that will ensure long-term and stable, quality biohydrogen.

Abbreviations

acetyl-CoA	acetyl coenzyme A
ADP	adenosine diphosphate
$\mathbf{A_{ox}}$	oxidized e^- acceptor
$\mathbf{A_{red}}$	reduced e^- acceptor
ATP	adenosine triphosphate
CE	coulombic efficiency
Chl	chlorophyll
COD	chemical oxygen demand
CODH	carbon monoxide dehydrogenase
CSTR	Continuous Stirred Tank Reactor
$\mathbf{D_{ox}}$	oxidized e^- donor
$\mathbf{D_{red}}$	reduced e^- acceptor
EMP pathway	Embden–Meyerhof–Parnas pathway
ΔG_o	Gibbs free energy
HRT	hydraulic retention time
NAD	Nicotinamide adenine dinucleotide
NADP	Nicotinamide adenine dinucleotide phosphate
MFC	microbial fuel cell
PE	photochemical efficiency
PNB	purple sulfur bacteria
PNSB	purple nonsulfur bacteria
SRT	solids retention time

References

[1] I. Akkerman, M. Janssen, J.M.S. Rocha, R.H. Wijfeels, Photobiological hydrogen production: photochemical efficiency and bioreactor design, Int. J. Hydrog. Energy 27 (11–12) (2002) 1195–1208.

[2] A.J.M. Stams, Metabolic interactions between anaerobic bacteria in methanogenic environments, Antonie. Van. Leewenhoeck 66 (1–3) (1994) 271–294.

[3] C.T. Gray, H. Gest, Biological formation of molecular hydrogen, Science 148 (1965) 186–192.

[4] R. Nandi, S. Sengupta, Microbial production of hydrogen: an overview, Crit. Rev. Microbiol. 24 (1) (1998) 61–84.

[5] P.-Y. Lin, L.-M. Whang, Y.-R. Wu, W.-J. Ren, C.-J. Hsiao, S.-L. Li, et al., Biological hydrogen production of the genus *Clostridium*: metabolic study and mathematical model simulation, Int. J. Hydrog. Energy 32 (12) (2007) 1728–1735.

[6] X. Ye, X. Zhang, E. Morgenroth, K.T. Finneran, Anthrahydroquinone-2,6-disulfonate increases the rate of hydrogen production during Clostridium beijerinckii fermentation with glucose, xylose, and cellobiose, Int. J. Hydrog. Energy 37 (16) (2012) 11701–11709.

[7] Q. An, J.-L. Wang, Y.-T. Wang, Z.-L. Lin, M.-J. Zhu, Investigation on hydrogen production from paper sludge without inoculation and its enhancement by *Clostridium thermocellum*, Biores. Technol. 263 (2018) 120–127.

[8] T. Tanaka, E. Ishimoto, Y. Shimomura, M. Taniguchi, S. Oi, Purification and some properties of raw starch-binding amylase of *Clostridium butyricum* t-7 isolated from mesophilic methane sludge, Agric. Biolog. Chem. 51 (2) (1987) 399–405.
[9] K. Trchounian, N. Muller, B. Schink, A. Trchounian, Glycerol and mixture of carbon sources conversion to hydrogen by *Clostridium beijerinckii* DSM791 and effects of various heavy metals on hydrogenase activity, Int. J. Hydrog. Energy 42 (12) (2017) 7875–7882.
[10] T. Miller, M.J. Wolin, Formation of hydrogen and formate by *Ruminococcus albus*, J. Bacteriol. 116 (2) (1973) 836–846.
[11] Y. Shi, P.J. Weimer, J. Ralph, Formation of formate and hydrogen, and flux of reducing equivalents and carbon in *Ruminococcus flavefaciens* FD-1, Antonie. Van Leeuwenhoek 72 (2) (1997) 101–109.
[12] I. Ntaikou, H.N. Gavala, M. Kornaros, G. Lyberatos, Hydrogen production from sugars and sweet sorghum biomass using *Ruminococcus albus*, Int. J. Hydrog. Energy 33 (4) (2008) 1153–1163.
[13] I. Ntaikou, E. Koutros, M. Kornaros, Valorisation of wastepaper using the fibrolytic/hydrogen producing bacterium Ruminococcus albus, Biores. Technol. 100 (23) (2009) 5928–5933.
[14] A.M. Egorov, T.V. Avilova, M.M. Dikob, V.O. Popotov, Y.V. Radionov, I.V. Berezin, NAD depended FDH from methylotrophic bacteria strain I, purification and characterization, Eur. J. Biochem. 99 (3) (1979) 569–576.
[15] S. Kawamura, J.G. O' Neil, J.F. Wilkinson, Hydrogen production by methylotrophs under anaerobic conditions, J. Ferment. Technol. 61 (2) (1983) 151–156.
[16] J.M. Blamey, M.W.W. Adams, Purification and characterization of pyruvate ferredoxin oxidoreductase from the hyperthermophylic archaeon *Pyrococcus furiocus*, Biochim. Biophys. Acta 1161 (1) (1993) 19–27.
[17] S. Mukund, M.W.W. Adams, The novel tungsten-iron-sulfur protein of the hyperthermophilic archaebacterium, *Pyrococcus furiosus*, is an aldehyde ferredoxin oxidoreductase: evidence for its participation in a unique glycolytic pathway, J. Biol. Chem. 266 (22) (1991) 14208–14216.
[18] S. Mukund, M.W.W. Adams, Characterization of a novel tungsten-containing formaldehyde ferredoxin oxidoreductase from the hyperthermophilic archaeon, *Thermococcus litoralis*: a role for tungsten in peptide catabolism, J. Biol. Chem. 268 (18) (1993) 13592–13600.
[19] M.W.W. Adams, Enzymes and proteins from organisms that grow near and above 100°C, Ann. Rev. Microbiol. 47 (1993) 627–658.
[20] L.H. Stickland, The bacterial decomposition of formic acid, Biochem. J. 23 (1929) 1187–1998.
[21] J. Yudkin, Hydrogenlyases: some factors concerned in the production of the enzymes, Biochem. J. 26 (1932) 1859–1871.
[22] A.C. Blackwood, A.C. Neish, G.A. Ledingham, Dissimilation of glucose at contrilled pH values by pigment and non pigment strains of *Escherichia coli*, J. Bacteriol. 72 (4) (1956) 497.
[23] H.W. Doelle, Bacterial Metabolism, Academic Press, New York, 1969. ISBN:9781483272740.
[24] N. Kumar, D. Das, Enhancement of hydrogen production by *Enterobacter clocae* IIT-BT 08, Process. Biochem 35 (6) (2000) 589–593.

[25] M.A. Rachman, Y. Nakashimada, T. Kakizono, N. Nishio, Hydrogen production with high yield and high evolution rate by self flocculated cells of *Enterobacter aerogenes* in a packed bed rector, Appl. Microbiol. Biotechnol 49 (1998) 450–454.
[26] F. Boshagh, K. Rostami, N. Moazami, Biohydrogen production by immobilized Enterobacter aerogenes on functionalized multi-walled carbon nanotube, Int. J. Hydrog. Energy 44 (28) (2019) 14395–14405.
[27] H. Masukawa, K. Nakamura, M. Mochimaru, H. Sakurai, Photobiological hydrogen production and nitrogenase activity in some heterocysteous cyanobacteria, in: J. Miyake, T. Matsunaga, A. San Pietro (Eds.), BioHydrogen II, An Approach to Environmentally Acceptable Technology, Elsevier, 2001.
[28] D.C. Howard, G.A. Codd, The uptake and production of molecullar hydrogen by unicellular cyanobacteria, J. Gen. Microb. 131 (7) (1985) 1561–1569.
[29] J. Miyake, Biological solar energy conversion, in: K. Miyamoto (Ed.), Renewable Biological Systems for Alternative Sustainable Energy Production, FAO, UN, 1998. FAO Agricultural Services Bulletin 128.
[30] A.V. Chalam, Ch Sasikala, Ch.V. Ramana, P.R. Rao, Effect of pesticides on hydrogen metabolism of *Rhodobacter sphaeroides* and *Rhodopseudomonas palustris*, FEMS Microbiol. Ecol. 19 (1) (1996) 1–4.
[31] K. Nath, M. Muthukumar, A. Kumar, D. Das, Kinetics of two-stage fermentation process for the production of hydrogen, Int. J. Hydrog. Energy 33 (4) (2008) 1195–1203.
[32] B. Uyar, I. Eroglu, M. Yücel, U. Gündüz, Photofermentative hydrogen production from volatile fatty acids present in dark fermentation effluents, Int. J. Hydrog. Energy 34 (10) (2009) 4517–4523.
[33] B. Uyar, I. Eroglu, M. Yücel, U. Gündüz, L. Türker, Effect of light intensity, wavelength and illumination protocol on hydrogen production in photobioreactors, Int. J. Hydrog. Energy 32 (18) (2007) 4670–4677.
[34] N. Basak, D. Das, Photofermentative hydrogen production using purple non-sulfur bacteria *Rhodobacter sphaeroides* O.U.001 in an annular photobioreactor: a case study, Biomass Bioenergy 33 (6–7) (2009) 911–919.
[35] M. Yetis, U. Gunduz, I. Eroglu, M. Yucel, L. Turker, Photoproduction of hydrogen from sugar refinery wastewater by *Rhodobacter sphaeroides* O.U. 001, Int. J. Hydrog. Energy 25 (11) (2000) 1035–1041.
[36] K. Seifert, M. Waligorska, M. Laniecki, Brewery wastewaters in photobiological hydrogen generation in presence of *Rhodobacter sphaeroides* O.U. 001, Int. J. Hydrog. Energy 35 (9) (2010) 4085–4091.
[37] X. Wang, Y. Fang, Y. Wang, J. Hu, A. Zhang, X. Ma, H. Yang, L. Guo, Single-stage photo-fermentative hydrogen production from hydrolyzed straw biomass using *Rhodobacter sphaeroides*, Int. J.Hydrog. Energy 43 (30) (2018) 13810–13820.
[38] J. Benemann, Hydrogen biotechnology: progress and prospects, Nat. Biotechnol. 14 (9) (1996) 1101–1103.
[39] J.R. Benemann, J.A. Berenson, N.O. Kaplan, M.D. Kamen, Hydrogen evolution by chloroplast-ferredoxin-hydrogenase system, Proc. Natl. Acad. Sci. USA 70 (1973) 2317–2320.
[40] C.J.P. Spruit, Simultaneous photoproduction of hydrogen and oxygen by Chlorella. Mededel. Landbouwhogeschool Wageningen, 58, *1958*, 1-17.
[41] P.C. Hallenbeck, J.R. Benemann, Biological hydrogen production: fundamentals and limiting processes, Int. J. Hydrog. Energy 27 (11–12) (2002) 1185–1193.

[42] J.Y.H. Kim, B.H. Jo, H.J. Cha, Production of biohydrogen by recombinant expression of [NiFe]-hydrogenase 1 in *Escherichia coli*, Microb. Cell Factories (2010) 9. art. no. 54.

[43] A. Scoma, D. Krawietz, C. Faraloni, L. Giannelli, T. Happe, G. Torzillo, Sustained H_2 production in a *Chlamydomonas reinhardtii* D1 protein mutant, J. Biotechnol. 157 (4) (2012) 613–619.

[44] M.L. Ghirardi, Implementation of photobiological H_2 production: the O_2 sensitivity of hydrogenases, Photosynth. Res. 125 (3) (2015) 383–393.

[45] K. Belafi-Bako, D. Bucsu, Z. Pientka, B. Balint, Z. Herbel, K. Kovacs, et al., Integration of biohydrogen fermentation and gas separation processes to recover and enrich hydrogen, Int. J. Hydrog. Energy 31 (11) (2006) 1490–1495.

[46] Y. Kung, C.L. Drennan, A role for nickel-iron cofactors in biological carbon monoxide and carbon dioxide utilization, Curr. Opin. Chem. Biol. 15 (2) (2011) 276–283.

[47] R.L. Kerby, S.S. Hong, S.A. Ensign, L.J. Coppoc, P.W. Ludden, G.P. Roberts, Genetic and physiological characterization of the *Rhodospirillum rubrum* carbon monoxide dehydrogenase system, J. Bacteriol. 174 (16) (1992) 5284–5294.

[48] M. Alfano, C. Cavazza, The biologically mediated water-gas shift reaction: structure, function and biosynthesis of monofunctional [NiFe]-carbon monoxide dehydrogenases, Sustain. Energy Fuels 2 (8) (2018) 1653–1670.

[49] D. Kim, J. Lee, S. Kang, P. Hallenbeck, E. Kim, J.K. Lee, et al., Enhanced photo-fermentative H_2 production using *Rhodobacter sphaeroides* by ethanol addition and analysis of soluble microbial products, Biotechnol. Biofuels 7 (1) (2014) 79.

[50] D. Das, T.N. Veziroglu, Hydrogen production by biological processes: a survey of literature, Int. J. Hydrog. Energy 26 (1) (2001) 13–28.

[51] G. Antonopoulou, I. Ntaikou, K. Stamatelatou, G. Lyberatos, Biofuels chemical and biochemical conversion processes and technologies: biological and fermentative production of hydrogen, in: R. Luque, J. Campelo, J. Clark (Eds.), Handbook of Biofuels Production: Processes and Technologies, Woodhead Publishing Limited, 2010. Woodhead Energy Series No. 15, 2010, ISBN 1 84569 679 4.

[52] A. Reungsang, N. Zhong, Y. Yang, S. Sittijunda, A. Xia, Q. Liao, Hydrogen from photo fermentation, in: Q. Liao, J. Chang, C. Herrmann, A. Xia (Eds.), Bioreactors for Microbial Biomass and Energy Conversion. Green Energy and Technology, Springer, Singapore, 2018.

[53] P.C. Hallenbeck, Photofermentative biohydrogen production, in: A. Pandey, J.-S. Chang, P.C. Hallenbecka, C. Larocce (Eds.), Biohydrogen, Elsevier, 2013.

[54] D.A. Bryant, N.U. Frigaard, Prokaryote photosynthesis and phototrophy illuminated, Trends Microbiol. 14 (11) (2006) 488–495.

[55] J.-H. Hwang, A. Kabra, J. Kim, B.-H. Jeon, Photoheterotrophic microalgal hydrogen production using acetate- and butyrate-rich wastewater effluent, Energy 78 (2014) 887–894.

[56] J. Miyake, S. Kawamura, Efficiency of light energy conversion to hydrogen by the photosynthetic bacterium *Rhodobacter sphaeroides*, Int. J. Hydrog. Energy 12 (3) (1987) 147–149.

[57] E. Nakada, S. Nishikata, Y. Asada, J. Miyake, Photosynthetic bacterial hydrogen production combined with a fuel cell, Int. J.Hydrog. Energy 24 (11) (1999) 1053–1057.

[58] R.K. Thauer, K. Jungermann, K. Decker, Energy conservation in chemotrophic anaerobic bacteria, Bacteriol. Rev. 41 (1) (1977) 100–180.

[59] G. Gottschalk, Bacterial fermentations, Bacterial Metabolism, Springer-Verlag, New York Inc, 1986.

[60] I. Ntaikou, H.N. Gavala, G. Lyberatos, Modeling of fermentative hydrogen production from the bacterium *Ruminococcus albus*: Definition of metabolism and kinetics during growth on glucose, Int. J.Hydrog. Energy 34 (9) (2009) 3697–3709.

[61] G. Antonopoulou, H.N. Gavala, I.V. Skiadas, G. Lyberatos, Modeling of fermentative hydrogen production from sweet sorghum extract based on modified ADM1, Int. J. Hydrog. Energy 37 (1) (2010) 191–208.

[62] D. Blankenhorn, J. Phillips, J.L. Slonczewski, Acid- and base induced proteins during aerobic and anaerobic growth of *Escherichia coli* revealed by two-dimensional gel electrophoresis, J. Bacteriol. 181 (7) (1999) 2209–2216.

[63] C. Kirkpatrick, M.L. Maurer, N.E. Oyelakin, Y.N. Yoncheva, R. Mauer, J.L. Slonczewski, Acetate and formate stress: opposite responses in the proteome of *Escherichia coli*, J. Bacteriol. 183 (21) (2001) 6466–6477.

[64] R.J. Zoetemeyer, P. Arnoldy, A. Cohen, C. Boelhower, Influence of temperature on anaerobic acidification of glucose on a mixed culture forming part of a two stage digestion process, Water Res. 16 (3) (1982) 313–321.

[65] I. Ntaikou, G. Antonopoulou, G. Lyberatos, Biohydrogen production from biomass and wastes via dark fermentation: a review, Waste Biomass Valor 1 (1) (2010) 21–39.

[66] H. Liu, H. Hu, J. Chignell, Y. Fan, Microbial electrolysis: novel technology for hydrogen production from biomass, Biofuels 1 (1) (2010) 129–142.

[67] A. Kadier, M.S. Kalil, W. Logroño, A. Mohamed, H.A. Hasan, Hydrogen production through electrolysis, in: T. Lipman, A. Weber (Eds.), Fuel Cells and Hydrogen Production, Springer, New York, 2019. Encyclopedia of Sustainability Science and Technology Series.

[68] B.E. Logan, D. Call, S. Cheng, H.V.M. Hamelers, T.H.J.A. Sleutels, A.W. Jeremiasse, et al., Microbial electrolysis cells for high yield hydrogen gas production from organic matter, Environ. Sci. Technol. 42 (23) (2008) 8630–8640.

[69] A. Bandyopadhyay, J. Stöckel, H. Min, L. Sherman, H. Pakrasi, High rates of photobiological H2 production by a cyanobacterium under aerobic conditions, Nat. Commun. 1 (9) (2010) 139.

[70] S. Cheng, D. Xing, D. Call, B. Logan, Direct biological conversion of Eelectrical current into methane by electromethanogenesis, Environ. Sci. Technol. 43 (10) (2009) 3953–3958.

[71] A. Adessi, R. Philippis, Photobioreactor design and illumination systems for H_2 production with anoxygenic photosynthetic bacteria: a review, Int. J. Hydrog. Energy 39 (7) (2014) 3127–3141.

[72] A. Adessi, G. Torzillo, E. Baccetti, R. Philippis, Sustained outdoor H_2 production with *Rhodopseudomonas palustris* cultures in a 50 L tubular photobioreactor, Int. J. Hydrog. Energy 37 (10) (2012) 8840–8849.

[73] E. Boran, E. Özgür, M. Yücel, U. Gündüz, I. Eroglu, Biohydrogen production by *Rhodobacter capsulatus* hup$^-$ mutant in pilot solar tubular photobioreactor, Int. J. Hydrog. Energy 37 (21) (2012) 16437–16445.

[74] P.A.M. Claassen, T. de Vrije, Non-thermal production of pure hydrogen from biomass: hyvolution, J. Clean. Prod. 18 ((Suppl. 1) (2006). S4.

[75] N. Basak, A. Jana, D. Das, D. Saikia, Photofermentative molecular biohydrogen production by purple-non-sulfur (PNS) bacteria in various modes: the present progress and future perspective, Int. J. Hydrog. Energy 39 (13) (2014) 6853–6871.

[76] C.N. Dasgupta, J. Gilbert, P. Lindblad, T. Heidorn, S.A. Borgvang, K. Skjanes, et al., Recent trends on the development of photobiological processes and photobioreactors

for the improvement of hydrogen production, Int. J. Hydrog. Energy 35 (19) (2010) 10218–10238.
[77] C.-Y. Chen, W.-B. Lu, C.-H. Liu, J.-S. Chang, Improved phototrophic H_2 production with *Rhodopseudomonas palustris* WP3-5 using acetate and butyrate as dual carbon substrates, Biores. Technol. 99 (9) (2008) 3609–3616.
[78] S. Venkata Mohan, Harnessing of biohydrogen from wastewater treatment using mixed fermentative consortia: Process evaluation towards optimization, Int. J.Hydrog. Energy 34 (17) (2009) 7460–7474.
[79] P.C. Hallenbeck, D. Ghosh, Advances in fermentative biohydrogen production: the way forward? Trends Biotechnol. 27 (5) (2009) 287–297.
[80] M. Zhou, H. Wang, D.J. Hassett, T. Gu, Recent advances in microbial fuel cells (MFCs) and microbial electrolysis cells (MECs) for wastewater treatment, bioenergy and bioproducts, J. Chem. Technol. Biotechnol. 88 (4) (2013) 508–518.
[81] M. Liu, M. Liu, D. Ding, K. Blinn, X. Li, L. Nie, Enhanced performance of LSCF cathode through surface modification, Int. J.Hydrog. Energy 37 (10) (2012) 8613–8620.
[82] P.A. Selembo, J.M. Perez, W.A. Lloyd, B.E. Logan, High hydrogen production from glycerol or glucose by electrohydrogenesis using microbial electrolysis cells, Int. J. Hydrog. Energy 34 (13) (2009) 5373–5381.
[83] J. Ditzig, H. Liu, B.E. Logan, Production of hydrogen from domestic wastewater using a bioelectrochemically assisted microbial reactor (BEAMR), Int. J.Hydrog. Energy 32 (13) (2007) 2296–2304.
[84] D. Call, B.E. Logan, Hydrogen production in a single chamber microbial electrolysis cell lacking a membrane, Env. Sci. Technol. 42 (9) (2008) 3401–3406.
[85] C.-H. Liu, T.-H. Ko, E.-C. Chang, H.-D. Lyu, Y.-K. Liao, Effect of carbon fiber paper made from carbon felt with different yard weights on the performance of low temperature proton exchange membrane fuel cells, J. Power Sources 180 (1) (2008) 276–282.
[86] A.W. Jeremiasse, H.V.M. Hamelers, C.J.N. Buisman, Microbial electrolysis cell with a microbial biocathode, Bioelectrochemistry 78 (1) (2010) 39–43.
[87] M.R. Rahimpour, F. Samimi, A. Babapoor, T. Tohidian, S. Mohebi, Palladium membranes applications in reaction systems for hydrogen separation and purification: a review, Chem. Engin. Process. Process Intensif. 121 (2017) 24–49.

Hydrogen energy

11

Zuo-Yu Sun
Hydrogen Energy and Space Propulsion Laboratory, School of Mechanical, Electronic and Control Engineering, Beijing Jiaotong University, Beijing, P.R. China

I believe that water will one day be employed as fuel, that hydrogen and oxygen will furnish an inexhaustible source of heat and light. — Jules Verne

11.1 Introduction

From the end of 17th century, the phlogiston theory made most people believe that an element, phlogiston, existed that makes fire. Henry Cavendish believed the *element* was the gas he found which could produce water during combustion with air. Antoine-Laurent de Lavoisier obtained the gas after decomposing water vapor and named it *hydrogen*. Albeit the phlogiston theory is invalid, obtaining hydrogen from water and again producing water via combustion are scientific truth.

All raw materials had been used as fuel (like biomass, fossil fuels) can be employed as sources to produce hydrogen [1]. Some natural energy sources (like wind, solar) can also be employed to produce hydrogen [2]. Furthermore, combustion using hydrogen as a fuel would produce virtually zero harmful emissions [3]. Therefore hydrogen is a promising sustainable fuel for the future.

11.2 Production technologies for hydrogen gas

Although hydrogen (H) is the most abundant element in the universe, the existence of two hydrogen atoms (H-H) is scare in nature which means hydrogen gas (H_2) is not a direct energy source and needs to be produced from other materials via various technologies.

11.2.1 From water: electrolysis and catalytic decomposition

Fig. 11.1 illustrates the process of hydrogen production from water using electrolysis. The experimental bench is composed by two electrodes, one current generator, and one water tank. Since pure water inhibits electrical conduction, the tank is filled up with an acidic electrolyte (like hydrochloric acid) or aqueous solution (like

Sustainable Fuel Technologies Handbook. DOI: https://doi.org/10.1016/B978-0-12-822989-7.00012-3

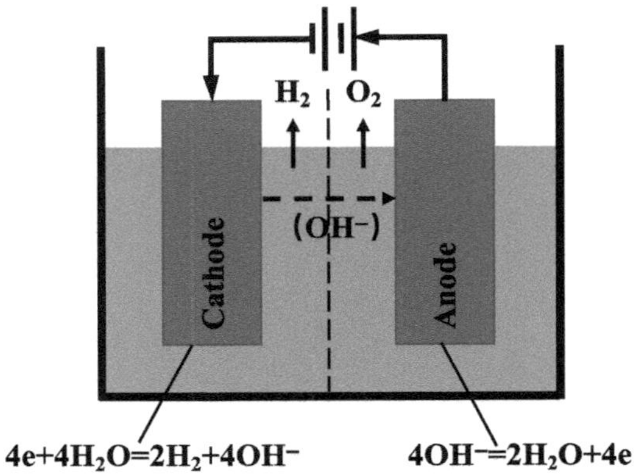

Figure 11.1 Illustration of water electrolysis to produce hydrogen gas.

sodium hydroxide, potassium hydroxide) for the delivery of electrical current. Such hydrogen occupies about 4% the current global market share [4].

Alkaline water electrolysis is a technology introduced by Troostwijk and Diemann in 1789 [5]. Electrolysis provides the megawatt range of H_2 for the current global commercialization [6]. Taking potassium hydroxide (KOH) as an electrolyte, for example, will ionize when it is dissolved in water, as:

$$KOH \Leftrightarrow K^+ + OH^- \tag{11.1}$$

the aqueous solution will be able to conduct electricity for the abundant ions (like K^+ and OH^-). When electricity is applied, the water molecules will be split to form unstable ions of hydrogen (H^+) and hydroxide (OH^-), as:

$$H_2O \Leftrightarrow H^+ + OH^- \tag{11.2}$$

Both positively charged ions (K^+ and H^+) migrate toward the cathode, where H^+ will first gain electrons to form stable hydrogen atoms since the electric potential of H^+ (about -1.71 V) is higher than that of K^+ (about -2.66 V), and individual hydrogen atoms combine to form hydrogen molecules (H_2). Meanwhile hydroxide ions (OH^-) migrate toward the anode where extra electrons are stripped away and drawn into electrical circuit, and oxygen (O_2) and water (H_2O) molecules will be left. These reactions could be expressed as:

$$\begin{aligned} &\text{Cathode}: 2H^+ + 2e^- \rightarrow H_2 \\ &\text{Anode}: 4OH^- - 4e^- \rightarrow O_2 + 2H_2O \end{aligned} \tag{11.3}$$

The total reaction of water electrolysis in potassium hydroxide solution can be expressed as:

$$H_2O - \Delta H^0 \rightarrow H_2 + 1/2O_2 \tag{11.4}$$

In an aqueous solution, water electrolysis happens at lower temperatures (300K–350K), nickel materials and asbestos diaphragm are the common electrodes [6], and the electrolyte concentration is about 20%–30% [7]. Potassium hydroxide is a popular solution since potassium has better vivacity for metal ions than other metals (like sodium, magnesium, etc.). However, hydrogen electrolysis has some weakness on the requirements of working pressure, energy efficiency and the corroding effects of potassium hydroxide on electrodes.

Most current commercial electricity is generated by the combustion of fossil fuels (mainly coal) in thermal-power plants; however, by-produced CO_2 emissions are unacceptable in long-term. Furthermore, compared to the hydrogen produced by vaporeforming hydrocarbons, the hydrogen produced by electrolyzing water has weaker attraction due to conversion efficiency and fuel cost, unless the electricity origins are from solar, wind, ocean energy, or even nuclear energy.

Besides providing electricity for electrolyzing water, nuclear reactors also could produce hydrogen from water [8] during the process of low-grade heat production, as:

$$8H_2O + 6FeCl_2 \rightarrow 2H_2 + 2Fe_3O_4 + 12HCl \tag{11.5}$$

The produced hydrogen gas also has higher purity (up to 97%), while this reaction needs a higher working temperature (about 1100K–1200K); moreover, an abundance of hydrochloric acid (HCl) would be by-produced, which makes the practical working difficult.

11.2.2 From fossil fuels: gasification and reforming

Producing hydrogen from coal occupies about 18% in the current global market. As early as 1833, coal-derived hydrogen gas had begun to be employed as fuel in the format of "water gas," which is a synthesized mixture (of H_2, CO, and CO_2) produced by alternately supplying air and steam to hot coke in a fixed-bed gasifier, as:

$$\begin{aligned} C + H_2O &\rightarrow H_2 + CO \\ C + 2H_2O &\rightarrow 2H_2 + CO_2 \end{aligned} \tag{11.6}$$

The endothermic reactions require the gasifier to be heated. Air is generally sent into the furnace first, part of the fuel is burned, and the heat is stored in the fuel layer and the regenerator. Then steam is introduced into the burning fuel layer for the reaction [9]. Because of the reaction heat absorption, when the temperature of the fuel layer and regenerator drops to a certain temperature, air is sent back to the furnace for heating up again. This cycle is illustrated in Fig. 11.2.

This production technology has a wide adaptative range for the gasification of raw materials, lignite coal, anthracite coal, and inferior coal [10]. Since the hydrogen will mixed with other materials in the production, the rate of steam decomposition will be less than 40%. Furthermore, such technology has serious environmental

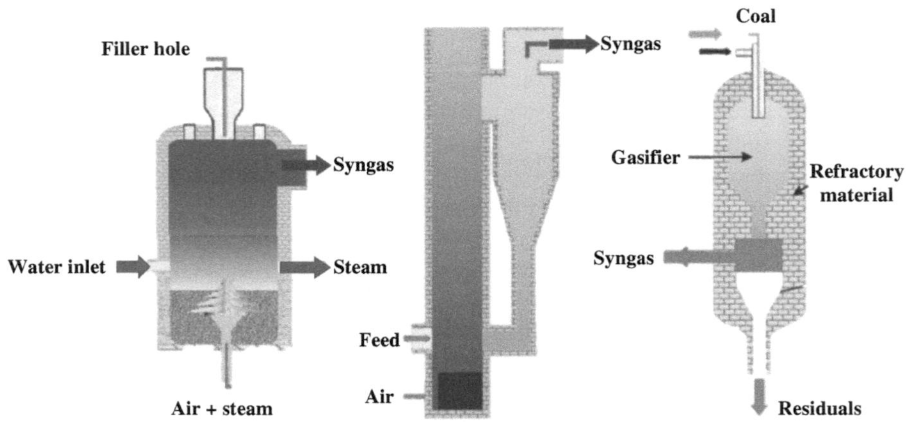

Figure 11.2 Flow chart of syngas production in a fixed-bed gasifier.

problems: (1) undecomposed steam results in lots of phenol water; and (2) a long-term operation will produce lots of pollutants (like hydrogen sulfide, tar, etc.).

Fluidized-bed gasification technology and entrained-flow gasification technology can also be used for syngas production from coal. Compared to fixed-bed gasification technology, fluidized-bed gasification technology is not limited to lump coal and mechanical transmission; and thus, the generated gas basically contains no coal volatile [11]; but it has a lower thermal efficiency and a lower carbon conversion rate of carbon attributed to the mix even stick of ash and unreacted carbon. Entrained-flow gasification technology has the widest adaptive range of coal, the largest gasification intensity, the highest conversion efficiency of coal (up to 99%), and few emissions (like tar, naphthalene and phenol water) [12]; however, the lower thermal efficiency in the furnace due to the parallel operation needs a huge heat-recovery system, while the lining material that can be used for reducing the corrosion of high temperature slag is limited.

Besides coal, natural gas and oil could also be reformed to produce hydrogen gas in various syngas methods. In fact, natural gas and liquid hydrocarbons, respectively, occupy about 48% and 30% share ratio of hydrogen sources in the current global market.

Like coal, natural gas could produce hydrogen gas via water steam, as:

$$CH_4 + H_2O \rightarrow 3H_2 + CO \tag{11.7}$$

This reaction is endothermic with the enthalpy change of about 253 kJ/mol (under standard conditions if the water is in the liquid phase), and the required heat is always provided by burning partial methane in an industrial steam reformer [13] as,

$$CH_4 + 2O_2 \rightarrow 2H_2O(\text{steam}) + CO_2 \tag{11.8}$$

During the production process, the practical amount of input steam is about double, or even triple, the requirement of stoichiometric reaction for further water–gas shift (WGS) reaction in which the surplus steam will be reacted with the produced carbon monoxide [14], as:

$$CO + H_2O \rightarrow H_2 + CO_2 \tag{11.9}$$

Combing (11.7) and (11.9), the gross reaction combined can be expressed as:

$$CH_4 + 2H_2O \rightarrow 4H_2 + CO_2 \tag{11.10}$$

The vaporeforming of methane happens at moderate pressure (from 2.0–3.0 MPa) under high-working temperature (i.e., 1200K). For avoiding the crack of methane under high temperatures (as demonstrated in the equation of 11.11) [15], the temperature should not be significantly raised albeit it may be benefit to enthalpy change:

$$CH_4 \rightarrow 2H_2 + C \tag{11.11}$$

Such carbon encapsulates catalyst atoms on the catalyst's surface to block the reaction of steam reforming; therefore more input steam is required to flush them away.

Although it is the major method to current hydrogen production, vaporeforming natural gas has been limited in larger scale applications due to the high production cost. First, the first-step reaction and WGS reaction happen in different reactors, thus the required two heat exchangers increase the cost of devices. Second, by-produced carbon dioxide needs to be flushed away for increasing fuel quality, which increases the cost. Third, compared to carbon dioxide whom can be removed by flushing, carbon monoxide needs higher-cost technologies to be removed.

Methane also can be oxidized into hydrogen, as:

$$2CH_4 + O_2 \rightarrow 4H_2 + 2CO \tag{11.12}$$

Compared to the vaporeforming, the oxidation of methane into hydrogen (as demonstrated in the equation of 11.12) has a faster reaction rate; furthermore, the insufficient oxygen in the partial oxidation reaction makes vehicles or ships can afford on-board hydrogen production [16].

$$2CH_4 + xO_2 + (2 - 2x)H_2O \rightarrow (6 - 2x)H_2 + 2CO \tag{11.13}$$

Besides natural gas, hydrocarbons are the second major source of hydrogen production. These produce hydrogen by reforming technologies, as:

$$2C_nH_m + 2nH_2O \rightarrow (2n + m)H_2 + 2nCO \tag{11.14}$$

At industrial scale, partial oxidation of hydrocarbons is more favored for the wide adaptive choices of feedstock, and such production requires a separate performer to strip carbon from higher hydrocarbons (since they have more noticeable differences in reaction rate and thermal cracking than methane) which means the hydrogen production becomes more complex and more expensive. Furthermore, besides carbon monoxide and carbon dioxide, lots of sulfur is contained in the products, which entails more cost and more pollutants.

11.2.3 From biomass: thermochemical and biological process

For the carbon neutral cycle, producing hydrogen from biomass is one promising sustainable route and it could be realized by thermochemical and biological processes [17]. Fig. 11.3 illustrates the classic pathways for hydrogen production from biomass.

The thermochemical process is mainly the gasification of cellulosic and sugar-rich materials. Such biomass materials could be transferred to carbon, like the cracking of sugar, as:

$$C_6H_{10}O_5 \rightarrow 6C + 5H_2O \tag{11.15}$$

And then carbon, similar to coal, could produce hydrogen with the existence of steam, as:

$$C + H_2O(\text{steam}) \rightarrow H_2 + CO \tag{11.16}$$

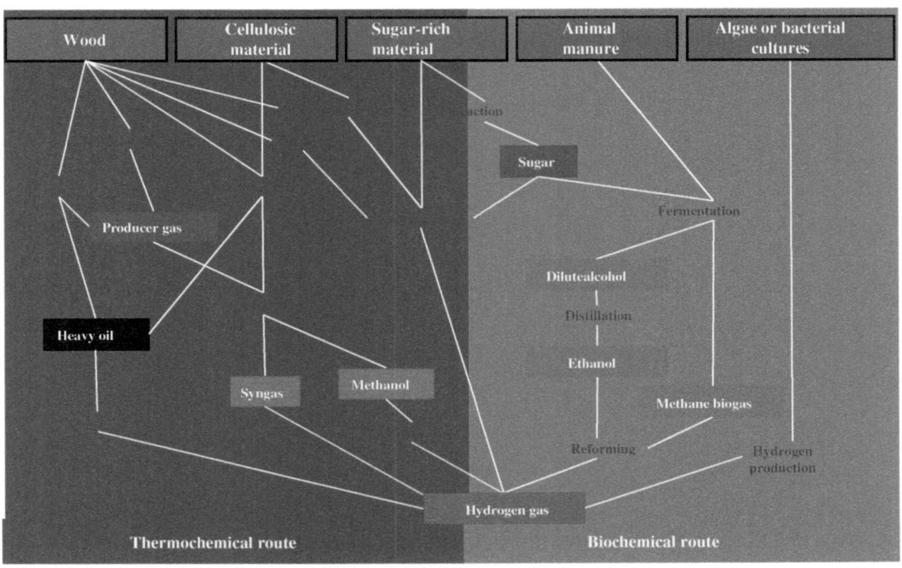

Figure 11.3 Pathways for hydrogen production from biomass.

In the gasification of lignin-containing biomass, there are also some additional processes like the oxygen gasification process, Boudouard process, and WGS reaction.

Since some gaseous by-products (like carbon monoxide, carbon dioxide, methane, etc.) exist together with hydrogen, the production is considered as "medium-quality" (the calorific value is less than 20 MJ/m^3, just about two-thirds of the quality required by pipeline gas) and the heat value does not meet the requirements of wider utilization. Furthermore, some other impurities (including ash, char, liquid wastewater, and tar) will be produced, the posttreatment toward environment-friendliness should be extra concerned.

The biological process mainly contains two routes: (1) the decomposition of water by algae or bacteria [18]; and (2) the fermentation of bio-wastes (like animal manure, etc.) [19].

Some algae or bacteria, like a bioreactor, could use water as raw materials to produce hydrogen under sunlight. The physical mechanism is due to the existence of enzymes related to hydrogen production in some organisms (light-biology and fermentation bacteria) [18]. Photosynthetic bacteria contain a photosynthetic system, and charge separation will be carried out once photons are captured and sent to the photosynthetic system to generate high-energy electrons and form proteins, as:

$$2H_2O \rightarrow O_2 + 4H^+ + 4e^- \tag{11.17}$$

Finally, H-reduction will be carried out under the action of nitrogenase to generate hydrogen gas, as

$$2e^- + 2H^+ \rightarrow H_2 \tag{11.18}$$

Nitrogenase is the main enzyme that catalyzes hydrogen production from photosynthetic bacteria [20]. Such biological hydrogen production could be realized under normal temperature and pressure if suitable enzyme catalysts are employed.

Photosynthetic organisms that produce hydrogen include photosynthetic bacteria and algae. Many algae (e.g., green algae, red algae, and cyanobacteria) are photosynthetic, hydrogen-producing microorganisms, and H_2 metabolism is mainly carried out by hydrogenase. Prokaryotes (like Rhodopseudomonas rubra and Rhodopseudomonas) has become a hot topic in the photosynthetic bacteria for hydrogen production since: (1) they are easier to cultivate on various waste organics with high conversion rates; (2) they have a wider absorption spectrum; (3) they have smaller free energy; (4) they have higher purity of hydrogen (more than 95%); and (5) no oxygen is produced, etc. However, the high dependence on a light source limits the practical development for the two main reasons: (1) more input energy is required for ensuring the intensity and the continuity of light; and (2) the pigments produced by photosynthetic bacteria and the turbidity of reaction solution reduce the light efficiency by influencing the spread and distribution light in the reaction system.

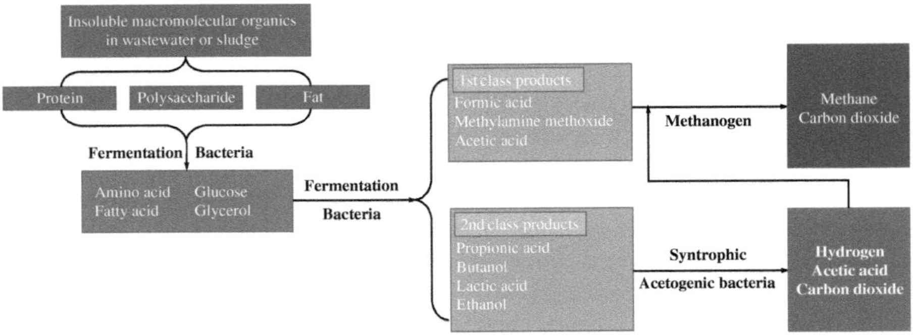

Figure 11.4 Some classic routes for hydrogen production by fermentation of biomass.

Fermentation is another significant biological method of hydrogen production from organic substrates under oxygen-free and dark conditions. Such hydrogen production is independent of light and easy to cultivate from pure or mixed bacteria, which allows large-scale production of hydrogen for the easy control of reactors [21]. The transformation process from glucose to acetic is a classic fermentation process (with higher conversion efficiency), as:

$$C_6H_{12}O_6 + 2H_2O \rightarrow 4H_2 + 2CO_2 + 2C_2H_4O_2 \quad (11.19)$$

Different bacterial can be employed to the fermentation of biomass for hydrogen production in different routes, as demonstrated in Fig. 11.4.

The conversion rate of hydrogen gas per unit substrate is very low (glucose is only about 2 mol/mol) as most hydrogen elements are still fixed in the fermentation products (e.g., propionic acid, butyric acid, lactic acid, and ethanol). To break down biomass in one obstacle for the hydrogen production by dark fermentation,so the hydrogen conversion rate per unit substrate is the bottleneck in the industrialization.

11.3 Combustion properties of hydrogen

As a potentially sustainable fuel, the combustion properties of hydrogen should be highly concerned. Hydrogen gas could be employed as the fuel to both spark-ignition power devices and compression-ignition power devices, but the most common combustion manner of hydrogen gas in internal combustion engines is spark-ignition.Table 11.1 provides a comparison of hydrogen gas with methane and iso-octane which, respectively, represent natural gas and gasoline (two major spark-ignited fuels).

The table indicates that hydrogen gas has the lightest molecule weight and the smallest density, which means it will be mobile on-board without the worries about extra loads. The highest capabilities of calorific value make hydrogen gas more

Table 11.1 Comparison of fundamental properties between hydrogen and others.

Properties	Hydrogen	Methane	Iso-octane	
Molecular weight (g/mol)	2	16	114	Lowest
Density (kg/m^3)	0.08	0.65	690	Lowest
Lower calorific value per mass (MJ/kg)	120	50	44.3	Highest
Higher calorific value per mass (MJ/kg)	142	55	48	Highest
Stoichiometric fuel/air ratio (kmol/kmol)	0.42	0.11	0.017	Highest
Mass diffusivity in air (cm^2/s)	0.61	0.16	0.07	Highest
Minimum ignition energy (mJ)	0.02	0.28	0.28	Lowest
Explosive range (in equivalence ratio)	0.10−7.10	0.50−1.67	0.66−3.85	Widest
Minimum quenching distance (mm)	0.64	2.03	3.50	Lowest

valuable as fuel. The high stoichiometric fuel/air ratio allows the same amount of air to combust more fuel to provide more energy and/or heat. The high mass diffusivity ensures a uniform fuel/air mixture for better combustion performance. The minimum ignition energy helps hydrogen-fueled engines lower disagree of spark-plug ablation. The wide range of explosive limits ensures hydrogen-fueled engines easily realize ultra-lean combustions for near-zero combustion emissions.

11.3.1 Flame structure and flame instabilities

Fig. 11.5 illustrates the flame structure of hydrogen-air premixed flames with different equivalence ratios under the same initial conditions (300K, 0.10 MPa, laminar). All the images were obtained in the Laboratory of Hydrogen Energy, Beijing Jiaotong University, China. The experimental methodology can be found in [22−24].

For different regimes of fuel concentration, the initial laminar premixed hydrogen flames are very smooth and propagate in a highly spherical shape; for the physical effects of electrodes, some cracks could be formed on the initial surface of hydrogen premixed flames, and such effects will be suppressed with the increase of equivalence ratio (φ). However, with continuous propagation, the laminar premixed hydrogen flames will express different behaviors for different regimes of equivalence ratio (as listed in Table 11.2), as:

1. for the stoichiometric and rich hydrogen flames (whose $\varphi \geq 1.0$), the flames can still outwardly propagate with highly smooth surface, the initial cracks continuously develop; with the decrease of φ, more cracks appear to form cellular structure on flame-front, the onset of new cracks and cellularity will be advanced (in the format of flame size);
2. for lean hydrogen flames ($0.4 \leq \varphi < 1.0$), the flames also can still outwardly propagates, and cells are uniformly formed on the flame-front, the oblivious cellularity makes the flame-front more and more wrinkled rather than maintaining the initial smooth; with the decrease of φ, the level of cellularity and wrinkled at a same flame size are enhanced, the onset of cellularity is advanced;
3. for the very lean hydrogen flames (like $\varphi = 0.30$), the flame propagation expresses a tendency toward upwardly from the initial highly outwardly manner when it develops to a certain size, and this tendency will become more and more obvious; furthermore, the cells

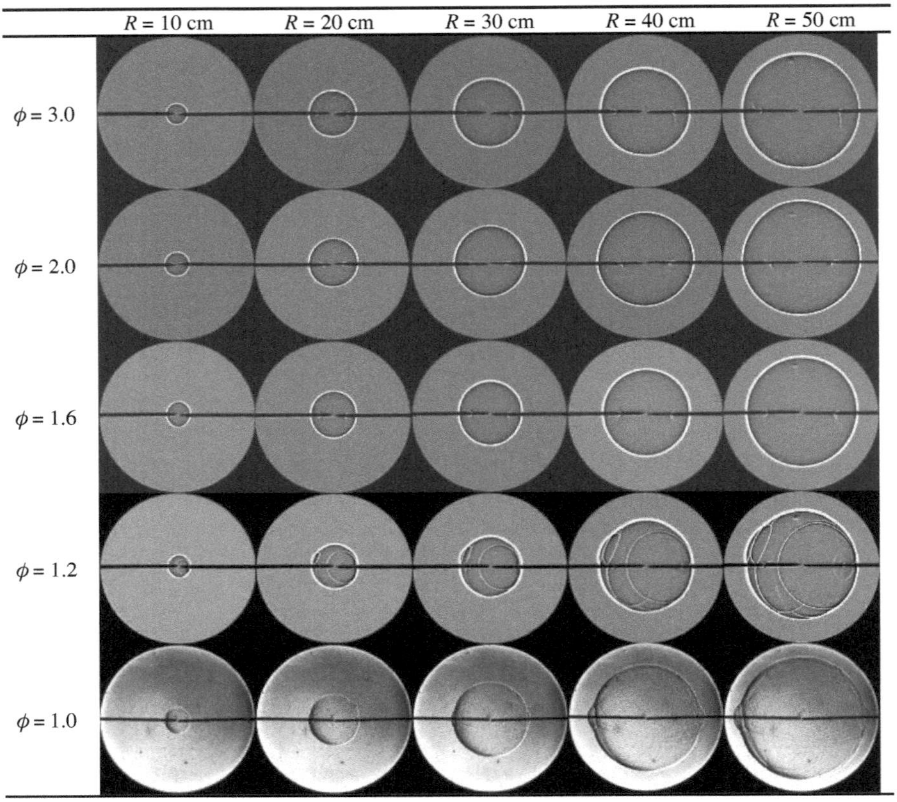

Figure 11.5 Flame images of laminar premixed hydrogen flames with different equivalence ratios under standard conditions.

at the lower partial flame-front (taking the geometrical center as the reference center) are continuously divided into small cells (as lean hydrogen flames) but the development of cells at the upper partial is manifested by enlargement (albeit split still exists), and the distribution and the size of cellularity become more and more uneven;

4. for ultra-lean hydrogen flames (like $\varphi = 0.20$), the flame propagation becomes upwardly from the initial outwardly manner, and little cellularity was found on the flame-front.

Fig. 11.6 illustrates the flame images of laminar premixed hydrogen flames whose equivalence ratio falls into the rage of 0.2–0.3 under standard conditions. All those flames express buoyancy behaviors during the propagation. To the ultra-lean hydrogen flames (whose $\varphi \leq 0.25$), they will be blowing upwardly during the development, and the cells just appear at the lower partial of flame-front. To the hydrogen flames whose $0.26 \leq \varphi \leq 0.30$, their geometrical centres will moved upwardly during the development, the cells will appear on the whole flame-front. With a decrease of φ, cellularity is significantly suppressed, buoyant behaviors are enhanced, and the onset of upwardly tendency (expressed as the difference on

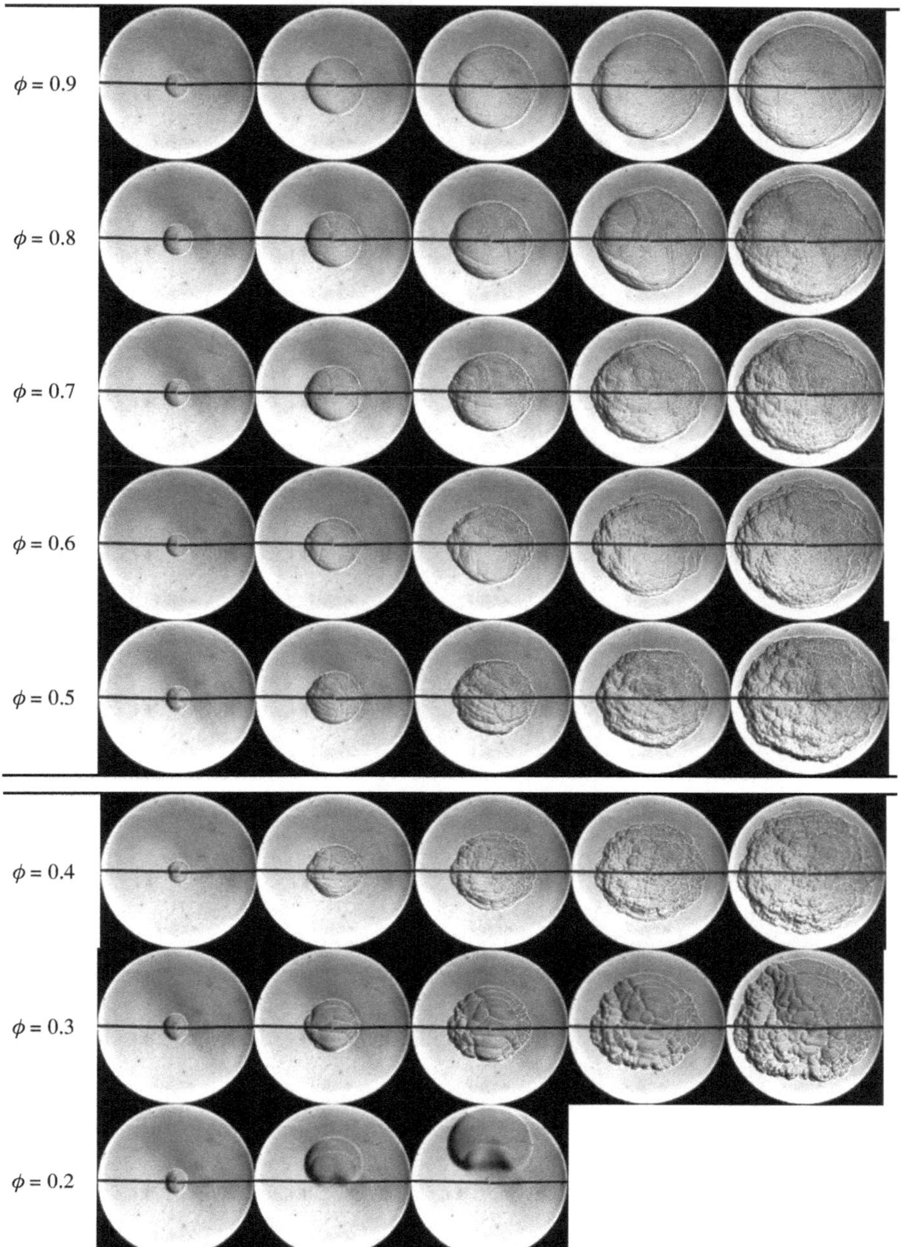

Figure 11.5 continued

Table 11.2 Structure evolution of hydrogen laminar premixed flames with different regimes of fuel concentration.

Range of equivalence ratio	Behaviors and variations with a decline of φ
Rich and stoichiometric ($1.0 \leq \varphi$)	• Highly spherical shape and highly smooth surface • Highly outwardly propagating • As φ decreases, cracks have been formed on the surface • As φ decreases, the onset of cracks is advanced
Lean ($0.4 \leq \varphi < 1.0$)	• Highly spherical shape and wrinkled surface • Highly outwardly propagating • As φ decreases, the wrinkled level is enhanced • As φ decreases, cells have been uniformly formed on the surface • As φ decreases, the onset of cellularity is advanced
Very lean (like $\varphi = 0.30$)	• Hardly spherical shape and wrinkled surface • Outwardly propagating with obvious buoyant tendency • The size and distribution of cellularity express uneven
Ultra-lean ($\varphi \leq 0.25$)	• Garlic-spherical shape and smooth surface • Upwardly propagating • As φ decreases, the wrinkled level is suppressed • As φ decreases, the onset of upwardly propagating is advanced

propagation speed between the upper partial and the lower partial) is advanced (in the format of flame size).

Fig. 11.7 illustrates the effects of initial pressure on hydrogen flames at 300 K. To the ultra-lean hydrogen flames (like $\varphi = 0.2$,), with the increase of initial pressure (p_{int}), the upward propagation is advanced while the cellularity tendency gives a few enhancements. To the lean and stoichiometric hydrogen flames, with the increase of p_{int}, the cellularity is obviously enhanced; to the hydrogen flame with $\varphi = 0.4$, a slight buoyant tendency still exists albeit cellularity has been expressed; while at subpressure conditions, the buoyant behaviors become more noticeable. The effects of p_{int} on hydrogen-rich flames are highly similar to those of stoichiometric hydrogen flames since both experience hydrodynamic instability [22].

Hydrodynamic instability (also known as Darrieus-Landau [D-L] instability) is an intrinsic instability that exists in all flames since it results from the density jump across the flame-front, and the intensity is commonly indicated by density ratio across flame-front (σ) and flame thickness (δ, which effects D-L instability via the baroclinic torque on the local zones of flame-front and the density gradient on flame-front). Fig. 11.8 demonstrates the variation regulations of the two parameters to equivalence ratio and initial pressure. Under the same thermodynamic conditions, with the increase of φ, the value of σ first

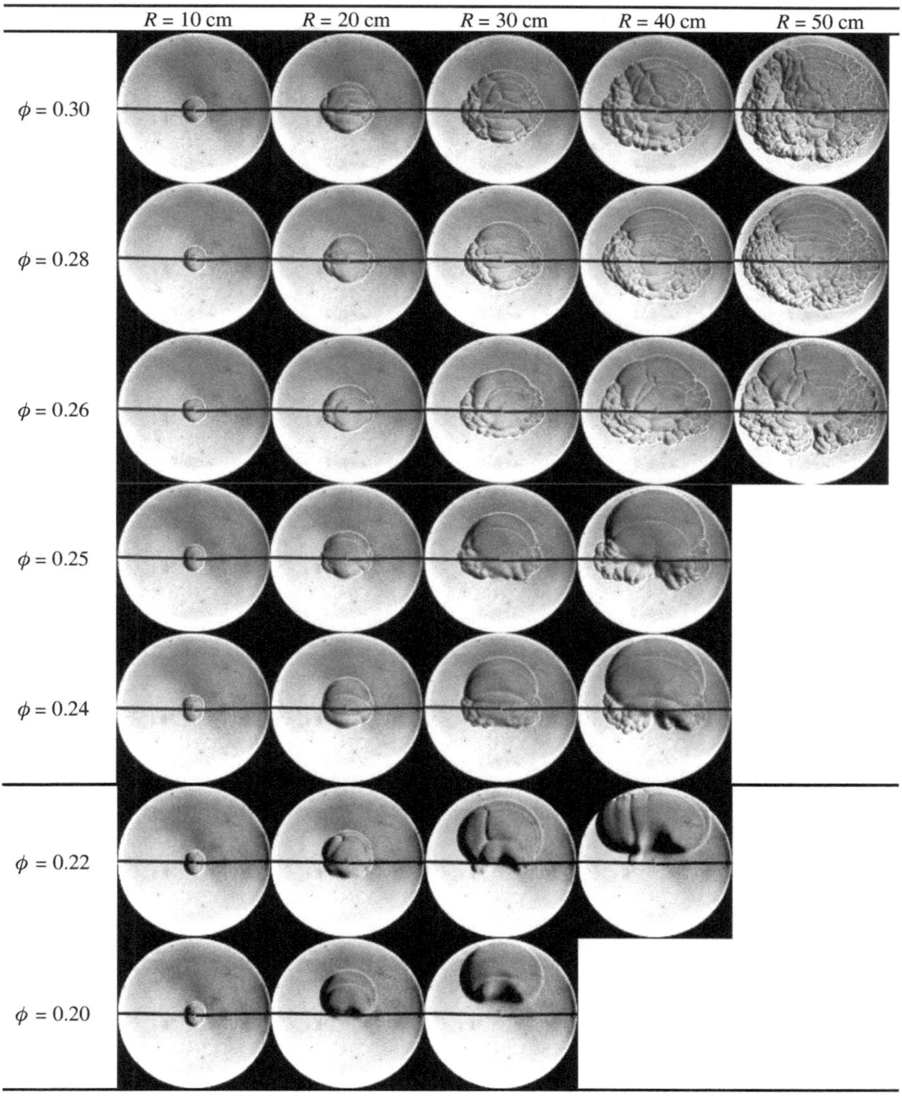

Figure 11.6 Flame images of laminar premixed hydrogen flames whose equivalence ratio falls in the range of 0.2−0.3 under standard conditions.

rises and then declines, the maximal value is attained at the stoichiometric, while the value of δ first declines and then rises, the lowest value is attained at $\varphi = 1.8$. Increasing p_{int}, the value of δ will be significantly reduced but the value of σ gives few variations. Such variations are well matched with the phenomena about cellularity induced by φ and p_{int} for stoichiometric and hydrogen-rich flames, namely, the flame structure of stoichiometric and rich laminar hydrogen flames is determined by D-L instability.

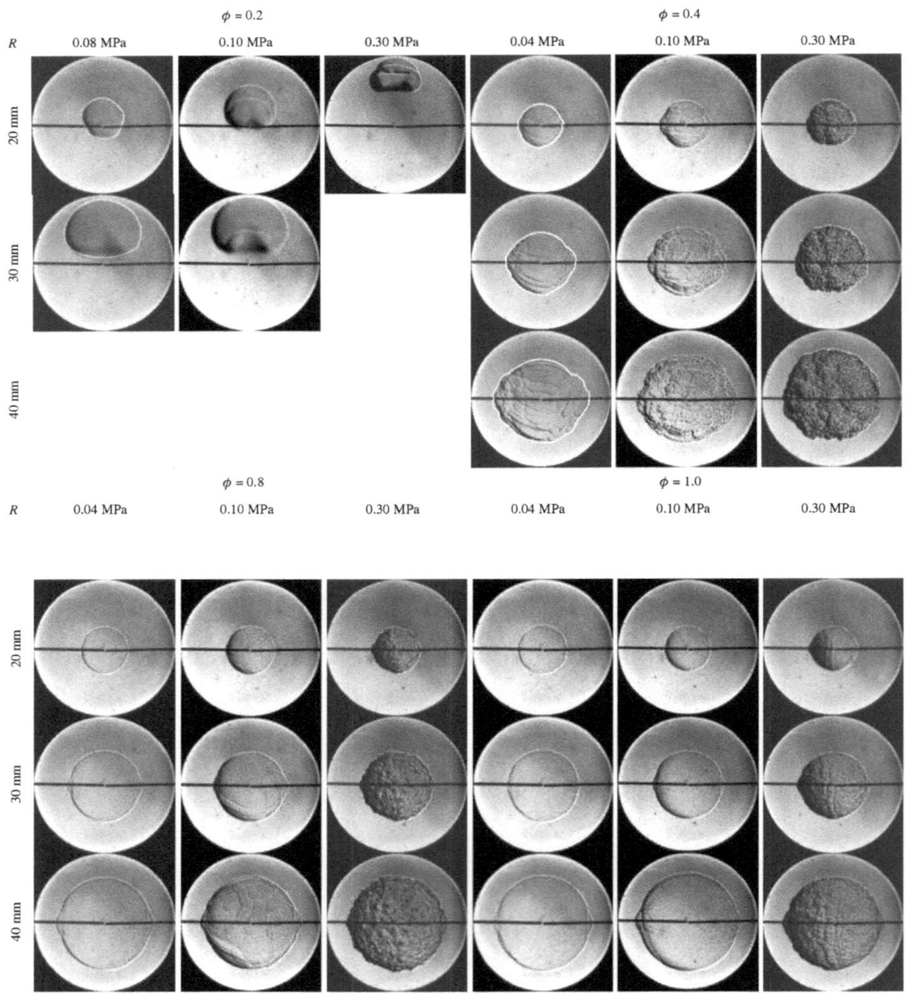

Figure 11.7 Effects of initial pressure on laminar premixed hydrogen flames at 300K.

Lean hydrogen flames have stronger cellularity behaviors than the stoichiometric, but the intensity of D-L instability is weaker than the stoichiometric for the reduced density ratio and the thickened flame thickness. Such phenomena are attributed to the noticeable difference on the relationship between heat conduction and mass diffusion on the flame-front. To stoichiometric and hydrogen-rich flames, the thermal diffusivity is no less than mass diffusivity; but to lean hydrogen flames, mass diffusivity is significantly enhanced which makes the flame-front more unstable. The related influence mechanism is named thermal-diffusive instability (also known as D-T instability). The intensity of D-T instability is always indicated by Lewis number (*Le*, the ratio of thermal diffusivity to mass diffusivity across flame-front).

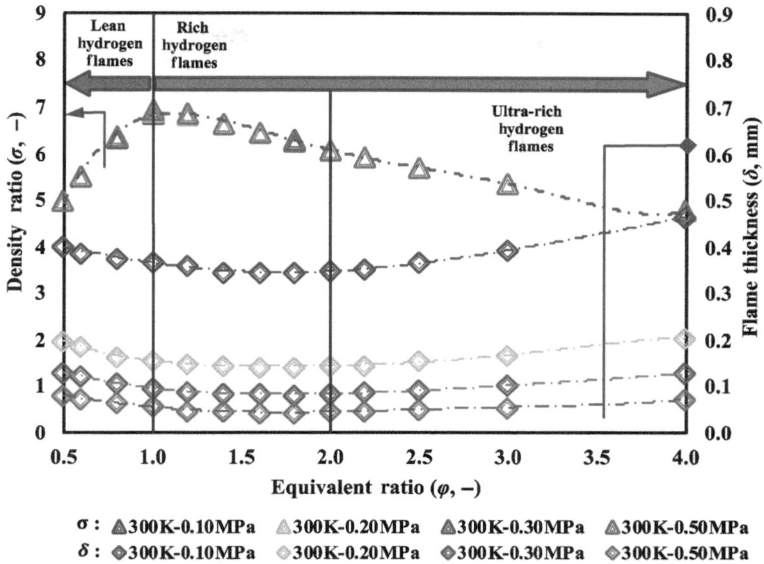

Figure 11.8 Density ratio and flame thickness of hydrogen flames.

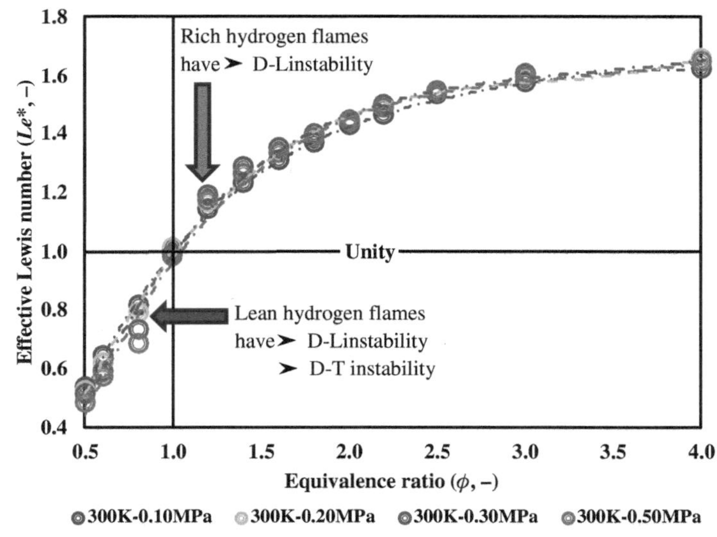

Figure 11.9 Lewis number of laminar premixed hydrogen flames.

Fig. 11.9 illustrates the Lewis number of laminar premixed hydrogen flames. With the decrease of φ, the Lewis number monotonously declines to be less than unity for all the lean conditions, which means just lean hydrogen flames suffer D-T instability and the instability intensity is enhanced with the increase of φ. Furthermore, with the increase of p_{int}, the Lewis number of lean hydrogen flames is reduced, namely rising p_{int} can enhance the intensity of D-T instability. Considering the detailed variation of cellularity behaviors illustrated in Fig. 11.6, it could be believed that the cellularity of lean hydrogen flames is mainly determined by D-T instability [22], but such conclusions cannot explain the evolutions of flame structure when φ further declines to ultra-lean.

By applying the Archimedes' principle, the upward propagation is driven by the difference between buoyancy and gravity. The buoyancy is proportional to the product of flame volume and the density of unburnt mixtures, while the gravity is proportional to the produce of flame volume of the density of burnt mixtures; therefore, the buoyant potential is related to flame volume and the density difference across flame-front ($\Delta\rho$) [23]. For a developing premixed flame, the flame volume is continuously raised, and thus, for hydrogen flames with similar size, the buoyant tendency would depend on the density difference. As demonstrated in Fig. 11.10, the maximal value of $\Delta\rho$ is attained at $\varphi = 0.4$, and the value of $\Delta\rho$ is noticeably raised with the increase of p_{int}. The variations of $\Delta\rho$ could explain the appearance of buoyant tendency at $\varphi = 0.4$, but it can not explain why the ultra-lean hydrogen flames more buoyant.

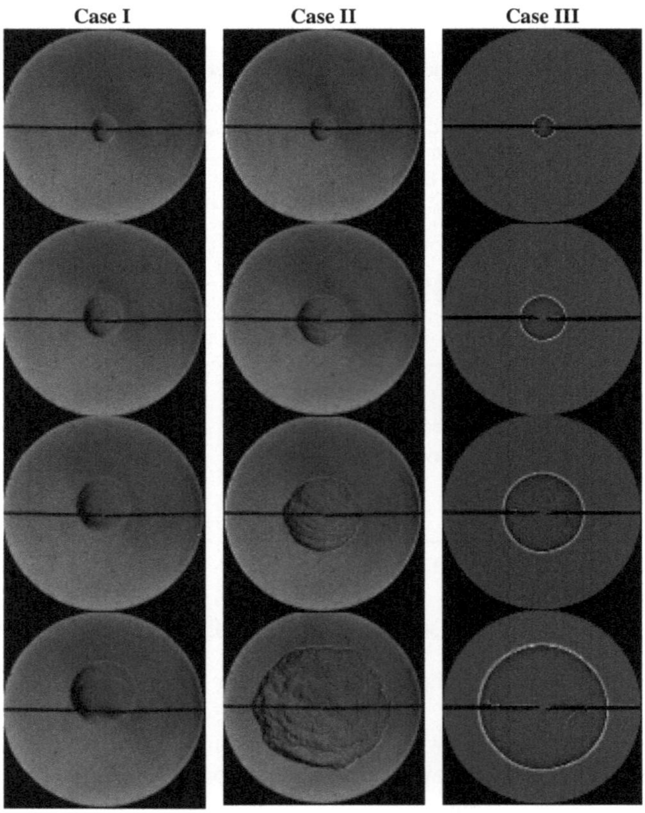

Figure 11.10 Density difference of hydrogen flames.

Besides gravity and buoyancy, an initial outwardly propagating flame also suffers stretch effects which are regarded as suppression effects on intrinsic instabilities [24]. As reported by Sun [24], to lean hydrogen flames in laminar, the value of stretch rate (a) on the flame-front with a same size is significantly reduced with the decrease of φ; furthermore, the lower the φ is, the faster the value of a declines to the converge. It could be proposed that the weaker stretch effects play more effective roles on the exhibitions of buoyant unstable to ultra-lean hydrogen flames.

Fig. 11.11 illustrates the effects of external turbulence on the evolution of hydrogen flames with different equivalence ratios at standard condition, all the images

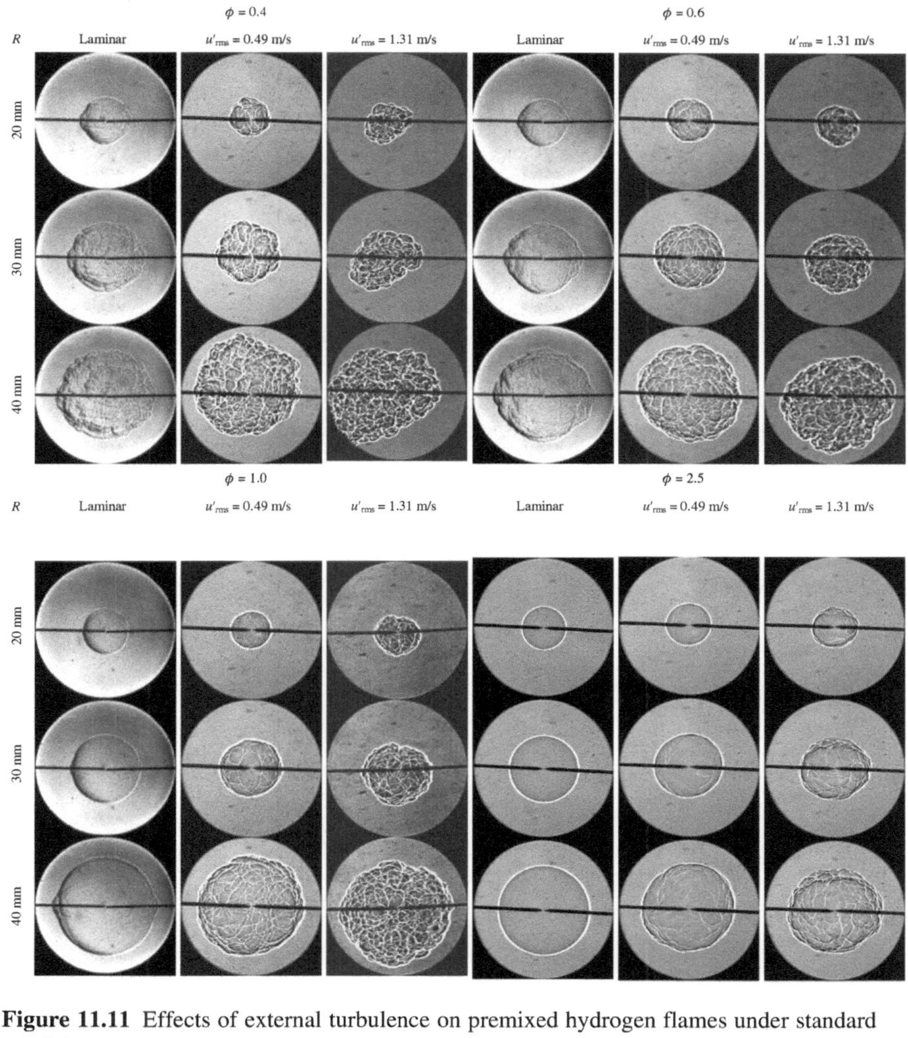

Figure 11.11 Effects of external turbulence on premixed hydrogen flames under standard conditions.

were obtained in the Laboratory of Hydrogen Energy, Beijing Jiaotong University, China, the experimental methodology can be found in [25–27]. The flame-front will be wrinkled by external turbulence (obviously in $\varphi = 2.5$) to induce extra cracks and then advances the development of cellularity. Such phenomena will be significantly enhanced once intrinsic instabilities have been expressed (seen in $\varphi = 1.0$). With the increase of turbulence intensity (u'_{rms}), the wrinkled level and the cellularity behaviors are significantly enhanced (like $\varphi = 0.6$); furthermore, to lean hydrogen flames, with the decrease of φ, the flame propagation direction express high randomness (i.e., seen from the comparison between $\varphi = 0.6$ and $\varphi = 0.4$).

11.3.2 Flame stretch and flame speed

Albeit that all hydrogen flames have intrinsic instabilities, the unstable behaviors won't be expressed at the initial period for the suppression of flame stretch until the stretch effects have been declined to a certain level or they are too weak to suppress the instabilities [22]. The stretch effects on the flame-front is commonly indicated by stretch rate (a), which is defined as the Lagrangian time derivative of the logarithm of surface area.

Fig. 11.12 demonstrates the effects of φ and p_{int} on the stretch rate of laminar premixed hydrogen flames. As the flame develops, the value of a decreases in exponential tendencies. To hydrogen flames with a same size, as φ rises from lean to rich, the value of a first rises and then decreases, and the maximum value is attained at rich mixture (the critical φ is about 1.8). p_{int} expresses different effects on flame stretch for hydrogen flames with different equivalence ratios. To lean hydrogen flames (like $\varphi = 0.6$), the initial value of a is few sensitive to p_{int}, but the decline gradient of a is generally reduced as p_{int} rises; to rich hydrogen flames (like $\varphi = 1.8$), the initial value of a monotonously raises with the increase of p_{int}, but the rise extent declines as flame continuously propagates; and to ultra-rich hydrogen flames (like $\varphi = 3.0$), as p_{int} rises, the initial value of a decreases while the decline extent nearly keeps a same level, resultantly the values of a always decline with as p_{int} rises in the whole propagation.

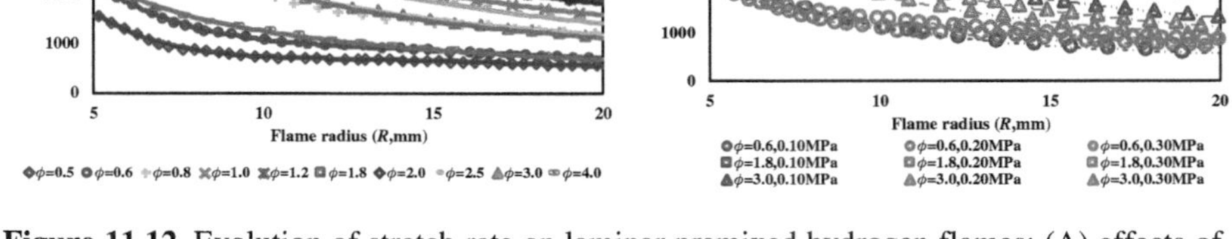

Figure 11.12 Evolution of stretch rate on laminar premixed hydrogen flames: (A) effects of equivalence ratio at standard condition; and (B) effects of initial pressure.

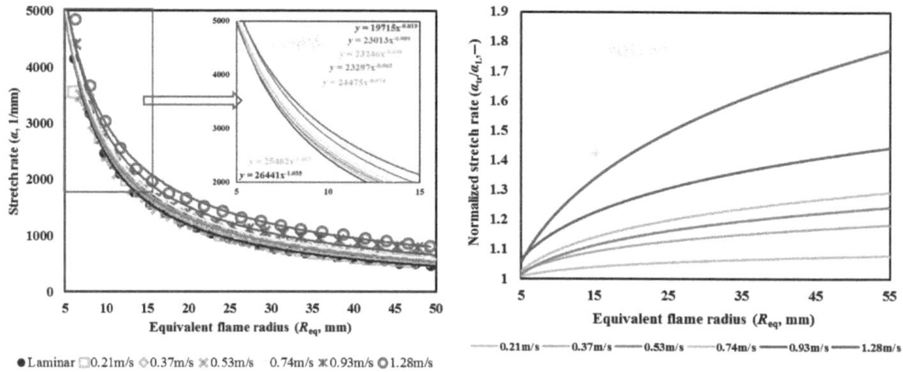

Figure 11.13 Effects of turbulence on the stretch effects of stoichiometric hydrogen flames under standard conditions: (A) evolution of stretch rate; and (B) normalized stretch rate.

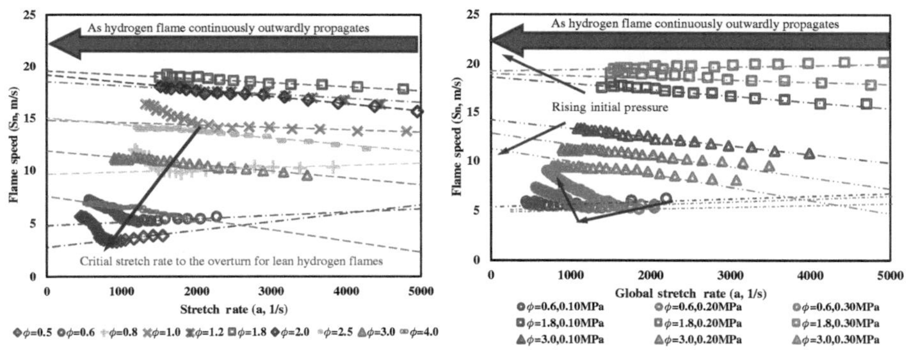

Figure 11.14 Nexus between propagation speed of laminar hydrogen flames and stretch rate: (A) effects of equivalence ratio under standard conditions; and (B) effects of initial pressures.

Fig. 11.13 illustrates the effects of external turbulence on the stretch rate of hydrogen flames (the case of $\varphi = 1.0$ at standard condition). For hydrogen flames with a same size, with the increase of u'_{rms}, the value of a rises since external turbulence enhances the interactive moment between flame and ambient flow. For a detailed viewpoint, the effects of turbulence on flame stretch are indicated by the normalized stretch rate (α_{tr}/α_l, the ratio of the stretch rate on turbulent flame to that on laminar flame with the same size) as demonstrated in Fig. 11.13B. With the increase of u'_{rms}, the value of α_{tr}/α_l rises during the whole propagation with increasing growth extents. Albeit that the enhanced stretch effects could suppress the expression of intrinsic instabilities, the enhancement of flame stretch is obviously weaker than the enhancement of flame curvature induced by external turbulence. Therefore the cellularity and the wrinkled level express significant enhancement under turbulence.

Fig. 11.14 illustrates the nexus between the propagation speed (S_n, time derivative of flame radius) of laminar hydrogen flames and stretch rate. For lean and stoichiometric hydrogen flames, the value of S_n first decreases with the decline of a;

when the value of a has been declined to a lower level, the stretch effects be too weak to continuously suppress the intrinsic instabilities, then the value of S_n will be raised. With the decease of φ, the critical stretch rate to the overturn variation of S_n is reduced. For hydrogen-rich flames, there exists linear relationship between S_n and a in which the fitting slop first declines and then rises with the increase of φ, and the critical status is attained at $\varphi = 1.8$. The effects of p_{int} on the nexus between S_n and a are different for hydrogen flames with different φ; in the detailed figure, as p_{int} rises: (1) for lean hydrogen flames, the initial value of S_n at a same value of a is reduced but such regulation will be reversed once S_n begins to rise with the further decline of a; (2) for rich hydrogen flames (like $\varphi = 1.8$), S_n at a same value of a will be raised but the variation extent of S_n during the whole propagation is significantly reduced; and (3) for ultra-rich hydrogen flames (like $\varphi = 3.0$), S_n at a same value of a will be noticeably reduced but the variation extent seems few sensitive to p_{int}.

Fig. 11.15 demonstrates the effects of turbulence on the nexus between the propagation speed and flame stretch of stoichiometric hydrogen flames at standard condition, the effects of turbulence on stretch effects are indicated by the normalized stretch rate (α_{tr}/α_l) and the effects of turbulence on flame propagation are indicated by the normalized flame speed (S_{tr}/S_n, the ratio of the turbulent flame speed to laminar flame speed at the same flame radius). As the value of α_{tr}/α_l rises, the value of S_{tr}/S_n first rises and then declines. At the same α_{tr}/α_l, with the increase of u'_{rms}, the value of S_{tr}/S_n rises, and the critical value of α_{tr}/α_l to the overturn of S_{tr}/S_n during the propagation also remarkably rises; moreover, as u'_{rms} rises, the growth extent of

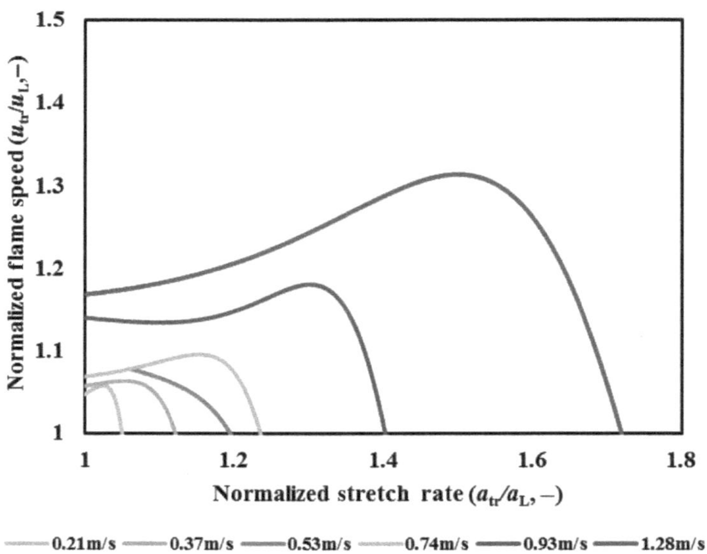

Figure 11.15 Effects of turbulence on the nexus between flame speed and stretch rate of stochiometric hydrogen flames.

α_{tr}/α_l during the propagation is enlarged. The results mean that a higher turbulence will induce stronger stretch effects on hydrogen flames, but the enhanced stretch effects won't always bring positive effects during the propagation process.

11.4 Problems relating to using hydrogen: Explosion potential and nox emissions

The explosion characteristics of hydrogen gas is double-edged, it makes hydrogen as a promising fuel but also a hazardous material; furthermore, the combustion of H_2 into air also produces NOx emissions for the reaction of nitrogen and oxygen under high temperature released by the reaction of hydrogen and oxygen.

11.4.1 Explosion characteristics of hydrogen gas

Fig. 11.16 illustrates the maximal explosion overpressure (p_{max}) of hydrogen-air mixtures under standard conditions in a constant-volume space of 28.73 L; the detailed experimental methodology can be found in [28–31]. As φ rises from stoichiometric to rich, the value of p_{max} obtained in both laminar and turbulent hydrogen explosion monotonously declines, and it rises with the increase of u'_{rms}; however, from the format of normalized maximal explosion overpressure ($p_{max,tr}/p_{max,l}$, the ratio of p_{max} in turbulent hydrogen explosion to that in laminar hydrogen explosion), with the increase of φ, the value of $p_{max,tr}/p_{max,l}$ first rises and then declines to convergence values, and the maximal value of $p_{max,tr}/p_{max,l}$ is attained at $\varphi = 1.2$.

The theoretical value of p_{max} is determined by adiabatic flame temperature, the practical value of p_{max} is lower than the theoretical value for the heat loss during explosion in nonadiabatic system. Albeit that hydrogen has a peaking adiabatic flame temperature off-stoichiometric [32], the corresponding equivalence ratio is

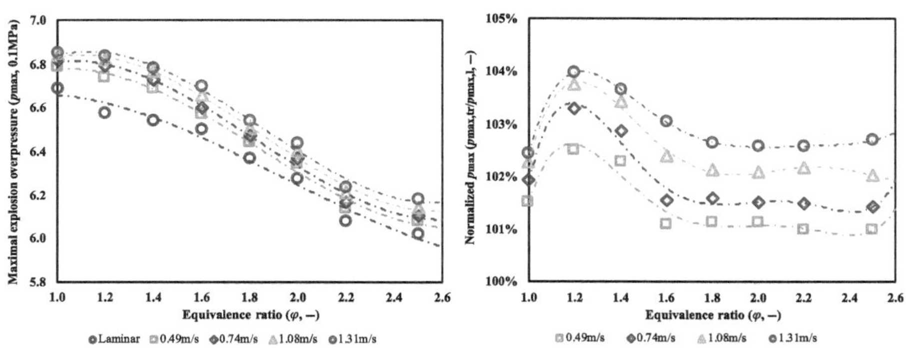

Figure 11.16 Maximal explosion overpressure of hydrogen in a 28.73 L space: (A) in the format of absolute values; and (B) in the format of normalized values.

closer to 1.0 (about 1.05) than the others in Fig. 11.16. Thus the stoichiometric hydrogen explosion could obtain the maximal value of p_{max} for the most heat release. With the increase of u'_{rms}, the temperature difference won't be noticeably changed for the constant adiabatic flame temperature to each specific φ, but the explosion duration will be significantly shortened for the increased flame speed. Therefore the accumulated heat loss during the explosion of hydrogen with specific φ will be reduced as u'_{rms} rises, which results in the decrease of pressure drop between practical p_{max} and the theoretical. A similar value of adiabatic flame temperature and a faster flame speed reasonably make the value of $p_{max,tr}/p_{max,l}$ attained at $\varphi = 1.2$; and it is also reasonably believed the convergence of $p_{max,tr}/p_{max,l}$ is caused by the equilibrium in the competition between adiabatic flame temperature and heat loss under different influence factors.

Fig. 11.17 illustrates the maximal rise rate of explosion pressure ($(dp/dt)_{max}$) and the corresponding deflagration index (K_G) of hydrogen at standard condition, the latter is defined as the product of $(dp/dt)_{max}$ and the cube root of explosion space's volume for removing the effects of space size.

As φ rises from the stoichiometric, the value of $(dp/dt)_{max}$ of hydrogen explosion first rises and then declines, the corresponding φ is 1.4 in laminar and will be reduced with the increase of u'_{rms}. The phenomena are mainly attributed to the competition on the variations of explosion capability and heat loss as mentioned in the previous. In all the cases demonstrated in Fig. 11.17, the value of K_G in hydrogen explosion is higher than 30 MPa·m/s (the criterion of Stand Level III defined by National Fire Protection Agency, the most dangerous level of explosion), which

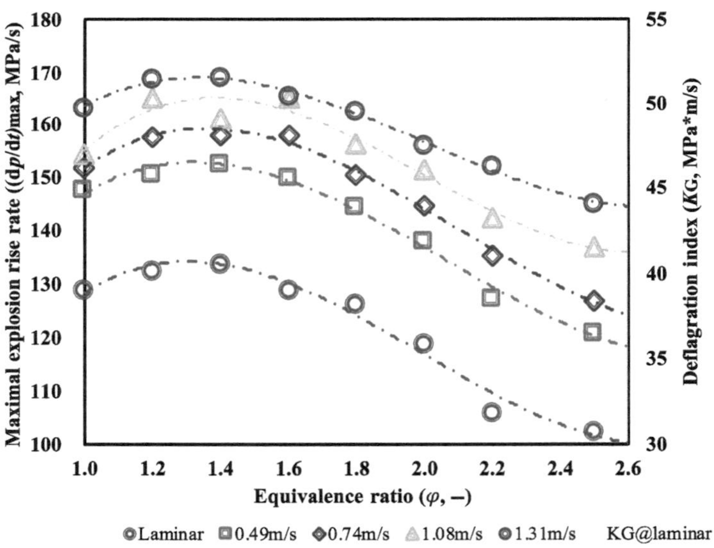

Figure 11.17 Maximal rise rate of explosion overpressure in a 28.73 L space and the corresponding deflagration index of hydrogen at standard condition.

means stoichiometric and rich hydrogen-air mixtures are highly hazardous and the hazardous level will be raised as u'_{rms} rises.

11.4.2 NOx emissions of hydrogen gas

Hydrogen flames have a high adiabatic flame temperature (about 2400K for the stoichiometric combustion at standard condition) that can activate the reaction to generate NOx emissions in the combustion of hydrogen and air [33]. As illustrated in Fig. 11.18, there are various pathways (including N_2-NO-NO_2, N_2-NH-NO-NO_2, N_2−N_2O-NO-NO_2) to oxidize N_2 during the combustion of hydrogen; and to the crucial intermediate NO, its formation has three major step-reactions (including $N + NO \rightarrow N_2 + O$, $NNH + O \rightarrow NH + NO$ and $N_2O + H \rightarrow NH + NO$) and the normalized sensitivity of $N + NO \leftrightharpoons N_2 + O$ is too high to reach about 0.73.

As an alternative fuel to on-road vehicles, NOx emissions should be within stringent emissions regulations. Adopting off-stoichiometric combustion strategies could significantly reduce the amount of NOx generated during hydrogen combustion [34]. Fig. 11.19 illustrates the effects of φ and EGR rate (exhaust gas re-circulation) on the NOx emissions in a port-fuel-injection hydrogen engine with a same operation speed 1000 rpm.

For a hydrogen-fueled engine the conditions of higher (like 50%−100% of the peak value) NOx emissions mainly fall into the lean combustion regime since abundant oxygen is left to take a part in the reaction during the formation of NOx. However, once the fuel concentration has further decreased to a certain level (like $\varphi < 0.7$ for the cases without EGR), the NOx emission will noticeably decline since the flame temperature has been significantly reduced that the reaction activity of NOx formation is corresponding suppressed. With the introduction of EGR, the NOx emissions also can be noticeably reduced, the fuel concentration regime of higher NOx emissions will be narrowed and the critical fuel concentration to the highest NOx emissions is reduced. The phenomena are mainly attributed to the reduction of air whose amount within the cylinder has been occupied by EGR and

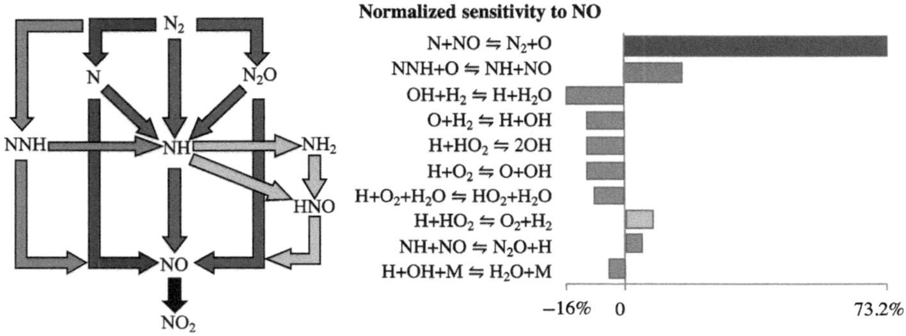

Figure 11.18 Pathways to NOx emissions and the sensitivity of NO during hydrogen combustion.

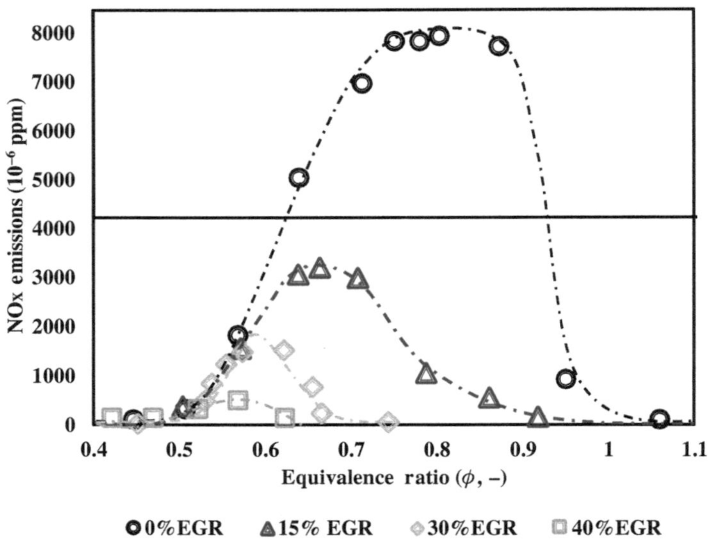

Figure 11.19 Effects of equivalence ratio and EGR on NOx emissions in a PFI-H_2ICE.

the reduction of temperature for the cooling effects of EGR. Albeit ultra-lean combustion and/or the combustion with higher EGR rate can significantly reduce NOx emissions, the maximal explosion pressure and the maximal rise rate of explosion will also be seriously reduced for less reaction and lower flame speed, namely the power performance of hydrogen-fueled engines will be worsened. Therefore adopting rich combustion may be a better strategy to hydrogen-fueled engines for the higher values of $(dp/dt)_{max}$ and lower NOx emissions, which is significantly different to the combustion strategies for fossil-fueled engines.

11.5 Conclusion

Any H-contained materials (like fossil fuels, biomass materials, etc.) can be employed as feedstock to hydrogen gas via thermochemical and/or biological pathways; and thus, hydrogen has been widely regarded as one promising alternative fuel.

Although hydrogen gas can be produced from water by electrolysis and catalytic reactions, the high energy cost and the source of electricity make water-derived hydrogen gas not a promising choice in the current. Gasifying and reforming fossil fuels are mature technologies to hydrogen production, but fossil fuels are still consumed as feedstock. Biomass-derived hydrogen gas will be a promising route for the supplement of hydrogen gas in the future for the independence of fossil resources and the sustainable carbon neutral cycle albeit some crucial issues are urgent to be solved.

The evolution of flame structure is different for hydrogen flames with different fuel concentrations, hydrogen-rich flames can outwardly propagate with smooth flame-fronts under standard conditions, lean flames will become self-wrinkled for the cellularity during the outward propagation and the cellularity will be enhanced with the decrease of equivalence ratio (φ). However, when the fuel concentration reduces to a certain level (like $\varphi = 0.4$), those ultra-lean hydrogen flames will express the tendency of upward propagation. The phenomena are mainly attributed to different mechanisms of instinct flame instabilities of hydrogen, the cellularity of hydrogen-rich flames is determined by D-L instability, that of lean hydrogen flames is dominated by D-T instability, the buoyant behaviors are believed related to the competition between buoyant force and flame stretch, and the existence of external turbulence will enhance the cellularity by the coupling effects. Suffering different mechanisms of flame instabilities and flame stretch, the variations of flame speed are also different according to fuel concentration, and the maximal flame speed are attained at rich ($\varphi = 1.8$) rather than the stoichiometric for hydrogen flames.

The explosion potential of hydrogen and the NOx emissions during hydrogen combustion are two major problems for the promotion of hydrogen gas as fuel. The maximal explosion pressure of hydrogen is attained at the stoichiometric level, but the maximal rise rate of explosion pressure is attained in rich explosion for the faster flame speed. Both are significantly raised with the existence of turbulence for the reduction of accumulated heat loss during explosion and the increase of flame speed. Furthermore, from the viewpoint of deflagration index, stoichiometric and rich hydrogen-air mixtures are very hazardous. For the existence of nitrogen and the high temperature released in the combustion of hydrogen in the air, NOx emissions exist in the hydrogen combustion which makes hydrogen-fueled vehicles still have tail gas. Higher NOx emissions are generated in lean hydrogen combustion for the abundant oxygen, but it can be significantly reduced in ultra-lean combustion (for the lower flame temperature) and rich combustion (for the insufficient oxygen left) and the combustion with EGR (for the both mentioned reasons). Considering ultra-lean combustion and the existence of EGR will also reduce the maximal rise rate of explosion overpressure and flame speed, rich combustion may be a promising strategy for hydrogen-fueled engines.

References

[1] C. Acar, I. Dincer, Comparative assessment of hydrogen production methods from renewable and non-renewable sources, Int. J. Hydrog. Energy 39 (1) (2014) 1–12.

[2] I. Dincer, C. Acar, Innovation in hydrogen production, Int. J. Hydrog. Energy 42 (22) (2017) 14843–14864.

[3] S. Verhelst, Recent progress in the use of hydrogen as a fuel for internal combustion engines, Int. J. Hydrog. Energy 39 (2) (2014) 1071–1085.

[4] G. Cipriani, V. di Dio, F. Genduso, D. la Cascia, R. Liga, R. Miceli, et al., Perspective on hydrogen energy carrier and its automotive applications, Int. J. Hydrog. Energy 39 (16) (2014) 8482–8494.

[5] M. Carmo, D.L. Fritz, J. Mergel, D. Stolten, A comprehensive review on PEM water electrolysis, Int. J. Hydrog. Energy 38 (12) (2013) 4901–4934.

[6] S.S. Kumar, V. Himabindu, Hydrogen production by PEM water electrolysis – a review, Mater. Sci. Energy Technol. 2 (3) (2019) 442–454.

[7] K. Zeng, D. Zhang, Recent progress in alkaline water electrolysis for hydrogen production and applications, Prog. Energy Combust. Sci. 36 (3) (2010) 307–326.

[8] R. Gharari, H. Kazeminejad, N.M. Kojouri, A. Hedayat, A review on hydrogen generation, explosion, and mitigation during severe accidents in light water nuclear reactors, Int. J. Hydrog. Energy 43 (4) (2018) 1939–1965.

[9] G. Perkins, Underground coal gasification – part II: fundamental phenomena and modeling, Prog. Energy Combust. Sci. 67 (2018) 234–274.

[10] M.F. Irfan, M.R. Usman, K. Kusakabe, Coal gasification in CO2 atmosphere and its kinetics since 1948: a brief review, Energy 36 (1) (2011) 12–40.

[11] G. Li, Z. Liu, T. Liu, J. Shan, Y. Fang, Z. Wang, Techno-economic analysis of a coal to hydrogen process based on ash agglomerating fluidized bed gasification, Energy Convers. Manage. 164 (2018) 552–559.

[12] T. Xu, A. Bhattacharya, Direct and two-step gasification behaviour of Victorian brown coals in an entrained flow reactor, Energy Convers. Manage. 195 (2019) 1044–1055.

[13] S.D. Angeli, G. Monteleone, A. Giaconia, A.A. Lemonidou, State-of-the-art catalysts for CH_4 steam reforming at low temperature, Int. J. Hydrog. Energy 39 (5) (2014) 1979–1997.

[14] S. Saeidi, F. Fazlollahi, S. Najari, D. Iranshahi, J.J. Klemeš, L.L. Baxter, Hydrogen production: perspectives, separation with special emphasis on kinetics of WGS reaction: a state-of-the-art review, J. Ind. Eng. Chem. 49 (2017) 1–25.

[15] S. Clarke, A. Dicks, K. Pointon, T. Smith, A. Swann, Catalytic aspects of the steam reforming of hydrocarbons in internal reforming fuel cells, Catal. Today 38 (4) (1997) 411–423.

[16] B.C. Enger, R. Lødeng, A. Holmen, A review of catalytic partial oxidation of methane to synthesis gas with emphasis on reaction mechanisms over transition metal catalysts, Appl. Catal. A-Gen. 346 (1-2) (2008) 1–27.

[17] Y. Liu, R. Lin, Y. Man, J. Ren, Recent developments of hydrogen production from sewage sludge by biological and thermochemical process, Int. J. Hydrog. Energy 44 (36) (2019) 19676–19696.

[18] M. Dębowski, M. Zieliński, A. Grala, M. Dudek, Algae biomass as an alternative substrate in biogas production technologies—review, Renew. Sust. Energy Rev. 27 (2013) 596–604.

[19] R. Łukajtis, I. Hołowacz, K. Kucharska, M. Glinka, P. Rybarczyk, A. Przyjazny, et al., Hydrogen production from biomass using dark fermentation, Renew. Sust. Energy Rev. 91 (2018) 665–694.

[20] M. Anwar, S. Lou, L. Chen, H. Li, Z. Hu, Recent advancement and strategy on biohydrogen production from photosynthetic microalgae, Bioresour. Technol. 292 (2019). No. 121972.

[21] P. Mishra, S. Krishnan, S. Rana, L. Singh, M. Sakinah, Z.A. Wahid, Outlook of fermentative hydrogen production techniques: an overview of dark, photo and integrated dark-photo fermentative approach to biomass, Energy Strateg. Rev 24 (2019) 27–37.

[22] Z.Y. Sun, F.S. Liu, X.C. Bao, X.H. Liu, Research on cellular instabilities in outwardly propagating spherical hydrogen-air flames, Int. J. Hydrog. Energy 37 (9) (2012) 7889–7899.

[23] Z.Y. Sun, G.X. Li, H.M. Li, Y. Zhai, Z.H. Zhou, Buoyant unstable behavior of initially spherical lean hydrogen-air premixed flames, Energies 7 (8) (2014) 4938–4956.
[24] Z.Y. Sun, G.X. Li, Propagation characteristics of laminar spherical flames within homogeneous hydrogen-air mixtures, Energy 116 (2016) 116–127.
[25] Z.Y. Sun, G.X. Li, Propagation speed of wrinkled premixed flames within stoichiometric hydrogen-air mixtures under standard temperature and pressure, Korean J. Chem. Eng. 34 (6) (2017) 1846–01857.
[26] Z.Y. Sun, Structure of turbulent rich hydrogen-air premixed flames, Int. J. Energy Res. 42 (8) (2018) 2845–2858.
[27] Y. Xie, Z.Y. Sun, Effects of the external turbulence on centrally-ignited spherical unstable CH_4/H_2/air flames in the constant-volume combustion bomb, Int. J. Hydrog. Energy 44 (36) (2019) 20452–20461.
[28] Z.Y. Sun, G.X. Li, Turbulence influence on explosion characteristics of stoichiometric and rich hydrogen/air mixtures in a spherical closed vessel, Energy Convers. Manag. 149 (2017) 1526–1535.
[29] Z.Y. Sun, Explosion pressure measurement of 50%H2-50%CO synthesis gas-air mixtures in various turbulent ambient, Combust. Sci. Technol. 190 (6) (2018) 1007–1022.
[30] Z.Y. Sun, Experimental studies on the explosion indices in turbulent stoichiometric H_2/CH_4/air mixtures, Int. J. Hydrog. Energy 44 (1) (2019) 469–476.
[31] Z.Y. Sun, Laminar explosion properties of syngas, Combust. Sci. Technol. 192 (1) (2020) 166–181.
[32] C.K. Law, A. Makino, T.F. Lu, On the off-stoichiometric peaking of adiabatic flame temperature, Combust. Flame 145 (4) (2006) 808–819.
[33] D.S. Ma, Z.Y. Sun, Progress on the studies about NOx emission in PFi-H_2ICE, Int. J. Hydrog. Energy 45 (17) (2020) 10580–10591.
[34] Z.Y. Sun, F.S. Liu, X.H. Liu, B.G. Sun, D.W. Sun, Research and development of hydrogen fueled engines in China, Int. J. Hydrog. Energy 37 (1) (2012) 664–0681.

Section IX

Emerging Technologies (Nanotechnology)

Multifaceted usage of miniaturized energy technologies for sustainable energy harvesting

12

Siddharth Thakur, Ashok Kumar Dasmahapatra and Dipankar Bandyopadhyay
Department of Chemical Engineering, IIT Guwahati, Guwahati, India

12.1 Introduction

The possibility of utilizing fossil fuels to power futuristic technologies have gradually been waning owing to a number of factors, including (1) an upsurge in average temperature across the globe due to fuel burning, (2) assimilation of noxious gases into the environment due to the toxic emissions, and (3) limited feedstocks with respect to ever-increasing demand. Measures taken to mitigate these problems prompted the use of renewable and nonconventional technologies [1,2]. For example solar, hydro, biomass, geothermal, wind, or tidal energy resources are renewable forms that have been tipped as potential supplements to fossil fuels [3]. However, technologies harnessing renewable resources are yet to attain grid parity due to the high-capital cost involved along with issues related to the integration with independent off-grid facilities for eventual power supply [4]. However, every year the energy contribution from renewable energy resources has been growing in the global energy consumption statistics, which is an encouraging sign [4].

Another silent paradigm shift has also taken place wherein the focus of the global market has been shifted to the modern portable engineering systems [5]. The state-of-art, energy-efficient, flat-panel televisions, cellphones, laptops, ipads, or google glasses demand the development of miniaturized high-density power sources [5]. The inventions of power banks—Li–ion, Na–ion, or Li–polymer batteries—are directed toward this end [5,6]. Emergence of tactile or touch sensors, drug-delivery capsules, or portable microelectronic gadgets over the past few decades have also necessitated the development of novel energy harvesting technologies [7–9]. The vision here is that these technologies must not only be capable of harnessing high-density energy from changes in their immediate environment at a lower cost-per-unit basis, but also be able to last for considerable periods of time without maintenance. Thus intensive research in this direction has resulted in a gamut of miniaturized energy technologies (MET) that are capable of harvesting power at a higher energy density from their surroundings [7].

In some ways the previous journey on the technological progress for large-scale (e.g., power plants) and localized (e.g., batteries) energy production ushered in a

Sustainable Fuel Technologies Handbook. DOI: https://doi.org/10.1016/B978-0-12-822989-7.00013-5

few important directions for an energy-efficient and sustainable world. These include (1) strategizing the use of energy resources specific to engineering applications; (2) minimizing energy loss and improving efficiency; (3) integrating a supply chain based on the quality of energy to enable reusability; (4) reversing the concept of localized generation, distribution, and switch to distributed generation and use; and (5) developing a technology-enabled economic model to minimize installation, operating, distribution, and maintenance costs. For example, instead of powering houses through a distribution of networks from a localized thermal power plant, making each house capable of harvesting power to become self-reliant like a TESLA home is such a gen-next solution [10].

For low energy–intensive, modern, miniaturized applications such as actuators, signal emitters, or light-emitting diodes (LED) that can be employed to perform a wide range of tasks in different resource settings such as at workplaces, it may be economically feasible to fulfill their energy requirement quota by the use of novel energy harvesting technologies. Therefore for powering such microelectronic devices a wide variety of MET has been conceptualized and integrated with the specialties of smart materials such as piezoelectricity, triboelectricity, or electromagnetism on the micro- to nanoscale [11]. Independent, real-time uses of these MET technologies or using them as a part of a "symbiotic" group of energy harvesters has also been tested to improve the supply and storage of energy to a specific device.

In the near future, as per a report by the Electronic Devices and Network Annex, an annex of the International Energy Agency, more than 50 billion devices (personal and nonpersonal) are going to be connected to the Internet [11]. The rapid rise of such integration of technologies with the online network has helped improve the utility of such microelectronic devices incorporating a variety of METs. Subsequently this process has laid down the foundation for the generation of a particular field of connected networking medium, namely, the Internet of things (IoT) [12]. Online connectivity among different modules ensures the working and operation of devices off-grid and independently, thereby improving the end-user interfacing of these devices and may eventually render the utility of complex machinery and manual supervision redundant [5,12].

In view of this background, this chapter has been dedicated to give a comprehensive picture of the METs available and their possible applications. First, different sustainable and self-sufficient energy harvesting technologies are introduced alongside a brief discussion of fundamental aspects related to the various types of MET. Thereafter, incorporation of these METs in different scenarios is described wherein various models or types of these devices have been developed and utilized for a large number of applications. Finally, we discuss the future outlook for each of the different energy-harvesting techniques wherein the opportunities for further improvements are (not ***after***) highlighted. Additionally, the nascent usage of IoT has also been discussed in the context of an array of integrated METs that can be utilized in the future to power a network of interconnected devices. A few interesting characteristics of such devices required for powering a network have been presented along with some design considerations as examples.

12.2 Types of miniaturized energy technologies

Energy harvesting by MET from ambient resources depends primarily on the scale of requirement. For example, heartbeat monitors or pacemakers require a few μW of energy to function, whereas LED lighting or actuators for buzzers require a few mW. Thus different techniques for energy harvesting have been designed and developed, tailor-made to cater to different technological settings. MET catering to various scales of energy production are discussed with some of their real-world applications. In particular, we consider four different types of MET that are widely employed: (1) triboelectric, (2) piezoelectric, (3) electromagnetic, and (4) thermoelectric.

12.2.1 Triboelectric energy harvesters

12.2.1.1 Introduction

The effect of triboelectricity has been observed since the early ages, many centuries ago. This phenomenon is a combination of tribology and charge transfer between the surfaces [13,14]. It is now well-known and documented that the process of charge generation occurs between different types of interfaces during friction, involving transfer of electrons [14,15]. This phenomenon happens when the interatomic distance between the two different materials is smaller than the equilibrium distance wherein a strong electron cloud overlap facilitates the transfer of charges. However, for liquid-phase interactions, the ion transfer process takes over after the first layer of charge transfer has occurred. A significant amount of charge transfer can be made possible between solid surfaces depending upon their selection based on the triboelectric series [16]. Thus this ubiquitous phenomenon can be extensively applied in different environments for recovering a large amount of energy from common activities.

A prime example can be found in the domestic setting where physical activities such as walking or running, or even while driving a car, can be utilized in a suitably modified manner to harvest energy employing this phenomenon. Interestingly, triboelectric nanogenerators (TENG) are facile, low-cost, easy-to-fabricate, and industrially scalable and can be deployed in different environments [1,7]. Broadly speaking there are four types of working modes for these devices: vertical contact-separation, lateral-sliding, single-electrode, and freestanding triboelectric layer modes, as shown in Fig. 12.1. The different modes can be operated as independently working units or can be employed in combination in order to enhance the durability and output capabilities of the device. This chapter highlights the different aspects of TENG phenomenon alongside highlighting a few fundamentals of this energy-harvesting technique [13,17].

12.2.1.2 Applications

The first energy harvesting application utilizing this principle in the form of TENG was shown in 2012, wherein the device construction consisted of two different

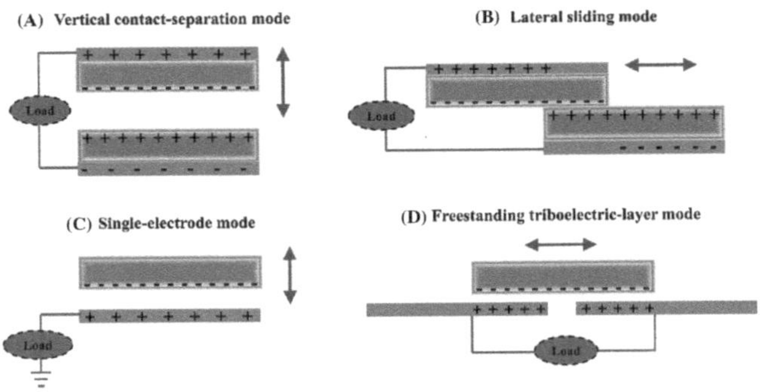

Figure 12.1 Schematic diagram of the different modes of triboelectric nanogenerators. (A) Vertical contact-separation mode, (B) lateral-sliding mode, (C) single-electrode mode, and (D) freestanding triboelectric-layer mode.

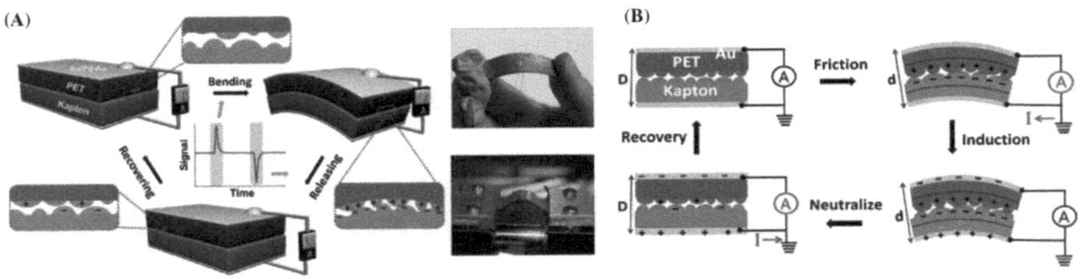

Figure 12.2 (A) Schematic diagram of the different stages of bending and releasing of a triboelectric nanogenerator. (B) Charge transfer schematic as a result of the bending and releasing processes.
Source: Reprinted from F.R. Fan, Z.Q. Tian, Z. Lin Wang, Flexible triboelectric generator. Nano Energy 1 (2012) 328–334. Copyright 2012, Elsevier.

polymer layers bound by two metal electrodes. The device architecture is represented in Fig. 12.2 [13]. Bending the entire structure resulted in development of friction between the two interfaces, which resulted in the formation of a collection of dipoles. Subsequently, the induced opposite charges at the respective electrodes facilitated charge transfer in the external circuit to reduce the potential build-up at the interface. This work showed that the potential developed as a result of simultaneous bending and leaving was capable of generating an output voltage of 3.3 V at a power density of 10.3 mW/cm^2. Incorporation of such cost-effective materials for the purpose of energy harvesting led to the inception of a new form of technology, and intense research has been witnessed in this field ever since. Different techniques have been tried in order to enhance the output characteristics of the device. Primarily, three factors that have been studied are surface roughness, material composition, and environmental effect. The procedure that has generally been followed to enhance the performance characteristics of a device includes the introduction of micro- or nanostructures on the materials' surface. This is chiefly done to increase the interfacial contact area between the layers, which results in increased surface-charge density.

Another approach involves the introduction of a liquid layer, which acts as a triboelectric layer and as an electrode [18,19]. This technique overcomes the problem of improper contact between solid—solid layers to a large extent by utilizing a rubber encasing filled with a conducting salt-water solution [19]. The surface of the rubber was etched to create nanorods in order to enhance the triboelectric effect. When the salt solution comes in contact with the rubber surface, due to the higher affinity of electrons in the rubber, the charges are transferred to the rubber surface because of triboelectric effect. Thus as the positively charged material approaches the rubber, the electric potential increases significantly. This effect also initiates the movement of electrons from the ground to the positively charged material under a short-circuit condition. The electric field potential developed attains a maximum value when the relative motion between the surfaces stops. Thereafter as the liquid material moves away from the surface, the electric potential reduces because the negative charges move back to the ground. Finally, the triboelectric charges again attaining the initial value when the system reaches its base configuration. This entire movement constitutes one cycle and thus an overall alternating current cycle has been generated due to the back-and-forth movement of the charges. The device was able to perform without any noticeable difference in output parameters even after withstanding ~55k cycles of deformation and could withstand tensile strain up to 300%. Materials composition has also been tested to select the best possible combination from the triboelectric series to enhance the charge trapping property. Under testing, at a load resistance of 300 MΩ, the device was able to produce ~225 mW/m^2 at a frequency of 3 Hz. Even with the use of distilled water, the device was capable of generating a decent output without electrolysis due to the low-potential difference developed across the electrodes. The device could be applied under different situations due to its deformability and stretchability.

The incorporation of high-dielectric materials such as $BaTiO_3$ (BTO) have also been studied, and have significantly improved the device characteristics. Fig. 12.3 is a schematic diagram showing the working principle of a novel design approach for construction of a high-energy density TENG [20]. The study utilized poled poly (vinylidenefluoride-*co*-polyethylene) [P(VDF-TrFE)] materials to enhance the charge-attraction property and BTO to improve the charge-trapping property. This study reported a very high output potential of 1130 V and current of 1.5 mA. Other modifications that have been undertaken to improve the output performance parameters of these devices include the surface enhancement of the materials by chemically treating the different interacting layers. For example, the usage of thiols and silanes to modify the Au electrode layer and SiO_2 dielectric layer. However, the best device performance was achieved via introduction of triboelectrically positive $-NH_2$ groups while use of triboelectrically negative $-Cl$ groups could not live up to expectations [21].

The atmospheric conditions are also found to strongly influence the working of the TENG device. For example, Lu et al., reported the effect of temperature by changing the operating temperature from −20°C to 20°C in the presence of their

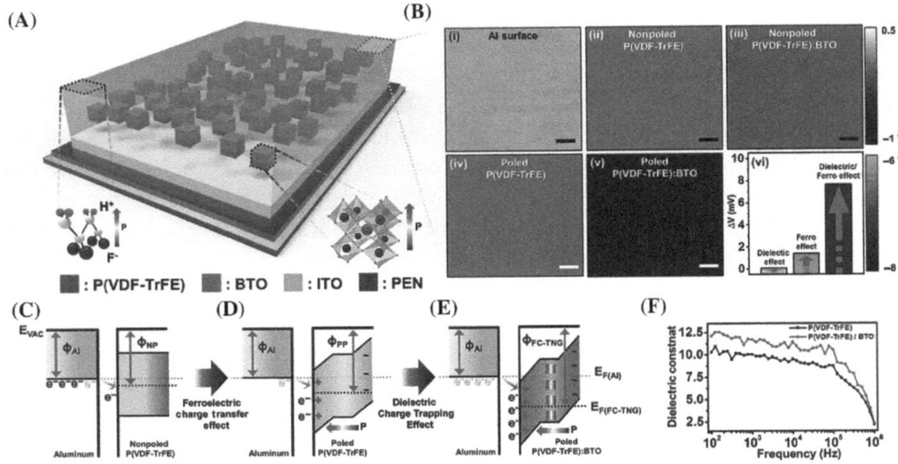

Figure 12.3 (A) Schematic diagram of the device composed of ferroelectric composite. (B) Kelvin force microscopy study of the different materials employed to fabricate the high-performing triboelectric device. (C–E) Band diagrams of the different materials depicting the significant enhancement in charge transfer upon dielectric and BTO additions. (F) Dielectric material characterization of the ferroelectric composite with, and without, the dielectric component.

Source: Reprinted from W. Seung, H.J. Yoon, T.Y. Kim, H. Ryu, J. Kim, J.H. Lee, et al. Boosting power-generating performance of triboelectric nanogenerators via artificial control of ferroelectric polarization and dielectric properties. Adv. Energy Mater. 7 (2017) 1600988–1600995. Copyright 2016, John Wiley and Sons.

device that was constructed utilizing Al and PTFE materials in a single-electrode mode configuration. The device displayed a reduction in performance values during this cycle although it gave a relatively stable output from 20°C to 100°C [22]. Thereafter a sharp reduction in output values with increasing temperature was noted. Higher temperatures facilitated faster reaction between the ambient oxygen groups and dangling bonds of carbon atoms in PTFE sheet, which reduced the electronegative character of the material. Moreover, inclusion of oxygen atoms also led to a reduction in relative permittivity of the material thereby weakening its charge-storage capabilities.

Different examples of TENG devices have found utility in applications ranging from wastewater flow to smart textiles to hearing aids [23–25]. A study on the incorporation of TENG-powered devices with clothing was reported by Tian et al. [23] The performance characteristics of the device has been illustrated in Fig. 12.4. Silicon and Ni–coated polyesters were the two triboelectric materials chosen to harness the human motion, which resulted in the transfer of charges from the skin to the textile material. The output voltage of 500 V and current of 60 μA could be attained via use of a single-layer triboelectric textile. The authors also reported that a double-layered triboelectric device with dimensions of $5 \times 5\ cm^2$ was able to

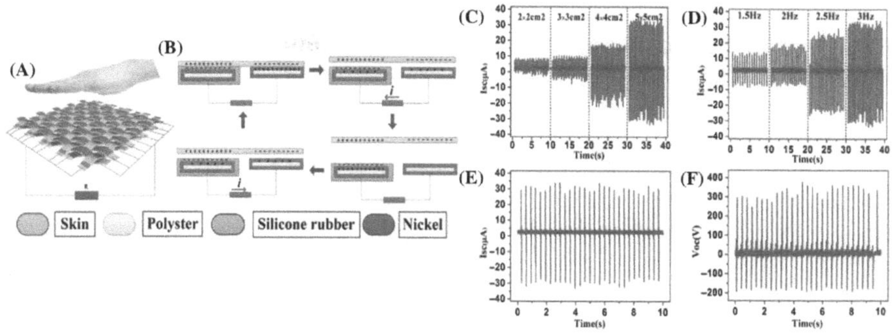

Figure 12.4 (A) Overall schematic diagram of a textile-based triboelectric nanogenerator. (B) Charge-transfer mechanism during the different stages of operation. (C–F) Output-device characteristics for the triboelectric device wherein short-circuit current density was observed as a function of time for different areas of devices, different input frequency of operation cycles, and at a fixed condition. Moreover, open-circuit voltage was also observed as a function of time.
Source: Reprinted from Z. Tian, J. He, X. Chen, Z. Zhang, T. Wen, C. Zhai, et al. Performance-boosted triboelectric textile for harvesting human motion energy. Nano Energy 39 (2017) 562–570. Copyright 2017, Elsevier.

generate an output voltage of 540 V and short-circuit current density of 140 μA. Thus incorporation of TENG devices into accessories utilized on a daily basis brings into focus an exciting field of technology applicable to the development of wearables-based, energy-harvesting devices.

In order to establish the versatility of TENG devices, researchers have also devised techniques to harvest ambient energies such as from wind and tides on a large scale. The concept of blue energy was put forward by Yi et al. [26] wherein techniques to harness the random low-frequency water flow energy in random direction along with the kinetic energy of the water waves was introduced. In order to demonstrate the resourcefulness of the TENG devices, researchers reported the recovery of waste energy from the environment, particularly from kinetic motion of the automobiles. Chang et al. [27] performed a study wherein they utilized a Al-PTFE designed device for extracting waste energy produced during the rotation and braking of the tires. Their model was able to generate a significant current density of 43 mA/cm^2 and output voltage >500 V for the braking case while 1.5 mA/cm^2 and 65 V could be achieved for the noncontact rotating case. The values of the output device were large enough to power small-scale devices such as LED arrays at various locations within the vehicle.

The field of TENG devices has also made significant strides in the field of human–machine interface systems by utilizing the different limbic movements to generate electrical signals producing desired electrical values. In one particular study by Pu et al., eye movement was utilized in order to generate tailor-made control commands [28]. The multilayered TENG device consisted of FEP and natural latex as triboelectric layers and employed a single-electrode TENG configuration via the usage of the ITO-coated electrode on the FEP side. During blinking, the movement from the eye muscles forces the separation distance to vary between the FEP and latex layers (the latex layer is attached to the skin besides the eye region). As a result a potential is developed and by integrating the device to a circuit a smart home-control system was developed. By suitably modifying the circuit and amplifying the output result, this hands-free technique has been utilized to operate electrical appliances such as a fan, doorbell, and table lamp. The group further designed a circuit for predetermined signals, such that the eye blinking movement could be utilized to type letters thereby offering a unique and convenient way of communication.

Environmental remediations have also been reported by several groups wherein they have utilized charge electrodes for the successful capture of particulate matter. Han et al., have proposed the development of a high-voltage TENG device by employing PTFE pellets between Al electrodes [29]. The schematic diagram along with the different device performance characteristics is shown in Fig. 12.5. The charges are distributed over the two electrodes because of the surface contact and polarization due to the movement of the highly-negatively charged pellets. The output values could reach as much as 6 kV near the bottom or top electrodes during the synchronous motion of the pellets, while their asynchronous motion could only generate open-circuit potential values close to 170 V. Hence the motion of the pellets made a significant change in the output parameters.

12.2.1.3 Future perspectives

The field of triboelectricity has significantly influenced MET technology over the past few years. It possesses versatile characteristics that makes it useful for the deployment in a number of different settings from homes to industrial locations. In particular, TENGs have been shown to be effective for converting low-frequency random energy dispersed in our environment. They have the ability to extract energy on different scales from waves in the ocean to mechanical displacement of machineries in industries to biomechanical movement in humans. Since the working principle is based on displacement current and the output voltage is independent of the drive frequency, this technology is better suited to harness energy from slow, arbitrarily moving structures. Further, owing to its nano- or microscale design, these devices can be operated at remote locations, working independently, and yet can be connected to form a network to constitute a self-powered system.

In order to enable these devices to have wider acceptance and achieve commercialization in the coming decades, it is important to work on selected targeted areas that can significantly improve its odds of developing into a mainstream technology [30]. Primarily, the theoretical understanding concerning this field of technology needs to be developed further not only to improve the fundamentals at the interface of the two triboelectric materials, but also to explain the different interactions occurring at the fluid–solid, solid–solid, or liquid–liquid interfaces. Although a noteworthy amount of work regarding this has been established [1,8,26,30], there is a wide scope for researching such a variety of TENG systems. On the other hand,

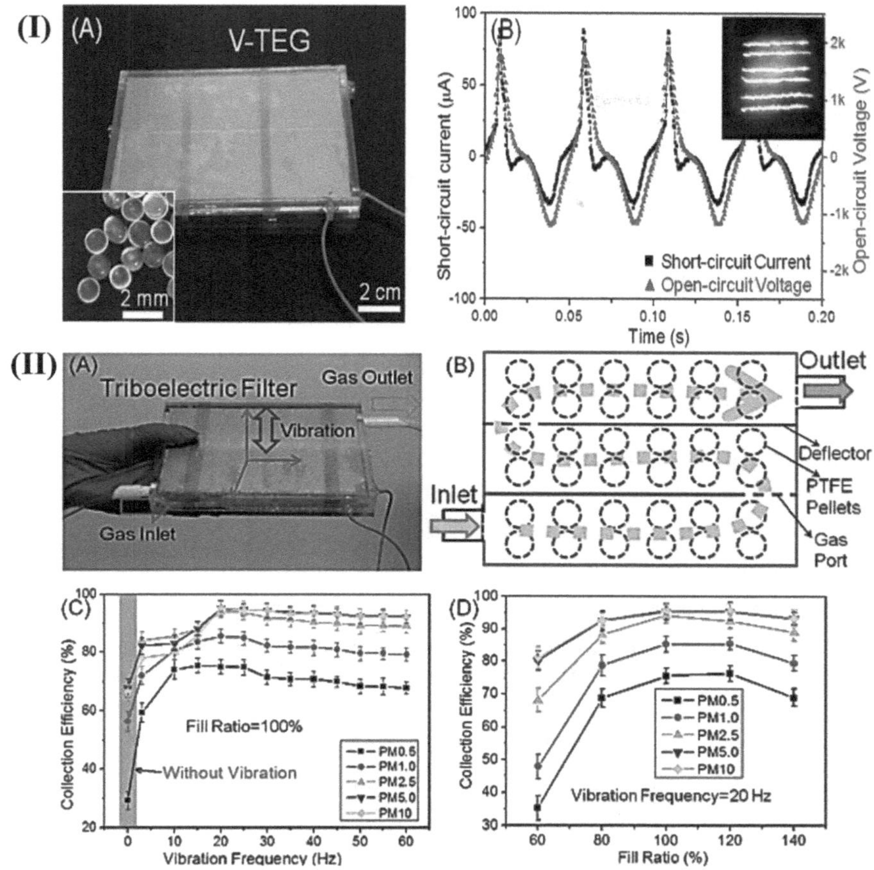

Figure 12.5 (I) (A) Picture of the vibrations based triboelectric device. Inset are the PTFE pellets. (B) Short-circuit current and open-circuit voltage values at 20 Hz vibration. Inset shows the lighting up of 150 LED's. (II) (A) Picture of the triboelectric device–based filter. (B) Schematic diagram of the working of the device. (C, D) Variation of collection efficiencies without and with varying vibration frequencies for different particle size and fill ratios.

Source: Adapted from C.B. Han, T. Jiang, C. Zhang, X. Li, C. Zhang, X. Cao, et al. Removal of particulate matter emissions from a vehicle using a self-powered triboelectric filter. ACS Nano 9 (2015) 12552–12561 Copyright 2015, ACS.

the improvement in the design and power management concerning their deployment of devices in harsh environments such as ocean waves need to be researched further. In order to enhance the output parameters of such systems, such as the overall voltage and power density, studies need to be undertaken to identify different materials and their combinations that possess suitable dielectric properties to generate the highest surface-charge densities. Lastly, the surface modifications of the triboelectric layers can also be performed in order to enhance the surface roughness via different cost-effective techniques such as chemical etching and soft lithography [7].

12.2.2 Piezoelectric energy harvesters

12.2.2.1 Introduction

In 1880 Jacques and Pierre Curie discovered the effect of piezoelectricity while measuring the surface charges on quartz crystals, Rochelle salt, and topaz when subjected to mechanical stress. Two years later, they were involved in the discovery of the reverse piezoelectric effect whereby they observed small deformations in the crystal structure in the presence of an external electric field. The former phenomenon has been largely incorporated as a kinetic-energy harvesting source. The piezoelectric materials are available in different structural phases such as single crystals (e.g., quartz), polycrystalline (e.g., PZT), thin film (sputtered or chemically grown ZnO), and polymeric forms (e.g., PVDF) [31].

The basic phenomenon of piezoelectricity can be explained with the help of Fig. 12.6A and B wherein insulator piezoelectric material is shown bound by electrodes on both ends. Upon application of mechanical stress, the piezoelectric polarization charges at the electrodes are generated, which increases with an increase in applied stress. In order to balance the potential developed at the electrodes, there is electron flow through the load in the external circuit. This results in the conversion of mechanical energy into electrical energy. In Fig. 12.6, ε represents permittivity of the material, $\sigma_{ps}(z)$ represents the piezoelectric polarization charges on the surface, which is a function of materials thickness, and $\sigma(t)$ represents the charge density of free electrons on the electrode [15]. Thus the displacement current density

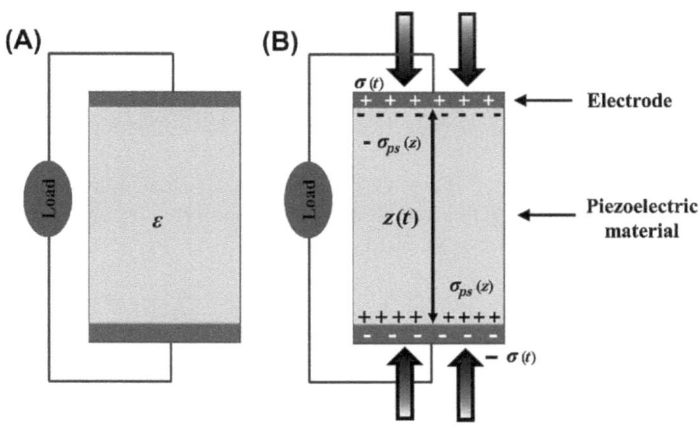

Figure 12.6 Schematic diagram representing piezoelectric phenomena under (A) no stress and (B) mechanically-applied stress situations.

(J_d)—that is, output current density—observed while using piezoelectric devices is related to the rate of change of surface polarization charges as given by Eq. (12.1).

$$J_d = \frac{\partial \sigma_{ps}(z)}{\partial t}, \tag{12.1}$$

where z represents the specimen thickness, while the magnitude of open-circuit voltage (V_{oc}) is given by Eq. (12.2).

$$V_{oc} = \frac{z\sigma_{ps}(z)}{\varepsilon}. \tag{12.2}$$

Therefore considering an external load R in the circuit, the overall relationship can be shown by Eq. (12.3).

$$RA\frac{d\sigma}{dt} = \frac{z\left[\sigma_{ps}(z) - \sigma(z)\right]}{\varepsilon}, \tag{12.3}$$

where A represents the area of the electrode. The output performance characteristics can also be found out using Eq. (12.3).

12.2.2.2 Applications

The earlier piezoelectric devices consisted of PZT material as the load-bearing medium; however, it had limited usage owing to its brittle nature which restricted the amount of strain it could withstand. Its usage in atomic force microscopy (AFM) was reported by Kong et al. [32] where different PZT-based configurations were utilized to create tips for scanning surface textures while having a resolution less than ~10 nm. The AFM technique has been extensively employed to characterize different piezoelectric materials down to a very small scale. In a study reported by Beach et al. [33], AFM was used to determine the impact of doping on the III-nitride heterostructures with regards to the piezoelectric field and charge distribution in strained layers. This facilitated the pathway for testing different types of arrangements that were then deployed to test the best possible geometries for piezoelectric materials utilization. Glynne-Jones et al. [34] developed a cantilever structure with piezo materials adhered to the top and bottom surfaces.

The cantilever structure worked preferably in the bending mode at low-resonant frequencies, thereby constantly applying strain to the piezo surfaces. The group used a tapered structure for the beam thereby providing it with a constant strain along the length for a given displacement. The piezo material consisted of PZT-5H powder blend, which was screen printed onto steel plate. During operation, the structure operated at a frequency of ~80 Hz that produced up to 3 μW of power into a load of 333 kΩ. Marzencki et al., performed a simulation study on a micromachined MEMS version of silicon cantilever 2 mm × 2 mm in dimensions, as shown in Fig. 12.7 [35]. Therein the 5 μm thick silicon layer which constituted the

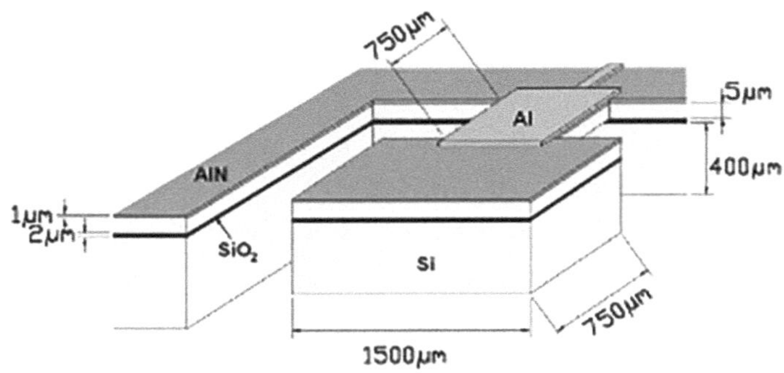

Figure 12.7 Schematic diagram representing the MEMS model undertaken to perform a simulation study for designing optimal output characteristics.
Source: Reprinted with permission from S. Basrour, B. Charlot, M. Marzencki, A. Grasso, M. Colin, L. Valbin, Design and fabrication of piezoelectric micro power generators for autonomous microsystems, in: Symposium on Design, Test, Integration and Packaging of MEMS/MOEMS (DTIP'05), 2005, pp. 99–302. ⟨hal-00009215⟩.

top surface, was utilized as the cantilever structure with a length of 750 μm. Simulation studies were also performed at 900 Hz resonant frequency which predicted an output voltage of ~100 nW and ~600 nW for AlN and PZT devices.

Studies in the literature also report on the combined effects of piezoelectric materials when incorporated in bundles of fibers. Mohammadi et al. [36] reported on the development of epoxy-based piezoelectric systems wherein PZT fibers of different diameters (15–250 μm) were aligned and molded into an epoxy, as shown in Fig. 12.8. A number of samples of dimensions 34 mm × 11 mm × (1.2–5.8 mm) were cut to form composites wherein fiber orientation was aligned with plate thickness direction. Performance tests were conducted by dropping 33.5 gm weight on the prepared samples. It was observed that specimens with smaller fiber diameters had the highest piezoelectric coefficient and smallest dielectric constant. A maximum voltage of 350 V and power of 120 mW was produced when the specimen employed was 5.85 mm thick with the fiber diameter of ~15 μm.

One-dimensional (1D) nanomaterials (e.g., nanowires and nanorods) have been the preferred choice of piezoelectric structures utilized in devices over the past few decades. Employing these structures in devices for kinetic or mechanical energy harvesting has certain advantages associated with it. Such as the generation of significant internal strain upon the application of an external stress, that results in enhanced piezoelectric properties due to flexoelectric effect [1,37]. Utilizing 1D nanomaterials as the base structure results in enhanced crystal alignment and lattice configuration that leads to superior flexibility, strain-loading capacity, and operational lifetime. Since the dimensions of the piezo elements are very small, the sensitivity of the device increases and allows a higher strain for a very small mechanical load in the range of nN or pN. Song et al. [38] reported an AFM study on the employment of 1D piezoelectric ZnO nanowires grown on a sapphire substrate. The AFM tip was composed of Pt coated Si operated in contact mode for the analysis.

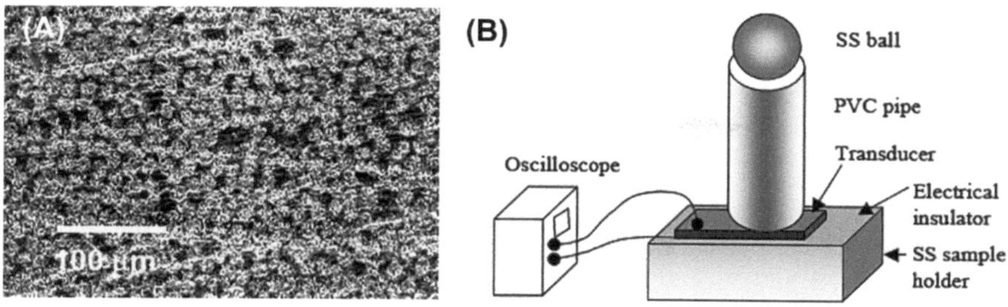

Figure 12.8 (A) PZT based composite microfiber distribution. (B) Schematic diagram of the impact-actuated, energy-harvesting setup.
Source: Reprinted from F. Mohammadi, A. Khan, R.B. Cass, Power generation from piezoelectric lead zirconate titanate fiber composites. Mater. Res. Soc. Symp. Proc. 736 (2002) 263–268. Copyright 2002, Cambridge University Press.

As the tip scanned over the surface, there was a deflection of the nanowires that experienced a positive strain on the outer surface (under tension) and a negative strain on the inner surface (under compression). Sharp output in the range of mV was observed during the AFM scanning and was attributed to the combined effects of piezoelectricity and semi-conducting nature of ZnO nanowires. Primarily, when the Pt tip was in contact with the ZnO nanowire, the tensile stresses within the piezo created a positive potential on the outer ZnO surface and resulted in a negatively biased metal-tip interface (as the Pt tip was at a nearly zero potential). This caused the formation of a reverse-biased Schottky diode, which led to the charge accumulation. Subsequently, when the AFM tip was in contact with the compressed side of ZnO nanowire, the metal-tip interface created a forward-biased Schottky diode that resulted in the transfer of charges from the nanowire to the AFM tip and gave an enhanced overall signal during the AFM scan. The different stages representing the various interactions are shown in Fig. 12.9.

In order to demonstrate the capability of nanowires for powering practical devices, Hu et al. [39] reported the development of a rational unipolar assembly of ZnO nanowires in a PMMA matrix for powering a liquid crystal display. They deposited multiple layers of conical ZnO nanowires with alternate layers of PMMA on a kapton tape and capped both ends with Cr/Au metallic electrodes. The conically shaped nanowires were deposited normal to the substrate. Thus the projected component of all the nanowires normal to the substrate added up along the vertical direction. This resulted in the development of significant piezoelectric polarization potential across the substrate. The device characteristics after five deposition cycles of conical ZnO nanowires and PMMA layers resulted in an output voltage of 1.5 V and current $\sim$35 nA, which was enough to power the small LCD display of a calculator.

In order to develop devices that could interface better with human beings and assist in operating self-powered devices, Yang et al., demonstrated the use of ZnO nanowires on flexible substrates and thereby demonstrated a potential avenue for harvesting energy from biomechanical movements [40]. The ZnO nanowires were

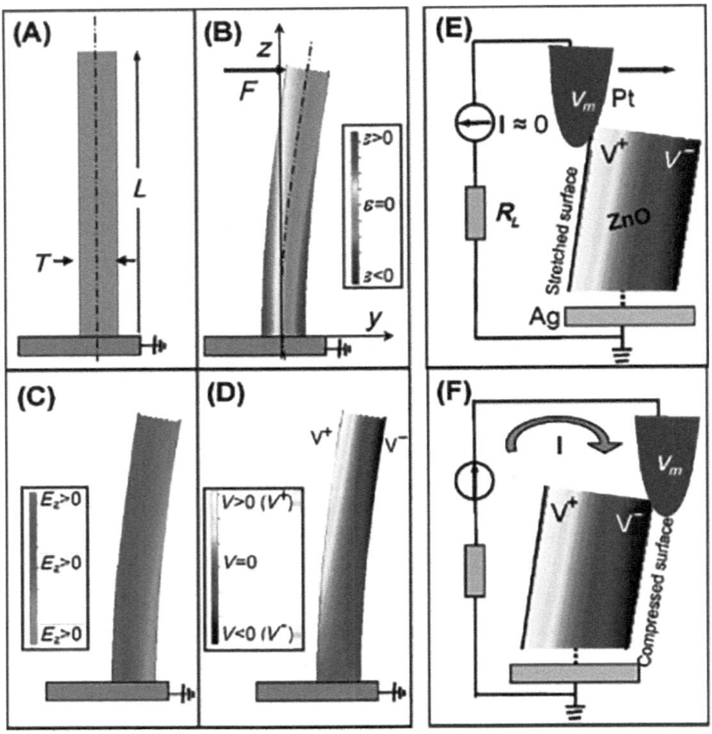

Figure 12.9 Schematic diagram representing the different stages of AFM tip interaction with the ZnO nanowire and subsequent build-up and release of charges in an external circuit. (A) Diagram representing the co-ordinate system followed in the study. (B) Occurrence of longitudinal strain in nanowire upon interaction with AFM tip. (C) Corresponding piezoelectric field distribution within the bent nanowire. (D) Positive and negative potential distribution in the nanowire, at two ends. (E) Presence of a Schottky barrier at the metal–semiconductor interface prevents the flow of charges and leads to charge accumulation. (F) Forward-biased Schottky barrier leads to charge flow across the interface and reduces the built-in potential with subsequent tip movement.
Source: Reprinted from J. Song, J. Zhou, Z.L. Wang, Piezoelectric and semi-conducting coupled power generating process of a single ZnO belt/wire: A technology for harvesting electricity from the environment, Nano Lett 6 (2006) 1656–1662. Copyright 2006, ACS.

placed between two electrodes on a flexible substrate. Upon bending the substrate, a uniaxial strain developed in the nanowires which caused the formation of piezoelectric potential that drove the charges to flow in the external circuit. Cyclic bending and release of the device resulted in a pulsed alternating current signal having an open-circuit voltage value of ~20–50 mV and short circuit current ~400–750 pA. In order to enhance the output characteristics of the device, large-scale integration of ~20,000 nanowires were performed in around 700 rows on a flexible substrate which resulted in a voltage output of ~1.3 V and current output of 28.8 nA. Fig. 12.10 shows the device configuration and output characteristics.

In another versatile application of piezoelectric systems, development of self-powered chemical sensors that could be deployed in harsh environmental conditions was reported. In a study performed by Lee et al., the group demonstrated the utility of employing a flexible ZnO nanowire-based nanogenerator wherein the piezo element was attached to a supercapacitor via a rectifier, and the output of the supercapacitor was utilized to power a carbon nanotube (CNT) based Hg^{2+} ion sensor [41]. The piezoelectric nanogenerator was composed on flexible kapton substrates on which, at one end, ZnO nanowires were grown using a hydrothermal technique and, on the other end, the Au electrode was coated. The kapton substrate was utilized because being flexible in nature it could follow the nonuniform, chemically grown ZnO nanowire morphology so that effective contacts could be maximized between the two materials [41,42]. Additionally it was also ensured that nanowire growth occurred along the polar direction of the multiple stacked layers. The device characteristics for a single-layered device with a dimension of 0.8 cm^2 was represented by an output voltage of ~350 mV and a current density of 125 nA/cm^2. Stacking multiple layers (e.g., 10) helped increase the output values to around

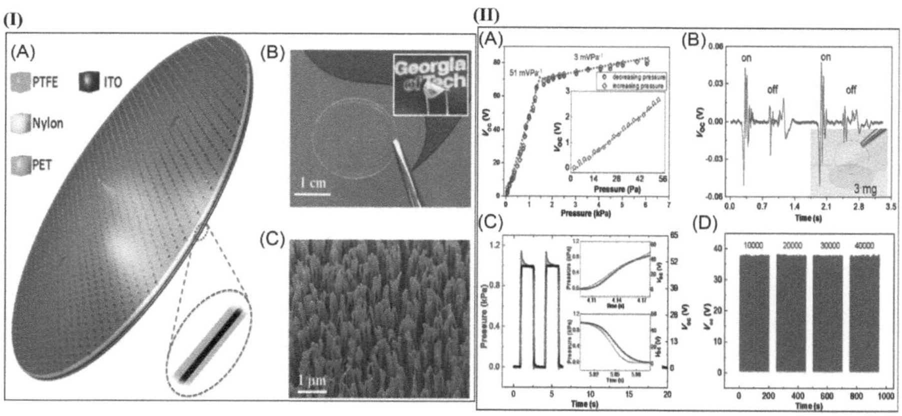

Figure 12.10 I (A) Schematic design of the bionic membrane sensor. (B) Image of as-fabricated piezo device. (C) SEM image of surface-etched PTFE nanowires. Scale bar is 1 μm. (II) Electro-mechanical analysis of the piezo device. (A) Effect of external pressure variation on the output voltage of the piezoelectric device. (B) Output voltage variation upon the placement and removal of human hair. (C) Periodic pressure application upon the device to determine its response time. Inset shows zoomed-in views of loading and unloading process for determination of response time. (D) Tests were undertaken to determine the mechanical stability of the device; 0.8V kPa applied for 40000 cycles at 2 Hz frequency resulted in a negligible change in output variation.
Source: Reprinted from J. Yang, J. Chen, Y. Su, Q. Jing, Z. Li, F. Yi, et al. Eardrum-inspired active sensors for self-powered cardiovascular system characterization and throat-attached anti-interference voice recognition. Adv. Mater. 27 (2015) 1316–1326. Copyright 2015, John Wiley and Sons.

2.1 V and 105 nA. This output was then used to charge a supercapacitor composed of multiwalled CNT-ITO electrodes and PVA/H_3PO_4 electrolyte. The supercapacitor was then eventually utilized to form a part of a self-powered, sustainable circuit designed to detect Hg^{2+} ions in the environment.

Deployment of devices in locations that are largely inaccessible requires powering mechanisms that can be operated remotely. Hence efforts have been concentrated toward devising novel piezoelectric systems which incorporate remote actuators. In one such study, Cui et al. [43] reported on a contactless piezoelectric device wherein magnets were employed to initiate the charge-transfer process. The top glass plate was initially coated with a 50 μm strip of Ag, onto which a ZnO microwire of length 500 μm was placed. Thereafter, a photoresist layer was spun casted onto it which after baking, fixed the wire in its location. Bottom glass plate was covered with a Pd electrode 50 μm in width. The two glass plates together formed a sandwich structure wherein the ZnO formed a Schottky barrier with the Pd metal. A magnetic composite cap was put on the free end of the ZnO microwire as it was dipped in a solution of Fe_3O_4 and paraffin. Fig. 12.11 shows the device representation along with the output characteristics. The approaching bar magnet induced a bending in the ZnO microwire wherein the strain generated caused the formation of piezo potential. The charges accumulated at the metal–semiconductor interface due to the Schottky barrier and later, as the magnet receded, the charges were allowed to flow in the external circuit. The forward and backward motions of the magnet produced output voltage, current values of ~30 μV, 12 pA, and 54 μV, 26 pA, respectively.

Besides the commonly utilized ZnO, the gamut of other piezoelectric materials utilized for designing flexible piezoelectric nanogenerators include PZT [44,45], $BaTiO_3$ [46,47], and PVDF [48,49]. For example, Park et al. [50] reported the

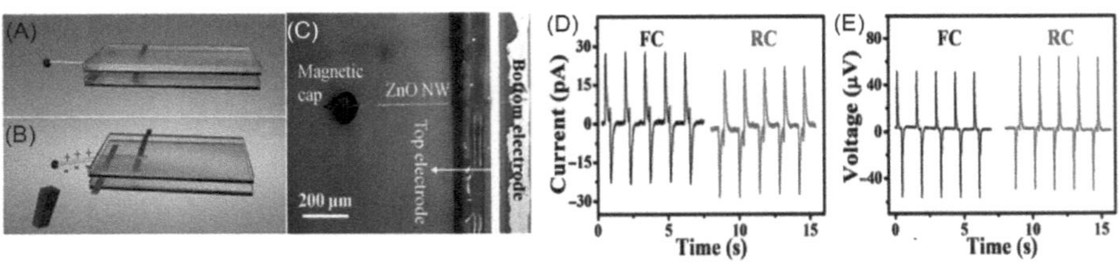

Figure 12.11 (A) Schematic diagram of the overall device. (B) Working mechanism of the piezoelectric device wherein the ZnO microwire is capped with a magnetic material. The cap undergoes movement under the influence of magnetic field which generates the piezo potential within the ZnO structure due to applied stress. (C) Picture of the overall device. (D, E) Output current and voltage characteristics of the device under forward connection (FC) and reverse connection (RC). FC means positive end of measurement probe connected with positive end of piezo device and RC means negative end of measurement probe connected with negative end of piezo device.
Source: Reprinted with permission from N. Cui, W. Wu, Y. Zhao, S. Bai, L. Meng, Y. Qin, et al. Magnetic force driven nanogenerators as a noncontact energy harvester and sensor. Nano Lett. 12 (2012) 3701–3705. Copyright 2012, ACS.

development of a flexible piezoelectric device composed of a perovskite $BaTiO_3$ film. The piezo element was sputter coated onto a $Pt/Ti/SiO_2/Si$ substrate and was subsequently made to undergo a poling process in order to align the different domains, at a field strength of 100 kV/cm. The metal-piezo (insulator)-metal sandwich structure was then transferred onto the kapton substrate and connected by interdigitated electrodes. Upon the application of a mechanical stress onto the flexible nanogenerator, an output voltage of 1 V and a current of 0.19 $\mu A/cm^2$ could be obtained.

Maintaining environmental awareness in work procedures also encouraged different groups to incorporate utilization of alkaline niobates, particularly sodium and potassium, as piezoelectric elements and hence they gained further importance during the period 2010–2015 and beyond. For example, Kang et al., formed Mn-doped [K, Na] NbO_3 fibers by electrospinning them in polyvinylpyrrolidone and annealing them at a very high temperature, that is ~800°C. They were then implanted in a PDMS layer and deposited on a polyether sulfone substrate where they were attached to it using Pt interdigitated electrodes. After the device was subjected to mechanical stresses, the open circuit voltage and short circuit current observed were ~0.3 V and 50 nA [51]. In another study that resulted in the development of high-performance piezo systems, Jeong et al. [52] developed a piezoelectric nanogenerator based on alkaline niobate–based particles. KNN-based alkaline niobate particles are lead-free piezo materials having high Curie temperature, electromechanical coupling factor, and piezoelectric coefficient besides being biocompatible in nature. The device construction included the use of ITO electrodes as current collectors and PDMS layers as the barrier against excessive mechanical deformation and layer breakdown during poling. During the application of mechanical stresses on the device, the embedded alkaline niobate particles experienced a strain as a part of the 250 μm thick composite layer, generating the potential gradient along a particular direction. This built-in potential caused the movement of electrons in the external circuit in a particular direction. The device was able to generate 12 V output voltage and 1.2 μA output current after undergoing a poling process at 150°C and 2 kV electric field strength.

12.2.2.3 Future perspectives

Conversion of mechanical energy into the electrical energy can be a fruitful technique and a potential gamechanger in the field of micro- and nanoscale devices. There are an abundance of examples in the environment wherein vibration energy generated under different situations have not been harvested effectively. Encouragingly, studies in the literature have demonstrated potential sources in the form of biomechanical locomotion, air flow, or vehicle movements as they usually generate random and low-frequency energy. Piezoelectric devices can play a massive role in filling the void in terms of standalone, environment-friendly, durable, affordable, and portable units. Such devices can also be deployed in remote and access-restricted areas while being connected to each other on large-scale, through, wireless networks. However, certain issues concerning their use needs to be addressed, which may lead to their limited use for on-field operations [1].

First, it should have the capability of being more sensitive to the external vibrations so that slight movements such as body pulse or breathing can be captured to generate meaningful output.

Second, utilization of biomaterials toward this end needs to be researched so that one could move toward construction of a fully biodegradable device with a minimal carbon-footprint.

Third, theoretical studies concerning the piezo potential development in soft biological structures needs to be undertaken so that its potential can be truly utilized [53].

Fourth, since the power output levels of this technology is low, that is, $\sim 1-7$ μW, studies have to be performed in order to enhance the output so that the devices cater to a diverse range of applications.

Finally, in order to gain a larger footprint among conventional green technologies, large-scale, cost-effective techniques for producing piezoelectric devices commercially need to be optimally designed.

12.2.3 Electromagnetic energy harvesters

12.2.3.1 Introduction

Electromagnetic-energy harvesting largely follows Michael Faraday's law of electromagnetic induction. The conversion of mechanical or kinetic energy into the electrical one takes place due to the movement of a conductor (coil) in a nonparallel direction to the external magnetic field or movement of the external magnetic field around a stationary conductor, in a closed loop. Fig. 12.12 is a schematic of one type of electromagnetic energy–harvesting device. Different studies have presented theories on designing the ideal systems focusing on factors such as narrow response frequency band, potential mismatch of resonant frequency of electromagnet energy

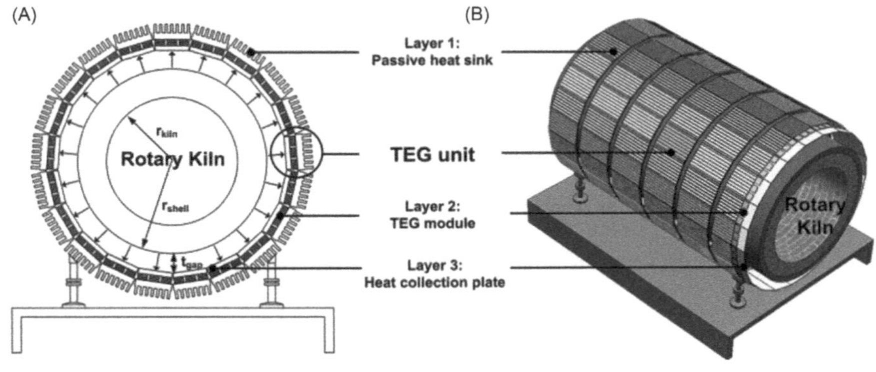

Figure 12.12 Schematic diagram of a general electromagnetic energy–harvesting setup.

harvester with external vibrations and material dimensions, and the properties of the electromagnetic core.

The expression used to determine the output characteristics of the device is,

$$V_{EMF} = -N\frac{d\phi}{dt}\frac{dx}{dt}. \tag{12.4}$$

where V_{EMF} refers to the magnitude of induced voltage generated, N refers to the number of turns of the coil, $\frac{d\phi}{dt}$ refers to the change of magnetic flux with time and $\frac{dx}{dt}$ refers to the change in displacement of the magnet with respect to the coil [54,55].

12.2.3.2 Applications

Design and development of small units for harvesting electrical energy from overhead power lines or transformers vicinities have attracted attention for a long time. Systems have been designed with appropriate power management circuits in order to minimize losses occurring during energy harvesting and power transmission. Since the external vibrations have a random and low-frequency nature, the output of the harvesting unit is significantly affected by the offset in matching between the different coupling units. In general the factors that are considered while designing this kind of circuit are the relationship between resonant frequency of the electromagnetic harvesting unit to the frequency of external vibration and the overall band for the output response frequency. Theoretical studies have been performed by Bernal et al. [56] to understand the significance of interaction between electromagnetic induction and associated damping alongside understanding the relationship between dimensions and output parameters. The interaction between the inductance, capacitance, and resistance of the associated units within the device are found to play pivotal roles in deterring the output performance of an MET. An extensive review by Naifar et al. [54] summarize the fundamentals and operating mechanisms of a number of different types of electromagnetic-energy harvesters.

In order to demonstrate the efficacy of these systems to harvest energy from real-world applications, Gupta et al. [57] reported the development of a device that could harvest power from running AC lines for driving wireless sensor networks with a low-duty cycle. They measured the voltage developed across inductors of different strengths when they were placed between two current carrying conductors connected to a load. Variation in distances between the conductors affected the magnitude of induced voltage developed across the inductor with V_{EMF} varying from 0.9 to 0.1 V over a separation distance $>$15 inches and $\sim$1 inch, respectively. Experiments relating the variation in the heights of the inductor from the conductor plane were also carried out. The study was able to report an overall power output of 1–2 mW utilizing a facile design with simple affordable materials.

In a similar vein, Ahola et al. [58] utilized the current transformer setup in order to build an energy-harvesting system that can extract energy from environments with alternate magnetic fields such as electric motors. The main design principles associated with the development of these kind of technologies is secondary load-carrying ability and linearity. However, effective measures can be taken to develop devices which employ commonly available materials for making cores. The group employed a switched-mode power converter in order to convert the AC power supply to a DC power supply. It utilized a series of diodes and capacitors to efficiently power the load when enough power was generated. Simultaneously, it allowed the charging of the capacitor and also prevented the discharge of secondary current. The group studied the usage of incorporating two transformers in parallel and reported that the output power almost doubled in the latter case, wherein a power output of 25 W was obtained for a 100 A primary source current when the supply frequency of the transformer was 50 Hz. The average output peak to peak voltage was around 24 V with a ripple of 1.5 V. The output parameters and efficiency of the overall device was also affected by the usage of different diodes such as Schottky diodes.

Applications such as pipeline monitoring require usage of wireless sensor networks in order to obtain independently, remote data with regards to the health and maintenance of the pipelines. In a study reported by Sadia et al. [59] simulation and modeling of a lumped parameter system was presented for an electromagnetic energy–harvesting device. They considered a flexible membrane with a coil attached to that it was placed on a channel through which there was a continuous movement of fluid. The movement displaced the coil and caused the distance to vary with a cylindrical permanent magnet, which was kept in the vicinity of the coil. The simulation was performed for a situation wherein the applied pressure was $\sim$630 Pa, and the device under frequency sweep was operated at a damping ratio of 0.01 and resonant frequency of $\sim$500 Hz. The model predicted an output load voltage of $\sim$2.5 V for a load resistance of 100 Ω. However, for an optimal load of $\sim$4 Ω, the output power was enhanced to $\sim$0.2 W.

Studies have also been reported that deal with recovering biomechanical energy spent while performing daily tasks. In a study presented by Halim et al. [60], the group demonstrated a compact electromagnetic energy harvester in the form of a dual-eccentric motor which could utilize the pseudo-walking movement, that is, movement of the arm during walking. The device consisted of a sprung eccentric motor to absorb the mechanical energy, and a transducer (fixed within the rotor) to convert it into electrical energy. The two sides of the rotor rotated together with a torsional spring. This setup also contained 5 magnet pole pairs with a back-iron shield and 10 copper coils have been placed in between the dual-rotor structure using a printed circuit board. The back-iron shield on either side helped enhance the magnetic flux density at the locations where the coils are placed.

Upon movement of the entire mechanical setup, the relative motion between the magnets and copper coils occurred and resulted in the generation of induced electromotive force according to Faraday's law. The study also developed a dynamical electromagnetic model of a sprung eccentric rotor and performed optimization tests to enhance output power under a pseudo-walking situation. The configuration of the dynamic system was optimized using finite element analysis. In order to demonstrate the practical capability of this device, a prototype was developed and tested

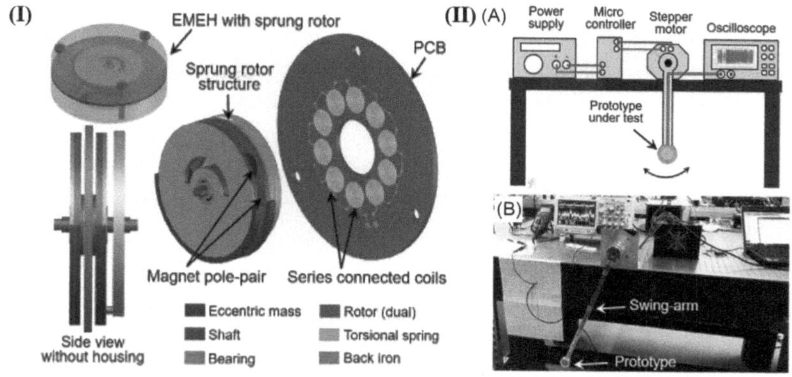

Figure 12.13 (I) Schematic of the dynamic electromechanical configuration utilizing a sprung eccentric rotor. (II) Schematic of the overall configuration.
Source: Adapted from M.A. Halim, R. Rantz, Q. Zhang, L. Gu, K. Yang, S. Roundy, An electromagnetic rotational energy harvester using sprung eccentric rotor, driven by pseudo-walking motion. Appl. Energy 217 (2018) 66–74. Copyright 2018, Elsevier.

on a mechanical swing-arm (mimicking a wrist-worn motion). The device was able to generate ~62 μW power at a rotation of 25 degrees and 1 Hz frequency. Fig. 12.13 depicts the schematic diagram of the prototype, along with the test setup.

Deployment of electromagnetic-energy harvesters in industrial settings has been touted to be a very useful innovation as significant amount of energies could be recovered from vibration movements of different industrial parts and electromagnetic radiations from towers. Toward that end, a study presented by Ren et al. [61], demonstrated a pneumatically actuated electromagnetic energy harvester that could be deployed in gas pressure fluctuation environments. The prototype consisted of a cylindrical casing wherein two permanent magnets were fixed at either end, a movable magnet with an attached spring structure exists, and two fixed-conducting coils are present. A gas inlet was provided that allowed the gas pressure to directly interact with the piston consisting of a movable magnet.

Fig. 12.14 depicts the schematic of an electromagnetic-harvester unit. It also shows the overall setup employed to harvest the variations in fluidic pressure and convert it into electrical signals. A finite element analysis of the developed model is also presented as the study was carried out to understand the fluidic, mechanical, and electromagnetic dynamics. The system was initiated with the gas pressure fluctuation–driven movement of the piston that was subsequently balanced by the repulsion forces generated due to the presence of permanent magnets, on either end of the casing. The oscillating magnet created the necessary electromotive force along with current on the conducting coils when they were operated under a load. The experiments were performed keeping the output pressure of the pneumatic source constant and duty cycle of the pulsed width modulated pneumatic signal at 50%. The output power density of the device was calculated as average power-per-unit volume and

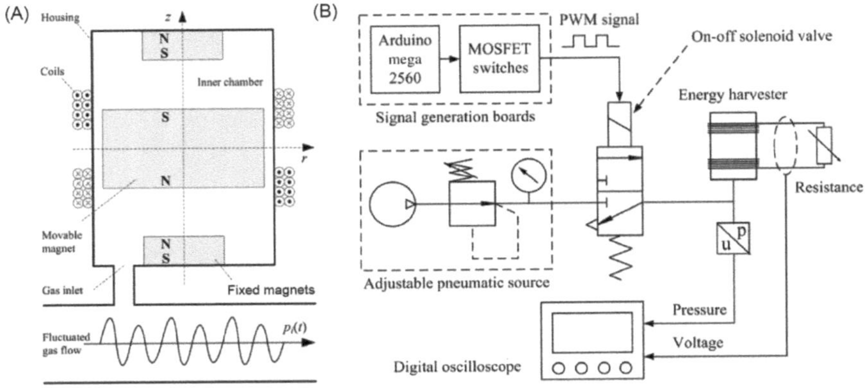

Figure 12.14 (A) Schematic diagram of the electromagnetic energy harvesting unit. (B) Schematic of the overall setup employed to utilize pneumatic gas pressure fluctuations and convert it into usable energy.
Source: Adapted from H. Ren, T. Wang, Development and modeling of an electromagnetic energy harvester from pressure fluctuations. Mechatronics 49 (2018) 36–45. Copyright 2018, Elsevier.

normalized power density as the ratio of the power density to the peak to peak pressure squared. The parameters of the developed prototype were around 145 and 0.02 W/m^3 kPa^2, respectively. The study concluded that the output parameters of the prototype relied significantly on the input pressure variation and its frequency and magnitude. The output parameters were comparable or better than the reported METs in the literature for similar applications.

In order to effectively capture the biomechanical movement of the arm during walking or running, another study developed toward creating a better human interface with electromagnetic energy–harvesting systems proposed the development of a curved oscillatory electromagnetic energy harvester. The work performed by Samad et al. [62] reported that the curved configuration had better output parameters in comparison with its linear counterpart since the curved design was able to harvest energy in two directions.

The polyethylene tube casing consisted of a freely-moving neodymium iron boron (NdFeB) magnet between two ends of the tube where dampeners were placed. An enameled copper coil with 1000 turns was wrapped near the center of the tube. The tube of a suitable diameter was chosen, which eventually reduced the air gap between the magnet and the outer casing such that the overall magnetic field experienced by the copper coils was enhanced. The inner surface of the tube was corrugated so as to reduce any friction generated during contact. It also assisted in making the tube bendable. Studies were performed with varying curvatures of the outer casing and an optimal curvature value of 0.25 m was observed. At this curvature, the highest value of induced voltage was observed probably because the gravity aided in the enhancing the relative movement of the magnet across the coils, thereby contributing to the induced voltage generated. For curvature values more or less than the optimal value (0.25 m), an inferior performance was observed when compared with the output generated using a straight tube. The possible reason attributed for this was the increased gradient experienced by the magnet on its return journey which might have affected its relative velocity. The output parameters for this device was observed at optimal curvature value of 0.25 m to be between 2.5–4 V under a no-load condition when the movement of the prototype was at a frequency between 2–5 Hz. A full-wave bridge rectifier was also incorporated in the device configuration in order to produce stable output DC voltages. The overall device weighed 0.14 kg. The power output was close to 5.2 mW when tested under actual running conditions after placing it on the arm. The device presented a pathway to harvest usable energy from simple daily activities such as arm movements. The concept can be extended for more efficient energy generation at a larger magnitude when integrated with sports such as tennis, boxing, football, badminton, or table tennis which involve rapid movements of limbs.

12.2.3.3 Future perspectives

Electromagnetic-energy harvesters have been utilized for energy production purpose since early 1900s although it came into limelight after the groundbreaking discovery of electromagnetic induction by Faraday almost a century ago [55]. The majority of applications today utilize the rotational and linear design of energy harvesting unit. This technology is capable of producing energy from the μW to mW levels by simply recovering stray kinetic energy or converting surrounding electrical or magnetic signals from a variety of appliances. They work at their optimality when deployed in hostile surroundings or difficult-to-reach areas such as electric poles or transformers, or in vicinity of industrial machineries. However, in order to garner wide-scale recognition and commercialization of these devices few improvements with regards to its design and operative parameters needs to be performed.

For example, the low frequency and narrow operating band limits their usage to few applications. Thus circuit models need to be developed to the next stage in order to expand the bandwidth of these technologies. Smart designs for the conductor coils need to be developed so that not only the increase in electrical resistance due to lengthy coils can be prevented, but also the size limitation of these devices can be overcome. This may enable the construction of the miniaturized energy harvesting units. A better compatibility with the other peripheral microelectromechanical systems in terms of matched bandwidth and loading resistances is certainly another area where lots of activities can be thought of. The models involving different mechanisms needs to be developed in order to generate a hybrid approach for achieving a better synchronization among different components of this technology. This may enable the enhancement of the energy output as well as restrict high losses associated with these systems.

12.2.4 Thermoelectric energy harvesters

12.2.4.1 Introduction

Thermoelectric energy harvesters convert thermal energy into useful electricity to power micro- to macroscale applications. This technique has many advantages associated with its operation such as the absence of moving parts, noise free, eco-friendly, and energy efficient [63–65]. Such devices convert the existing temperature difference across the junction of two dissimilar materials into potential difference following the principles of the Seebeck effect. Fig. 12.15 depicts the schematic diagram of a typical thermoelectric device wherein the hot junction is at a temperature T_H and cold junction at T_C. The thermal motion drives the negatively charged particles of an *n*-type semiconductor and positively charged particles of a *p*-type semiconductor to travel toward the opposite ends in the presence of thermal gradient. The potential difference across the junction can be represented as Eq. (12.5):

$$V = (\alpha_A - \alpha_B)(T_H - T_C), \tag{12.5}$$

where, α_A and α_B are the Seebeck coefficients for the two dissimilar materials.

The Seebeck coefficient depends on the thermal conductivity of the materials and electrical conductivity along with the nature of charge carriers such as the electrons or holes. The flow of charges in a conductor due to a temperature difference also leads to the Thomson effect for which the heat generated can be evaluated from Eq. (12.6),

$$Q = \rho J^2 - \mu_{AB} J \Delta T, \tag{12.6}$$

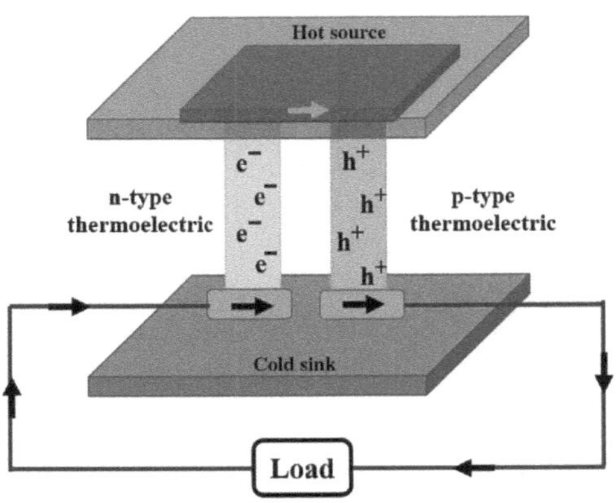

Figure 12.15 Schematic diagram of a typical thermoelectric energy harvesting device with two dissimilar materials attached to the ends at different temperatures.

wherein, ρ represents the electrical resistivity, J is the current density, μ_{AB} is the Thomson coefficient, and ΔT represents the temperature gradient along the conductor. The relationship between the Seebeck coefficient and Thomson coefficient is given by Eq. (12.7),

$$m_{AB} = T \times \frac{da_{AB}}{dT}. \tag{12.7}$$

Another dimensionless thermoelectric figure-of-merit used to characterize the energy harvester performance is ZT. This parameter depends on the thermal conductivity, k, electrical conductivity, σ, and Seebeck coefficient, α, according to the Eq. (12.8),

$$ZT = \frac{\alpha^2 T \sigma}{k}. \tag{12.8}$$

Researchers have been studying techniques to enhance the thermoelectric properties of materials. A few studies have reported that efficient thermoelectric materials should have (1) lower thermal conductivity to maintain a higher thermal gradient at the junction, (2) higher electrical conductivity to facilitate charge transfer, and (3) higher ZT value, greater than 1.

12.2.4.2 Applications

Thermoelectric energy harvesting systems are capable of serving a multitude of applications ranging from biologically implanted devices to wireless sensors to powering aerospace applications [64,65]. The main advantage in such devices lies in tapping naturally abundant thermal radiations from the industrial flue gases, cooling towers, waste burning, automobile exhaust, or human body. The working principle behind this technique opens up the possibility to generate energy on different scales from a few μW to the level of kW. Recently the drive toward designing self-powered electronics capable of being operated by miniaturized but effective power sources have gained traction. Aligned with this objective, different studies have been documented in the literature that show significant potential toward developing thermal energy harvesters that are capable of recovering heat energy from biomechanical movements. For example, in a study reported by Wang et al. [66], a wearable, rationally designed thermoelectric generator was reported. The device consisted of a thermopile structure made out of poly-SiGe, wherein each chip contained 4700 thermocouples connected electrically via series connection and thermally via parallel connection. Silicon substrate was utilized to be the base out of which a rim structure thermopile was developed. Simulation was also carried out in order to validate the design criteria and output characteristics. The open-circuit voltage of the device resulted in 12.5 V/(K cm^2) while the output power was ~0.03 μW/(K cm^2). Upon the usage of the device on the human body, the output voltage was observed to be 0.15 V with an output power ~0.3 nW for a matched load.

The introduction of different METs linked with online networks has ushered in an era of wearable technology that has significantly reduced the inconvenience and economic burden of individuals. Different self-powered devices have been oriented toward this direction that are capable of driving different sensors which are able to sense different pathological signatures produced by the human body. In another work directed toward the objective of designing continuous monitoring systems, Kim et al. [67], reported the development of a wearable electrocardiography (ECG) system wherein the ECG system was powered by the thermoelectric generator drawing the energy from the human body. For this work they utilized a polymer-based heat sink. The device reported was flexible in nature and comprised of a super-absorbent polymer along with fiber, which facilitated liquid evaporation. Studies were conducted by optimizing the different parameters of the system. The output power from the device was high enough to drive not only the ECG system, but also the sensors and other power management circuits. The device produced an output power density of 38 $\mu W/cm^2$ for 10 min. Thereafter, more than 13 $\mu W/cm^2$ for over 22 h of continuous running, as shown in the Fig. 12.16. Thermoelectric energy harvesting devices have also been applied to domestic settings in order to salvage ambient thermal energy.

In order to enhance the applications domain of this kind of energy-harvesting device, a study presented by Hashim [68] reported the deployment of a device that could be utilized in a domestic setting. The thermal energy harvesting device developed was attached to the side wall of an electric oven from where it was able to recover the excess heat energy lost by the system. The schematic diagram of the setup along with the actual setup is depicted in Fig. 12.17. In the study, two K-type thermocouples were utilized to measure the temperature at the two ends of the thermoelectric material whereby, the top thermocouple was attached to the oven surface while the bottom thermocouple was attached to the heat sink. A different material

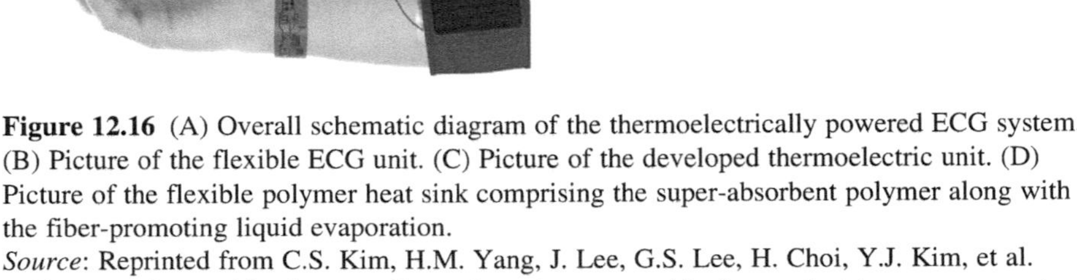

Figure 12.16 (A) Overall schematic diagram of the thermoelectrically powered ECG system. (B) Picture of the flexible ECG unit. (C) Picture of the developed thermoelectric unit. (D) Picture of the flexible polymer heat sink comprising the super-absorbent polymer along with the fiber-promoting liquid evaporation.
Source: Reprinted from C.S. Kim, H.M. Yang, J. Lee, G.S. Lee, H. Choi, Y.J. Kim, et al. Self-powered wearable electrocardiography using a wearable thermoelectric power generator. ACS Energy Lett. 3 (2018) 501–507. Copyright 2018, ACS.

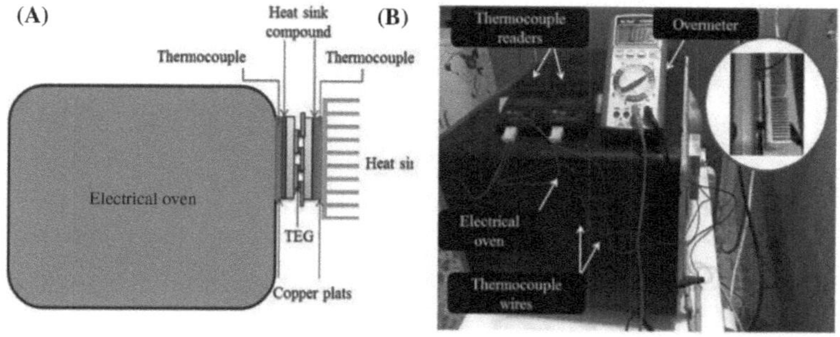

Figure 12.17 (A) Schematic diagram of the thermoelectric energy harvesting unit deployed in a domestic setting. (B) actual snapshot of the device in operation.
Source: Adapted from H.T. Hashim, Energy harvesting from the waste heat of an electrical oven via thermoelectric generator. J. Phys. Conf. Ser. 1032 (2018) 012024–012031.

acting as heat sink with a definite thermal conductivity was placed between the oven and thermoelectric on one side, while the same material was placed between the heat sink and thermoelectric on the other side. The energy harvester design consisted of 256 legs of temperature-sensitive material.

The output characteristics of the device resulted in the output voltage of 1 V and power of 30 mW, in an open-circuit condition whereas, a 20°C temperature difference in a closed-circuit condition resulted in a voltage generation on 0.5 V, and power output of 25 mW. This power output was claimed to be enough to charge a mobile phone or light LEDs. In order to overcome the hurdle of mass printing of flexible thermoelectric devices, Shin et al. [69] reported a scalable, flexible screen printed thermoelectric device on glass fabric surface. The authors carefully optimized the amounts of thermoelectric inks (*p*-type/*n*-type composed of BiSbTe and BiTeSe alloys, respectively), binder ratio's and organic solvent. Identification of a binder namely methyl cellulose was reported which at low concentrations (0.5 wt. %) offered adequate viscosity values of the solutions for screen printing while hardly affecting the electrical transport of charges themselves. The binder material was then burnt off to result in the reduction of thermal conductivity in *n*-type material due to excessive phonon interactions. After the heat-treatment, the powdered-binder mixture was loosely placed on the fabric and thus uniaxial hot pressing was applied to increase the adhesion to the underlying substrate. Thus the enhanced electrical conductivity (<200 S/cm for *n*-type material) and reduced thermal conductivity (0.77 W/mK) resulted in the device's ZT values of around 0.81 for the *n*-type material and 0.65 for the *p*-type material. Fig. 12.18 depicts a schematic diagram of the methodology followed to prepare the screen-printed thermoelectric energy harvester unit.

The thermal energy in the form of automotive exhausts seriously contributes to the inefficiency of the combustion engine, with reported thermal energy losses up to 40% of the energy produced during combustion [70,71]. Therefore energy-

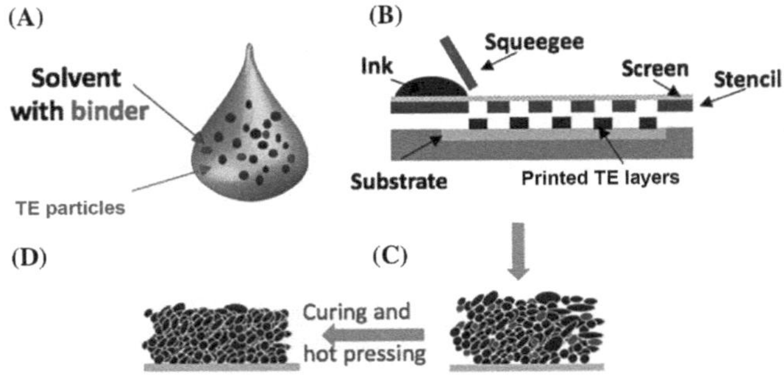

Figure 12.18 Schematic diagram of the constituents, and methods employed to prepare the screen-printed thermoelectric energy harvester device. (A) Printing ink, depicting the TE particles along with the solvent and binder. (B) Screen printed methodology followed to obtain different layers. (C) A particular screen-printed layer comprising of thermoelectric elements. (D) Hot-pressed technique used to compress the different layers so as to increase contact between them.

Source: Reproduced under the terms of CC-BY 4.0 license S. Shin, R. Kumar, J.W. Roh, D.S. Ko, H.S. Kim, S.I. Kim, et al. High-performance screen-printed thermoelectric films on fabrics. Sci. Rep. 7 (2017) 7317–7325. Copyright 2017, Nature.

harvesting systems developed to tap this source of energy may aid the overall fuel efficiency. In a study reported by Thacher et al. [72] a thermoelectric energy harvesting unit was developed and attached to the exhaust pipe of a 1999 GMC Sierra pick-up truck. The overall system consisted of the energy harvesting unit, power conditioning unit and other accessories. Tests were performed under three different situations: (1) initially the thermoelectric unit was deployed solely onto the exhaust pipe; (2) then the thermoelectric energy harvesting unit was deployed along with a portion of exhaust pipe made insulated; and (3) finally, the unit was deployed with insulated upstream exhaust pipe fitted with a precooling heat exchanger. The results obtained suggested that utilizing the insulated pipe configuration along with the lowered coolant temperature aided in enhancing the overall efficiency of the system as the temperature difference across the thermoelectric unit increased. The thermoelectric module consisted of 71 serially connected units wherein one leg was made of *p*-type and another of *n*-type BiTe based material. The two ends of the thermoelectric units were connected to the heat source and sink through aluminum oxide layer that acted as thermal conductors and electrical insulator. It was reported that the increment in power was from 25 W to 125 W for an increase in vehicle speed from 50 to 110 kph. This increment in the vehicle speed resulted in a temperature increase of almost 180°C.

In an extensive study carried out by Aranguren et al. [73], a thermoelectric energy harvesting prototype was built for recovering lost thermal energy in an industrial setting from a combustion chamber utilizing natural gas as the fuel. The setup consisted of 48 different microscale thermoelectric modules, along with heat exchangers, finned heat sinks, and pipes. The group reported that at higher temperatures and gas glow rates, more power could be produced by the energy harvesting units. Another study was also carried out wherein the group studied the optimal design of a chimney where they reported a loss of ~42% of thermal energy in the absence of thermoelectric units. Therefore the overall thermoelectric device when deployed near the combustion chamber was able to harvest a net power of ~22 W with an effective surface area of 0.25 m^2. This study clearly indicated the effectiveness of employing energy harvesting units at locations wherein significant temperature differences resulted in large thermal energy losses. Luo et al., developed a mathematical model to calculate the energy savings from the utilization of thermoelectric modules in rotary kilns present in the cement industry [74]. It has been observed that approximately 10%–15% of the energy consumed for cement production gets dissipated into the atmosphere from the rotary kiln surface. Since the overall cement production process consumes 110–120 kWh/ton energy; the fraction of thermal energy lost can be significant.

Hence this study proposed a deployment of miniaturized hybrid Bi_2Te_3-PbTe thermoelectric modules for the process of energy recovery. Fig. 12.19 shows a schematic diagram of the proposed thermoelectric module design in 2D and 3D configurations, respectively. The model was tested assuming a 1D heat-transfer problem under steady-state conditions. The properties of thermoelectric modules are assumed to be constant and electrical, thermal resistance was negligible. The simulation results proposed that for an ~5 m × 72 m long rotary kiln, ~211 kW of electrical energy could be produced and 3283 kW of thermal energy loss could be conserved which accounted for around 33% of energy savings due to the presence of the thermoelectric energy harvesting unit.

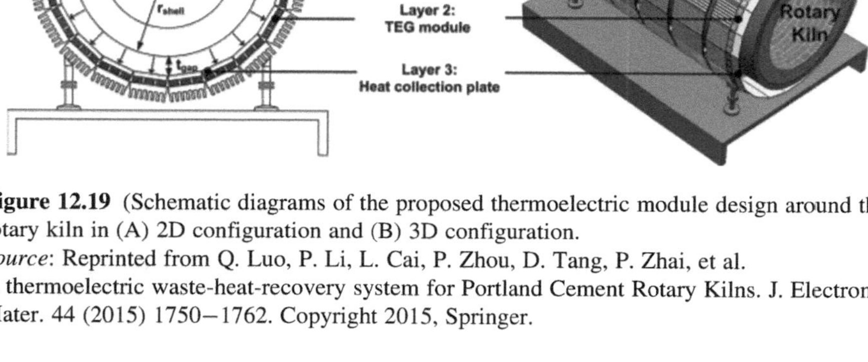

Figure 12.19 (Schematic diagrams of the proposed thermoelectric module design around the rotary kiln in (A) 2D configuration and (B) 3D configuration.
Source: Reprinted from Q. Luo, P. Li, L. Cai, P. Zhou, D. Tang, P. Zhai, et al. A thermoelectric waste-heat-recovery system for Portland Cement Rotary Kilns. J. Electron. Mater. 44 (2015) 1750–1762. Copyright 2015, Springer.

12.2.4.3 Future perspective

Thermoelectric generators have the potential to harvest significant quantities of lost thermal energy when deployed in conventional settings such as industries, automobile sector, or cooling towers. They can also be integrated with rather nonconventional settings that experience a lower temperature gradient such as human body, LED light sources, computers, domestic settings, or workplaces. However, in order to widen its applicability, further improvements for device performance could be achieved by designing materials with enhanced electrical and thermal properties, fabricating thermopiles on high-aspect ratio structures, or optimizing structural design for tailor-made resource settings. Incorporation of nanoparticles of different geometries within thermoelectric modules can open up a potential avenue of developing composite structures with enhanced physical properties. Further, the development of organic thermoelectric materials alongside hybrid ones with the inorganic materials, are yet to be explored in detail. In particular, hybrid devices can be considered for future use as they can combine the advantages of flexibility, lightweight, long lifetime, and a high-output voltage. These devices have the capability of delivering power on different scales and thereby can be considered as prime candidates for powering wireless sensors and other peripherals for the IoT sector.

12.3 Internet of things

Transportation through IoT-enabled cabs have revolutionized the operation, monitoring, and maintenance of diverse application over the past few years. Using the same principle, a host of microdevices can be deployed, operated, monitored, and maintained while dynamically addressing the issues related to the rainwater harvesting, LPG transportation, surveillance for children or women's safety, crack propagation inside ventilations, or targeted drug delivery. The vision here is to use energy-efficient technologies that are of micro- to nanoscale dimensions, portable, self-powering, low-energy intensive, low-maintenance, robust, and affordable. The prevalence of such devices is expected to create effective communication between different objects enabling them to form part of a larger network and eventually form an ecosystem where all are in a virtual connection with each other. Such an ecosystem of devices is called the Internet of Things (IoT). IoT devices find applications perhaps in every engineering sector and have a significant influence on the global market. Reports have suggested that not only the number of devices related to the IoT may reach ~20 billion by the end of 2020, but the associated market will also amount to USD ~5 billion in the foreseeable future [11,12,75]. In order to tap the market on the global scale commercially, IoT devices should be able to fulfill the basic characteristics of emerging technologies, such as:

1. low capital cost of implementation so that it does not affect the revenue generation negatively and minimum requirement for maintenance so that the operational costs remain below a threshold;

2. they should be versatile and be available in variety of sizes so that they will be able to cater to a large market;
3. techniques need to be designed, which can the make the volume of devices scalable so that a vast deployment can be made possible;
4. devices should be sustainable and be able to harvest energy from the surroundings, thereby not only requiring minimum human intervention but also lowering their dependencies on batteries and other sources that may drain the natural resources.

Many different technologies have been pushed as primary energy sources for this ever-growing field of technology. A few of these have been discussed in this chapter such as triboelectric nanogenerators, piezoelectric nanogenerators, electromagnetic wave harvesting, thermoelectric. Table 12.1 provides an overview of the different METs with their associated features. The different energy-harvesting techniques possess capabilities of powering devices across many energy levels. A few considerations that need to be understood while designing IoT-based systems are [12,75,81]:

- *Effective power management*—The devices powering the IoT technologies are to operate in remote areas. Therefore appropriate power-management modules for varying load resistances should be chosen carefully. Moreover, since the device need not require 24 × 7 delivery of data to the network, the update rates have also to be considered since the transition of devices from sleep to active cycles may consume more power.
- *Device protection*—The devices need to be encased in proper coatings so that the internal circuitry is not affected due to any environmental factors. Moreover, unauthorized access to the device may result in tampering with its functionality. Therefore proper mechanisms need to be installed so that free access is limited.

Table 12.1 Comparison of various METs for powering IoT devices.

Energy source	Power output	Advantages ([1,16,98])	IoT applications	References
Triboelectric	0.1 μW–1 MW	Facile device construction, cost-efficient, high-power density, robust	Wireless networks, sensors, actuators, smart-home devices	[19,26,28,76]
Piezoelectric	1 μW–100 W	Scalable, compact architecture, high-power density	Sensors, smart-home devices, actuators	[5,43,47,53,77]
Electromagnetic wave harvesting	1 μW–10 mW	High efficiency, low cost, scalable	Wireless networks, sensors	[54–57,78]
Thermoelectric	10 μW–100 W	Reliable, high-power density, compact	Wireless networks, miniaturized motors, actuators	[5,64,65,79,80]

- *Energy storage*—Appropriate storage units such as capacitors or batteries need to be chosen for various energy harvesting systems when catering to different applications. Applications may require power throughout the day on varying scales; therefore, a secondary unit is required for storing the energy harvested from the environment and delivering it to the device.
- *Working spectral range*—It has been reported that the operational bandwidth of devices is inversely proportional to the power output. This makes the devices more pertinent to narrow-bandwidth data. Thus proper care needs to be given to MET selection since the frequency-to-occurrence rate of external actuations powering the device is often unpredictable.
- *Hybrid technology*—Utilization of ambient dissipated energy may be enhanced by the utilization of alternate METs working in tandem. Studies need to be performed so that the output energy gained from the hybrid functionality comes at a cost that is rational and financially viable.

12.4 Summary

This chapter reviews four primary energy harvesting technologies that have been studied for powering the next generation of nano- or microscale devices intended for usage toward Internet of Things applications. The different METs have the advantage of being portable, compact, robust, environment-friendly, and, most importantly, have the capability of producing output power over a wide range of scales. These devices are capable of harvesting energy from the environment, which not only makes them almost maintenance-free, but their usage also discourages the deployment of a large number of batteries for powering different systems. The chapter also discusses the versatility of these METs in terms of their wide range of service sectors. These sustainable technologies are expected to empower the future of miniaturized devices by allowing the different sensors, actuators to be powered independently at remote locations while allowing them to form a part of a larger interconnected network. However, for commercialization of these devices on a large-scale further research needs to be performed in order to develop higher efficiency devices capable of working in different modes and with different technologies so as to form hybrid systems. Additional improvements in material selection can be carried out by the incorporation of appropriate nanomaterials. Efficient design of power management circuits needs to be performed so that the devices are able to operate under optimal conditions and provide improved DC supply for longer operational periods. The chapter also highlights a few drawbacks and possible remedies for enhancing the output of different energy harvesting mechanisms.

References

[1] S. Chandrasekaran, C. Bowen, J. Roscow, Y. Zhang, D.K. Dang, E.J. Kim, et al., Micro-scale to nano-scale generators for energy harvesting: self powered piezoelectric, tribo-electric and hybrid devices, Phys. Rep. 792 (2019) 1−33.

[2] Z.L. Wang, W. Wu, Nanotechnology-enabled energy harvesting for self-powered micro-/nanosystems, Angew. Chem. Int. Ed. 51 (2012) 11700−11721.
[3] P. Moriarty, D. Honnery, Can renewable energy power the future? Energy Policy 93 (2016) 3−7.
[4] M.A. Islam, M. Hasanuzzaman, N.A. Rahim, A. Nahar, M. Hosenuzzaman, Global renewable energy-based electricity generation and smart grid system for energy security, Sci. World J. (2014). Article ID 197136.
[5] K.V. Selvan, M.S. Mohamed Ali, Micro-scale energy harvesting devices: review of methodological performances in the last decade, Renew. Sustain. Energy Rev. 54 (2016) 1035−1047.
[6] B. Scrosati, J. Garche, Lithium batteries: status, prospects and future, J. Power Sources 195 (2010) 2419−2430.
[7] S. Rathore, S. Sharma, B.P. Swain, R.K. Ghadai, A critical review on triboelectric nanogenerator, IOP Conference Series: Materials Science and Engineering, 377, Institute of Physics Publishing, 2018, p. 012186.
[8] J. Luo, Z.L. Wang, Recent advances in triboelectric nanogenerator based self-charging power systems, Energy Storage Mater. (2019) (In press).
[9] K.S. Dhathathreyan, N. Rajalakshmi, R. Balaji, Nanomaterials for fuel cell technology, Nanotechnology for Energy Sustainability, Wiley-VCH Verlag GmbH & Co. KGaA, Weinheim, Germany, 2017, pp. 569−596.
[10] C. Truong, M. Naumann, R. Karl, M. Müller, A. Jossen, H. Hesse, Economics of residential photovoltaic battery systems in germany: the case of Tesla's powerwall, Batteries 2 (2016) 14.
[11] Task 6—Energy Harvesting, https://edna.iea-4e.org/tasks/task6 (accessed 31.10.19).
[12] S.M. Reisch, Powering the internet of things, Chem. Eng. N. 95 (2017) 16−17. Available from: https://cen.acs.org/articles/95/i32/Powering-internet-things.html (accessed 31.10.19).
[13] F.R. Fan, Z.Q. Tian, Z. Lin Wang, Flexible triboelectric generator, Nano Energy 1 (2012) 328−334.
[14] Z.L. Wang, A.C. Wang, On the origin of contact-electrification, Mater. Today (2019) (In Press).
[15] Z.L. Wang, On Maxwell's displacement current for energy and sensors: the origin of nanogenerators, Mater. Today 74−82 (23) (2017).
[16] H. Zou, Y. Zhang, L. Guo, P. Wang, X. He, G. Dai, et al., Quantifying the triboelectric series, Nat. Commun. 10 (2019) 1427−1435.
[17] J. Shao, M. Willatzen, Y. Shi, Z.L. Wang, 3D mathematical model of contact-separation and single-electrode mode triboelectric nanogenerators, Nano Energy 60 (2019) 630−640.
[18] W. Tang, T. Jiang, F.R. Fan, A.F. Yu, C. Zhang, X. Cao, et al., Liquid-metal electrode for high-performance triboelectric nanogenerator at an instantaneous energy conversion efficiency of 70.6%, Adv. Funct. Mater. 25 (2015) 3718−3725.
[19] F. Yi, X. Wang, S. Niu, S. Li, Y. Yin, K. Dai, et al., A highly shape-adaptive, stretchable design based on conductive liquid for energy harvesting and self-powered biomechanical monitoring, Sci. Adv. 2 (2016) 1501624−1501633.
[20] W. Seung, H.J. Yoon, T.Y. Kim, H. Ryu, J. Kim, J.H. Lee, et al., Boosting power-generating performance of triboelectric nanogenerators via artificial control of ferroelectric polarization and dielectric properties, Adv. Energy Mater. 7 (2017) 1600988−1600995.

[21] S. Wang, Y. Zi, Y.S. Zhou, S. Li, F. Fan, L. Lin, et al., Molecular surface functionalization to enhance the power output of triboelectric nanogenerators, J. Mater. Chem. A 4 (2016) 3728–3734.
[22] C.X. Lu, C.B. Han, G.Q. Gu, J. Chen, Z.W. Yang, T. Jiang, et al., Temperature effect on performance of triboelectric nanogenerator, Adv. Eng. Mater. 19 (2017) 1700275–1700282.
[23] Z. Tian, J. He, X. Chen, Z. Zhang, T. Wen, C. Zhai, et al., Performance-boosted triboelectric textile for harvesting human motion energy, Nano Energy 39 (2017) 562–570.
[24] C.R.S. Rodrigues, C.A.S. Alves, J. Puga, A.M. Pereira, J.O. Ventura, Triboelectric driven turbine to generate electricity from the motion of water, Nano Energy 30 (2016) 379–386.
[25] H. Guo, X. Pu, J. Chen, Y. Meng, M.-H. Yeh, G. Liu, et al., A highly sensitive, self-powered triboelectric auditory sensor for social robotics and hearing aids, Sci. Robot. 3 (2018). eaat2516.
[26] Y. Xi, H. Guo, Y. Zi, X. Li, J. Wang, J. Deng, et al., Multifunctional TENG for blue energy scavenging and self-powered wind-speed sensor, Adv. Energy Mater. 7 (2017) 1602397–1602402.
[27] C. Bao Han, W. Du, C. Zhang, W. Tang, L. Zhang, Z. Lin Wang, Harvesting energy from automobile brake in contact and non-contact mode by conjunction of triboelectrication and electrostatic-induction processes, Nano Energy 6 (2014) 59–65.
[28] X. Pu, H. Guo, J. Chen, X. Wang, Y. Xi, C. Hu, et al., Eye motion triggered self-powered mechnosensational communication system using triboelectric nanogenerator, Sci. Adv. 3 (2017) e1700694.
[29] C.B. Han, T. Jiang, C. Zhang, X. Li, C. Zhang, X. Cao, et al., Removal of particulate matter emissions from a vehicle using a self-powered triboelectric filter, ACS Nano 9 (2015) 12552–12561.
[30] Z.L. Wang, Triboelectric nanogenerators as new energy technology and self-powered sensors - principles, problems and perspectives, Faraday Discuss. 176 (2014) 447–458.
[31] R.J. Dewhurst, Measurement science and technology: a historical perspective, Measurement Science and Technology, Institute of Physics Publishing, 2013.
[32] L.C. Kong, Integrated electrostatically resonant scan tip for an atomic force microscope, J. Vac. Sci. Technol. B Microelectron. Nanom. Struct. 11 (1993) 634–672.
[33] R.A. Beach, T.C. Mcgill, Piezoelectric fields in nitride devices, J. Vac. Sci. Technol. B 17 (1999) 1753–1759.
[34] P. Glynne-Jones, S.P. Beeby, N.M. White, Towards a piezoelectric vibration-powered microgenerator, IEEE Proc. Sci. Meas. Technol. 148 (2001) 68–72.
[35] S. Basrour, B. Charlot, M. Marzencki, A. Grasso, M. Colin, L. Valbin, Design and fabrication of piezoelectric micro power generators for autonomous microsystems, in: Symposium on Design, Test, Integration and Packaging of MEMS/MOEMS (DTIP'05), 2005, pp. 99–302. ⟨hal-00009215⟩.
[36] F. Mohammadi, A. Khan, R.B. Cass, Power generation from piezoelectric lead zirconate titanate fiber composites, Mater. Res. Soc. Symp. Proc. 736 (2002) 263–268.
[37] X. Wang, Piezoelectric nanogenerators-harvesting ambient mechanical energy at the nanometer scale, Nano Energy 1 (2012) 13–24.
[38] J. Song, J. Zhou, Z.L. Wang, Piezoelectric and semiconducting coupled power generating process of a single ZnO belt/wire: a technology for harvesting electricity from the environment, Nano Lett. 6 (2006) 1656–1662.

[39] Y. Hu, Y. Zhang, C. Xu, G. Zhu, Z.L. Wang, High-output nanogenerator by rational unipolar assembly of conical nanowires and its application for driving a small liquid crystal display, Nano Lett. 10 (2010) 5025–5031.
[40] J. Yang, J. Chen, Y. Su, Q. Jing, Z. Li, F. Yi, et al., Eardrum-inspired active sensors for self-powered cardiovascular system characterization and throat-attached anti-interference voice recognition, Adv. Mater. 27 (2015) 1316–1326.
[41] M. Lee, J. Bae, J. Lee, C.S. Lee, S. Hong, Z.L. Wang, Self-powered environmental sensor system driven by nanogenerators, Energy Environ. Sci. 4 (2011) 3359–3363.
[42] S. Xu, Y. Qin, C. Xu, Y. Wei, R. Yang, Z.L. Wang, Self-powered nanowire devices, Nat. Nanotechnol. 5 (2010) 366–373.
[43] N. Cui, W. Wu, Y. Zhao, S. Bai, L. Meng, Y. Qin, et al., Magnetic force driven nanogenerators as a noncontact energy harvester and sensor, Nano Lett. 12 (2012) 3701–3705.
[44] L. Gu, N. Cui, L. Cheng, Q. Xu, S. Bai, M. Yuan, et al., Flexible fiber nanogenerator with 209 v output voltage directly powers a light-emitting diode, Nano Lett. 13 (2013) 91–94.
[45] J.M. Wu, K.H. Chen, Y. Zhang, Z.L. Wang, A self-powered piezotronic strain sensor based on single ZnSnO3 microbelts, RSC Adv. 3 (2013) 25184–25189.
[46] K.I. Park, M. Lee, Y. Liu, S. Moon, G.T. Hwang, G. Zhu, et al., Flexible nanocomposite generator made of BaTiO3 nanoparticles and graphitic carbons, Adv. Mater. 24 (2012) 2999–3004.
[47] K.I. Park, S.B. Bae, S.H. Yang, H.I. Lee, K. Lee, S.J. Lee, Lead-free BaTiO3 nanowires-based flexible nanocomposite generator, Nanoscale 6 (2014) 8962–8968.
[48] M. Ataur Rahman, B.C. Lee, D.T. Phan, G.S. Chung, Fabrication and characterization of highly efficient flexible energy harvesters using pvdf-graphene nanocomposites, Smart Mater. Struct. 22 (2013) 085017–085026.
[49] Y. Mao, P. Zhao, G. McConohy, H. Yang, Y. Tong, X. Wang, Sponge-like piezoelectric polymer films for scalable and integratable nanogenerators and self-powered electronic systems, Adv. Energy Mater. 4 (2014) 1301624–1301630.
[50] K.I. Park, S. Xu, Y. Liu, G.T. Hwang, S.J.L. Kang, Z.L. Wang, et al., Piezoelectric BaTiO3 thin film nanogenerator on plastic substrates, Nano Lett. 10 (2010) 4939–4943.
[51] H.B. Kang, J. Chang, K. Koh, L. Lin, Y.S. Cho, High quality Mn-Doped (Na,K)NbO3 nanofibers for flexible piezoelectric nanogenerators, ACS Appl. Mater. Interfaces 6 (2014) 10576–10582.
[52] C.K. Jeong, K.I. Park, J. Ryu, G.T. Hwang, K.J. Lee, Large-area and flexible lead-free nanocomposite generator using alkaline niobate particles and metal nanorod filler, Adv. Funct. Mater. 24 (2014) 2620–2629.
[53] S. Guerin, S.A.M. Tofail, D. Thompson, Organic piezoelectric materials: milestones and potential, NPG Asia Mater. 11 (2019) 10–14.
[54] S. Naifar, S. Bradai, C. Viehweger, O. Kanoun, Survey of electromagnetic and magnetoelectric vibration energy harvesters for low frequency excitation, Meas. J. Int. Meas. Confed. 106 (2017) 251–263.
[55] S.P. Beeby, T. O'Donnell, Electromagnetic energy harvesting, Energy Harvesting Technologies, Springer US, 2009, pp. 129–161.
[56] A.G. Bernal, L.E. Linares García, The modelling of an electromagnetic energy harvesting architecture, Appl. Math. Model. 36 (2012) 4728–4741.

[57] V. Gupta, A. Kandhalu, R. Rajkumar, Energy harvesting from electromagnetic energy radiating from ac power lines, in: HotEmNets '10 Proceedings of the 6th Workshop on Hot Topics in Embedded Networked Sensors, 17, 2010, pp. 1–6.

[58] J. Ahola, T. Ahonen, V. Särkimäki, A. Kosonen, J. Tamminen, R. Tiainen, et al. Design considerations for current transformer based energy harvesting for electronics attached to electric motor, in: SPEEDAM 2008—International Symposium on Power Electronics, Electrical Drives, Automation and Motion, 2008, pp. 901–905.

[59] S. Bakhtiar, F.U. Khan, Analytical modeling and simulation of an electromagnetic energy harvester for pulsating fluid flow in pipeline, Sci. World J. (2019) 1–9.

[60] M.A. Halim, R. Rantz, Q. Zhang, L. Gu, K. Yang, S. Roundy, An electromagnetic rotational energy harvester using sprung eccentric rotor, driven by pseudo-walking motion, Appl. Energy 217 (2018) 66–74.

[61] H. Ren, T. Wang, Development and modeling of an electromagnetic energy harvester from pressure fluctuations, Mechatronics 49 (2018) 36–45.

[62] F.A. Samad, M.F. Karim, V. Paulose, L.C. Ong, A curved electromagnetic energy harvesting system for wearable electronics, IEEE Sens. J. 16 (2016) 1969–1974.

[63] I. Petsagkourakis, K. Tybrandt, X. Crispin, I. Ohkubo, N. Satoh, T. Mori, Thermoelectric materials and applications for energy harvesting power generation, Sci. Technol. Adv. Mater. 19 (2018) 836–862.

[64] G.J. Snyder, Thermoelectric energy harvesting, Energy Harvesting Technologies, Springer US, 2009, pp. 325–336.

[65] D. Enescu, Thermoelectric energy harvesting: basic principles and applications, Green Energy Advances, IntechOpen, 2019, pp. 1–16.

[66] Z. Wang, V. Leonov, P. Fiorini, C. Van Hoof, Realization of a wearable miniaturized thermoelectric generator for human body applications, Sens. Actuators A Phys. 156 (2009) 95–102.

[67] C.S. Kim, H.M. Yang, J. Lee, G.S. Lee, H. Choi, Y.J. Kim, et al., Self-powered wearable electrocardiography using a wearable thermoelectric power generator, ACS Energy Lett. 3 (2018) 501–507.

[68] H.T. Hashim, Energy harvesting from the waste heat of an electrical oven via thermoelectric generator, J. Phys. Conf. Ser. 1032 (2018) 012024–012031.

[69] S. Shin, R. Kumar, J.W. Roh, D.S. Ko, H.S. Kim, S.I. Kim, et al., High-performance screen-printed thermoelectric films on fabrics, Sci. Rep. 7 (2017) 7317–7325.

[70] K.M. Saqr, M.K. Mansour, M.N. Musa, Thermal design of automobile exhaust based thermoelectric generators: objectives and challenges, Int. J. Automot. Technol. 9 (2008) 155–160.

[71] J. Yang, F.R. Stabler, Automotive applications of thermoelectric materials, J. Electron. Mater. 38 (2009) 1245–1251.

[72] E.F. Thacher, B.T. Helenbrook, M.A. Karri, C.J. Richter, Testing of an automobile exhaust thermoelectric generator in a light truck, Proc. Inst. Mech. Eng. Part D J. Automob. Eng 221 (2007) 95–107.

[73] P. Aranguren, D. Astrain, A. Rodríguez, A. Martínez, Experimental investigation of the applicability of a thermoelectric generator to recover waste heat from a combustion chamber, Appl. Energy 152 (2015) 121–130.

[74] Q. Luo, P. Li, L. Cai, P. Zhou, D. Tang, P. Zhai, et al., A thermoelectric waste-heat-recovery system for Portland Cement Rotary Kilns, J. Electron. Mater. 44 (2015) 1750–1762.

[75] M. Shirvanimoghaddam, K. Shirvanimoghaddam, M.M. Abolhasani, M. Farhangi, V. Zahiri Barsari, H. Liu, et al., Towards a green and self-powered internet of things using piezoelectric energy harvesting, IEEE Access. 7 (2019) 94533–94556.
[76] Y. Wang, Y. Yang, Z.L. Wang, Triboelectric nanogenerators as flexible power sources, NPJ Flex. Electron. 1 (2017) 10–19.
[77] Z.L. Wang, Towards self-powered nanosystems: from nanogenerators to nanopiezotronics, Adv. Funct. Mater. 18 (2008) 3553–3567.
[78] R. Amirtharajah, A.P. Chandrakasan, Self-powered signal processing using vibration-based power generation, IEEE J. Solid-State Circuits 33 (1998) 687–695.
[79] N. Wang, L. Han, H. He, N.-H. Park, K. Koumoto, A novel high-performance photovoltaic–thermoelectric hybrid device, Energy Environ. Sci. 4 (2011) 3676–3679.
[80] S. Khan, R.S. Dahiya, L. Lorenzelli, Flexible thermoelectric generator based on transfer printed Si microwires, in: 2014 44th IEEE European Solid State Device Research Conference (ESSDERC), 2014, pp. 86–89.
[81] K. Yallup, L. Basiricò, K. Iniewski, Sensors for Diagnostics and Monitoring, CRC Press, 2018, ISBN: 9780815370208, . ISBN.

Section X

Design, Scale-up, and Commercialization of Sustainable Fuel Technologies

Communicating energy consumption: developing a low-cost, real-time, SMS feedback tool for off-grid household electricity consumption

13

Tonny Kukeera[1,2,3,5], *Noara Kebir*[3], *Setu Pelz*[3,4] *and Sebastian Groh*[5]
[1]MicroEnergy Systems Research Group, Technical University of Berlin, Berlin, Germany, [2]Pan African University Institute of Water and Energy Sciences (Including Climate Change), Tlemcen, Algeria, [3]MicroEnergy International GmbH, Berlin, Germany, [4]Department of Energy and Environmental Management, Europa-Universität Flensburg, Berlin, Germany, [5]School of Business and Economics, North South University, Dhaka, Bangladesh

13.1 Introduction

Universal electricity access is critical in economic development and without it, poverty eradication and, hence, the achievement of Sustainable Development Goal 7 is close to impossible. This followed the formation of Sustainable Energy for All (SEforAll) initiative with the primary aim to: (1) address global access to modern energy services; (2) double the global rate of improvement in energy efficiency (EE); and (3) double renewable energy share in global energy production [1]. In the background is the cognizance that energy is the engine for economic development and hence poverty alleviation [2]. The framework of this is not only limited to the extension of energy access, but also doing it sustainably while putting other factors into consideration like environmental protection and greenhouse gas (GHG) emissions. Notably there has been an increase in concerns over climate change and global warming over recent years that has seen the issue surfacing on top of almost every agenda concerning environmental conservation and effects mitigation [3,4]. The role household energy use plays in this regard cannot be underrated [5–7]. Increased EE and reduction in energy demand and use is a crucial step in curbing GHG emissions and climate change [8]. It also presents an opportunity for households to cut down the family budget and boost income through less spending on energy and, hence, improved standards of living. Furthermore, in off-grid systems, users profit instantaneously from EE because they have more energy available for use.

A significant percentage of African countries continue to suffer from the lowest rates of electrification with countries like Burundi, Chad, and Liberia having only

Sustainable Fuel Technologies Handbook. DOI: https://doi.org/10.1016/B978-0-12-822989-7.00014-7

1.25% grid access [9]. This coupled with low grid access (See Fig. 13.1), has been one of the biggest motivations for the development of the off-grid industry in the region. Among other factors is the arrangement of homesteads (i.e., many households are still dispersed and located in remote rural areas) and this favors off-grid energy use as shown in Fig. 13.2, and falling prices of stand-alone renewable energy systems. Products like solar mobile phone chargers, solar lamps, and solar home systems (SHS) can provide tier 1−3 energy services (as per the global tracking framework tier-based system) [11,12] for between 4%−20% of the required cost for grid extension [10]. There has been an increase in funding focused on initiatives involved in developing off-grid energy services and products through the advancement of standards and EE, with a specific focus on off-grid appliances [13−15]. Saving a unit Watt from an off-grid appliance's load results in lower initial solar package costs, improved service, or both [16]. The stand-alone electricity product market is expanding rapidly, and some models [17] estimate the market for SHS products to develop from $550 million in 2014 to $2.4 billion by 2024.

It is important to note that the success of off-grid technologies for providing energy solutions in recent years is mostly attributable to the availability of energy-efficient appliances [10]. Fig. 13.3 shows the effect EE can have on usage in terms of the number of appliances and hours of usage.

On the other hand, however, off-grid systems still suffer from capacity limitation, high maintenance and servicing costs, and sustainability issues [18,19], among others. To boost the diffusion of SHSs and off-grid systems in general among rural households, EE is an indispensable part that cannot be ignored in emerging economies. EE means doing more with less, so performance is maintained while saving energy and money; it saves, allows, and extends energy to be utilized elsewhere, hence providing more energy services to consumers.

Figure 13.1 Global population share without grid access [9].

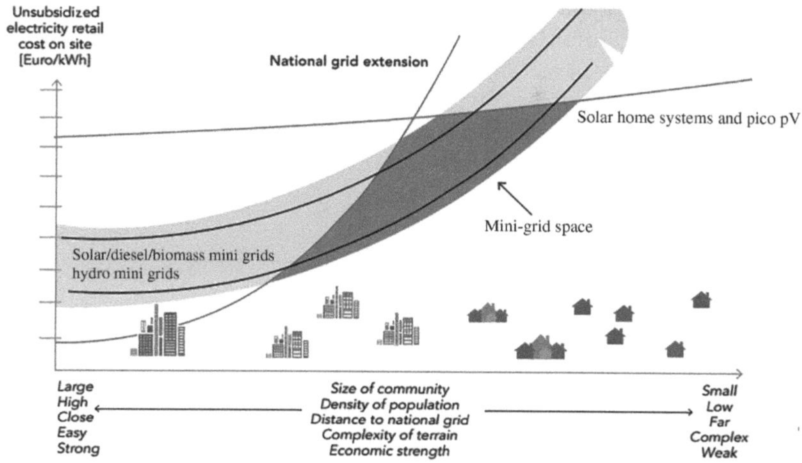

Figure 13.2 Variation of least-cost option for energy access with load and distance [10].

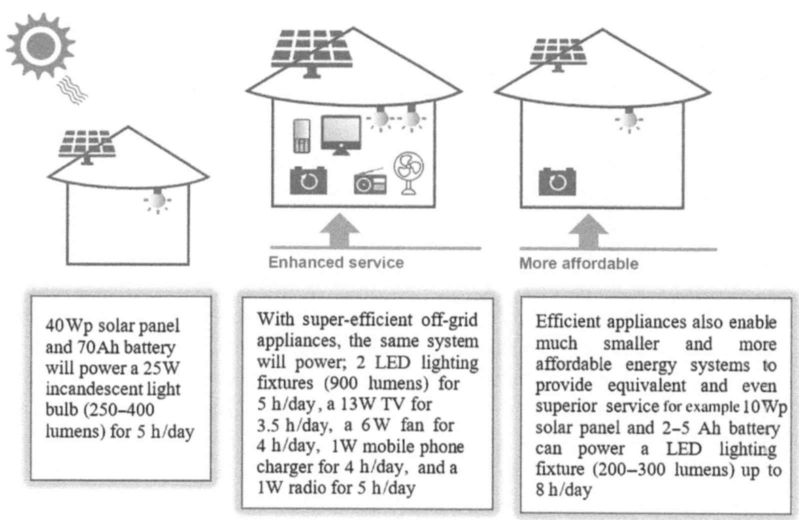

Figure 13.3 Energy efficiency in terms of appliance and usage time [10].

It is therefore imperative that as countries in subSaharan Africa (SSA) continue to grapple with low-energy supply, more supportive measures that encourage understanding, and efficient use of the available energy resources being employed. One of the ways by which this can be promoted is through behavior-based energy strategies that aim to change how people perceive their energy use [20]. This can affect electricity-use behaviors and hence help achieve EE and savings [21,22]. By providing off-grid household energy users with timely, understandable, and straightforward consumption information, energy feedback of a subset of behavior-based energy strategies discloses energy use information to consumers and supports them to access more usable energy by influencing their behavioral patterns [21,23]. Furthermore, it sheds more light on energy use and improves people's understanding of investing in energy-efficient materials in their homes.

13.2 Consumption information access in off-grid households

In many off-grid households in Africa, where over 600 million people have no access to electricity [24], no consumption information is available to users, hence denying them a chance to monitor their energy-use patterns adequately. The situation is worse in households that are off the grid; users are often charged a fee in bulk without provision of their actual energy consumption. This leaves a gap as consumers are challenged to monitor their consumption patterns which can be rather complicated without timely, understandable information. This denies consumers an opportunity to keep track of their household electricity consumption and to save energy. The effects of this in low-income households are adverse, ranging from underdevelopment to a vicious cycle of abject poverty among the masses as they are deprived of good energy service quality.

Furthermore, users are challenged by system failures, costly servicing and maintenance, as well as capacity limitation. The general lack of access to consumption information coupled with improper servicing infrastructures as well as the asymmetric nature of information flow surrounding solar technologies [25], exacerbate the situation. Additionally, it contributes to general system underperformance and practices like system misuse, capacity mismatch, battery over-discharge, and overcharges among others [26,27]. The result is a general loss of morale and trust in solar systems in communities [28]. Off-grid systems have for long suffered from a problem of poor-servicing infrastructures, and this has been identified as the weakest link in the off-grid systems development chain [19,26]. It is against such a background that this study proposes a low-cost, real-time, SMS tool that would support consumers to track their consumption by providing real-time consumption information on mobile phones through text messages. Additionally, by drawing from literature of behavioral models and experiences in household solar off-grid use, the study further gives insights on how better consumption could be translated and communicated to end-users. This study lays a foundation for better community engagement in off-grid households.

13.2.1 *Investment growth in the off-grid industry*

An investment growth in off-grid companies has contributed significantly to the success of the off-grid industry; through innovation and development of more

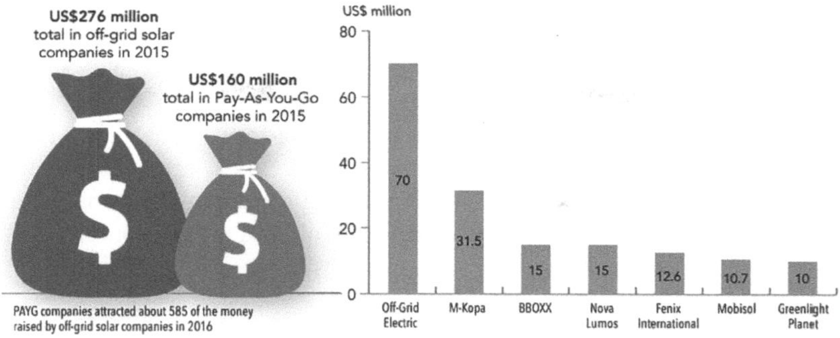

Figure 13.4 Increasing investments in off-grid and share of PAYG in 2015 ([29], p. 21).

efficient business models like PAYG among others. PAYG companies were allocated the highest share of the 2015 capital investment in off-grid development as shown in Fig. 13.4. PAYG is the model where customers can overcome the bulk cost of SHSs by paying in small installments. In the event of nonpayment by the user, the system functionality is locked. When the total payment of the solar system is completed over time, the user owns the solar system perpetually. The model mimics the traditional users' behavior of having to buy traditional fuels like kerosene every day for lighting whenever it is needed. PAYG relies on mobile phone technology to manage SHS payments. Like PAYG, a number of off-grid industry delivery models have focused much of their efforts on the business side such as compelling and making payment plans easier for users [30] with not as much regard or consideration of the users' knowledge, skills, and behavior in relation to the solar systems [25]. The gap created renders solar systems inefficient with many failures and hence low life spans. A sustainable model that not only compels users to pay, but also ensures that they understand what they are paying for and how to get the best out of it, would play a significant role in improving the sustainability, EE, and reducing the maintenance problems associated with SHS. A thorough understanding of the consumption of a particular appliance can be essential in determining the operational costs (where electricity is for a fee-for-service) and the system size (for stand-alone solar systems).

13.2.2 Feasibility of an SMS-based feedback tool in rural Africa

The global usage of mobile phones has grown enormously over the past few years, with over 93% of the people having one in the developing countries as reported by [31]. In Africa alone, it has been reported that over 93% of people have access to cellphones [32]. "[The] mobile phone, a device that was once considered a luxury of the affluent has now become a necessity" [33]. The number of mobile phones in some African countries like Kenya has been reported to be higher than the number of adults in the whole country—38 million mobile phones in a population of

45 million people [34]. While mobile phones are utilized for many uses, a study [35] found out that the most common usage for phones is sending text messages. The ascent of mobile phone usage continues to remove ineffectiveness in conventional business models and improve customers' access to services.

Moreover, the advancement of mobile money platforms in Uganda and other SSA countries (over 130 running mobile money services) has many companies integrating this into their business models [36]. Many off-grid companies have incorporated this in their systems too [37–39]. Such technologies piggyback on the success of mobile phone usage.

13.2.3 Mode of communication and message interpretation

Feedback messages require an understanding of the user's consumption in terms of appliances and behavior patterns so that specific messages tailored to specific user needs are sent. Different users have different motivations, but the end goal is to create and ensure sustainable energy use behavior among users. According to a study by He et al. [40], electricity users go through different stages of behavior transition from the first message received. It is over time and through a series of reminders in the form of messages that sustainable energy use behaviors become part-and-parcel of an individual's every day behavior. Various studies have found that messages tailored to specific individuals produce more positive results over time than a "one-size-fits-all" type of message that involves availing the "*same feedback to differently motivated individuals at different stages of readiness, willingness and ability to change*" [8,20,22]. Different motivation techniques or models are used, including attitude, rational economic, information, positive reinforcement, and elaboration likelihood models [40].

13.3 Solar power system

The solar system was broken down into three sections, namely power generation, storage, and consumption.

13.3.1 Power generation

In the power generation part, the focus was on how much power could be generated at a specific time of the day, taking into consideration the weather conditions. Photovoltaic (PV) forecasting based on the physical approach (see Fig. 13.5), where numerical weather prediction models are employed [42], was used for a fictitious location. The PV system characteristics from the manufacturer's datasheets were used to determine the irradiance. Irradiance is one of the most crucial factors that affect PV power output.

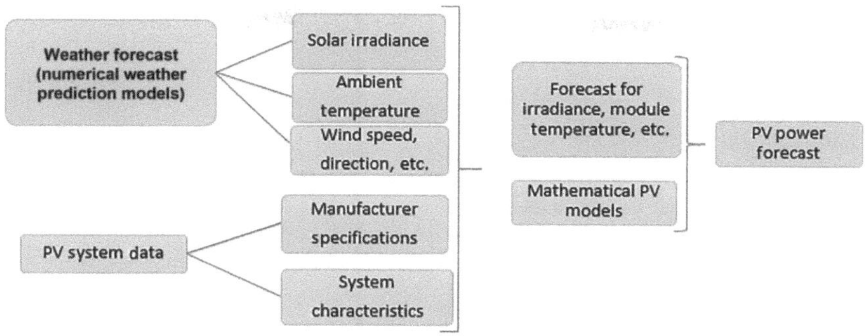

Figure 13.5 PV forecasting based on the physical approach [41].

13.3.2 Power storage

Storage is a vital component of energy systems in general. In SHSs, batteries guarantee energy availability at night and other times when the sun is not shining. Batteries in solar systems handle various charge and discharge processes throughout the day. This is especially important since a battery's capacity, performance, and life span is dependent on the number of cycles (i.e., on the discharge/charge speed and the charge events that occurred in the near past). In order to put such effects into account, the kinetic battery model (KiBaM) by Maxwell and McGowan was chosen to model the batteries' performance [43]. The model assumes that a battery has two types of charge, that is, available charge and bound charge. This allows obtaining the battery state of charge (SOC) at any specific time.

13.3.3 Consumption

Consumption was mainly focused on power-usage monitoring. A device known as the Karana was used for monitoring general solar-system performance. Karana is an open-source, cheap, and easily extendable hardware and software framework for data acquisition based on the STM32[1] microcontrollers. The device was developed at Micro Energy International GmbH.[2] It was developed as an alternative to conventional monitoring devices and relies entirely on open-source software and hardware. The device has embedded abilities to monitor and communicate with the back-end system at the server point where data in time-series format is stored to the database—Influx DB.[3] The data consists of the measurements, tag keys (i.e., host and region) with the tag values, field key, field value, and a timestamp. The device also provides a data visualization platform that can be observed via an open-source, real-time graphing software (Grafana) for further inspection (see Fig. 13.6). The device has three ports: for load connections, solar panel, and battery. The data

[1] https://www.st.com/en/microcontrollers-microprocessors.html
[2] https://mekarana.jimdo.com/
[3] https://www.influxdata.com/products/influxdb-overview/

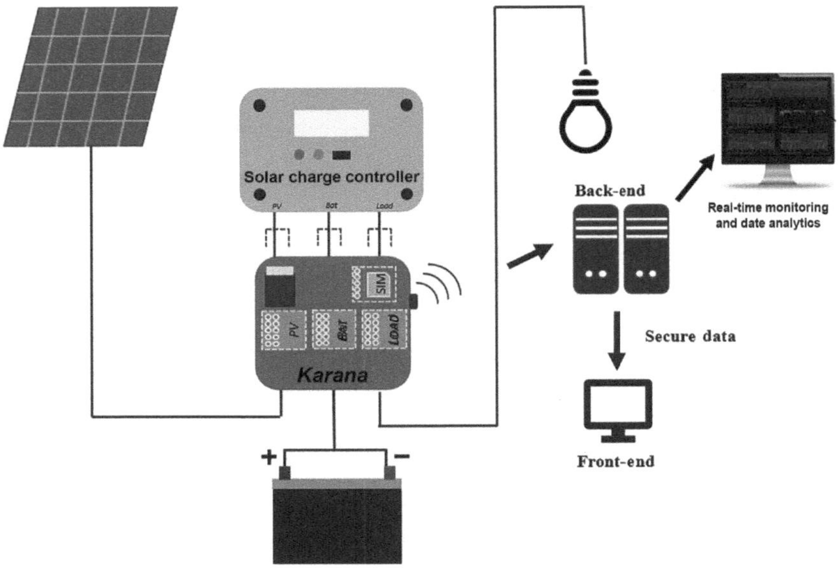

Figure 13.6 Sample of web access to the current status of a solar home system installed at the office (Grafana visualization).

collected ranges from load current and voltage, PV current and voltage, plus battery current and voltage. This allows for efficient monitoring of power production, storage, and consumption. The device is GSM enabled, which means it can be deployed anywhere and accessed from almost everywhere. Fig. 13.7 shows the setup of the monitoring system with the device.

13.3.4 Message sending and integration

The algorithm was written using python as the coding language. Python is a widely used high-level programming language for general-purpose programming. The lightweight Application Programming Interface (API) written in Python with hug[4] was used in integrating the SMS sending algorithm with the Karana back-end system. The hug API lays a platform which enables the different system components, that is, the influx database, Karana device, and Grafana, to communicate. The Karana back-end system is the set of code procedures and algorithms that run in the background of the Karana device described in Section 3.2 of chapter 3. This set of code forms the entire back-end API. For message sending, the Twilio[5] text message sending platform was leveraged.

[4] http://www.hug.rest/
[5] https://www.twilio.com/sms

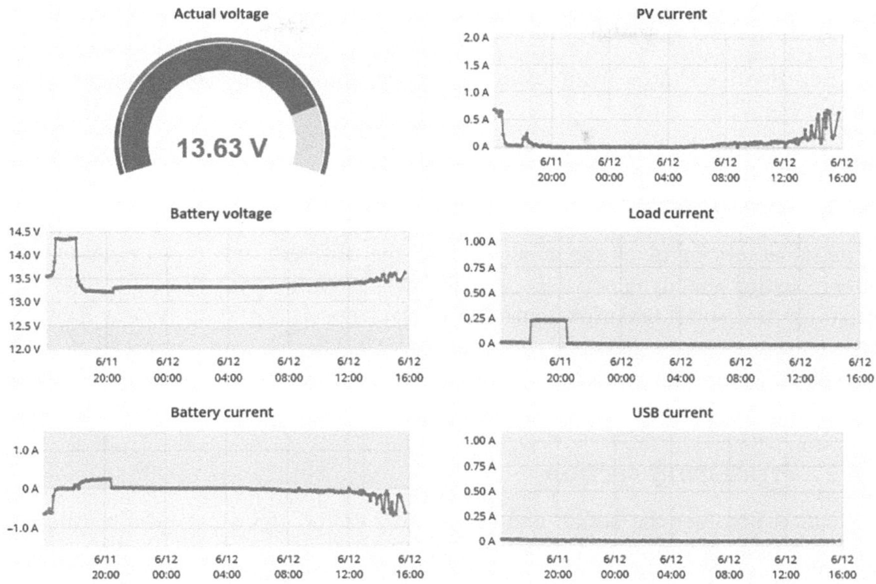

Figure 13.7 Monitoring system setup.

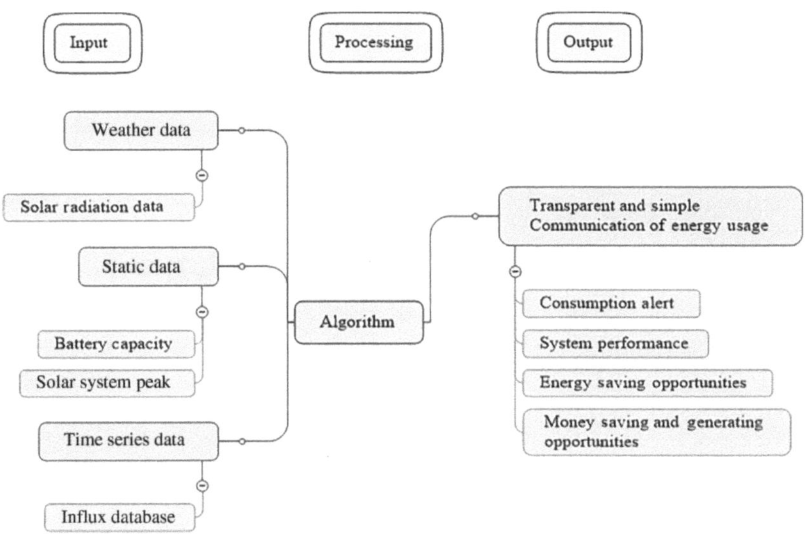

Figure 13.8 General process setup.

13.4 General process setup

The setup involved three main sections, namely input, processing, and output, as shown in Fig. 13.8.

13.4.1 Input section

The input section involves the acquisition of the different data types ranging from weather, static, and dynamic data, also known as time-series data. Weather data consists of the solar radiation data. Static data consists of PV system data that entails battery capacity and solar system peak. Time-series data comprises the real-time data from the influx database.

13.4.2 Processing section

This section involves the action and functioning of the algorithm, which follows a systematic flow as shown in Fig. 13.9. P_1M, P_2M, P_3M, and P_4M are the different types of message sending schemes which are triggered by the events as described in the flow chart.

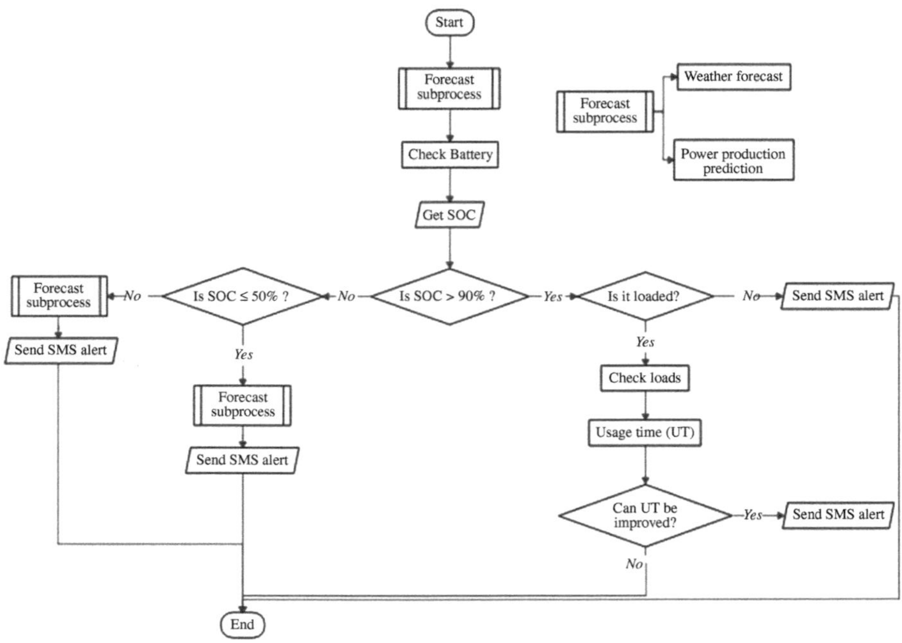

Figure 13.9 A process flow chart of the algorithm with the different message categories.

The starting point initiates the processing section. The command proceeds to the forecast subprocess. The forecast subprocess involves two steps; a weather forecast followed by determination of solar-power production and prediction. At this stage, the expected solar radiation is attained, and this is used to calculate the expected power in a day, and hence the estimated time it would take the battery to charge. The next step is to check the battery in order to ascertain the level of charge. This is done by acquiring the SOC, which is analyzed further in order to determine the next operation. If the SOC is above the high threshold mark, for example, 90%, a command is prompted to check whether there are any loads or appliances in use.

If there are no loads, text messages sending function is invoked (P_3M). At this point, the text message to be sent is to alert the user of the surplus energy availability and how this could be utilized for productivity. The message is also supposed to offer protection to the battery against overcharging through demand-side management.

On the flipside, if there are loads or appliances in use, the algorithm proceeds to check the load rating and consumption. This is followed by estimating how much the battery power would last at the current consumption rate; or for how long the battery would support the appliances until the SOC reaches the low threshold point, that is, 50%. If the appliance usage time can be improved or increased, a message sending function is invoke, and an SMS sent to a user suggesting how best usage time can be improved (P_4M). This message entails information suggesting ways of how the usage time can be increased, such as by turning off redundant lights or replacing the appliances with more energy-efficient ones, among others.

If the SOC is not above 90%, the algorithm proceeds to check whether it is below the low threshold point (i.e., 50%). If SOC is below 50%, the forecast subprocess is revisited to determine how long it may take the battery to charge to an adequate SOC. A text message is sent to the user informing them how long it would take for the battery to charge (P_2M) and hence the recommended time at which appliances can be connected. On the other hand, if the SOC is not below 50%, the algorithm proceeds to the forecast subprocess to determine how long it would take the battery to charge (i.e., general system performance) and then sends a text message informing the user of these revelations (P_1M) and the recommended time for appliance connection.

13.4.3 Output section: message categories, details, and sending criteria

This section encompasses the SMS interpretation and sending part of the algorithm.

Guided by [40] models, messages to be sent to the users were categorized into priorities and sample messages drafted. There are mainly four message categories labeled according to the priority number given (P_1M, P_2M, P_3M, P_4M). Priorities were given depending on the frequency a particular message type is sent, that is, SMSs that ought to be sent every day were categorized priority 1 (P_1M). The categorization of messages was done to avoid sending many unnecessary messages that could turn out redundant. It also helps to lessen the costs involved in message sending. On the users' side, it helps to avoid the unnecessary notifications in the form of inbox messages on their phones, which has been reported to be a contributing factor to disinterest and general lack of response to feedback systems [22].

13.4.3.1 Priority 1 message

This category of SMS aims at informing and advising users of the most appropriate time for connecting appliances. This message is also aimed at promoting and ensuring load shifting among users by alerting them of the times when excess energy is expected. Moreover, in the periods when the weather forecast shows a period of low irradiance from the sun, adjustments that would ensure prolonged usage and maximum utilization without harming the solar system would be communicated. This message category is sent at least once daily in the morning, depending on the state of the solar system and usage. For what more message samples would look like see Fig. 13.10 and there are others such as:

To have more lighting time, at night, consider connecting your appliances at 2:00pm.

To have enough TV time with your mates, consider doing your laundry at 1:30pm.

To have more lighting time for your children's homework, consider charging your phones at 1400 hrs.

It is examinations time, to ensure more reading hours at night for your children, consider connecting appliances at 2:00pm

13.4.3.2 Priority 2 message (P_2M)

This message category is invoked when the battery SOC is below or equal to the low-threshold point. This message type is aimed at protecting the battery as well as informing the user of the appropriate time for connecting his or her desired appliances. This message would be sent regardless of whether priority 1 message was sent or not. However, a second message of the same category cannot be sent on the same day. Message samples would look like this (See Fig. 13.11):

To enjoy your system tomorrow and the days to come, consider disconnecting appliances and connecting later at 3:00pm.

13.4.3.3 Priority 3 message (P_3M)

This message category aims at making the users realize the wasted and missed energy usage opportunities and how this can be avoided the next time [44]. By

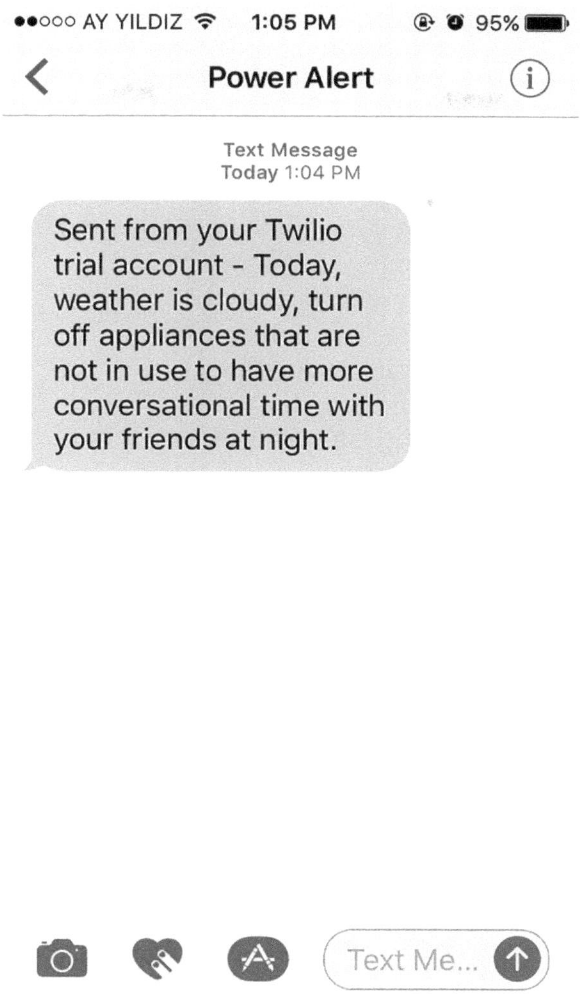

Figure 13.10 Screenshot of sample P_1M messages generated and sent by the developed algorithm.

informing the user of their past consumption and how this could be improved in different ways, the message points the user to the energy-awareness benefits and how the system can be utilized optimally. This message type also includes short texts that provide encouragement for small-energy actions to promote more substantial energy-efficient actions in the future. This is expected to influence user behavior rather than solely focusing on appliance efficiency. The message also aims at alerting the user in case the battery SOC goes beyond the maximum threshold value. Messages of this category would be weekly and dependent on the time a previous

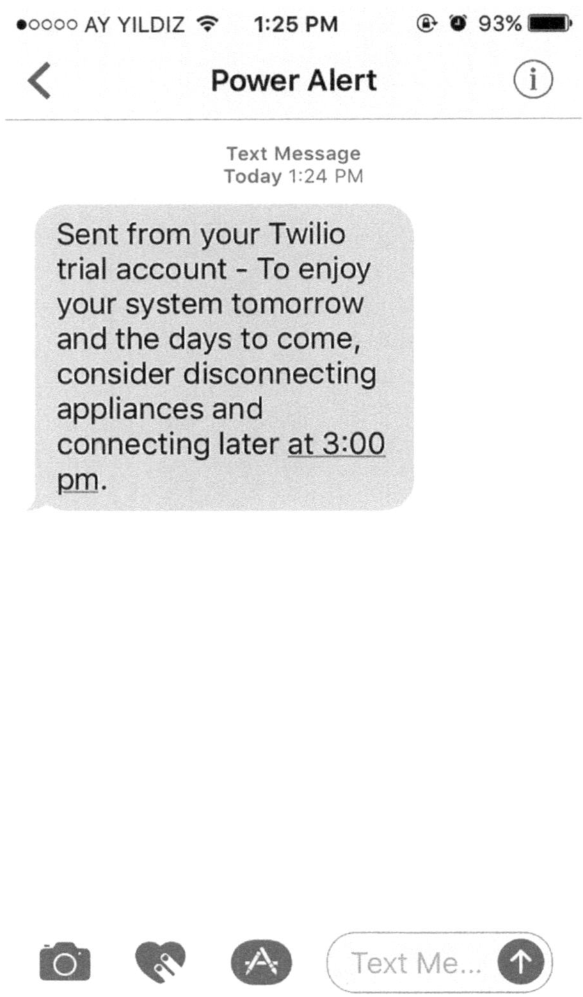

Figure 13.11 Screenshot of sample P_2M message generated and sent by the developed algorithm.

message of the same category was sent. Message samples (see Fig. 13.12) would look like this:

> *Last week you missed 10 hours of lighting time because of high consuming appliances, replace 10 W bulbs with 3 W bulbs for more reading time for your children*

> *Thanks for changing your lights, you were able to achieve 2 hours more lighting time for your children's homework*

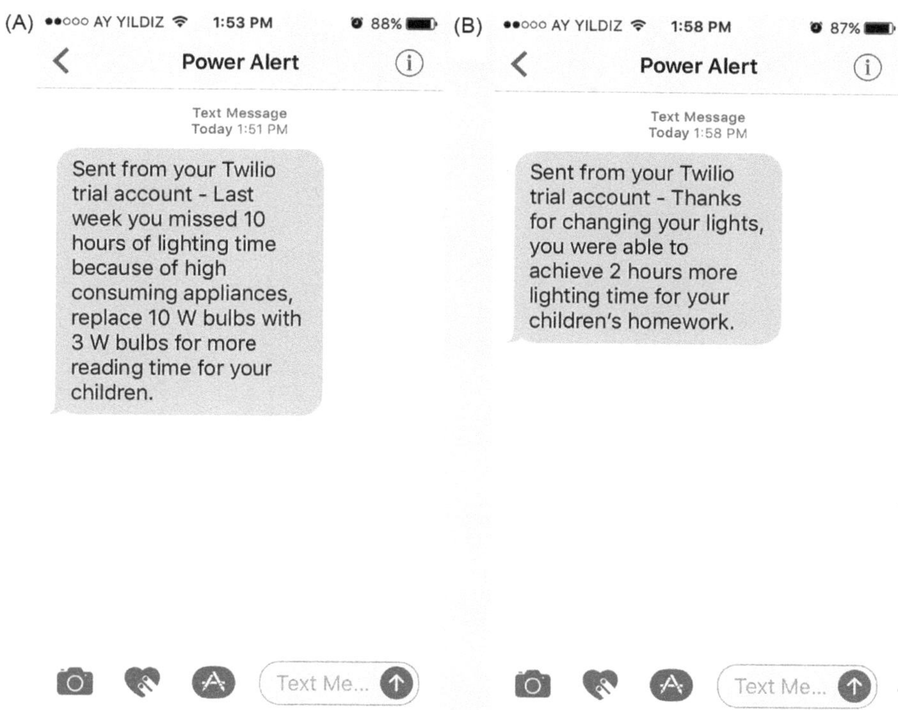

Figure 13.12 Screenshots of sample P_3M messages generated and sent by the developed algorithm.

13.4.3.4 Priority 4 message (P_4M)

This message category focuses on informing users of the appliance usage times and how this can be increased by replacing with more energy-efficient appliances plus other recommendations. These messages aim at educating the user how well the system would perform if more energy-efficient appliances were used. This is communicated through tangible ways like citing lighting time, amount of laundry that can be done, and the number of phones that can be charged. Also, this message category addresses the general system routine preventive maintenance. There could be a fall in the system performance caused by dust on the panels, loose connections, corroded battery terminals, or any other breakages in the system [44,45] due to poor maintenance. Since such incidences are not so common in the system, messages of this category are sent after 2–3 weeks from the previous time a message of the kind was sent. Message samples (see Fig. 13.13) would look like this:

To have more un-interrupted TV time, consider cleaning your solar panels

To ensure more lighting time consider replacing your 4 Watt bulbs with 2 Watt bulbs

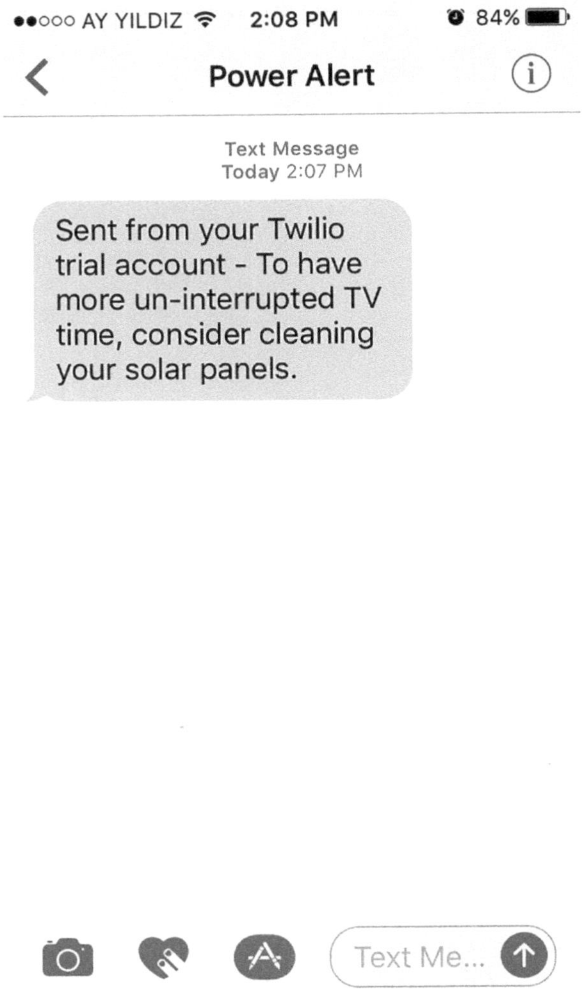

Figure 13.13 Screenshot of sample P_4M messages generated and sent by the developed algorithm.

13.4.4 Effective communication strategy

The information metrics is as important as the feedback system itself. Conversion into currency values has been well thought out as more desirable by some researchers [21,46]. In one study [47] cost-based feedback was found to be preferred by the consumers. However, in other research no positive results were yielded when cost-based feedback was employed [47]. Environmental metrics have also been used in many cases [48]. Given the current climate change and global warming discourse, environmental metrics would play a leading role in stimulating personal norms and connections concerning the environment [46,47,49]. A study [50] revealed no significance in the results for electricity use efficiency when environment-based feedback was used [51] postulated that the link between environmental concerns and electricity usage is not absolute even in households that seem to be aware of the environment. The standard delivery system is the computerized system where users interface with applications on display screens and an extension usually added through emails [21,52]. Email delivery has been cited to not only provide personal delivery but also can be integrated to websites that provide more interactive options than the email feedback alone. The use of timely feedback coupled with an SMS and email system was used in another study and an average yearly consumption reduction of 3% was noted. However, the authors further argued that this kind of feedback is most likely to yield better results in another country setting [53]. The feedback that allows users options to choose and interact was cited to be effective [52]. In some cases, paper-based feedback has been considered environment-unfriendly [54], while in others, they are preferred over other feedback systems [55]. Whereas internet-based feedback is more comfortable and consequently more feasible for utilities, demographics and other location factors are a critical point that cannot be ignored.

The decision of what to send and how well this can be communicated is mainly dependent on the user's day-to-day activities, social behavior, perspective, and understanding. The user perspective is among the most critical aspects of feedback systems because the actions that proceed after receiving the information are dependent on whether the information is understood or not. We would recommend having the messages communicated in the local language of the intended user. For the effectiveness of communication, it would be essential to involve users in the development process by engaging and sensitizing them on the importance of such a system and how to interpret the messages received. This would help to draw a line on what matters to the users and how well this can be incorporated into the messaging translation system. In some studies, feedback systems have been coupled with incentives to make them more efficient [21,23]. Incentivizing the whole process would have an impact on the results achieved and has been reported in some studies.

13.4.5 Frequency and time of notification

The frequency of notification is dependent on the message priorities that were set, that is, the algorithm sends a minimum of one message every day (P_1M) depending on the system performance and general usage. The messages were prioritized depending on the period of delivery and on how frequent specific information would be required to be passed on to the user in order to maintain the positive usage behaviors repetitively.

However, the user has a pivotal role to play in deciding how frequent he or she should be notified and deliberating on the appropriate time to do this. This is because the messages are dependent on how well the system is being used and maintained.

13.4.6 Potential benefits

A system such as this would not only benefit the users by ensuring better system performances, reduction in system failures and hence economic boost, but would also ensure energy awareness among off-grid energy users. The different messages sent are meant to capture the scope of the entire solar system in use, hence ensuring an understanding of what users are paying for when it comes to metering and other platforms like PAYG. Also, communicating consumption to users would be vital in reducing maintenance costs involved and system breakages. These sent messages can play a role in stimulating users to think of other business avenues like community phone charging and other appliances due to the availability of surplus energy.

Some studies have mentioned the failure of many donor-funded off-grid projects due to a complete detachment of the communities from the projects [18,19]. Feedback systems like the one demonstrated in this chapter would be vital in stimulating community participation and ownership in such projects.

13.5 Conclusion

This chapter attempts to answer the question of how consumption information can be made accessible and communicated in a transparent, understandable, and straightforward way in off-grid rural households. Furthermore, it elaborates on how this information can be leveraged to improve the performance of solar systems by influencing user behaviors. The study argues that in order to achieve this, already existing platforms such as mobile phones could be an advantage in this regard because they are common in rural households. The algorithm developed in the study uses weather data, static data like battery capacity and solar system size, plus real-time data series from the influx database. In the weather data, solar irradiation data is targeted thereby predicting the expected power production from the solar system. The battery capacity data and solar system data are used to set threshold parameters for monitoring the entire system. The battery, solar PV, and loads or appliances are the basis over which the SMS-sending scheme depends. The algorithm was tested and screenshots of messages are presented in Section 13.4.

In the face of slow growth of grid electricity among isolated and dispersed communities, off-grid electricity has experienced steep growth. However, the literature as far as measures to address EE and customer understanding in off-grid systems is concerned is still minimal. In this regard, therefore, this chapter tackles the problem of invisibility of consumption information in off-grid households, which is a stronghold for influencing user behaviors and hence promoting EE. This study could, therefore, be used by solar supplier companies to develop sustainable models that not only persuade users to pay, but also encourage understanding and ensure better performance of solar systems.

Acknowledgment

This study was carried out with the assistance of funding from the Alexander von Humboldt foundation under the framework of the international climate protection fellowship.

Sincere appreciation goes to Mr. Rolex Muceka, Mr. Yunus Alokore, Mr. Michael Goette, and Ms. Ogechi Nwadiaru Vivian for the advice and technical support they provided throughout the study.

References

[1] World Bank, Global tracking framework 2017—progress toward sustainable energy (WWW Document), World Bank. <https://www.worldbank.org/en/topic/energy/publication/global-tracking-framework-2017>, 2017a (accessed 15.10.19).
[2] S. Groh, The role of energy in development processes—the energy poverty penalty: case study of Arequipa (Peru), Energy Sustain. Dev. 18 (2014) 83–99. Available from: https://doi.org/10.1016/j.esd.2013.12.002.
[3] K. Zhou, S. Yang, Understanding household energy consumption behavior: the contribution of energy big data analytics, Renew. Sustain. Energy Rev. 56 (2016) 810–819.
[4] J.J.C. Da Silva, A.M. Ganhao, Training Program About Energy Efficiency in Existing Buildings, PLEA, 2013.
[5] K.B. Janda, Buildings don't use energy: people do, Archit. Sci. Rev. 54 (2011) 15–22.
[6] M.K. Mattinen, P. Tainio, M. Salo, M. Jalas, A. Nissinen, E. Heiskanen, How building users can contribute to greenhouse-gas emission reductions in Finland: comparative study of standard technical measures, user modifications and behavioural measures, Energy Effic. 9 (2016) 301–320.
[7] L. Collins, S. Natarajan, G. Levermore, Climate change and future energy consumption in UK housing stock, Build. Serv. Eng. Res. Technol. 31 (2010) 75–90.
[8] W. Abrahamse, L. Steg, C. Vlek, T. Rothengatter, The effect of tailored information, goal setting, and tailored feedback on household energy use, energy-related behaviors, and behavioral antecedents, J. Environ. Psychol. 27 (2007) 265–276.
[9] I. Orlandi, N. Tyabji, J. Chase, Off-Grid Solar Market Trends Report 2016, New Energy Finance World Bank IFC Global Off-Grid Lighting Association, Bloomberg, 2016.
[10] World Bank, State of Electiricity Access Report 2017 (Vol. 2): Full Report (No. 114841), The World Bank, 2017b.
[11] S. Groh, S. Pachauri, N.D. Rao, What are we measuring? An empirical analysis of household electricity access metrics in rural Bangladesh, Energy Sustain. Dev. 30 (2016) 21–31.
[12] International Energy Agency, Sustainable Energy for All 2013–2014: Global Tracking Framework Report, The World Bank, 2014.
[13] A.P. Heynen, P.A. Lant, S. Sridharan, S. Smart, C. Greig, The role of private sector off-grid actors in addressing India's energy poverty: an analysis of selected exemplar firms delivering household energy, Energy Build. 191 (2019) 95–103.
[14] S.C. Bhattacharyya, Financing energy access and off-grid electrification: a review of status, options and challenges, Renew. Sustain. Energy Rev. 20 (2013) 462–472.

[15] H. Olk, J. Mundt, Photovoltaics for productive use applications, A Catalogue of DC-Appliances Dtsch. Ges. Für Int. Zusammenarbeit GIZ GmbH, 2016.

[16] R.V. Buskirk, Doubling the global pace of progress for energy efficiency (EE): applying a "Moore's Law" of EE to technology innovation for off-grid applications, 2015.

[17] T. van Gevelt, J. Holmes, Business models for home-based electricity services, Smart Villages Initiative Technical Report, 2015.

[18] P. Díaz, C.A. Arias, M. Gomez-Gonzalez, D. Sandoval, R. Lobato, Solar home system electrification in dispersed rural areas: a 10-year experience in Jujuy, Argentina Prog. Photovolt. Res. Appl. 21 (2013) 297−307.

[19] E. Mills, The specter of fuel-based lighting, Am. Assoc. Adv. Sci. 308 (5726), (2005) 1263−1264. https://doi.org/10.1126/science.1113090.

[20] A. Mahone, B. Haley, Overview of Residential Energy Feedback and Behavior-based Energy Efficiency, Energy and Environmental Economics Inc., 2011.

[21] S. Darby, The effectiveness of feedback on energy consumption. A review for defra of the literature on metering, billing and direct displays, Sci. Res. 486 (2006) 26.

[22] D. Vine, L. Buys, P. Morris, The effectiveness of energy feedback for conservation and peak demand: a literature review, Open J. Energy Effic. 2 (2013) 7−15.

[23] S. Darby, Smart metering: what potential for householder engagement? Build. Res. Inf. 38 (2010) 442−457.

[24] S.C. Bhattacharyya, D. Palit, Mini-Grids for Rural Electrification of Developing Countries: Analysis and Case Studies from South Asia, Springer, 2014.

[25] N.N. Opiyo, Impacts of neighbourhood influence on social acceptance of small solar home systems in rural western kenya, Energy Res. Soc. Sci. 52 (2019) 91−98.

[26] C.A. Friebe, P. von Flotow, F.A. Täube, Exploring the link between products and services in low-income markets—evidence from solar home systems, Energy Policy 52 (2013) 760−769.

[27] F.D.J. Nieuwenhout, A. Van Dijk, P.E. Lasschuit, G. Van Roekel, V.A.P. Van Dijk, D. Hirsch, et al., Experience with solar home systems in developing countries: a review, Prog. Photovolt. Res. Appl. 9 (2001) 455−474.

[28] I.P. Da Silva, G. Batte, J. Ondraczek, G. Ronoh, and C.A. Ouma, Diffusion of solar energy technologies in rural Africa: Trends in Kenya and the LUAV experience in Uganda. 2015.

[29] R. Ren21, Global Status Report. REN21 Secr. Paris, 2016.

[30] J. Amankwah-Amoah, Solar energy in sub-Saharan Africa: The challenges and opportunities of technological leapfrogging, Thunderbird Int. Bus. Rev. 57 (2015) 15−31.

[31] International Telecommunication Union, 2018. Global ICT developments (2001−2018) [WWW Document]. <https://www.itu.int/en/ITU-D/Statistics/Pages/stat/default.aspx> (accessed 16.10.19).

[32] W. Mitullah, R. Samson, P. Wambua, S. Balongo, AD69: Building on progress: Infrastructure development still a major challenge in Africa | Afrobarometer [WWW Document]. <http://afrobarometer.org/publications/ad69-building-progress-infrastructure-development-still-major-challenge-africa>, 2016 (accessed 16.10.19).

[33] P. Stalin, S.B. Abraham, K. Kanimozhy, R.V. Prasad, Z. Singh, A.J. Purty, Mobile phone usage and its health effects among adults in a semi-urban area of southern India, J. Clin. Diagn. Res. JCDR 10 (2016) LC14.

[34] M. Mutiga, Z. Flood, Africa calling: mobile phone revolution to transform democracies, The Guardian, 2016.

[35] Pew Research Center, Cell phones in Africa: communication lifeline. Pew Research Center, Global Attitudes Project <https://www.pewresearch.org/global/2015/04/15/cell-phones-in-africa-communication-lifeline/>, 2015 (accessed 16.10.19).
[36] GSMA, The Mobile Economy Sub-Saharan Africa, 2019.
[37] Fenix International, ReadyPay solar power—lease-to-own off-grid systems [WWW Document]. Fenix Int. <https://www.fenixintl.com/product/>, 2019 (accessed 2.8.19).
[38] BBOXX, Our technology. BBOXX. <https://www.bboxx.co.uk/our-technology/>, 2019 (accessed 2.8.19).
[39] M-KOPA Solar, Our technology, M-KOPA Corporate [WWW Document]. <http://solar.m-kopa.com/about/our-technology/>, 2019 (accessed 2.8.19).
[40] H.A. He, S. Greenberg, E.M. Huang, One size does not fit all: applying the transtheoretical model to energy feedback technology design, in: Proceedings of the SIGCHI Conference on Human Factors in Computing Systems, ACM, 2010, pp. 927–936.
[41] I. Koumparou. The development of a forecasting tool for predicting the energy yield from PV systems in Cyprus, 2014.
[42] A.C. Lorenc, Analysis methods for numerical weather prediction, Q. J. R. Meteorol. Soc. 112 (1986) 1177–1194.
[43] J.F. Manwell, J.G. McGowan, Lead acid battery storage model for hybrid energy systems, Sol. Energy 50 (1993) 399–405.
[44] A. Reinders, A. Sudradjat, V.A.P. Van Dijk, R. Mulyadi, W.C. Turkenburg, Sukatani revisited: on the performance of nine-year-old solar home systems and street lighting systems in Indonesia, Renew. Sustain. Energy Rev. 3 (1999) 1–47.
[45] A. Kalirasu, S.S. Dash, Simulation of closed loop controlled boost converter for solar installation, Serbian J. Electr. Eng. 7 (2010) 121–130.
[46] C. Fischer, Feedback on household electricity consumption: a tool for saving energy? Energy Effic. 1 (2008) 79–104.
[47] G. Fitzpatrick, G. Smith, Technology-enabled feedback on domestic energy consumption: articulating a set of design concerns, IEEE Pervas. Comput. 8 (2008) 37–44.
[48] K. Ehrhardt-Martinez, A. John, K.A. Donnelly, Beyond the meter: enabling better home energy management, Energy, Sustainability and the Environment, Elsevier, 2011, pp. 273–303.
[49] J. Schleich, M. Klobasa, S. Gölz, M. Brunner, Effects of feedback on residential electricity demand—findings from a field trial in Austria, Energy Policy 61 (2013) 1097–1106.
[50] G. Brandon, A. Lewis, Reducing household energy consumption: a qualitative and quantitative field study, J. Environ. Psychol. 19 (1999) 75–85.
[51] B. Gatersleben, L. Steg, C. Vlek, Measurement and determinants of environmentally significant consumer behavior, Environ. Behav. 34 (2002) 335–362.
[52] C. Fischer, Influencing electricity consumption via consumer feedback: a review of experience, in: Proceedings, European Council for an Energy-Efficient Economy, ECEEE, 2007, pp. 1873–1884.
[53] M. Gleerup, A. Larsen, S. Leth-Petersen, M. Togeby, The effect of feedback by text message (SMS) and email on household electricity consumption: experimental evidence, Energy J. (2010) 113–132.
[54] S. Roberts, W. Baker, Towards effective energy information: improving consumer feedback on energy consumption, Cent. Sustain. Energy 3 (2003) 20.
[55] M.S. Martinez, C.R. Geltz, Utilizing a pre-attentive technology for modifying customer energy usage, in: Proceedings, European Council for an Energy-Efficient Economy, 2005.

Recent developments in hydrogen fuel cells: Strengths and weaknesses

14

Natarajan Rajalakshmi, Rengarajan Balaji and Sundararajan Ramakrishnan
Centre for Fuel Cell Technology, International Advanced Research Centre for Powder Metallurgy and New Materials, IIT Madras Research Park, Chennai, India

14.1 Fuel cells based on hydrogen

14.1.1 History and utilization

Hydrogen and fuel cells are among various technologies that could pave the way to deploy renewable energy sources for all the energy related applications with CO_2-free energy cycles. A fuel cell is an electrochemical device that converts the chemical energy of a fuel (hydrogen, alcohols, and hydrogen-rich hydrocarbons) into electrical energy. Fuel cells in principle are not limited by the Carnot cycle efficiency like the internal combustion (IC) engines. In case of fuel cells, efficiency largely depends on the ability of an electrocatalyst for fuel oxidation and oxygen reduction. The thermodynamic efficiency of the fuel cell which is the ratio of change in Gibbs free energy to the change in total enthalpy ($\Delta G/\Delta H$) is 83% for a simple H_2 and O_2 reaction. The ideal thermodynamic voltage and the effect of reactant concentration on the efficiency of the fuel cells can be found in detail elsewhere [1–3]. The free energy of any redox reaction could theoretically be completely converted to electrical energy. However, there are certain factors that inhibit the efficiency of the real fuel cell.

Thermodynamically an ideal fuel cell would supply any amount of current maintaining a constant voltage as long as it is supplied with sufficient fuel. In practice, however, the measured voltage output of a real fuel cell is less than the ideal thermodynamically predicted voltage due to irreversible losses. The more the current that is drawn from the fuel cell the greater will be the losses. The real voltage output for a fuel cell can thus be written by starting with the thermodynamically predicted voltage output of the fuel cell and then subtracting the voltage drops due to various losses:

$$V = E_{\text{thermo}} - E_{\text{act}} - E_{\text{ohmic}} - E_{\text{conc}} \tag{14.1}$$

Since a single fuel cell can produce only low voltage at operating current, several cells are connected to build a stack of cells to produce the required power output. A fuel

Sustainable Fuel Technologies Handbook. DOI: https://doi.org/10.1016/B978-0-12-822989-7.00015-9

cell stack can be configured in series, parallel, series–parallel, or as single units, depending on the type of application. The number of fuel cells in a stack determines the total voltage, and the surface of each cell gives the total current. Fuel cells can be classified primarily by the kind of electrolyte they employ, but they all work on the same principle. Different types of fuel cells are:

- polymer electrolyte membrane fuel cells (PEMFC);
- alkaline fuel cells;
- phosphoric acid fuel cells;
- molten carbonate fuel cells; and
- solid oxide fuel cells (SOFC).

Fuel cells can also be classified primarily based on their operating temperature as well with the type of fuel they use. Ion-conducting species also change depending on the fuel cell and the materials that are employed [4]. They are summarized in Table 14.1.

PEMFC are approaching technical maturity and economies of scale. Commercial deployment for number of applications has started in several countries like Japan, United States, and in Europe. Energy conversion with a PEMFC from hydrogen to electricity would essentially result in water as the only emission by-product and low-quality heat, with the low temperature providing high tolerance for cycling operation. The operation requires pure hydrogen and the operating temperature is quite low compared to all the other types of fuel cells. The relative maturity of the technology can lead to relatively low cost. Green house gas emissions from aviation sector is less than road transports still aircraft manufacturers are vying on increased efficiency with reduced emissions. Aircraft architecture is undergoing a trend toward "more electric aircraft" which is replacing the hydraulically or pneumatically driven components. Fuel cell systems are a promising technology solution regarding the enhancement of energy efficiency for cruise and ground operations. For many years, the manufacturers of commercial aircraft Boeing and Airbus have realized that fuel cells may offer advantages for commercial aircraft operation. Apart from emission reductions and thermal efficiency, they can constitute distributed power systems, enabling locating the power near the point of use (reducing wiring) and also reducing the power drawn from the engines. The efficiency is moderate, around 50%–60%, and with the low operating temperature, heat recovery is not feasible. The modules currently have an output of up to 120 kW, and the physical size is small, which is beneficial for many applications. The PEMFC technology has received the highest ranking in spite of its major challenges like durability, cost, sensitivity to impurities, and the complex water management system (both gas and liquid). This technology also is characterized by highly sensitive operating conditions and a high system mass.

14.1.2 Hydrogen fuel

Utilization of hydrogen as a fuel dates back some centuries to when balloons were used to provide lift as shown in Fig. 14.1 and in airships during the late and early

Table 14.1 Classification of fuel cells.

Fuel cell type	DMFC	PEMFC	AFC	PAFC	MCFC	SOFC
Electrolyte type	Polymeric ion-exchange membrane	Polymeric ion-exchange membrane	Immobilized aq. KOH solution/ anion-exchange membrane	Immobilized liquid phosphoric acid/ phosphoric acid doped PBI membrane	Immobilized liquid molten carbonate	Ceramic (mostly YSZ)
Operating temperature (°C)	20–90	30–100	50–100	150–210	~650	500–1000
Charge carrier	H^+	H^+	OH^-	H^+	CO_3^{2-}	O^{2-}
Fuel	Methanol	Hydrogen	Hydrogen	Hydrogen	Natural gas/ hydrogen	Natural gas/ hydrogen
Applications and main advantages	*Electronic devices*: Mobile phone, laptop chargers etc. with methanol cartridges	*Automobile and transport applications*: PEMFC stacks are used with compressed H_2 cylinders	*Space applications*: Requires high-pure fuel and oxidant	*Stationary power generation*: Can tolerate CO_2 and even low concentrations of CO; at low temperatures ionic conductivity is very low	*Distributed power generation, CHP*: High conversion efficiency and flexibility of fuels (especially in case of SOFC)	

AFC, Alkaline fuel cells; *MCFC*, molten carbonate fuel cells; *PAFC*, phosphoric acid fuel cells; *PEMFC*, polymer electrolyte membrane fuel cells; *SOFC*, solid oxide fuel cells.
Source: From K.S. Dhathathreyan, N. Rajalakshmi, R. Balaji, Nanomaterials for PEM Fuel Cells, Wiley Publishers, 2017.

Figure 14.1 First Manned Hydrogen Balloon Flight, Paris, 1783. Courtesy: Creative Commons/Wikimedia
Source: From https://commons.wikimedia.org/wiki/File:Jacques_Charles_Luftschiff.jpg.

part of 17th and 18th centuries [5]. Prior utility was not documented until combustion engines started to capitalize on hydrogen as a fuel at the start of industrial revolution followed by fertilizer industries in the production of synthetic ammonia. At the behest of the industrial revolution, global demand for oil rose as did hydrogen, which is used for refining the oils.

Given the population explosion, the demand for energy rose to gargantuan levels and this resulted in over exploitation of natural resources, in particular coal. Other energy generation sources, for example, nuclear energy, entail heavy input costs and are associated with health risks. Solar energy conversion appears promising but inefficiency limits its utility. When it comes to energy carriers, electricity has been predominant and with the advent of electrolyzer and fuel cells, hydrogen assumes the second place. While the former converts water/water methanol mixture into useful hydrogen, the later utilizes it and generates electricity. Utilization of fuel cell technology in NASA Apollo missions in 1960s triggered further application of fuel cells across many fields [6]. Though it appears as an energy intensive process, electrolysis produces hydrogen without much harm to the atmosphere and electrolyzers are based on this principle. Apart from various factors, like the gradual depletion of oil-based natural resources, safety concerns with nuclear fuels, efficiency limits of solar cells, and given the quest for greener energy sources, hydrogen assumes importance in the current scenario.

A hydrogen council has been formed by many energy companies, industrial gas companies, original equipment manufacturers for vehicles, and other industry stakeholders to take advantage of the potential of hydrogen fuel. Initial efforts could focus on large-scale applications so as to rapidly generate economies of scale with minimal infrastructure requirements, as well as for sectors where hydrogen from renewables stands out as the best-performing option to meet climate targets and comply with local emission regulations [7]. Such applications include large-scale industry (e.g., petrochemicals, steel) and medium- to heavy-duty transport (medium to large passenger vehicles and commercial vehicles, large fleets of buses, trucks, trains, maritime, aviation, etc.).

Large-scale deployment of fuel cells is also hampered by the high cost and issues of durability under different operation conditions. Cost computation of fuel cell involves materials, processing, assembly and testing, including the man-hours involved in all associated processes. While in the case of low-temperature fuel cells (LTFC), the cost is dominated by the type of electrocatalyst and electrolyte used, the processing cost and wastage dominate the high-temperature fuel cell (HTFC) systems. Enhancing the efficiency of the cells at low electrocatalyst loading, the development nonnoble metal catalysts have been the focal point of research with LTFC. Recent developments are in the area of oxygen reduction reactions and electro-oxidation of small hydrocarbon fuels such as methanol, ethanol, and formic acid which are considered logistic fuels [8]. Synthetic fuels based on hydrogen produced from renewable electricity (through water electrolysis) are fed together with CO_2 into a reactor forming a synthesis gas (CO and H_2) that is then liquefied and further refined to become, for example, e-diesel or e-kerosene. Such fuels can be blended with existing fuels in IC engine vehicles and are also referred to as drop-in fuels.

14.1.3 Paramount importance

Hydrogen and fuel cell technologies play a big role in the transition to a low-carbon economy due to their similar performance, operation, and consumer experience to conventional fossil-fuel technologies. They also provide valuable opportunities such as carbon capture and storage, bioenergy and hybrid heat pumps. Significant changes from the last hydrogen "hype cycle" in the 2000s is that manufacturing scale-up and cost decreases mean hydrogen and fuel cells are being commercialized in several sectors, from portable electronics and backup power to fork-lift trucks. Many international corporations have formed the hydrogen councils to position hydrogen as a key solution for energy transition that involves many challenges due to its complexity and diversity.

Nearly half of the energy demand for global transport is for light-duty vehicles and the number of passenger cars worldwide is expected to rise from 1 to 2.5 billion by 2050. Deployment of hydrogen fueled vehicles will result in significant reduction of GHG emissions into the atmosphere. It has been observed that over half a million premature deaths per year across Europe are due to particulates and NO_x emissions. The direct cost of air pollution due to illness-induced loss of production, healthcare, crop-yield loss, and damage to buildings is around $28 Billion per year across Europe with external costs estimated to be euro 330–940b per year. More than 92% of the world's population are exposed to poor air quality. Internationally diesel-powered cars and trucks will be banned by 2025, and pure combustion vehicles are expected to be banned by 2040 [9].

Hydrogen and fuel cell technologies can support climate change, energy security in terms of transport, industry, buildings, power sector, and so forth. Hydrogen can also be connected to different energy sectors and energy networks, thereby increasing the operational flexibility. The major advantages lie in achieving very low-carbon individual motorized transport, integrate variable renewable energy (VRE) into the energy system, and contribute to the decarbonization of the industry and building sectors.

14.1.4 Fuel cells for stationary applications

For a very long time PEM fuel cells have been considered as a future technology for stationary applications. The main objective of this application area is to develop a PEM fuel cell system with reduced stack cost, and improved performance to the level required by the combined heat and power (CHP) and stationary power generation markets. It can be achieved only by reducing the gap existing between laboratory prototypes and precommercial systems.

The deployment of PEMFC systems for large-scale stationary applications in a specific region can also be influenced by the presence of a PEMFC system manufacturer in the area that increase the awareness with potential users regarding the benefits of the technology they provide versus competing technologies and the presence of end-users that require, steady electricity supply (e.g., data centers) in regions with poor grid reliability. The application of PEMFCs for the stationary sector shows an increasing trend of technology transfer from the manufacturer to end user and is being recognized as a reliable option compared to conventional technologies such as IC engines or batteries. The major commercially available stationary PEM-based fuel cell systems for various ranges is listed in Table 14.2.

The Navigant report indicates that over the past few years the positive growth of stationary fuel cell industry is due to a surge in US and foreign governments' interest in flexible and reliable energy sources, and the global market potential could increase to more than 50 GW. Nearly 2.5 million telecom towers will be supported by fuel cell-based back-up power systems across the globe by the end of 2020 [10].

Table 14.2 List of fuel cell manufactures for stationary application around the globe and their product description.

Manufacturer	Country	Product	Output Range in kW
Ballard power systems	Canada	ClearGen	Multi-500 kW power banks
ENEOS Cell Tech	Japan	Ene-Farm	0.250–0.700
Hydrogenics	Canada	MW-power plant	1 MW
Nedstack	Netherlands	HP	2–10 (scalable)
Panasonic	Japan	Ene-Farm	0.200–0.750
Toshiba	Japan	Ene-Farm	0.250–0.700
Acta S.p.A	Italy	Acta power	2 and 4
Ajusa	Spain	NOIL 5000AC UP	5
Altergy Systems	United States	Freedom Power	0.500, 1, 5, and 7.5
Axane	France	CommPAC 500	0.500–10
Ballard power systems	Canada	FCgen-1020ACS	1.5–3.6
		FCgen-1300	2–11
		ElectraGen-ME	2.5 and 5
		ElectraGen-H2	1.7, 2.5, and 5
PowerCell	Sweden	S1	3.0
Electro Power Systems S.p.A.	Italy	Electroself, Electro	1.5–10
First Element Energy	United States	Air-cooled/water-cooled	2–25
Horizon FCT	Singapore	H-Series	0.010W–5kW
		Ecobox-MR	1–10
		GreenHub Powerbox	0.500–2
		MFC Mini	0.050
Hydrogenics	Canada	HyPM XR Power Modules	4.5–12
		HyPM Rack	2–200
Intelligent Energy	United Kingdom	Air-cooled	5.0
Nedstack	The Netherlands	HP	2–10
		XXL	2–9.5
ReliOn, a Plug Power Company	United States	E-1000x	1–4
		E-1100	1.1–4.4
		E-200	0.175–0.525
		E-2200x	2.2–17.5
		E-1100v	1.1
		E-2500	2.5–20

Ballard Power Systems agreed to provide a 1 MW fuel cell distributed power generation system for Hydrogène de France that will be installed at a sodium chlorate chemical plant in Bordeaux Métropole, utilizing its by-product hydrogen. Further, Nedstack signed an agreement to deliver a 2 MW PEMFC power plant using the by-product hydrogen obtained from chlor-alkali process at Ynnovate Sanzheng Fine Chemicals in China with the aim of lowering electricity consumption by 20%. The commercial operation of a 1 MW fuel cell power system was started by hydrogenics at a Hanwha-Total oil refinery site in Daesan, South Korea. Horizon Fuel Cell Technologies of Singapore has sold a 200 kW PEM fuel cell system to Ulsan in South Korea as part of the Ulsan Techno Park (UTP) project. This project is part of the Hydrogen Town initiative which aims to achieve 1 MW of electricity generation using "waste" hydrogen in the industrial city. The fuel cell developer Nedstack has aimed for developing a PEMFC system at a CAPEX of <1500 euro/kW, as the costs of the current MW scale units is seen as too high. The optimization efforts are focused toward improvement of MEAs, stack design, and overall system balance of the plant [11].

The use of PEMFCs for providing reliable backup power for communication networks is increasing steadily in rural and remote areas that may not have access to electrical grids or infrastructure. In India, providing backup power for telecom towers using a fuel cell system is expanding and several key players are making substantial sales. Ballard Power has delivered about 3000 PEM-based fuel cell system for backup power application for telecommunication sectors. Among these systems, about 100 systems were installed by Reliance Jio Infocomm Ltd., and about 50 systems were installed at Aditya Birla's 'group Idea Cellular network in India.

The history of stationary fuel cell development and its application has culminated in a strong growth during the past decade and the implementation of large-scale fuel cells is currently dominated by the United States and South Korean market (almost 95% of installed capacity). There are also major differences in the approach taken at state level within the United States, with the majority of the capacity installed in two states, California and Connecticut. In order to implement large-scale stationary fuel cells activities, fuel cell manufacturers are still largely dependent on public funding, whether through technology push or market pull measures. The main obstacles at this stage seem to be reliability and the cost of PEMFCs. A lack of fundamental understanding of electrochemical processes has been cited as one of the causes of the many setbacks suffered by the fuel cell industry.

A breakthrough with respect to cost and lifetime of PEMFC systems is essential to successfully introduce the PEMFC technology in at least a part of this huge market. Therefore the constant efforts targeted at improved PEFC stacks and their key components, which are mainly responsible for the high cost, is of highest priority. The higher operating temperatures and substitution of perflourinated membranes are ongoing research tasks. The improvement of the electrocatalyst activity of platinum or nonnoble metal electrodes is of similar importance to reduce material cost and simultaneously increase the power density of the stack.

14.1.5 Fuel cells for transport applications

Based on an earlier study, the majority of greenhouse gas (GHG) emissions is likely due to the transportation industry which accounts for roughly 17% of total emissions per year [12]. Key changes in the way propulsion systems are handled in the sector, if addressed, will bring larger implications in controlling air pollution. Looking at the larger picture, it will help to adhere to Kyoto protocol with reference to the global warming. GHG emissions and petroleum usage estimates for future mid-sized passenger cars that utilize various propulsion systems are depicted in Fig. 14.2 [13].

Fuel cells with near-zero harmful emissions will come to the fore to address the above issue in the transportation sector with twice as much efficiency than conventional gasoline-based propulsion, say ICEs [14]. Although a fuel cell with no

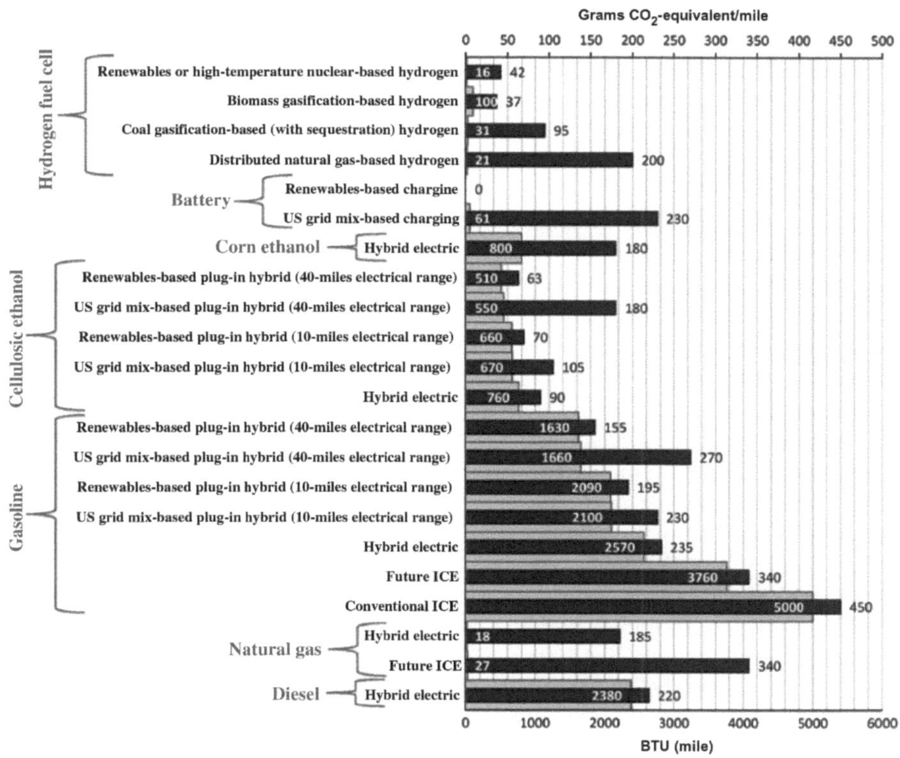

Figure 14.2 GHG emissions and petroleum use estimates for future (2035-2045) mid-size passenger car with various propulsion systems [13].
Source: From O.Z. Sharaf, M.F. Orhan, An overview of fuel cell technology: fundamentals and applications, Renew. Sustain. Energy Rev. 32 (2014) 810–853, Data sourced from T. Nguyen, J. Ward, Well-to-wheel greenhouse gas emissions and petroleum use for mid-size light-duty vehicles. US Department of Energy. Record#1001; 2010.

moving parts offers clean energy and efficient operation with little maintenance, the major handicap is cost. The membrane electrode assembly, which is where the energy conversion happens, contributes to the majority of the cost of the fuel cell since it often consists of noble metals like platinum. However, reduction in the cost of the fuel cell is due to the outcome of alternative and feasible solutions by researchers worldwide coupled with the subsidies from the governments. Absence of real-time application test data in the public domain is another shortfall this industry is facing as the companies which test them (i.e., majority of high-end passenger car manufacturers) prefer to patent them thus leaving little to fetch for. It may for their own benefit rather for collective growth.

Among all the types of fuel cells, PEMFCs are ideal for transportation applications due to their light weight, low operating temperature, and high-power density with better efficiency. Also the modularity it can offer, the freedom to change the size and shape according to the vehicle and the compactness of the system, is ideal for any transportation vehicle. Lack of hydrogen refuelling infrastructure is still a bottleneck that needs to be addressed. Due to the limited number of refueling stations available across the world, fuel cell–powered propulsion is yet to occupy its deserved place. Steps are required for sustained supply of high-purity hydrogen for commercialization.

The poisoning of the catalyst is a serious concern and requires further attention in the case of LTFC like PEMFC. SOFC, a HTFC, can offer fuel flexibility. For PEMFCs to be a candidate for transportation application, it should meet two critical aspects, volumetric and gravimetric power density. Durability of the fuel cell is also a serious concern to compete with gasoline or battery technology. Fast start-up is expected for any transportation vehicle as well as the sustainability of the fuel cell for repeated discontinuous operations [15]. Also the technology has to address uncertainty concerning safety regulations.

Most of the fuel cell development and application in the transportation application, initially was restricted to auxiliary power units. Consider the case of heavy-duty trucks where idling alone constitutes to 40% of running time, which will spew enough unspent gasoline into the atmosphere contributing to global warming and real waste of fuel. A fuel cell utilized along with a conventional propulsion system can supply the required power during idling and save energy. Two-wheelers outnumber passenger cars in terms of transportation vehicles and they are a major source of pollution. In future, both two-wheelers and passenger cars bound to adopt cleaner technologies like fuel cell to curb the menace it created so far in terms of GHG emissions. Though PEMFC is best suited for passenger cars, direct methanol fuel cell (DMFC) is gaining pace along due to the possibility of on-board production of hydrogen from methanol by steam reforming. However, the cleanup procedure is a bottleneck in this particular technology.

Automobile manufacturers including Hyundai, Toyota, Honda, General Motors, Daimler AG, Nissan, and Volkswagen have introduced fuel cell–powered passenger cars with Honda being at the forefront in delivering the model FCX Clarity in 2007. More recently Toyota came up with its version, called MIRAI (meaning “dream” in Japanese), which offered unparallel driving experience to its consumers. Automotive vehicle manufacturers around the world poised to introduce fuel-cell powered vehicles within the next decade in almost all the categories of vehicles[16]. Adoption of fuel cell–driven automobiles is not restricted to developed countries, and based on recent data, it is the developing countries where the technology is still nascent it is the developing countries where the technology is still nascent which are now in forefront to adopt cleaner technologies at large scale [17]. In India, the TATA group, one of the biggest conglomerates, operates a few buses to generate real-time data and automobile manufacturer Mahindra & Mahindra have tested their fleet of vehicles incorporating fuel cells. The objective is to understand the technology and modifying it according to the local requirements. Real-time data is all these automobile manufacturers require to adopt fuel cell technology.

Over the past few years there has been a sharp increase in heavy-duty vehicles that operate on hydrogen fuel cells, including buses, utility trucks, and locomotives. Fuel cell–driven buses have caught enormous attention in the US city of California and in Western Europe. However, fuel cell technology is yet to reach maturity for it to be adopted at a commercial level. The impediment is the high cost, which is also the reason for the absence of mass production. To make it competitive along with conventional transport vehicles, farsighted R&D and policy changes are required. Again, lack of real-time data restricts its utility, Again, lack of real-time data restricts its utility in marine applications in comparison to automobile applications. Factors like fuel cell stack volume it occupies, the weight it will add up and corrosion of components exposed to marine environment need to be addressed immediately. The recent surge in research on this technology, driven mainly by the pollution generated by the conventional automobile, lead to application of fuel cells in marine applications starting from the year 2003, when commercial passenger ship is Germany adopted fuel cell as APU. This led to induction of the technology in submarines also.

When NASA adopted fuel cell technology in its Apollo space program, other countries followed up and this generated a lot of interest among researchers worldwide in a technology that was dormant for a century after its invention. What makes this technology attractive for space application in particular is the generation of water, the by-product in the fuel cell system, which can be consumed by the cosmonauts and astronauts after some simple treatment procedures. The silent operation of fuel cell without obnoxious emissions is favored in unmanned aerial vehicles (UAV) but the reliability, durability, and sustainability of the fuel cell are the reasons for the hesitant approach by defense equipment manufacturers.

From the developers' perspective, the cost trade-off can be realized by adopting the correct choice/type of fuel cell for the fuel cell vehicles (FCV) and the best fuel-processing technology [18]. Methanol can be compacted better than hydrogen when used as fuel and requires a reformer to convert it into useful hydrogen, which at times may increase the weight and consumes power. To develop fuel cell infrastructure, cost analysis is the key which many fail to address without taking into account of well-to-wheel cost breakup and refilling infrastructure.

The following challenges are the major impediment for the commercialization of PEMFC technology in transportation applications:

1. hydrogen refilling infrastructure;
2. cost of the fuel cell; and
3. technology transformation.

The amortized nature of the existing gasoline infrastructure offers stiff resistance for fuel cell entry and this demands a change in the mindset. All the advantages, cost implications of the fuel cell needed to be taken into account with a focus to change entire infrastructure for the betterment of the planet. Huge cost implications at entry-level is unavoidable but that should not be the bottleneck for the adoption of this clean technology. PEMFC technology looks promising given the power density it delivers that can meet the requirements of most automobiles. When adopted, it will bring down the GHG emissions.

14.1.6 Techno commercial barriers

The major barrier is the absence of transportation and refuelling infrastructure which prevents the utilization of hydrogen in fuel cells. Production and distribution pathways like gasification and electrolysis need only incremental steps and do not require a major infrastructure transformation. However, it requires more time on par with demand. Hydrogen production using low-carbon hydrogen is the greatest barrier for developing the hydrogen energy system. Large-scale deployment of hydrogen technologies within every 100 km network will reduce the transportation costs considerably, which is currently pegged at $50/MWh.

Current PEMFC technology is based on perflourinated proton-exchange membranes which require fully hydrated conditions. There is considerable interest in lowering the membrane cost and extending the operating window of PEMs. The system complexity could also be reduced with the development of electrolytes that do not require hydration. The various parameters of membranes like proton conductivity, electronic conductivity, permeability to fuel, electro-osmotic drag coefficient, chemical/thermal stability, good mechanical properties, and reduced cost are classified as the "iron triangle" of performance, durability, and cost.

The various types of catalysts that are in use are Pt, Pt alloys/Pt dealloy, core shell, shape-controlled nanocaytalysts, nanoframes, nanocages, nonprecious metal catalysts, and others. All these catalysts are at various stages of maturity in terms of manufacturing capability, cost benefits, and higher mass activity. However, they also suffer from a few drawbacks like long-term availability, durability, Pt-loading targets, dissolution of core, scale-up, low-current density, and others.

The cost estimates for a fuel cell and its components are mostly based on statistical data obtained from automobile industries working on FCV programs. The key players in these programs are from the United States, Canada, Germany, Japan, and Korea. The targets set by the US Department of Energy (DoE) is mostly considered by manufacturers as well as OEMs for their estimates. However, the cost estimates of fuel cells are confidential due to the heavy competition in this field.

In general the system designs do not change with production rate, but material costs, manufacturing methods, and business-operational assumptions vary. Cost estimations at very low manufacturing rates (below 1000 systems per year) presents particular challenges. Traditional low-cost, mass-manufacturing methods are not cost-effective at low manufacturing rates due to high per-unit setup and tool costs, and lower manufacturing line utilization. Therefore less-defined and less-automated operations are typically employed. For some repeat parts within the fuel cell stack [e.g., the membrane electrode assemblies (MEAs) and bipolar plates], such a large number of pieces are needed for each system that even at low-system production rates (1000/year), hundreds of thousands of individual parts are needed annually. Thus for these parts mass-manufacturing cost reductions are achieved even at low-system production rates. However, other fuel cell stack components (e.g., end plates and current collectors) and all FCS-specific balance of plant (BOP) equipment manufactured in-house do not benefit from this manufacturing multiplier effect because there are fewer of these components per stack (e.g., only two end plates per stack, etc.).

System costs were projected at a number of manufacturing rates from 1000 to 500,000 per year (Fig. 14.3) [19].

Single variable sensitivity analyses at 100,000 and 500,000 systems per year is illustrated in Fig. 14.4.

The main difference between the two manufacturing volumes is that the GDL has not yet reached economies of scale, where at 500,000 systems per year it has. Both volumes are shown because a manufacturing volume of 100,000 units per year is comparable to manufacturing volumes for alternative advanced automotive technologies, and the 500,000 units per year volume shows the relative impact of the cost of constituent components on total system costs when at economy of scale. In

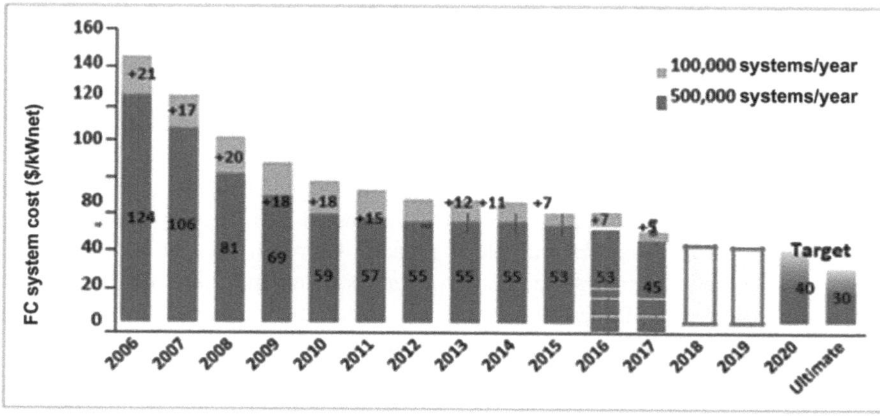

Figure 14.3 Modeled cost of an 80-kW PEM fuel cell system based on projection to high-volume manufacturing (100,000 and 500,000 units/year). Error bars represent the 90% confidence interval from a stochastic uncertainty analysis and reflect manufacturing uncertainty in the modeled system (19)

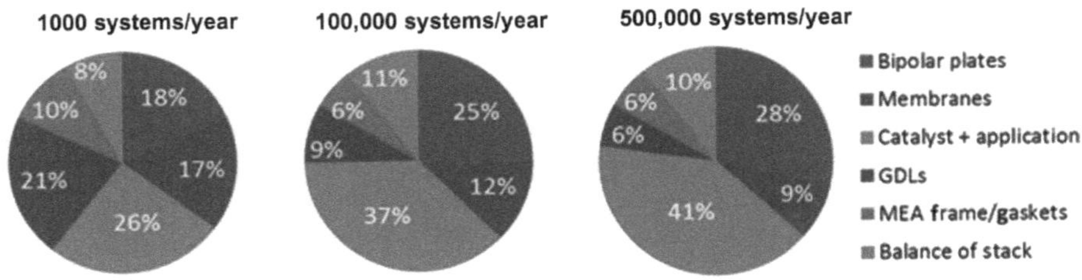

Figure 14.4 Breakdown of the 2017 projected fuel cell stack cost at 1000, 100000, and 500000 systems per year. (19)

both scenarios, air-loop cost is the highest contributor to system cost variability; enhanced MEA performance would be one way to relax air management requirements (i.e., the need for higher pressure) and lower CEM cost.

14.1.7 Fuel cell; world view

Hydrogen is a flexible energy carrier that can be produced from any primary energy source. And it can be effectively transformed into any form of energy for diverse end-use applications. Hydrogen is particularly well-suited for use in fuel cells that efficiently use hydrogen to generate electricity. As an energy carrier, hydrogen can provide energy system flexibility by enabling linkages between energy supply and demand for centralized and decentralized power. Although this technology is highly promising, the development of hydrogen generation, transmission and distribution, and retail infrastructure is challenging. The current performance of hydrogen in the transportation sector is provided in Table 14.3 [20].

Being both electric and hybrid, fuel cell vehicles benefit from technological advancement in both technologies as they have a significant number of parts (such as batteries and power electronics) in common. There are about 550 passenger cars and buses running in several demonstration projects across the world—specifically in Europe there are 192, Korea and Japan have around 100 each, and the United States has around 140. The number of fuel cell driven electric vehicles (FCEV) is expected to be around 8 million around 2030T. A small number of fuel cell heavy freight trucks (HFTs) are currently being used in a demonstration project at the port of Los Angeles which is testing the usability of range extenders with electric trucks. Currently 1 kg of hydrogen gives around 100 km, and demonstration cars have ranges of around 500–650 km. Vehicle costs are still high (~USD60,000) due to early market introduction. The fuel cell system cost is very high leading to higher vehicle costs, therefore the challenge lies in reducing the fuel cell stack and BOP costs with increased durability [20].

14.1.8 Road ahead

This technology still in the early stages of commercialization and is expensive, and therefore is struggling to compete with alternative technologies, including other

Table 14.3 Hydrogen in transportation sector.

Application	Power or energy capacity	Energy efficiency	Investment cost	Lifetime	Maturity
Fuel cell vehicles	80–120 kW	Tank-to-wheel efficiency 43%–60% (HHV)	USD 60,000–100,000	150,000 km	Early market introduction
Hydrogen retail stations	200 kg/day	~80%, incl. compression to 70 MPa	USD 1.5 million–2.5 million	–	Early market introduction
Tube trailer (gaseous) for hydrogen delivery	Up to 1000 kg	~100% (without compression)	USD 1,000,000 (USD 1000/kg payload)	–	Mature
Liquid tankers for hydrogen delivery	Up to 4000 kg	Boil-off stream: 0.3% loss per day	USD 750,000	–	Mature

Source: IEA, 2015. Technology Roadmap: Hydrogen and Fuel Cells, http://ieahydrogen.org/pdfs/TechnologyRoadmapHydrogenandFuelCells-(1).aspx,

low-carbon options. More R&D is required to assess the durability of the process as well the product, that is, fuel cell. Parameters such as performance, cost reduction, enhanced durability, and thermal and electrical management also need to be assessed depending on the application sectors (e.g., transportation or industrial).

FCV have shown a durability of 5000 h on a longer time interval, while DoE demands 8000 h of durability with 10% drop in performance. As the technology improvements are happening continuously, sufficient data on durability is not available. In addition most fuel cells use metallic flow-field plates in order to meet the targets of reduction of weight of the cell or cell stack. However, most flow-field plates are coated and researchers are trying to achieve better durability of such coatings to increase the lifetime of fuel cells. Durability, being the pivotal point, determines the sustainability of technology.

Governments can help accelerate the development and deployment of this technology by ensuring continued research, development and demonstration (RD&D) funding for hydrogen generation and conversion technologies, such as electrolyzers and fuel cells, and stimulate investment and early market deployment of hydrogen and fuel cell technologies and their infrastructure through effective policy support to bring down costs. They can prove on-road practicality and economics across the supply chain of FCEVs by putting the first tens of thousands of vehicles on the road, along with hydrogen generation, T&D and refueling infrastructure, including at least 500 to 1000 stations in suitable regions around the world, and cross-border projects. Governments can also increase the number of hydrogen-based energy storage systems suitable for integrating VRE and collect and analyze performance data under real-life conditions.

14.1.9 Infrastructure and commercialization

Similar to compressed natural gas (CNG) transport through pipelines, hydrogen can be synthesized and transported through pipelines to access larger distances; however, this technology is expensive. Dedicated onsite production or transport in the form of compressed tanks or cylinders are cheaper, although the ideal would be through pipelines. Centralized and distributed production have advantages and disadvantages. It is easier to opt for existing underground CNG pipelines to transport hydrogen than building a dedicated one that will increase the input cost. Nations which have such infrastructure in place already can switch over to hydrogen with ease. The world's largest hydrogen production plant with a capacity generation of 10 MW commenced its operation from March 2020 in Fukushima Hydrogen Energy Research Field (FH2R), Japan [21]. The entire plant is operated with solar energy.

Though the development of hydrogen as fuel for automobile industries gained momentum worldwide, one of the biggest bottlenecks toward realizing hydrogen-based transport is that the world is still driven by gasoline. Most low-end vehicles still utilize petrol and the transport vehicles rely on diesel with very few vehicles running on CNG owing to pollution norms in some places. The shift from gasoline to hydrogen-based transport requires extensive public–private partnership.

Of the total world energy utilization, 25% is for transport. Petroleum and other liquid fuels accounts for 96% of the world transport sector of which motor gasoline is 40%, diesel 36%, and jet fuel 12% based on 2012 estimates. Projected estimates for 2040 include that bio-diesel is estimated to increase its share as fuel and there will be drop in motor gasoline with little increase in other sources including EV and hydrogen [22]. This dependence on oil will not die soon until the infrastructure for production and distribution of hydrogen is in place. The policy makers related to the automotive industries, while drafting the policies for future automobiles should include subsidies, tax-free regimes to the industries which will come forward to adapt hydrogen technology.

At the beginning of 2020, there were about 432 hydrogen refilling stations operating worldwide with the majority being in California (United States), Germany, Japan, Korea, and China [23]. In 2019, 83 hydrogen refuelling stations added up to the existing 349 and from the start of 2020, there were about 432 stations operating worldwide with the majority being in California, Germany, Japan, Korea and China. The major bottleneck in establishing refilling stations stems from the cost involved and the supply of hydrogen.

The number of hydrogen refuelling stations meager in comparison with gas stations. Based on the data given by the Ministry of Petroleum & Natural Gas, Government of India, there were about 64,625 gas stations as of April 2014 [24]. Such is the penetration of gasoline, infrastructural framework, chain connectivity which goes in tandem and still enjoy as the undisputed leader in transport category. However, government and industries need to take ambitious and real-world actions to curb GHG emissions.

The upfront cost for setting up a hydrogen refilling station is on an average between USD2–3 million [25] whereas an EV recharging station with a single 350 kW charger will cost only a 10th of the above. Sustained efforts will bring down the cost or will be in a position to get the return on investments at a faster pace if the station operates at 80% of its capacity every day. Until the deployment of hydrogen refueling stations in sufficient numbers, the sale of vehicles driven by hydrogen will not pick up. To get these vehicles on road, significant initial upfront cost is inevitable. With the kind of expansion happening around the world, with much focus on greener fuels, such cost is reasonable from operable point of view. By scaling up the technologies, the cost of hydrogen fuel will be brought down. Countries approach in this direction demands key analysis, pragmatic and reasonable recommendations to envisage cleaner technologies.

Owing to the huge upfront cost, anywhere between USD2–5 million to set up a single refill station, creating infrastructure for the production and distribution of hydrogen is not an easy task. In certain regards, the cost involved in setting up the refill station is something akin to a "show-stopper" for hydrogen vehicles [26]. In the interest of making this technology viable, it is obvious we should have a near, mid-, and long-term action plans. Apart from the tried and tested near-term objectives like mixing a certain amount of hydrogen with CNG or its use along with methane, centralized hydrogen production and distribution to retail outlets, new norms should be in place to bring down pollution and cost. Some of the initial explorations in this direction may not be successful, such as when hythane is utilized in a compression ignition engine which eventually increased the torque and thermal efficiency, but NO_x and CO emissions could not be curbed [27]. Mid-term objectives may include decentralized hydrogen production near the refilling spot itself by means of electrolysis of water or steam reforming using bio-fuels. Also, on-board storage technologies need extensive improvements so that the space required is kept at a minimum. Excess hydrogen that goes untapped can be utilized. Production of hydrogen on the large scale should be best attempted through electrolysis assisted with solar energy. One example of this is being executed in Fukushima, Japan, where they have been successful in commissioning a hydrogen production facility which has the capacity to deliver 1200 Nm^3/h.

14.1.9.1 Hydrogen generation by electrolysis of water

To accelerate the adoption of hydrogen, it must be produced from a clean source and at rates and efficiencies competitive to gasoline. In the context of the hydrogen economy, hydrogen production methods based on fossil fuels are considered transitional technologies. Water is widely agreed to be the most interesting source of sustainable hydrogen of the future mainly because the process is easy to integrate with renewable energy sources.

Electrolysis of water is a mature technology based on the generation of hydrogen and oxygen by applying a direct electric current to water to dissociate the two elements. Water electrolysis is attracting increasing interest since it is considered the only method that may obtain large amounts of sustainable hydrogen without the emission of pollutant gasses or requiring the consumption of fossil or nuclear resources.

The phenomenon of electrolytic splitting of water into hydrogen and oxygen to the development of various versions of industrial technique to meet the hydrogen demands of various applications, the water electrolysis development has gone through several landmarks. Some historical events of water electrolysis are listed in Table 14.4 [28].

Despite the fact that the discovery of electrolytic water decomposing was first observed in acidic water, in industrial plants alkaline medium is preferred because corrosion is more easily controlled and cheaper construction materials can be used compared to acidic electrolysis technology. Depending on the type of electrolyte employed in the electrolysis process, it was further classified as PEM electrolysis, and solid oxide water electrolysis, which have been developed recently. The types of electrolyzers are listed in Table 14.5.

Table 14.4 Timeline of water electrolyzer technology.

Year	Event
1800–1920	Discovery of electrolytic splitting of water by Nicholson and Carlisle
1920s–70s	Worldwide manufacturing of Several large 100 MW size plants
1948	Development of first pressurized electrolyzer built by Zdansky/Lonza
1966	Development of first PEM system was built by General Electric Company
1970s	Development of first solid oxide water electrolyzer

Table 14.5 Different types of water electrolyzer technology.

Type	Alkaline	Acid	Polymer electrolyte	Solid oxide
Electrolyte type	Potassium or sodium hydroxide	Sulfuric or phosphoric acid	Polymer	Ceramic
Charge carrier	OH^-	H^+	H^+	O_2^-
Reactant	Water	Water	Water	Steam
Electrode	Nickel based	Platinum based	Platinum, iridium oxide based	composite of yttria stabilized zirconia (YSZ)
Cathode reaction	$2H_2O + 2e^- \rightarrow H_2 + 2OH^-$	$2H^+ + 2e^- \rightarrow H_2$	$2H^+ + 2e^- \rightarrow H_2$	$H_2O + 2e^- \rightarrow H_2 + O^{2-}$
Anode reaction	$2OH^- \rightarrow H_2O + ½O_2 + 2e^-$	$H_2O \rightarrow ½O_2 + 2H^+ + 2e^-$	$H_2O \rightarrow ½O_2 + 2H^+ + 2e^-$	$O_2^- \rightarrow ½O_2 + 2e^-$
Operating temperature	40°C–90°C	150°C	25°C–150°C	700°C–1000°C

14.1.9.2 International scenario

The global water electrolysis market is projected to be valued at USD 12,975 million by 2025 at an estimated CAGR of 7.58% during the forecast period. Water electrolysis for hydrogen production finds its application in various industries, such as chemicals, petroleum, glass, refineries, automobile, pharmaceuticals, metals and mining, and others. The key players of the global water electrolysis market are, Air Products and Chemicals, Inc., Siemens AG, Proton Onsite, Teledyne Energy Systems Inc., AREVA H_2Gen, and Hydrogenics Corporation. The global water electrolysis market has been categorized based on product category, application, and region. Based on the product category, the global market has been segmented into a PEM and alkaline water electrolysis. Among these products, the alkaline water electrolysis segment is projected to be dominant throughout the forecast period, whereas the PEM segment has been registered as the fastest growing technology. On the basis of region, the Asia-Pacific has accounted for the largest market share in the water electrolysis market followed by Europe and North America.

R&D on alkaline electrolyzers in India was initiated in early 1980s. BARC and CSIR-CECRI were active in developing materials for this type of electrolyzer as well as for stacks. Compact alkaline electrolyzers have been designed and demonstrated in BARC. The concept of renewable energy integrated water electrolyzer was demonstrated by the energy research and development association (ERDA), Vadodara, using wind energy with a commercial alkaline water electrolyzer (AWE) for a practical distribution generation system in 2013. In addition, NISE, Gurgaon has also installed a 120 kW solar photovoltaic system to produce electricity for the generation of hydrogen through an AWE. The generated hydrogen stored in high-pressure cylinders and used through a dispenser unit for fueling the demonstration vehicles. In addition a few companies are engaged in manufacturing AWE ranging from 1–50 Nm^3/h production capacity for industrial applications.

In the area of PEM water electrolyzers, CSIR-CECRI has demonstrated up to 5.0 Nm^3/h capacity. The electrolyzer was designed using a circular type platinum coated titanium flow-field plate, platinum black cathode, and iridium oxide anode. This electrolyzer stack can deliver hydrogen at 5–10 bar pressure. Recently the process know-how was transferred to India for commercialization. In addition, the concept of using depolarizers in a PEM water electrolyzer has also been reported by ARCI-CFCT and demonstrated up to 2.5 Nm^3/h capacity with a much lower energy consumption of 1.40 kWh/Nm^3. It also demonstrated for the first time the use of carbon-based materials in its construction and thus reducing the capital cost tremendously. JNTUH, Hyderabad, has also developed an indigenous PEM-based water electrolyzer with lower hydrogen capacity of 36 L/h. Indian Oil Corporation Ltd. recently commissioned a Hydrogen-CNG fuel dispensing station at its R&D center in Faridabad. This pilot station provides a hands-on experience with onsite hydrogen production using an imported PEM water electrolyzer with a storage, distribution, and supply system. BARC is working on developing material for a high-temperature steam electrolyzer of 1.0 Nm^3/h hydrogen production capacity for technology demonstration purposes.

14.1.10 Cost reduction and hydrogen neutrality

At present the cost per kilogram of hydrogen is pegged at \$6–8 and this will vary from place to place owing to the method adopted for its generation and the transportation cost involved. The US DoE estimates that the cost of hydrogen can be brought down to less than \$3/kg within a few years [29]. Significant cost reduction can be achieved if the generation is carried onsite through electrolysis assisted by solar energy. The various probabilities of production, storage and utilizing hydrogen for transportation applications offers wide possibilities to reduce the cost of the hydrogen.

Often termed as "tomorrow's oil," hydrogen is the best energy carrier poised to transform the entire transportation sector. The right term would be carbon neutrality which can be achieved by adopting green hydrogen production through solar energy–assisted water electrolysis. Although it can be used for a longer duration to store energy in comparison with other forms, the reluctance to adopt, lies in the complexity of the technology for safe handling, storing and distribution of hydrogen.

According to a report [30], countries like Japan, Korea, and Singapore are importing hydrogen for commercial use to meet the internal demands. In most of the cases, until recently, space for setting up a large plant to generate renewable energy was a distant dream owing to the geographical landscape prevalent in all these countries. Japan has commissioned the single largest hydrogen production hub in the Fukushima prefecture with initial capacity to generate 1200 Nm^3 of H_2/h. China is poised to take over as the largest producer of hydrogen, thanks to the policy decisions it took to shift to greener economy. Other countries like Australia, Canada, France, Germany, Norway, and the United Kingdom are already producing more hydrogen than the existing demand. Developing nations like India are hydrogen-neutral, where they do not need to import hydrogen to meet internal production.

Virtually all the hydrogen produced in the world is by steam methane reforming and coal gasification and very little by electrolysis. The oil refining sector consumes the majority of hydrogen, followed by ammonia and methanol production, while the rest is in the iron and steel industry [31]. Hydrogen has been utilized in the transportation sector as an energy carrier from the beginning of this century. The market potential of hydrogen is poised to increase from \$120 billion with a compound annual growth rate (CAGR) of 5% over the next decade [32]. This increase is due to the adoption of hydrogen energy by various industrial sectors and government policies which benefit hydrogen consumers. An insight into the study indicates that 18% of global energy demand will be met with hydrogen fuel by 2050 and most of it will be utilized in the transportation sector.

Though the near-term potential market is largely associated to heavy-duty vehicles like trucks and buses, for a longer term it is expected to replace all gasoline-driven, light-duty, passenger vehicles. These industries will switch to greener hydrogen fuel, thus reducing the carbon footprint which now stands at 37 billion tonnes of CO_2 by 2019 estimates [33].

Given the number of hydrogen powered vehicles sold in 2017 in the United States alone, more than Europe and Japan combined, and Japan installing 98% of the world's residential fuel cell systems which run on hydrogen in its soil, Europe adopting better policy for switching to hydrogen, the market potential for hydrogen in blooming and promising [34].

14.2 Strategic energy plan

Among the various energy forms that are available for consumption, the share of electricity should be increased to 40% by 2050 from the current level to achieve the decarbonized energy world envisaged by the Paris agreement. In this context, low-carbon electricity from renewables may become the preferred energy carrier. This challenge could be addressed by hydrogen from renewables, which allows large amounts of renewable energy to be channeled from the power sector into the end-use sectors.

Devising plans to achieve long-term goals will drive the hydrogen economy and will bring it close to realization on the commercial scale. However, it requires long-term commitment and support from public and private partnership as well as continuous monitoring. The opportunity before us should be tapped to envision a greener environment.

The plan must include a broad vision with realistic goals and should create an environment for utility of hydrogen energy. Due to geographical and economic differences, the strategy template is not unitary [35]. Each country must create, adopt, and modify strategies according to their experience or policies. Deployment of hydrogen refill spots will push the market for vehicles run with hydrogen. Perpetual demand will push the hydrogen economy further. Continued R&D activities will help to iterate the technology for improvement and to bring down the costs involved without sacrificing performance aspects. Economic incentives to attract private capital are critical for sustaining the technology on a longer run. Such strategies should be adopted to make hydrogen technology a reality.

The onus of making hydrogen technology available rests with the government's mandate. Unless the public office comes into participation with tax incentives, subsidies, or penalties for continued emissions that will pollute the environment, no industry will come forward to change its norms. The US state of California and the union territory of Delhi, India, are classic examples, where the governing councils pitched in to take cautionary measures to curb the pollution due to gasoline-driven automobiles and switch over to technologies which will not induce harm.

Coordination from public and private entities, standardization of practices to be followed, and key regulation measures are required for successful transition to a hydrogen economy. The technology roadmap should be in phases so that everyone can adopt within a flexible time frame without causing too much burden during switch over.

14.3 Technology barriers

The global hydrogen demand stands at 90 million metric tonnes per year, and it will increase in the forthcoming years. Though industries consume more hydrogen, a positive shift in the adoption of hydrogen in the transport sector is due to certain attributes like light weight, high-energy density, and ease of synthesis at industrial scale. Also, the possibility of deriving it from various carbon energy sources and widely prevalent resources such as water, make it a viable alternate source of energy.

In certain industries, hydrogen is available as by-product, for example, in chlor-alkali industries where it can be tapped extensively and used. Hydrogen can also be complimented where there is a possibility to replace the excess electricity is used, such as in steel production, transport, and heating applications. Hydrogen can bridge the deficit within the current energy demand if it is put in place with the required technology domain.

Currently global hydrogen production stands at 70 million metric tonnes. The majority of the production is done through steam reforming of natural gas, gasification of the coal, by partial oxidation of oil, and the rest is through electrolysis of water. Around 2.8 million metric tonnes of hydrogen is produced by Indian fertilizer and petroleum refineries and 0.4 million metric tonnes is available as by-product from chlor-alkali industries [36]. These values are based on 2011 estimates and the recent figures list India among hydrogen-neutral countries, in that it is self-sufficient (i.e., hydrogen demand is met with supply within the country). Many countries fall within the hydrogen-neutral category or where supply is more than the demand. Export of excess hydrogen is carried out to countries where the demand is higher.

Existing prominent barriers for the widespread adoption of hydrogen technology can be summarized as follows:

1. Energy density of compressed hydrogen gas is still less than the liquid gasoline used in transport vehicles, before taking into account of the fuel loss.
2. Existing infrastructure is meager and there are serious safety concerns associated with handling hydrogen.
3. Compressing hydrogen in cylinders as in gaseous or liquid form for use in transportation is an energy-intensive process. This increases the cost of hydrogen.
4. Shortage of skilled labor to handle hydrogen and associated technologies.

14.4 Conclusion

Hydrogen as a fuel plays a significant role in a future energy sector based on two energy carriers: hydrogen and electricity. It can be easily coupled with renewable energy sources to produce clean and sustainable hydrogen. The major challenges that need to be addressed are government support, infrastructure development, reducing the cost of hydrogen generation, need for less carbon-based methods, increased manufacturing capability of stack components, increased demand, and so forth.

Water electrolysis has the potential to play a significant role in the future energy sector as it can be easily coupled with renewable energy sources to produce clean and sustainable hydrogen. Currently alkaline and PEM water electrolyzers are commercially available and are both are safe and reliable, but alkali-based electrolytes are preferred over acid-based ones for larger capacity due to the use of nonnoble catalysts. The considerable reductions in investment and operating costs are also necessary for alkaline and PEM electrolyzers. The various results obtained globally from number of demonstrations programs is evidence of the suitability of water electrolysis to become the main route for producing renewable hydrogen. However, in India, industry participation is the most essential factor for the successful implementation as well as utilization of the hydrogen produced using the electrolysis method.

Acknowledgments

The authors would like to thank Dr. G. Padmanabham, Director, ARCI, and Dr. R. Gopalan, Regional Director, ARCI, for their constant support and encouragement. Technical Research Centre (TRC) (AI/1/ 65/ARCI-2014), Department of Science and Technology, Government of India, for financial assistance is also acknowledged herewith.

References

[1] N. Rajalakshmi, K.S. Dhathathreyan, Present Trends in Fuel Cell Technology Development, Progress in Fuel Cell Research, NOVA Publishers, New York, USA, 2007.

[2] N. Rajalakshmi, K.S. Dhathathreyan, Present Trends in Fuel Cell Technology Development, NOVA Publishers, New York, USA, 2008.

[3] K.S. Dhathathreyan, N. Rajalakshmi, Challenges in PEM fuel cell development, Indian Assoc. Nucl. Chem. Allied Sci. Bull. 3 (2009) 214.

[4] K.S. Dhathathreyan, N. Rajalakshmi, R. Balaji, Nanomaterials for PEM Fuel Cells, Wiley Publishers, 2017.

[5] J. Charles, M.N. Roberts, Departure of Jacques Charles and Marie-Noel Robert's 'aerostatic globe' balloon from the Jardin des Tuileries, Paris, December 1, 1783 <https://www.loc.gov/pictures/item/94508310/> (accessed 10.11.15).

[6] M. Warshay, P.R. Prokopius, The Fuel Cell in Space: Yesterday, Today and Tomorrow, Grove Anniversary (1839–1989) Fuel Cell Symposium, London, Great Britain, September 18–21, 1989.

[7] I. Staffell, D. Scamman, A.V. Abad, P. Balcombe, P.E. Dodds, Paul Ekins, et al., The role of hydrogen and fuel cells in the global energy system, Energy Environ. Sci. 12 (2019) 463–491.

[8] N. Rajalakshmi, R.I. Jafri, K.S. Dhathathreyan, Recent Advances in Low Temperature Fuel Cells, Wiley Publishers, 2017.

[9] L. Antoni, N. Garland, J. Han, M. Rex, R.C. Samsun, AFC TCP 2019 survey on the number of fuel cell vehicles, hydrogen refueling stations and targets, IEA Technology collaboration programme, Adv. Fuel Cell (2019).

[10] R.A. Felseghi, E. Carcadea, M.S. Raboaca, C.N. Trufin, C. Filote, Hydrogen fuel cell technology for the sustainable future of stationary applications, Energies 12 (2019) 4593–4621.
[11] E. Weidner, R.O. Cebolla, J. Davies, Global deployment of large capacity stationary fuel cells, European Commission, JRC Technical Report, 2019.
[12] R.H. Henne, K.A. Friedrich, Applications – transportation auxiliary power units: fuel cells, in: J. Garche (Ed.), Encyclopedia of Electrochemical Power Sources, Elsevier, Amsterdam, 2009.
[13] O.Z. Sharaf, M.F. Orhan, An overview of fuel cell technology: fundamentals and applications, Renew. Sustain. Energy Rev. 32 (2014) 810–853.
[14] J.W. Plunkett, Plunkett's Automobile Industry Almanac 2011, Plunkett Research, Houston TX, 2010.
[15] J.H. Wee, Applications of PEMFC systems, Renew. Sustain. Energy Rev. 11 (2007) 1720–1738.
[16] B.G. Pollet, I. Staffell, J.L. Shang, Current status of hybrid, battery and fuel cell electric vehicles: From electrochemistry to market prospects, Electrochim. Acta 84 (2012) 235–249.
[17] D. Carter, M. Ryan, J. Wing, The fuel cell industry review 2012, Fuel Cell Today 2012.
[18] W.P. Teagan, J. Bentley, B. Barnett, Cost reductions of fuel cells for transport applications: fuel processing options, J. Power Sources 71 (1998) 50–85.
[19] W. Adria, K. Gregory, P. Dimitrios, Department of Energy (DOE) hydrogen and fuel cell program record 17007, September 2017. <https://www.hydrogen.energy.gov/pdfs/17007_fuel_cell_system_cost_2017.pdf>
[20] Technology Roadmap Hydrogen and Fuel Cells OECD/IEA, International Energy Agency, 2015.
[21] New Energy and Industrial Technology Development Organization (NEDO), Toshiba Energy Systems & Solutions Corporation, Tohoku Electric Power Co., Inc., Japan, March 7, 2020 <https://www.nedo.go.jp/english/whatsnew_00239.html> (accessed 14.03.20).
[22] Transportation Sector Energy Consumption, International Energy Outlook 2016, U.S. Energy Information Administration, DOE/EIA-0484, May 2016.
[23] https://fuelcellsworks.com/news/in-2019-83-new-hydrogen-refuelling-stations-worldwide/ (accessed 19.02.20).
[24] State Wise Retail Outlets (2012–2019), Petroleum Planning & Analysis Cell's (PPAC) Snapshot of India's oil & gas data, abridged ready reckoner, ministry of petroleum & natural gas, Government of India, February 2020.
[25] M. Melaina, M. Penev, Hydrogen Station Cost Estimates; Comparing Hydrogen Station Cost Calculator Results with Other Recent Estimates, National Renewable Energy Laboratory, Colorado, U.S., September 2013.
[26] J.M. Ogden, Infrastructure for hydrogen fuel cell vehicles: a Southern California case studyJanuary Center for Energy and Environmental Studies, Princeton University, Princeton, NJ, 1997.
[27] T. Sandalci, O. Isın, S. Galata, Y. Karagoz, I. Guler, Effect of hythane enrichment on performance, emission and combustion characteristics of a CI engine, Int. J. Hydrog. Energy 44 (2019) 3208–3220.
[28] K. Zeng, D. Zhang, Recent progress in alkaline water electrolysis for hydrogen production and applications, Prog. Energy Combust. Sci. 36 (2010) 307–326.

[29] Multi-Year Research, Development, and Demonstration Plan, Office of Energy Efficiency and Renewable Energy, U.S. DOE Hydrogen and Fuel Cells Program, 2012.
[30] Opportunities for Australia from Hydrogen Exports, Australian Renewable Energy Agency (ARENA), August 2018.
[31] IRENA, Hydrogen from Renewable Power: Technology Outlook for the Energy Transition, International Renewable Energy Agency, Abu Dhabi, 2018.
[32] M.W. Melaina, D. Steward, M. Penev, S. McQuee, S. Jaffe, C. Talon, Hydrogen Infrastructure Market Readiness: Opportunities and Potential for Near-term Cost Reductions, NREL/BK-5600-55961, August 2012.
[33] Giuseppe Etiope, et al., Gridded maps of geological methane emissions and their isotopic signature, Earth Syst. Sci. Data 11 (2019) 1–22.
[34] I. Staffell, The role of hydrogen and fuel cells in the global energy system, Energy Environ. Sci. 12 (2019) 463–491.
[35] A. Hugo, P. Rutter, S. Pistikopoulos, A. Amorelli, G. Zoi, Hydrogen infrastructure strategic planning using multi-objective optimization, Int. J. Hydrog. Energy 30 (2005) 1523–1534.
[36] Compressed Natural Gas and Hydrogen Fuels, Lessons Learned for the Safe Deployment of Vehicles, U.S Department of Transport and U.S. Department of Energy Workshop, December 10–11, 2009, Workshop Notes.

Energy efficiency in buildings

15

Janmejoy Gupta[1] *and Manjari Chakraborty*[2]
[1]Department of Architecture, School of Planning and Architecture (SPA), Vijayawada, India, [2]Birla Institute of Technology, Mesra, Ranchi, India

15.1 Background: building in the wake of an energy crisis

An energy efficient building offers an appropriate environment for habitation with minimal energy consumption and wastage of energy, thereby maximizing energy conservation. It focuses on increasing the efficiency of resources like energy, water, and materials—while reducing its negative impact on human health and the environment during the building's lifecycle, through environment-friendly placement of the building on site, its design, construction, operation, and maintenance. The goal of energy-efficient buildings is to accomplish the same, or more, tasks and functions in the buildings as before while using less energy. The energy-efficient building acts as a smooth-functioning machine in which the equipment produces as much, or more, energy capacity than needed to operate an ideal human environment, including air quality, moisture conditions, air temperature, and lighting. The ideal energy-efficient building retains the best environment for human habitation while minimizing the need and cost of energy.

With the cost of energy rising and availability decreasing, the need to provide energy-efficient building designs becomes more important. Escalating energy costs are changing our design practices. It is no longer possible to design structures without specific reference to the projected energy consumption. An energy-efficient building balances all aspects of energy use in a building—lighting, space conditioning, and ventilation—by providing an optimized mix of passive solar design strategies, energy-efficient equipment, and renewable sources of energy. Use of materials with low-embodied energy also forms a major component in energy-efficient building design. In the field of architecture, the task is to link the built environment with nature. Buildings should be so designed as to consume high technology but low energy. Architectural design must be global enough in spirit to ensure global longevity and also local enough in expression to remain sustainable.

In the wake of regional vernacular wisdom of architecture being lost to uniform glass facades and aluminium composite paneled elevations due to so-called "globalization of architecture," awareness in the field of energy-efficiency has become a necessity.

Sustainable Fuel Technologies Handbook. DOI: https://doi.org/10.1016/B978-0-12-822989-7.00016-0

Imbalance in energy consumption is prevalent around the world. Energy consumption is high in most developed countries. On the other hand, the developing countries need to consume more energy to ensure economic growth. According to estimates, energy consumption in developing countries is only 10% of that in the developed countries. The economic development of many countries is hindered due to "energy poverty."

Natural gas has been the energy source with the highest rates of growth in recent years. The high end-use efficiency of natural gas has made it a popular choice for power generation projects. Renewable sources of energy are gaining popularity. However, fuel prices and regulatory policies of different countries play an important role in the development of renewables. Hydroelectricity has been a major use of hydro sources of energy around the world. The consumption of electricity is growing across the globe.

Energy is a necessity and clean renewable energy is an obligation to the Earth and our future. Due to the industrial revolution CO_2 emission and global warming have become serious global issues. Most of the countries' governments have set targets to reduce CO_2 emission in order to prevent global warming further. A transition from conventional energy systems to those based on renewable resources is necessary to meet the ever-increasing demand for energy and to address environmental concerns. Global demand for energy is rising fast as the population increases and developing countries such as China and India undergo dramatic economic growth.

Based on data from the International Energy Agency [1] (Table 15.1 and Fig. 15.1), the world's energy needs could be 50% higher in 2035 than they are today. Yet the fossil fuels on which the world still depends are finite and far from environment-friendly. Serious thought needs to be given to creating viable alternatives.

15.2 Energy use in buildings: energy modeling and energy audits

This section deals with the issues related to energy use in buildings by discussing aspects like the pattern and magnitude of energy consumed by buildings, and the various heads under which building design and construction processes consume energy. Factors influencing energy consumption in buildings have been identified and major methods of energy conservation are discussed. In India, as per energy statistics data obtained for 2017 from the Central Statistics Office, Ministry of Statistics and Program Implementation, and government, buildings account for the second-highest consumption of energy in India after industries, the highest consumer of energy. The energy consumed by buildings can be reduced significantly by proper design. With increasing cost of energy and availability of energy resources decreasing, the need to provide energy-efficient building designs becomes all the more important. The escalating energy cost is forcing us to change our design habits. It is no longer possible to design buildings without specific reference to the projected energy consumption. The two most important things to be discussed

Table 15.1 Evolution of the world energy mix (a projection).

Consumption	2005 Consumption (mill TOE)	2020 Consumption (mill TOE)	Gain or loss (mill TOE)	Percentage change
Total (sum of components below)	11,409	15,544	4135	36.2
Oil	3678	4300	622	16.9
Natural gas	2420	3600	1180	48.8
Coal (conventional)	2778	3193	415	14.9
Traditional biomass and waste	793	1400	607	76.5
Nuclear fission	624	790	166	26.6
Hydro	624	790	116	18.3
Other biomass, methanol, ethanol	370	388	18	4.9
Unconventional oil from tar, sands, shale	88	350	262	297.7
Coal processes total from liquefaction, oxygenation, gasification	–	500	–	–
Solar (photovoltaics, solar power towers, solar thermal, space solar)	11	100	89	809.1
Wind	8.5	100	92	1076.5
Nuclear fusion	0	0	0	–
Methane gas hydrates	0	22	22	–
Geothermal	4.8	50	45	941.7
Tides	1	1	1	900

Source: The Millineum Project, Retrieved from International Energy Agency. <https://www.iea.org/>, 2011.

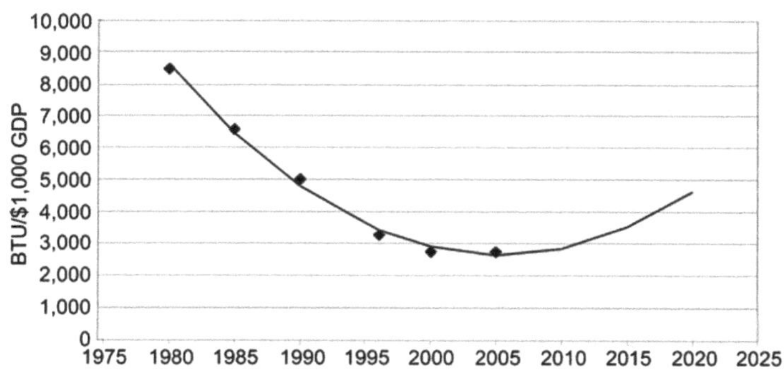

Figure 15.1 The trends of global energy efficiency.
Source: The Millineum Project, Retrieved from International Energy Agency. <https://www.iea.org/>, 2011.

with respect to projected building energy-consumptions are energy models and audits. The two are the bedrock on which buildings' energy consumptions are studied.

15.2.1 Energy model

An energy model is a representation of a building for the purposes of building energy simulation. The energy model can be used for analyzing a building's energy use. It consists of all the design and operating parameters associated with the energy consumption of that building.

An integrated-design approach targets energy and other performance issues at the most fundamental conception of design and relies on an ongoing holistic design philosophy and interdisciplinary communications to ensure optimum performance. It is in the integrated design approach where building energy simulations are essential. A truly integrated design cannot proceed without a better understanding of the type of building at hand. Design parameters include wall and roof constructions, window performance values, installed electricity for lighting and user equipment, occupant numbers, ventilation requirements, air and water distribution system descriptions, and cooling and heating equipment details.

Operating parameters include operational schedules for the lights, user equipment, occupants, thermostats, and fans as well as heating, ventilation, and air-conditioning (HVAC) controls and utility data. The simulation takes in the energy model data, combines it with historic weather data, and estimates the performance of the entire building and system components for every hour of the year.

As a design tool, energy models provide a comparative analysis between options. This can occur during conceptual stages to determine the energy impacts of an architectural form or can occur later in the design to evaluate the impact of a specific glazing choice.

As an accounting tool, energy models are most often used to compare a proposed design against a benchmark. The Leadership in Energy and Environmental Design (LEED) rating system, state energy codes, and federal tax incentives are examples of using energy models in this capacity.

Energy modeling tools, such as e-QUEST (a front end to DOE-2), Energy-Plus, IES Virtual Environment, Trane TRACE, and Design-Builder, among others, provide three distinct functions: (1) a design tool throughout all phases of design, (2) an accounting tool for the end of design, and (3) a financial justification tool after occupancy and calibration. Building components can be broken down into "spaces" and "systems." Spaces refer to the building as a whole and the rooms that comprise the building. Spaces are made up of walls, windows, roofs, doors, floors, and shade. Implicit in this list are factors like orientation, construction, and dimension. Also included are lighting levels and plug loads. Buildings may be viewed as a closed-energy system. Temperature within the building only changes as a result of the flow of energy from several sources. The categories in a building that utilize or consume energy are the heating systems, cooling systems, ventilation, and water-heating requirements. Other energy consumers are cooking equipment, computer equipment, and other machinery used within a building. It is also important to be able to identify the time span when maximum energy consumption occurs. Energy consumption is tailored to the months or season of the year to account for the changing climatic conditions and the building system that service the structure. Energy consumption is plotted along with the ordinate and the time frames in months of year are plotted along the abscissa. Building energy simulation will never replace good design judgment but it will always calibrate and inspire it.

15.2.2 Energy audit

To make a building is to create an efficient performing system linked to its surrounding environment and subject to a range of interactions affected by the climate and requirements of its occupants. Thus the main objective of this building system is to perform efficiently to provide the desired and required comfort level. This comfort of users of a building may be categorized as the thermal, visual, and acoustical comfort as well as indoor environmental quality. To achieve these comfort levels the building requires energy supply through passive and active strategies.

The energy audit in a building is a feasibility study. It serves to identify energy use among the various services and also identify the opportunities for energy conservation. It is basically a very first step in establishing an energy-management program. The audit will produce the data on which such a program is based. The study should reveal to the owner, manager, or management team of the building the options available for reducing energy waste, the costs involved, and the benefits achievable from implementing energy-conserving opportunities (ECOs). The energy management program is a systematic on-going strategy for controlling a building's energy consumption pattern. It is to reduce waste of energy by taking different measures like identifying the climate where the building is located, its functions, occupancy schedules, envelope retrofitting, and other such aspects. It establishes and maintains an efficient balance between a building's annual functional energy requirements and its annual actual energy consumption. The energy audit may range from a simple walk-through survey at one extreme to one that may span several phases, followed by monitoring of energy use in the building services, and then model analysis using computer simulation of the building's operation. The complexity of the audit is therefore directly related to the stages or degree of sophistication of the energy-management program and the cost of the audit exercise. The first stage is to reduce energy use in areas where energy is wasted. It begins with a detailed, step-by-step analysis of the building's energy use factors and costs, such as insulation values, occupancy schedules, chilling efficiencies, lighting levels, and records of utility and fuel expenditures. It includes the identification of specific ECOs, along with the cost-effective benefits of each one. The completed study would provide the building owner with a thorough and detailed basis for deciding which ECOs to implement, the magnitude of savings to be expected, and the energy-conservation goals to be established and achieved in the energy-management program. However, the ECOs may yield modest gains. The second stage is to improve the efficiency of energy conversion equipment and to reduce energy use by proper operations and maintenance. For these reasons it is necessary to reduce the number of operating machines and operating hours according to the demands of the load, and fully optimize equipment operations.

ECOs include the following:

- Building equipment operation
- Building envelope
- Air-conditioning and mechanical ventilation equipment and systems
- Lighting systems
- Power systems
- Miscellaneous services

The first two stages can be can be implemented without remodeling buildings and existing facilities. The third stage would require changes to the underlying functions of buildings by remodeling, rebuilding, or introducing further control upgrades to the building. This requires a certain degree of monetary investment, recoverable later in the form of equivalent of monetary worth of energy-savings.

The last stage is to carry out large-scale energy reducing measures when existing facilities have passed their useful life, or require extensive repairs or replacement because of obsolescence. In this case higher energy savings may be achieved. For these last two stages, the audit may be more extensive in order to identify more ECOs for evaluation, but at an increased need for heavier capital expenditure to realize these opportunities.

15.3 Passive solar techniques

Passive solar techniques are the most cost-effective means of making buildings energy efficient. Passive solar applications, when included in initial building design, adds little or nothing to the cost of a building, yet has the effect of realizing a considerable reduction in operational costs and reduced equipment demand. It is reliable, mechanically simple, and a viable asset to a building.

In passive solar design the system is integrated into the building elements and materials. The windows, walls, floors, and roof are used as the heat collecting, storing, releasing, and distributing system. These very same elements also act as major elements in passive cooling design but in a very different manner. It should be understood that passive solar design does not necessarily mean the elimination of standard mechanical systems.

Two elements must be present in all passive solar heating designs: (1) a south facing exposure of transparent material (glass, plastic) to allow solar energy to enter; and (2) material to absorb and store the heat (or cool) for later use. With these two basic elements in mind a number of approaches to design a passive solar architecture are possible.

15.3.1 Passive solar heating design considerations

15.3.1.1 Site consideration

The performance of any solar-energy building, especially one that uses passive design strategies, is strongly impacted by the site and the siting of the building in relation to its surroundings. In choosing a particular design approach, site and climate conditions must be evaluated carefully so that the best approach or combination of approaches is incorporated. No one particular passive design approach is most advantageous in all climates or on all sites and situations; it differs as per the site, climate, and design context.

15.3.1.2 Building shape and orientation

Different building shapes and orientations can be designed to perform efficiently by combining effective glazing, solar exposure, and shading into the building form. Buildings oriented along an east-west axis are generally more efficient for winter heating and summer cooling. An example of optimum orientation for Ranchi (capital of Jharkhand, India) is shown in Fig. 15.2.

This orientation allows for maximum solar glazing (windows) to the south for solar capture for heating. This orientation is also advantageous for summer cooling

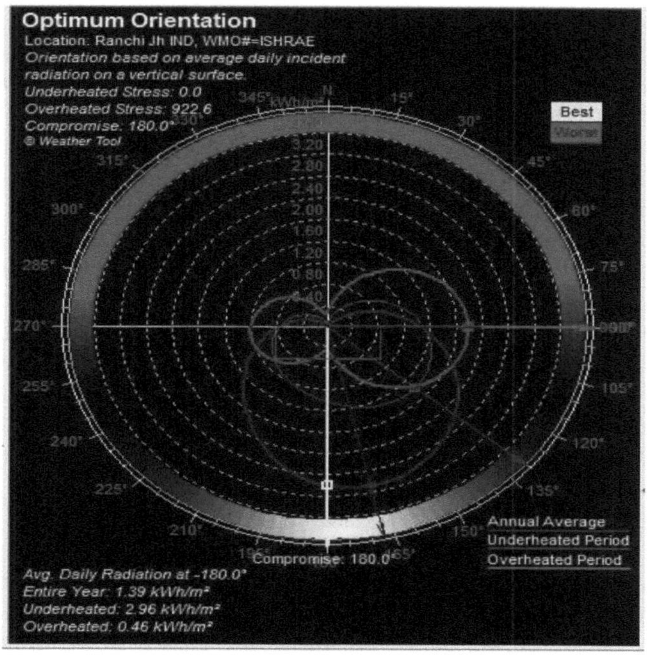

Figure 15.2 Optimum orientation for least possible exposure to sun in summer and maximum possible exposure to sun in winter, for Ranchi, the capital of Jharkhand, India, in the northern hemisphere. The longer side of the dwelling, shown in red aligned along east–west axis, with minimum exposure to east and west sides (Through which Ranchi, the Tropic of Cancer line, i.e., 23.5 degrees north imaginary line, passes).
Source: Simulated by the Author, Generated in Ecotect-Autodesk Software with Ranchi's detailed climate file, 2018 [2].

conditions since it minimizes east–west exposure to morning and afternoon summer sunlight.

15.3.1.3 Indoor space planning

Along with proper location, orientation, and shape of the building, another very important issue to be considered is the interior space planning. Habitable spaces that are mostly occupied should be arrayed along the south face of the building to access summer winds in the summer and solar radiation in winter in the northern hemisphere. Rooms that are least used (closets, storage areas, garages) should be placed along the north wall where they can act as a buffer between high-use living space and the cold north side. (The scenario reverses in the southern hemisphere.)

15.3.1.4 Openings

The major expanse of openings in a passive solar energy structure will be south-facing solar windows. Whole design planning should include the following considerations: the impact of heat gain in summer; views; natural lighting; and privacy requirements in determining the placement and size of windows in the building.

Passive heating techniques are used by architects in building design to achieve thermal comfort conditions in a cold climate. This technique allows maximum winter solar gain and prevents summer overheating. Heat energy is trapped by the building elements first then it is reradiated and distributed to where it is needed. Direct reradiation is the most effective means. Heat is also conducted through building materials and distributed by air movement. It is necessary to design floor plans to ensure that the most important rooms face south for the best solar access. Designing passive solar heating is about keeping out summer sun and letting in winter sun in the building. This will maximize winter heat gain, minimize winter heat loss, and concentrate heating where it is most needed. Climate classification is necessary to design buildings for passive solar heating, as the type of passive-design strategy to be used depends on climate type. For example, in tropical regions, built forms are maintained cool by planting tall canopy trees and attached overhead shading structures as shown in Fig. 15.3.

The functioning of passive solar heating basically works on the greenhouse effect. Solar radiation is trapped by the greenhouse action through windows exposed to sun. Trapped heat is absorbed and stored by materials with high-thermal mass (usually masonry) inside the building. It is released at night when it is needed to offset heat losses to lower outdoor temperatures. Passive design relies on greenhouse principles to trap solar radiation. Heat is gained when short-wave radiation passes through glass, where it is absorbed by building elements and furnishings and reradiated as long-wave radiation. Long-wave radiation cannot pass back through glass and thus is trapped inside. It is necessary to locate thermal mass where it is exposed to direct solar radiation or radiant heat sources. Thermal mass will also absorb reflected radiant heat. Openings should be sun-shaded in such a way that shading above openings excludes high-angle summer sun, but admits lower-angle

Figure 15.3 An example of built forms in tropical regions.

winter sun. Passive solar heating systems can be broadly classified into direct gain systems and indirect gain systems.

15.3.2 Direct gain

Direct gain is the most common passive solar system. Sunlight enters the room through windows to warm the interior space. South-facing glass admits solar energy into the building where it strikes thermal storage materials, such as floors or walls made of adobe, bricks, concrete, stone, or concrete-blocks. The interior thermal mass increases the temperature during the day by absorbing heat. At night the thermal mass radiates heat into the living space, thus warming the spaces. Direct gain can be achieved by various forms of openings such as clearstories and skylight windows designed for optimum heating.

15.3.3 Indirect gain

In an indirect gain system, thermal mass is located between the sun and the living space. The thermal mass absorbs the sunlight that strikes it and transfers to the living space. The indirect gain system uses 30%–45% of the sun's energy striking the glass adjoining the thermal mass. A few commonly used indirect gain systems are:

- *Trombe Wall*: A Trombe wall is a thermally massive wall with vents provided at the top and bottom. It may be made of concrete, masonry, and adobe and is usually located on the southern side (in the northern hemisphere) of a building in order to maximize solar gains.
- *Water Wall*: Water walls are based on the same principal as that for Trombe walls, except that they employ water as the thermal storage material.

- *Roof-Based Heating System*: In this technique, incident solar radiation is trapped by the roof and is used for heating interior spaces.

15.3.4 Sunspaces

A sunspace or solarium is the combination of direct and indirect gain systems. In the northern hemisphere, the basic requirements of sunspaces are: (1) a glazed south-facing collector space attached, yet separated, from the building; and (2) living space separated from the sunspaces by thermal storage wall. Solar radiation heats up the sunspace directly, which, in turn, heats up the living space (separated from the sun-space by a mass wall) by convection and conduction through the mass wall.

Notwithstanding the importance of passive-solar techniques for a building to be deemed energy efficient, at a certain point to make buildings and groups of buildings energy efficient, active solar techniques also need to be used. The basic active solar techniques are discussed in Section 15.4.

15.4 Active solar techniques

One of the ways in which the sun's energy can be utilized is by trapping the sun's energy through different types of solar collectors and using it for various purposes like heating water and space heating. This same solar energy can also be used for active cooling techniques wherein absorption cooling systems transfer a heated liquid from the solar collector to run a generator or a boiler, activating the refrigeration loop, which, in turn, cools a storage reservoir from which cool air is drawn into the space. Nowadays centralized district heating and cooling systems based on active solar techniques are becoming popular in the West. Active solar systems use solar panels for heat collection and electrically driven pumps or fans to transport the heat to the required spaces. Electronic devices are used to regulate the collection, storage, and distribution of heat within the system.

In active systems, solar collectors are used to convert the sun's energy into useful heat for hot water, space heating, or industrial processes. Flat-plate collectors are typically used for this purpose. These most often use light-absorbing plates made of dark-colored material such as metal, rubber or plastic that are covered with glass. The plates transfer the heat to a fluid; usually air or water flowing below them, and the fluid is used for immediate heating or stored for later use. There are two basic types of liquid based active systems, namely open loop and closed loop. An open-loop system circulates potable water through the collector. In closed-loop systems, the circulating fluid is kept separate from the system used for potable water supply. This system is mainly used to prevent the freezing of water within the collector system. Thermo-siphon systems are more convenient and simpler.

In thermo-siphon systems the water circulates from the collector to the storage tank by natural convection and gravity. As long as the absorber keeps collecting heat, water keeps being heated in the collector and rises into the storage tank, placed slightly above (at least 50 cm). The cold water in the tank runs into the collector to replace the water discharged into the tank. The circulation stops when there is no incident radiation. Thermo-siphon systems are simple, relatively inexpensive, and require little maintenance and can be used for domestic applications. Solar ponds have been developed that harness the sun's energy and can be used for various purposes including production of electricity. Other devices such as solar cookers, water distillation systems, solar dryers, and so forth have been developed which can be used to reduce energy requirements in domestic households and in industrial applications.

Among active systems, two main families can be identified according to the medium used for the heat transport:

- air-collector systems
- hydraulic-collector systems

Air collector systems are characterized by lower costs, but also lower efficiency than hydraulic ones, mainly due to air's low thermal capacity. Solar thermal gains are generally used immediately and without storage for preheating the fresh air needed for building ventilation. The heat can also be stored by forcing the air to circulate in a stone bed underneath the ground or by using the solar air as cold source in a heat pump air/water; such applications can be quite expensive, and are therefore rare. Air systems can only be used for space heating.

15.4.1 Solar collectors

Solar collectors absorb the sun's light energy and changes it into heat energy. This energy is then transferred to a fluid or air that is used to warm buildings, heat water, generate electricity, dry crops, or cook food. Solar collectors are used for any process that requires heat.

The main applications of solar collectors are:

- Hot water preparation in households, commercial buildings and industry.
- Water heating in swimming pools.
- Space heating in buildings.
- Drying crops and houses.
- Space cooling and refrigeration.
- Water distillation.
- Solar cooking.

15.4.2 Typical solar collectors

Typical solar collectors gather the sun's energy with rooftop arrays of piping and net-metal sheets painted black to absorb as much radiation as possible. They are encased in glass or plastic and angled toward the south to catch maximum sunshine. The collectors act as miniature greenhouses, trapping heat under their glass plates. Because solar radiation is diffuse, the collectors must have a large area. In active systems, solar collectors are used to convert the sun's energy into useful heat for hot water, space heating, or industrial processes.

There are three solar heating applications:

- Solar space heating
- Solar water heating
- Solar pool heating

Solar water heating systems use water tanks for the storage of solar energy. Both passive and active solar water heating use water tanks. Active indirect systems use a heat-exchanger to transfer the heat from the heat-transfer fluid. The other solar water heating systems use the actual household water and therefore do not need water tank with a heat-exchanger. Solar pool heating uses the swimming pool water for solar-energy storage. By circulating the swimming pool water through solar pool collectors, it is possible to extend the swimming season.

15.4.3 Active cooling

Absorption cooling systems transfer a heated liquid from the solar collector to run a generator or a boiler activating the refrigeration loop that cools a storage reservoir from which cool air is drawn into the space. Rankin steam turbines can also be powered by solar energy to run a compressed air-conditioner or water cooler. Solar refrigeration is independent of electric supply and without any moving parts, for example, the Zeolite refrigerator.

15.4.4 District heating and cooling systems

Sustainable energy supply systems, and mainly the use of district heating and cooling systems based on the use of renewable energies like solar and biomass or the use of waste heat, are the major tool to introduce clean and sustainable energy in cities. Produced energy may be supplied to the residential sector, industry, urban agriculture, and any other sector requiring hot or cold water. District heating and cooling installations using renewable energies are constantly increasing in Europe. In many European countries the potential for use of renewables for district heating systems is very high, while the number of settlements supplied by district heating networks is continuously increasing. In Sweden more than 50% of the energy is delivered by waste heat and renewable energy sources.

15.5 Energy-efficient landscaping

Efficient landscape design can be integrated with building design to save energy. Effectively done, landscape architecture affects the microclimate, vegetation of all types, and water bodies in its various forms. The use or manipulation of these natural elements or processes will eventually result in an overall supportive setup for energy-efficient buildings. Landscaping helps to sustain the environment.

A well-designed landscape will:

- Cut summer and winter energy costs dramatically.
- Protect a building from winter wind and summer sun.
- Protect the existing vegetation through efficient and careful planning.
- Help control noise and air pollution.

Landscaping is an easy and inexpensive way to improve the energy efficiency of a building, enhance the appearance and value of a property, and provide screening for privacy. Selected types of plants can be strategically positioned to protect the building from the harsh extremes of summer sun and chilling winter winds, improve comfort inside and outside the home, and reduce the need for supplementary heating and cooling. Tree planting and climate-appropriate landscaping offer substantial environmental benefits. Trees and vegetation control erosion, protect water supplies, provide food, create habitat for wildlife, and clean the air by absorbing carbon dioxide and releasing oxygen. Deciduous trees and vines can be "designed" to shade partially exposed outdoor living areas in summer, while still allowing the sun through in winter. Low shrubs and ground cover around the home can reduce reflected heat and glare. Dense trees and shrubs can be positioned to deflect strong winds and channel cooling summer breezes. The landscaping of a site can be designed to modify temperatures in and around the home. Shade from trees to roofs and/or windows can reduce indoor temperatures by 6°C−12°C in summer. Conversely, ensuring winter sun penetration through windows can save a substantial amount of heating energy. External temperatures can be reduced up to 6°C by using ground cover or lawn instead of hard paving. These temperature differentials improve both comfort and the energy efficiency of the home so that heating and cooling requirements are reduced. Too often the landscape is not considered until after a building is completed. Landscape design ideally is part of the wider design process, allowing interaction between the building orientation, building design, site conditions, and proposed landscape development. This presents the best opportunity for maximizing the landscape benefits for the building and its users. A carefully considered design of the external environment can be extremely beneficial in terms of thermal comfort. Appropriate selection and placement of trees and shrubs, trellises, and pergolas to shade walls and windows can greatly moderate the temperature inside the building. Equally, the selection and placement of an appropriate paving material in and around a building can dramatically have an impact on the internal temperature of a building.

15.5.1 Climate and appropriate landscaping

A well-designed landscaped garden enhances the appearance of the building, assists in modifying the indoor air temperature, and increases the resale value of the property. The tropical belt can be divided into four approximate climatic regions: warm humid, hot dry, composite, and cool dry. Energy-conserving landscape strategies should depend on the climate of the region. Here landscaping strategies are listed by region and in order of importance.

Warm-Humid Climate:

- Need to channel summer breezes toward the home.
- Maximize summer shade with trees that still allow penetration of low-angle winter sun.
- Avoid locating planting beds close to the home if they require frequent watering.

Hot-Dry Climate:

- Provide shade to cool roofs, walls, and windows.
- Allow summer winds to access naturally cooled homes.
- Block or deflect winds away from air-conditioned homes.

Composite Climate:

- Maximize warming effects of the sun in the winter.
- Maximize shade during the summer.
- Deflect winter winds away from buildings.
- Funnel summer breezes toward the home.

Cool Dry Climate:

- Use dense windbreaks to protect the building from cold winter winds.
- Allow the winter sun to reach south-facing windows.
- Shade south and west windows and walls from the direct summer sun, if summer overheating is a problem.

While there are broad climate zones throughout a country, each site will have its own microclimate and conditions that will influence the house design and landscape development. A site and microclimate analysis provides information to assist the designer to decide on the best placement of the building on the site, and to identify what types of landscape protection are needed. The analysis should consider:

- site size, topography, slope, soil, and drainage;
- prevailing seasonal winds, temperature, and humidity;
- relationship to sun and shade patterns in summer and winter;
- existing vegetation and any special features; location of other buildings and fences;
- location of vehicular and pedestrian access; views;
- any legal setbacks and building restrictions;
- adjacent site conditions.

15.5.2 Landscaping methods and their appropriateness

The climate immediately surrounding the building is called its microclimate. The designed building's microclimate may be more sunny, shady, windy, calm, rainy, snowy, moist, or dry than average local conditions. These factors all help determine what plants may or may not grow in that microclimate.

15.5.2.1 Selecting and planting trees and shrubs

Trees and shrubs come in all shapes and sizes. Selection of the trees and shrubs and the plantation technique directly affect the building occupants' comfort and energy efficiency. Trees and shrubs have a life span of many years and can become more attractive and functional with age. But poor planning of landscape improvements often creates trouble. That is why it's necessary to ensure proper plant placement and minimal maintenance.

15.5.2.2 Shape characteristics

Tree shapes are very diverse. The density of a tree's leaves or needles is important to consider. Dense evergreens make great windbreaks for winter winds. If it's necessary to impede summer winds, it's required to choose a tree or shrub with more open branches and leaves. Such trees are also good for filtering morning sun from the east, while denser trees are better for blocking harsh afternoon summer sun as shown in Fig. 15.4.

15.5.2.3 Growth

It is necessary to carefully select trees between slow-growing or fast-growing species. Although a slow-growing tree may require many years of growth before it shades your roof, it will generally live longer than a fast-growing tree. Also, because slow-growing trees often have deeper roots and stronger branches, they are less prone to breakage by windstorms or heavy snow loads. And they can be more drought-resistant than fast-growing trees. Considering growth rate, strength, and brittleness when locating trees near walkways or structures, it is necessary to check whether the mature tree's root system is likely to damage sidewalks, foundations, or sewer lines. The smaller the yard or open area, the more important it is to select a tree with manageable roots.

When planting trees, shrubs, hedges, or bushes it is necessary to find out how large the mature specimen will grow. For those plants close to the building, it is necessary to plan for at least 1 foot (30 cm) of extra clearance between the full-grown shrub and the wall of the building. This will prevent heavy pruning or damage to building siding in the future. Deciduous trees with high, spreading crowns (i.e., leaves and branches) can be planted to the south of the building to provide maximum summertime roof shading. Trees with crowns lower to the ground are more appropriate to the west, where shade is needed from lower afternoon sun angles. Trees should not be planted on the southern sides of solar-heated homes in cold climates because the branches of these deciduous trees will block some winter sun. A 6-foot to 8-foot (1.8-m to 2.4-m) deciduous tree planted near the building will begin shading windows the first year. Depending on the species and the building, the tree will shade the roof in 5–10 years. Trees, shrubs, and groundcover plants can also shade the ground and pavement around the building. This reduces heat radiation and cools the air before it reaches walls and windows. Vines can shade walls during their first growing season. A lattice or trellis with climbing vines, or a planter box with trailing vines, shades the building's perimeter while admitting cooling breezes to the shaded area. Shrubs planted close to the house will fill in rapidly and begin shading walls and windows within a few years. However,

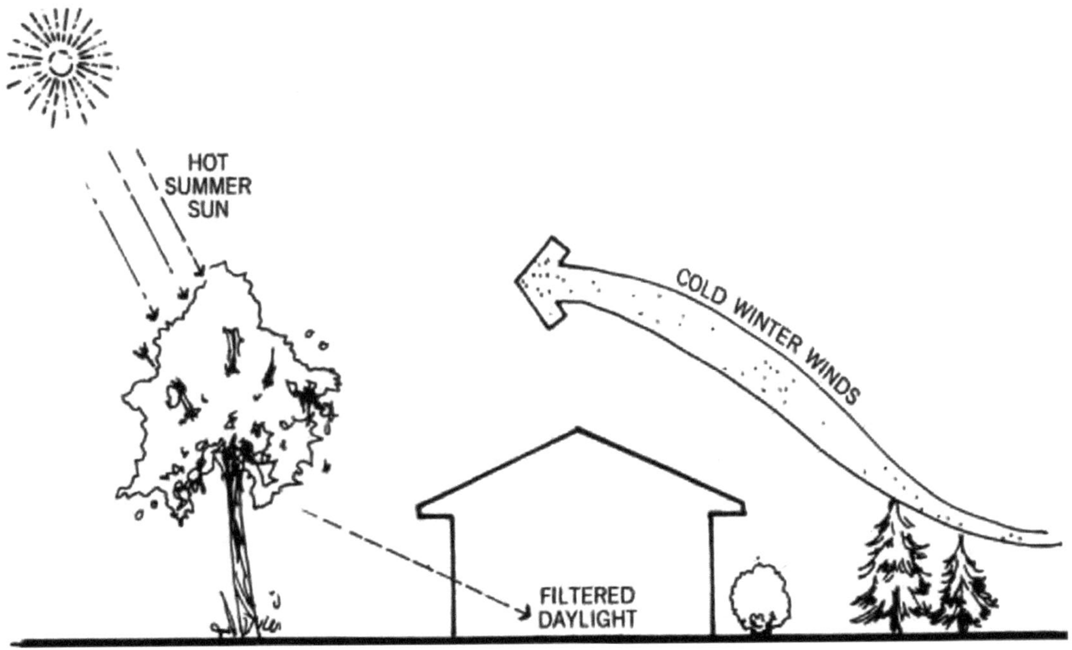

Figure 15.4 Plants can reduce winter heating loads and summer cooling loads by as much as 25%. Plants can also improve the quality of daylight by filtering and diffusing the light.
Source: The AIA Research Corporation, Regional Guidelines for Building Passive Energy Conserving Homes Catalogue, Washington, DC, 2018 [3].

dense foliage must not be allowed to grow immediately next to the building where wetness or continual humidity are problems. Well-landscaped buildings in wet areas allow winds to flow around the home, keeping its surrounding soil reasonably dry. Modifying summer temperatures is a high priority in most climate zones. Windows facing south and west will need protection from the hot summer sun. Landscaping can provide most of this protection economically and attractively. For example, a tree shading a window can reduce a room's temperature by up to 12°C.

15.5.3 Landscaping for comfort in summer and winter

The selection of suitable plants for shading depends on the part of the building being shaded (e.g., walls, outdoor living areas, windows). Different types of plants (trees, shrubs, vines) can be selected on the basis of their growth habit (tall, low, dense, light permeable) to provide the desired degree of shading for various window orientations and situations.

15.5.3.1 Points to be followed to achieve summer shading

- Deciduous trees and shrubs provide summer shade yet allow winter access.
- Trees with heavy foliage are very effective in obstructing the sun's rays and casting a dense shadow. Dense shade is cooler than filtered sunlight.
- Trees with light foliage filter the sunlight and produce a dappled shade.
- High-branching canopy trees can be used to shade the roof, walls and windows.
- Vertical shading is best for east and west walls and windows in summer, to protect from intense sun at low angles, for example, screening by dense shrubs, trees, deciduous vines supported on a frame, shrubs used in combination with trees
- Shading and insulation for walls can be provided by plants that adhere to the wall, such as English ivy, or by plants supported by the wall, such as jasmine.

15.5.3.2 Reducing glare and ground temperature

Glare is produced when strong sunshine is reflected from a surface such as light-colored paving, walls, water, or a shiny object. Lawns, ground cover, and low-growing shrubs absorb more sunlight and reradiate less heat than a paved surface. Ground cover planting not only reduces glare but can lower the temperature near the ground by approximately 6°C. In summer, an asphalt surface can be twice as hot as grass. By increasing the proportion of nonpaved areas, there is the added environmental benefit of increased storm water absorption.

To reduce glare:

- use low-growing shrubs, ground cover or grass to absorb reflected glare;
- avoid large areas of paving near the building to reduce reflected glare and ground temperature, especially south-facing windows;
- shade large areas of paving with trees or pergolas and planting.

15.5.3.3 Capturing summer breezes

Capturing summer breezes is possible by careful positioning of windbreak planting that can encourage the entry of desirable summer breezes and planting low shrubs and establishing lawns and ponds to the windward side will help cool hot summer winds.

15.5.3.4 Achieving winter sun penetration

Shading the building from direct sun is essential for summer comfort, but care should be taken that winter solar gain to windows and outdoor living areas is not excluded. To protect solar access in winter, the following strategies can be applied:

- Use deciduous trees and plants, particularly to the south of the building, to allow the access of winter sun to windows.
- Tall, low-branching evergreen trees should be kept at sufficient distance from south-facing windows to avoid overshadowing in winter.
- Where evergreen trees are used for shading the building, they should have branches high enough to permit the entry of as much sunlight as possible in winter.

15.5.3.5 Blocking cold winter winds

Cold winds can be blocked by following these steps:

- Windbreaks are most effective when located at 90 degrees to the direction of the wind.
- A vertical, somewhat permeable windbreak with 50%–60% density is generally more effective than a solid one, such as a wall, which can create turbulence.
- Wind velocity can be reduced on the windward and leeward side of a windbreak.
- Although wind velocity can be reduced by 50% for a distance of 10–20 times the height, the maximum protection on the leeward side of a windbreak is provided for a distance of 3–7 times the height.
- Earth mounds may be utilized to deflect winds.
- Large dense shrubs can be used as windbreaks to the windward side to counter cold winter winds and channel cooling summer breezes.
- Medium to large-sized shrubs or trees clipped to form a hedge can provide useful still air insulation and shading when grown close to a wall.
- Planting can be utilized to promote natural ventilation by positioning to deflect air flow through the building.
- Properly selected and placed landscaping can provide excellent wind protection that will reduce heating costs considerably. Furthermore, these benefits will increase as the trees and shrubs mature. The best windbreaks block wind close to the ground by using trees and shrubs that have low crowns. Evergreen trees and shrubs planted to the windward side are the most common type of windbreak. A windbreak will reduce wind speed for a distance of as much as 30 times the windbreak's height.
- Permeable barrier reduces wind speed without turbulence.
- Trees, bushes, and shrubs are often planted together to block or impede wind from ground level to the treetops. Evergreen trees combined with a wall, fence, or earth-berm (natural or artificial walls or raised areas of soil) can deflect or lift the wind over the building.

15.5.4 Reducing mechanical heating and cooling costs

Solar heat passing through windows and being absorbed through the roof is the major reason for mechanical conditioning. Shading is the most cost-effective way to reduce solar heat gain and cut air-conditioning costs. Using shade effectively requires knowing the size, shape, location, and the moving shadows that the shading device casts. But a building in cool regions may never overheat and thus may not require shading. Trees can be selected with appropriate sizes, densities, and shapes for almost any shading application. To block solar heat in the summer but let much of it in during the winter, deciduous trees should be used. To provide continuous shade or to block heavy winds, use evergreen trees or shrubs. Trees and shrubs can be selected and positioned to moderate the chilling effects of penetrating winter winds, particularly on large or exposed sites, and can also assist in capturing cooling summer breezes. In general, planting trees and shrubs as windbreaks, windscreens, and breezeways may control airflow over, around, and through buildings.

15.5.5 Water and its role in energy-efficient landscaping

Water is usually the least-used of all the landscaping elements, yet it is perhaps the most important. Water attracts fauna essential for a balanced ecosystem. Water features are also a great way to add a calming element to a landscape. The soft sound of running water is well-known for its relaxing properties.

Water has a profound impact on climate control, especially in the utilization of solar radiation and energy conservation. Water stores a large percentage of solar-insolation striking the water surface and a small percentage is radiated off. At the same time, while a small percentage of the solar energy strikes the land surface, a large percentage of the same energy is radiated back into the atmosphere. Including ponds, waterfalls, and fountains adds a pleasing kinetic dimension to any landscape plan. Environmental concerns and wilderness appeal have made water features more popular. Fountains, ponds, and waterfalls can be dramatic or restful; they can add beauty to stark walls, invite wildlife, and provide a sense of serenity after a long day.

Water features made of natural or artificial materials used for landscaping are:

15.5.5.1 Fountains

Flowing water adds delightful sounds and a restful ambience to any garden. Solar-powered fountain pumps are increasingly popular because they require no wiring or plumbing and cost nothing to operate. Breeze flows from the water body onto the shore during the day and vice versa at night. This natural airflow pattern may be utilized and controlled for natural ventilation and energy conservation.

15.5.5.2 Water gardens

Water is a popular element in Japanese landscape design. The soothing nature of water adds to the contemplative and peaceful feel of a Japanese garden. Water may be quietly flowing or still in a Japanese-style landscape, but never chaotic or loudly bubbling. Adding a water element is one of the easiest ways to incorporate Japanese design into an existing landscape.

15.5.5.3 Rainwater harvesting

Rainwater harvesting may be defined as the process of augmenting the natural infiltration of rainwater or surface run off into the ground by some artificial methods. The method generally adopted recharge through pits, trenches, and bore wells shafts by directly diverting runoff water into existing or disused wells or conserving the rainwater by artificial storing and using the same for human use.

Rainwater harvesting is a way to capture rainwater when it rains, the water should be either stored above ground or ground-water recharging should be done. This happens naturally in open, rural areas. In congested, over-paved metropolitan cities, we need to create methods to capture rainwater.

One of the tangible outputs of making buildings energy-efficient are creating buildings that are green. In fact, the terms green buildings and energy-efficient buildings are loosely used as synonyms. Though they are not synonyms, it is a fact that to a certain extent at least, all green buildings should employ energy-efficient building features in their design and construction. In Section 15.6 green buildings and how they have contributed to furthering the idea of energy-efficient buildings are discussed.

15.6 Concept of green buildings

Buildings have major environmental impacts over their entire life cycle. Resources such as ground cover, forests, water, and energy are depleted to give way to buildings. A green building depletes natural resources minimally during its construction and operation. The aim of green-building design is to minimize the demand on nonrenewable resources, maximize the utilization efficiency of these resources when in use, and maximize the reuse, recycling, and utilization of renewable resources. It maximizes the use of efficient, locally available building materials, and construction practices; optimizes the use of onsite sources and sinks by bioclimatic architectural practices. An example is shown in Fig. 15.5.

A green building uses minimum energy to power itself; uses efficient equipment to meet its lighting, air-conditioning, and other needs; maximizes the use of renewable sources of energy; uses efficient waste and water management practices; and provides comfortable and hygienic indoor working conditions. It is evolved through a design process that requires all concerned—the architect and landscape designer and the air conditioning, electrical, plumbing, and energy consultants—to work as a team to address all aspects of building and system planning, design, construction, and operation. They critically evaluate the impacts of each design decision on the environment and arrive at viable design solutions to minimize the negative impacts and enhance the positive impacts on the environment. In summary, the following

Figure 15.5 Water and trees as landscape elements used in a built form.

aspects of the building design are looked into in an integrated way in a green building:

- Site planning
- Building envelope design
- Building system design HVAC (heating, ventilation and air conditioning), lighting, electrical, and water heating
- Integration of renewable energy sources to generate energy on site
- Water and waste management
- Selection of ecologically sustainable materials (with high recycled content, rapidly renewable resources with low emission potential, etc.)
- Indoor environmental quality (maintains indoor thermal and visual comfort and air quality)

Hence the need to design a green building would be to address all these issues in an integrated and scientific manner. It is a known fact that it costs more to design and construct a green building. However, it is also a proven fact that it costs less to operate a green building that has tremendous environmental benefits and provides a better place for the occupants to live and work in. Thus the challenge of a green building is to achieve all its benefits at an affordable cost. A green building can have tremendous benefits, both tangible and intangible, over its life cycle. The immediate and most tangible benefit is in the reduction in operating energy and water costs right from day one, through the entire life cycle of the building. The energy savings could range from 25%–40% depending on the extent of green specifications. Other tangible savings would be reduction in first costs and enhanced asset values. Intangible benefits of green buildings include increasing productivity of occupants' health, safety benefits, and better indoor environmental quality.

15.6.1 Green rating systems prevailing in India

The LEED green building rating system evaluates environmental performance from a whole building perspective over a building's life cycle, providing a definitive standard for what constitutes a green building. LEED is a measurement system designed for rating new and existing commercial, institutional, and high-rise residential buildings. The LEED rating system is the most versatile and widely adopted rating system in the world. Around 15–20 countries the world over have adopted this rating system. Different buildings fall under different rating programs. Green buildings registered in India include corporate office complexes, IT parks, hospitals, government offices, educational institutions, airport, hotels, and so forth.

Parallel to LEED India, GRIHA was developed as an indigenous building rating system, particularly to address and assess non-air conditioned or partially air-conditioned buildings. GRIHA has been developed to rate commercial, institutional, and residential buildings in India emphasizing national environmental concerns, regional climatic conditions, and indigenous solutions. GRIHA stresses passive solar techniques for optimizing visual and thermal comfort indoors, and encourages the use of refrigeration-based and energy-demanding air conditioning systems only in cases of extreme thermal discomfort. GRIHA integrates all relevant Indian codes and standards for buildings and acts as a tool to facilitate implementation of the same. The envelope component–based compliance approach, namely the Energy Conservation Building Code (ECBC), sets requirements for exterior roof and ceilings, cool roofs, opaque walls, vertical fenestration, and skylights.

The ECBC building envelope requirements are based on the climatic zone in which the building is located. ECBC defines five climatic zones (hot-dry, warm-humid, composite, temperate, and cold). Based on the characteristics of climate, the thermal comfort requirements in buildings and their physical manifestation in architectural form are also different for each climatic zone. These physical manifestations, in turn, dictate the ECBC requirements for the envelope, as well as other building components that are applicable to the buildings. Last, but not the least, many indigenous ways of bioclimatic or green architecture to ensure thermal comfort for occupants is hidden in the various vernacular types of architecture practiced all over India. These strategies for climate-responsive architecture should be studied, documented, and adapted after necessary modifications for incorporation in modern buildings as well.

Urban Heat Islands are detrimental to outdoor and in some cases indoor thermal comfort in cities. People in cities often feel too hot in the summer due to the higher than average temperatures. To reduce the negative impact of the urban climate on people, we need to increase the number of green spaces in cities. Green spaces have a significant role in modulating the climatic elements. Green areas "filter" the air, they can produce oxygen, purify air and water, regulate microclimate, reduce noise, protect soil and water, maintain biodiversity, increase air humidity, reduce thermal stress, enhance local air circulation, and have recreational, cultural, and social value and improve our quality of life. In summer, green areas decrease air temperature by shading the ground. As a result of shading, soil surface temperatures in a park may be 12°C lower than the temperature of a street surface. It was found that urban tree planting can account for a 25% reduction in net cooling and heating-energy usage in urban landscapes. Urban shade trees offer significant benefits in reducing building air-conditioning demand and improving urban air quality.

15.6.2 Green roofs

Roof gardens, the precursors of contemporary green roofs, have ancient roots. The earliest documented roof gardens were the hanging gardens of Mesopotamia, considered one of the seven wonders of the ancient world. Green-roof habitats contribute to local biodiversity conservation and provide esthetic and psychological benefits for people in urban areas.

Also adopting strategies of reroofing and repaving in lighter color and planting shade trees can affect substantial energy savings directly and indirectly. Green roofs (roofs with a vegetated surface and substrate as shown in Fig. 15.6) provide ecosystem services in urban areas, including improved storm-water management, better regulation of building temperatures, reduced urban heat-island effects, and increased urban wildlife habitat. Green roofs are becoming increasingly popular in hot countries to combat the urban heat island effect.

Improved microclimatic conditions have major implications for the development of cities. The strong relationship between microclimatic and comfort conditions

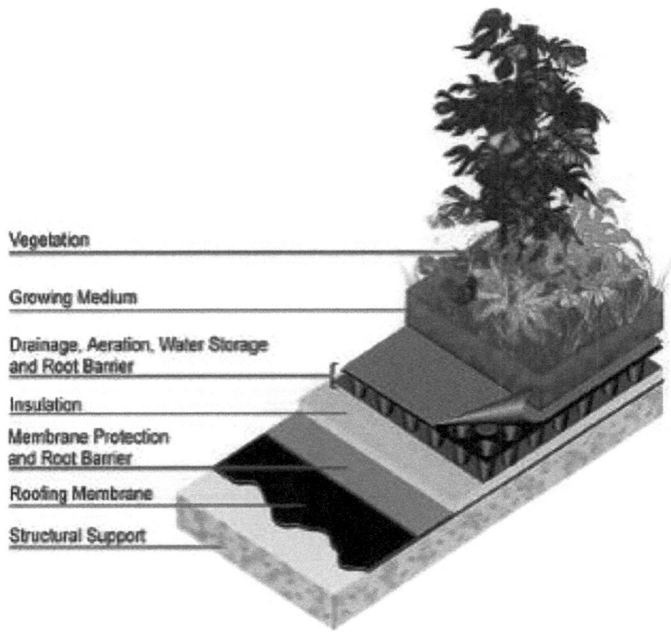

Figure 15.6 Schematic cross section of a green roof.
Source: Redrawn by Author, reference from Greenroofs, American Wick Drain. <https://www.awd-usa.com/drainage-applications/green-roof>, 2018 [4].

demonstrate that careful design can allow for the use of open spaces, balancing exposure, and protection from the different climatic elements.

15.7 Conclusion

To sum up, in the present situation the architect and his/her team's task to develop energy-efficiency in buildings is to

1 Examine the given conditions (site conditions, macroclimate, radiation, daylight, precipitation, microclimate);
2 Establish the limits of desirable or acceptable? conditions (temperatures, lighting and acceptable?noise levels);
3 Understand the local climate, relate the local climate to the effect it has on the inhabitants of that area, (which is essentially referred to as bio-climatic study), in which a lot of tools including but not restricted to Mahoney-tables, psychrometric-chart study of exact comfort-range required for biological thermal-comfort, and software tools. The subsequent design-interventions which can be incorporated inside the buildings would increase energy-efficiency of buildings. Nowadays, a lot of computer soft-wares which aid climate-responsive designs, like Ecotect (Autodesk-Revit), Climate-Consultant (developed by the University of California, UCLA) , Design-Builder and E-Quest, to name a few, are there which detail out the exact comfort-zone for a particular climate, and also go on to spell out the broad-level architectural design strategies that can be incorporated to improve energy-efficiency in buildings through appropriate energy-modelling.
4 Attempt to control these variables (heat, light and sound) by passive and active means (by the building itself) as far as practicable.
5 Based on the bio-climatic requirements, take appropriate technological solutions and implement them in tune with the existing architectural fabric and manifest the required technological solutions required to make the indoor habitable?spaces more comfortable?as well as energy-efficient, harmoniously via practical, functional and elegant architectural solutions.
6 Provide for energy-based services (heating, cooling, electric lighting, controlling sound), when needed, with the aid of energy-modeling and energy-auditing tools before and after construction respectively.
7 Conduct Life Cycle Assessment Checks and Operation and Maintenance Monitoring Post-Occupancy as well.

References

[1] The Millineum Project, Retrieved from International Energy Agency. <https://www.iea.org/>, 2011.
[2] Simulated by the Author, Generated in Ecotect-Autodesk Software with Ranchi's detailed climate file, 2018.
[3] The AIA Research Corporation, Regional Guidelines for Building Passive Energy Conserving Homes Catalogue, Washington DC, 2018.
[4] Greenroofs, American Wick Drain. <https://www.awd-usa.com/drainage-applications/green-roof>, 2018.

Section XI

Economic & Lifecycle assessment of Sustainable Fuel Technologies

Toward sustainable long-term energy planning for cities: an economic and environmental assessment of sustainable fuel technologies in the city of Donostia-San Sebastián

16

Eneko Arrizabalaga, Diego García-Gusano and Patxi Hernandez
TECNALIA, Basque Research and Technology Alliance (BRTA), Astondo bidea Edificio 700, Derio, Spain

16.1 Introduction

From a global perspective, growing trends in urbanization are pushing governments to look for solutions regarding the sustainability of economic growth. Cities, as main suppliers of services to citizenship, have started leading an energy transition (ET) process that will last decades. In this sense, the expectations derived from the Paris Agreement in COP21 (IPCC, 2015) should be transformed into investment plans within countries. Furthermore, those national objectives should consider the relevance that urban areas have as main consumers of energy in the present; a situation that should be modified to deploy measures the other way around. For example, self-consumption with solar PV systems and storage, demand–response schemes, smart controls, renewable heating/cooling systems as well as changes in citizens' behavior regarding energy habits, should serve as driving forces in the design of new forms of cities and governance; the so-called "citycentric" approach.

Due to the advances in the role of the population in the creation of policies, a sort of decentralization is expected when talking about supra-level schemes such as renewable targets or emission reductions at the national level. In general, it is accepted that a higher level of coordination among governmental bodies is not only desirable, but also required. Municipal governments—especially in large metropolitan areas—must work in coordination with the regional government and be aligned with the national interests for the highest good of all. Consequently, it seems crucial to develop quantitative tools to support such alliances under a solid basis of knowledge. The best decisions are those that minimize risks, while at the same time, maximize benefits for all the agents without endangering the future sustainability of the system.

The ET to fight or alleviate climate change consequences, together with social concerns such as inequality and poverty, is a matter of debate in most countries and

Sustainable Fuel Technologies Handbook. DOI: https://doi.org/10.1016/B978-0-12-822989-7.00017-2

cities around the world. Thousands of experts in many topics are advancing in their research to deploy solutions and create new ways of living that pursue—broadly speaking—the same, that is, to increase welfare without causing damages to nature. In that sense, the role of the citizen, as user/consumer, is crucial since they are the final agent and creator of awareness. Most modern states, based on democratic schemes of representativeness, behave as controllers of the system's evolution and must internalize the social awareness to keep working. In that, cities act as key actors due to the huge amount of people living in them.

To date, it is not so frequent that cities have energy and/or climate planning agencies. Traditionally, that was a concern of countries focused on ministerial tasks. That is changing significantly, especially in Europe. The Covenant of Mayors and the SECAP document that cities have been reporting as committed mechanisms to reduce climate change are interesting tools to inspire changes in society. However, the Paris Agreement commitments along with the establishment of the 17 Sustainable Development Goals (SDGs) included within the 2030 Agenda of United Nations have acted as stimuli to move cities into taking leadership. To do so, many cities have started to build their own goals in fighting climate change by pursuing huge amounts of electric vehicles in their streets, imposing taxes on pollution, or investing in renewables (mostly solar PV), among other measures.

Accordingly, evaluations to select among technologies or to determine to what level a measure should be deployed are founded on the utilization of analytical tools such as life cycle assessment (LCA), techno−economic assessment (TEA), energy system modeling (ESM), scenario analysis (SA), and multicriteria decision analysis (MCDA), among others. These tools or methodological approaches are frequently used in specific realms to evaluate concrete aspects of products or systems. For example, LCA is useful to evaluate the environmental performance of systems, TEA allows for economic behavior to be understood, whereas ESM facilitates the energetic performance of a system when looking at the future using scenarios. Moreover, MCDA is useful to make decisions among different options. Depending on the scope (global, continental, national, regional, or local), the interests (public, private), the system itself (a factory, a product, a mechanism, a process, or a city, etc.), these tools are useful in the search for the "best" solutions, that is, cost-effective, socially acceptable, environment-friendly, etc.

From an environment point of view, LCA is usually recognized as a well-known method to evaluate the impact on nature and human health of products and/or systems. The LCA rationale is easily transposable to life cycle costing (LCC) and even social LCA (SLCA), to consider aspects other than the environment. Hence a multiple approach to tackle a problem may be faced using a combination of those (LCA, LCC, SLCA), known as a life cycle sustainability assessment (LCSA). While there are thousands of LCA studies in scientific and business literature, LCSA works are not so common.

For a sustainability assessment of complex systems, including the exploration of multiple derivatives or future scenarios, joint methods integrating various of the mentioned methodologies and tools might be needed.

This multifaceted tactic of using several methodological frameworks to deal with a problem is a complex task from the side of the analysts. Different methods usually differ in the nature of the objects they include so a sort of harmonization is required to interpret parameters. This requires specific knowledge from the modelers to merge different parts of the system, and it is hard to automatize. In the case of cities, there are not so many holistic approaches encompassing energy, economy, and emissions in addition to sustainability assessments. Moreover, there is a lack of satisfactory, that is, useful, reliable, and complete, inputs (data) in energy planning for policy makers at the city level. Considering that decisions must be made anyway, every rigorous method that could assist policy makers in their work is beneficial.

In this chapter, the main motivation of the work is dual. On the one side, it is to deepen sustainability assessment methods by proposing a novel conceptual approach prior to evaluation. Such advance goes to the favor of understanding the suitability of LCSA in analyzing the future of cities by enriching the discussion about the definition of system boundaries in city planning. The approach has potential replicability regarding improvements in the evaluation of long-term energy plans aimed at sustainability in other systems or scales. On the other side, the chapter proves the application of the proposed approach through a case study for the city of Donostia-San Sebastián, located in the Basque Country, in Spain.

16.2 Conventional approaches in city energy planning

As previously explained, there are many studies on city planning and energy policy making for cities. Those studies use different approaches and methods to face the many concerns of sustainability, that is, technical, economic, social, and environmental dimensions.

Regarding LCA studies, holistic approaches in evaluating cities are not so frequent. City-related LCA works present specific evaluations of systems that may be considered "urban." For example, García-Guaita et al. [1] developed a multidisciplinary evaluation of a Spanish city using three consolidated environmental methodologies, namely LCA, material flow analysis, and urban metabolism. These authors evaluated the sustainability performance of the city to point out the sources of problems in terms of emissions, therefore, proposing measures to alleviate the situation. At a different level, the work carried out by Lausselet et al. [2] concerning the environmental performance of zero-emission neighborhoods during early planning stages is interesting. From a geographical point of view, Loiseau et al. [3] discussed the importance of the territorial aspects of urban areas when developing LCA studies.

In particular, the works carried out by Albertí et al. [4,5] are highly relevant to the present work. These authors are advancing LCA rationale by adapting the nature of the methodology, thus, adapting several substep definitions according to the specificity of cities. For instance, they developed the so-called city prosperity index to understand the functional unit (FU) of a city in terms of LCA. Besides, they make use of the number of inhabitants to proceed with normalization. Moreover, in the work of Albertí et al. [4], a thorough discussion on allocation procedures for LCA in cities is presented. Three methods for establishing the boundaries for an urban LCA are considered, namely (1) administrative-based limits, (2) density-based boundaries, defined by the physical nature of the population density, and (3) service-based boundaries, defined by considering (4) in addition to those services required by the city. In addition, four allocation methods are presented by the authors, namely (1) monetary-based allocation, (2) producer-based allocation (impacts are assigned to the city holding the production of the good), (3) consumer-based (impacts are assigned to the city holding the consumers of the good), and (4) category-based allocation (global impacts are assigned to the city where consumption occurs, whereas local impacts are allocated to the city where the impacts take place).

None of the previous articles discuss prospective terms. Frequently, the lack of "vision" when using LCA studies is a problem for many analysts who want to obtain benefits from the results obtained through LCA scenarios. LCA is usually taken as a static approach of a system that serves to point out the hotspots and aims at proposing alternatives. When one aims to understand how a system (in this case an entire city) will evolve into the future, some other tools such as ESM are required.

The combination of ESM and LCA approaches is an innovative area of research in the field of sustainable long-term energy planning. For instance, the works carried out by García-Gusano et al. [6,7] include a methodological framework to evaluate several energy systems (national electricity mix) with a time horizon up to 2050. These works evaluate in a prospective manner some social and environmental concerns, such as human health or the global warming potential (GWP), which are associated with the transformation of the energy system.

In terms of policy making, it seems convenient to highlight the Communication on Integrated Product Policy [COM (2003)302] of the European Commission (EC) [8]. It concluded that LCA provide the best framework for assessing the potential environmental impacts of products currently available. In the document, the need for more reliable data and consensus on LCA methodologies was emphasized. Besides, it was announced that the EC would deliver a platform to facilitate an initiative involving both ongoing data collection efforts in the European Union and existing harmonization initiatives. In a similar way, it is remarkable the work undergone by the California Energy Commission [9], where the state administration published a document on the incorporation of LCA to complement urban metabolism in sustainable community assessments.

16.3 A new conceptual framework for long-term energy scenarios for cities

It has, therefore, become clear that sustainability has an undeniable presence in conventional approaches to city energy planning, standing as one of the main motivators of such plans. In spite of this, its presence is mainly concentrated during the phases of defining objectives and monitoring results, and to a lesser extent, in the configuration of potential decarbonization scenarios. This reflects the fact that sustainability has not yet fully become the concept that underpins the bulk of the activities oriented to the planning processes.

Using sustainability, and in particular exante impact assessment as the principal connecting element, the framework presented in this section aims to lay the groundwork for the prioritization of scenarios to guide the ET of cities toward a low-carbon future. Fig. 16.1 outlines how the concept of sustainability assessment (highlighted in blue) should be involved in different ways in each of the phases of the energy planning process to maintain the coherence of the analysis.

Phase I: City contextualization

In the beginning of the study, during the phase of a city's contextualization (diagnosis), there are many indicator-based frameworks capable of evaluating the sustainability of cities for a specific year. Frameworks such as the one developed in the CITYKEYS project [10], City Protocol [11], Siemens Green City Index [12], PLEEC [13] project, European Smart Cities [14], or the international standard for a city sustainability assessment ISO 37120 [15], among many others, usually include various indicators related to sustainability. These metrics can be used as a reference to start a discussion with the municipality regarding its capacities, strengths, weaknesses, and needs.

Phase II: City vision and goal and scope definition for the planning

In the second phase, a long-term city vision is performed, and the goal and scope of the study are defined as well. In this case, sustainability usually appears when defining the objectives of the city. However, it should be fully integrated in the

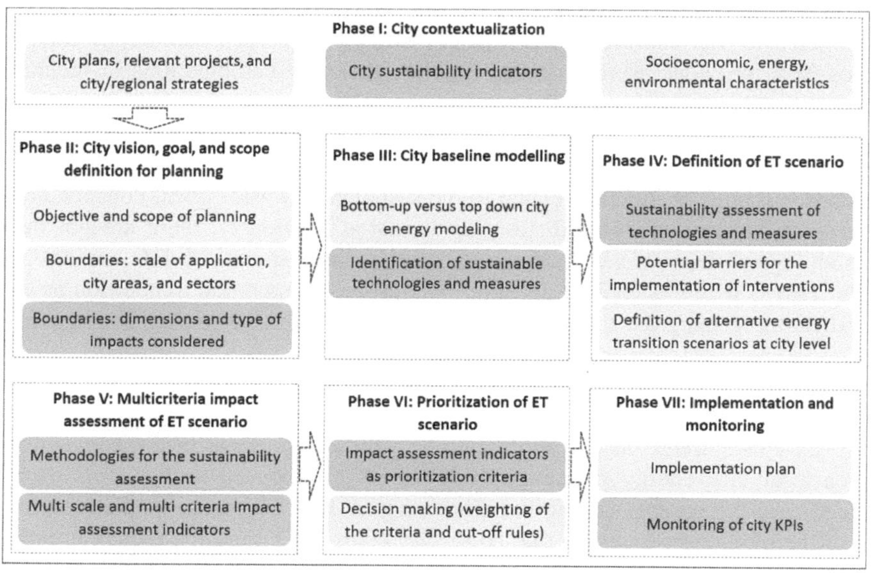

Figure 16.1 Conceptual framework for the prioritization of scenarios to guide the energy transition of cities toward a low-carbon future.

definition of the scope and boundaries of the analysis, specifying for each of the scales (district, city, region) both the dimensions considered, for example, economic, environmental, and social, as well as the type of impacts evaluated, for example, direct, indirect, and induced.

Phase III: City baseline modeling

The third phase is focused on city baseline modeling, and it is founded on the analysis of the current situation of the city in terms of energy consumption and generation.

This first step can be addressed from both the bottom-up and the top-down perspectives. The first one—bottom-up—is approached by aggregating detailed energy modelings. For the building sector, the evolution of the energy use of existing buildings and expected new constructions are modeled in detail, from the individual building level to the district level. The second perspective—top-down—can be approached using the general and aggregated energy consumptions of the city as a starting point as well as other types of socioeconomic parameters as drivers in order to explore the evolution of those consumptions. In any case, during this first step, sustainability mainly appears when these energy consumptions are transformed into different environmental impact categories such as the GWP or primary energy (PE).

The second step of the city baseline modeling focuses on the identification of sustainable fuel technologies and measures that are suitable to be included in the alternative ET scenarios for the city.

Phase IV: Definition of energy transition scenarios

The definition of the potential ET scenarios for the city is precisely the aim of the fourth phase. Comparison and prioritization of sustainable fuel technologies and measures is carried out through LCSA methods. The results of this assessment, combined with the analysis of the barriers associated to each technology for their deployment, provides useful criteria to define different combinations of technologies that finally establish the ET scenarios proposed.

Phase V: Multicriteria impact assessment of energy transition scenarios

Phase five is focused on the multicriteria impact assessment of ET scenarios. This phase should be directly related to the sustainability assessment concept and to the impact assessment. Nevertheless, as discussed previously, there are not many methodological frameworks addressing impacts assessment of ET scenarios in depth. This is why this chapter discusses and proposes a method in addition to a set of indicators to integrate the sustainability dimension for the impact assessment of long-term energy scenarios at the city level.

Phase VI: Prioritization of scenarios

Once the impact assessment of the proposed scenarios has been carried out, phase six concentrates on the prioritization of those scenarios. In the conventional approach of city energy planning, the number of proposed scenarios is usually reduced, and the process of prioritization is based, in most cases, on the minimization of the total costs, ensuring that politically committed emission reduction targets are achieved. Therefore the impact assessment of scenarios is a step that usually appears to provide additional information about the effects that can be expected from the implementation of the final scenario selected. This means that exante impact assessments are not commonly used for the generation of criteria that can be used by municipalities for the prioritization of alternative scenarios. In the method proposed, impact assessment indicators are precisely used as prioritization criteria. These criteria can be weighted by the municipality according to their specific interest and city objectives so that the different scenarios can be fully compared.

Phase VII: Implementation and monitoring

Finally, phase seven focuses on the implementation and monitoring of the plan. Here, the impact assessment indicators used in the previous phase for the exante impact assessment can be part of the indicators used for monitoring the performance of the city.

16.3.1 The sustainability dimension in long-term energy scenarios at the city level

Sustainability in its broadest sense is increasingly becoming the common goal of transformation processes in many countries and cities. Proof of this is the previously mentioned 2030 Agenda for Sustainable Development [16], adopted by the United Nations in 2015. At the heart of this agenda, 17 SDGs with 169 associated targets are intended to guide the actions to be adopted by all countries during the subsequent years. Within this agenda, SDG 11 says, "*Make cities and human settlements inclusive, safe, resilient and sustainable*," and proposes several targets and indicators for cities. Some of these targets highlight aspects such as the need for reducing the adverse per capita environmental impact of cities or the need for supporting positive economic, social, and environmental links at different scales to reinforce national and regional planning.

However, most of the studies carried out in the field of LCSA for cities focus on environmental and economic impact assessments of small and medium scale interventions such as buildings and sustainable fuel technologies. Due to this fact, the possible connection between exante impact assessment and strategic decision making at the city and regional levels is largely blurred.

When tackling the problem from the point of view of the municipality, the fact of integrating the analysis of the expected impacts as a central pillar of the analysis, may provide useful criteria to assist in decision-making processes. This opens up a new range of possibilities in the field of energy planning in cities. One of the difficulties in this regard is that there are multiple solutions when it comes to promoting the transformation of cities toward a low-carbon future. The municipality should be able to evaluate potential scenarios considering direct, indirect, and induced effects that their implementation will have on the city; effects that are difficult to foresee. To this end, the impact assessment methodology should offer the necessary sensitivity to measure the different effects that would arise once the decision is taken to opt for the deployment of certain sustainable fuels and technologies rather than other alternatives.

Unfortunately, the full integration of sustainability assessment into the modeling of long-term ET scenarios for cities is still a great challenge. This section, which further develops phase five of the framework described in Section 16.3 for city energy planning, aims to make some progress toward the harmonization of the way of integrating sustainability assessment within the conventional energy modeling of scenarios. Keeping this in mind, the methodology developed combines different approaches to impact assessment, which, through their interrelationship, allow for the effect of potential ET scenarios of cities to be evaluated from a multicriteria and multiscale point of view.

Fig. 16.2 shows graphically the proposed methodology, which places sustainability assessment as the core of the analysis. This proposal agglutinates various dimensions, scales of action, time horizons of the strategy, and scopes for the impact assessment that should be considered when dealing with long-term sustainable city planning.

It is relevant to note that the methodology does not attempt to assess city sustainability for a specific year. What it really seeks is to assess and compare the sustainability of the potential long-term scenarios proposed for the city, so that the optimum or at least a compromise solution scenario is identified. In this case, the scenario is understood as the agglutination of technologies and measures aimed at reducing energy consumption and the dependency on fossil fuels, and increasing the proportion of renewable energies, while ensuring the well-being of citizens and promoting the socioeconomic development of the city as a whole. These technologies and measures are deployed in different areas of the city with different intensities and temporal rhythms throughout the transition period defined.

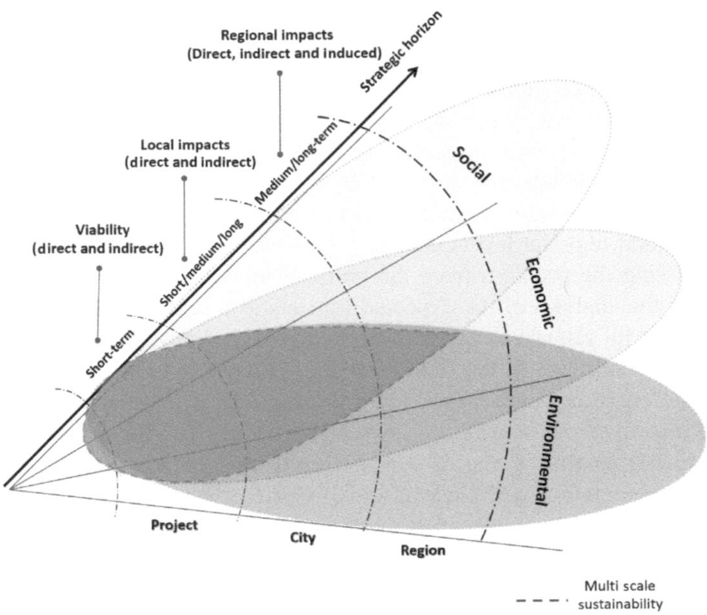

Figure 16.2 Conceptual framework for the multicriteria and multiscale impact assessment of energy transition scenarios of cities.

Fig. 16.2 shows that the impact assessment methodology covers the three pillars of sustainability by assessing the economic, social, and environmental impacts of the energy transition scenarios. In turn, for the evaluation of each of these dimensions, different waves can be distinguished that represent the different scales of the analysis. In this way, it is possible to distinguish between the project level (often associated with interventions at the building and district scales), the city level, and the regional level. Finally, it can be seen that each of the mentioned waves is associated with a time horizon of the strategy and a depth or scope of the impact assessment. Therefore a distinction can be made between short-term impact assessment (corresponding mainly to the project scale), medium-term impact assessment, and long-term impact assessment (corresponding to both the city and regional scales).

Project—district scale:

This first scale responds to the necessity of assessing impacts from a feasibility point of view, and provides relevant information for the company, entity, or citizen to make an investment. In this way, aspects such as the profitability or the payback period of the investment become relevant aspects for the economic dimension. In the environmental dimension, the quantification of environmental impacts from a purely informative point of view and for compliance with environmental requirements are usually the focus of the assessment, while for the social dimension, several aspects such as social acceptance or the working conditions can be of interest. This level of analysis is the most studied so far using impact assessment methodologies such as LCA, LCC, and SLCA.

City—energy transition scenario scale:

The second wave (Fig. 16.2) corresponds to the city scale. This includes various time horizons of the strategy, from isolated short-term actions at the project level to the implementation of transition scenarios that aim to transform the city in the medium- and long-term. Nevertheless, the assessment of the effects created in the society, economy, and environment of the city as a whole become more relevant. In this case, the evaluation of aspects such as the influence of the transition scenario on the self-sufficiency levels of the city, diversification of the energy matrix, creation of local employment, energy poverty, or the accomplishment of the established emission reduction targets should be addressed. These aspects are related to particular and inherent characteristics of the city such as the structure of production and consumption of goods and services or the quality of life. For their evaluation, the methodologies mentioned based on the life cycle approach must be adapted. An additional challenge of the proposed adaptation is to provide a method with sufficient simplicity and flexibility to allow its connection with other methodologies and modeling tools commonly used for city energy planning. In this regard, linking ESM (which provides the basic required inputs for the impact assessment) with impact assessment methodologies in such a way that the long-term ET scenarios for cities can be designed and assessed is precisely one of the main difficulties.

City—regional interactions:

Finally, the third wave of Fig. 16.2 is intended to establish the relationship between the actions carried out at the city level throughout the ET period and their effects at the regional level. In this case, the aim is to evaluate the influence that the direction adopted for the transformation of the city will have on more strategic aspects such as:

- The promotion and diversification of regional industry in order to reduce its vulnerability to possible economic crises.
- Improving the quality of life and the income level of the different social strata.
- The generation of stable economic activity with high added value to improve competitiveness in foreign markets.
- Alignment with climate change adaptation and mitigation policies that contribute to limiting both local and global impacts.

Exante impact assessment enables the most appropriate medium- and long-term transition strategies for the city to be explored, considering not only the direct and indirect effects, but also the induced effects at the macroeconomic level. Effects such as growth in regional gross domestic product (GDP), in sectoral value added (VA), in household disposable income, and in job creation should be assessed at this point.

It is clear that for this purpose, methodologies based on the life cycle perspective, as commonly used, fail to provide the necessary methodological framework. However, by means of some adaptations such as establishing a connection between the analysis of the life cycle costs of these scenarios with other types of socioeconomic impact assessment methods such as those based on input—output tables may be of interest. In this sense, using supply chain analysis of the technologies and measures included in the scenarios is a direction to explore further, which can serve to create an exogenous shock that represents the entire scenario, and that can be used as the main input for the socioeconomic impact assessment at the regional scale.

16.3.2 A focus on sustainability assessment: an innovative life cycle assessment adaptation for city planning

Despite the vast literature focusing on life cycle impact assessment, it is difficult to find a method designed expressly and integrally for the impact assessment of ET scenarios for cities. This means that the most basic aspects for life cycle impact assessment are still to be discussed and agreed upon. Aspects such as the definition of the goal and scope, the FU or reference unit, the lifetime, and the system boundaries or the impact assessment indicators, all of them must be defined specifically for this emerging problem. The definition of these aspects for the evaluation of the life cycle impact assessment of cities is becoming the focus of new studies such as that of Albertí et al. [17]. This publication represents a relevant progress in this field; however, the method proposed in this chapter does not attempt to assess the sustainability of the actual situation in the city, but it focuses on the sustainability assessment of the proposed ET scenario. In other words, the focus is placed on the second wave observed in Fig. 16.2, which represents the transition scenario for the city scale. More precisely, an adaptation of the LCA methodology for city energy planning is proposed.

Goal and scope, functional unit, and lifetime proposed:

The authors propose as the goal and scope of the assessment the comparison of competing long-term energy scenarios that are viable to be adopted by a specific city to guide its transition toward a low-carbon future. Note that the concept of time is included from the very first moment. Regarding the definition of the FU, it should be noted that this is a crucial aspect to ensure a fair comparison between different alternatives. As defined in the International Reference Life Cycle Data System Handbook [18], the FU should "name and quantify the qualitative and quantitative aspects of the function(s) along the questions 'what,' 'how much,' 'how well,' and for 'how long.'" In this case, the proposed FU can be described as, *an energy transition scenario that allows the city assessed to achieve in the most sustainable way possible an X% reduction in carbon dioxide (CO_2) emissions at the end of the transition period with respect to base year CO_2 emissions*.

Note that the target of CO_2 emission reduction is proposed as a limiting aspect in the form of a cut-off rule. Besides, it is established that it is not enough to comply with this requirement, but it is also necessary to consider the way in which this has been achieved by evaluating the level of sustainability of each alternative scenario. The lifetime is also implicitly defined in the description of the FU as it refers to the transition period evaluated.

System boundaries, life cycle phases, and information modules proposed:

Regarding system boundaries, and considering that the purpose of the study is to evaluate the different scenarios that will transform the way in which the city generates and consumes energy, the limits of the system are extended from the final energy uses to an environment where the different centralized and decentralized generation systems are also integrated. This is why the limits of the system will include all the infrastructures and technologies necessary for the generation of local energy within the physical limits of the city, including the generation of local energy by means of nonlocal fuels. In addition, all the phases involved between energy generation and consumption in the city must be considered, including transformation, distribution, storage, and supply.

On the other hand, for the definition of the phases and information modules to be considered, the perspectives proposed by both the Intergovernmental Panel on Climate Change [19] and the standard *EN 15978 Sustainability of construction works—Assessment of environmental performance of buildings—Calculation method* have been taken into account.

The two frameworks mentioned are of interest when assessing the sustainability of individual buildings and energy generation systems implemented in a given year. However, they do not establish how to consider the different phases when assessing the suitability of a city's transition scenario from a life cycle perspective, which will progressively implement different groups of sustainable fuel technologies and measures throughout the transition period. It is, therefore, necessary to adapt this framework in such a way as to assess the interaction between several interventions that will be implemented simultaneously and progressively in the city. This allows for the influence of aspects such as the order, the simultaneity, and the rhythm of the implementation of technologies and measures to be determined. These aspects have a relevant effect for impact assessment as they have a direct influence on how and when to consider, among others, the price escalators, discount factors, or energy technology cost trends.

Fig. 16.3 shows a schematic visualization of the adaptation of the framework for the definition of the phases and information modules proposed in this work. The phases of the life cycle included in the framework are phases A1–A3, which include the extraction of raw materials, transport, and manufacturing, phases A4–A5, which include transport, installation, and construction, phase B2, which includes maintenance, phase B4, which includes replacement, phase B6, which includes operation, and phase C (C1–C4), which includes the end of life of the products and components.

The need to perform analysis on an annual basis is evidenced. This enables the detailing of the year in which each of the sustainable fuel technologies and

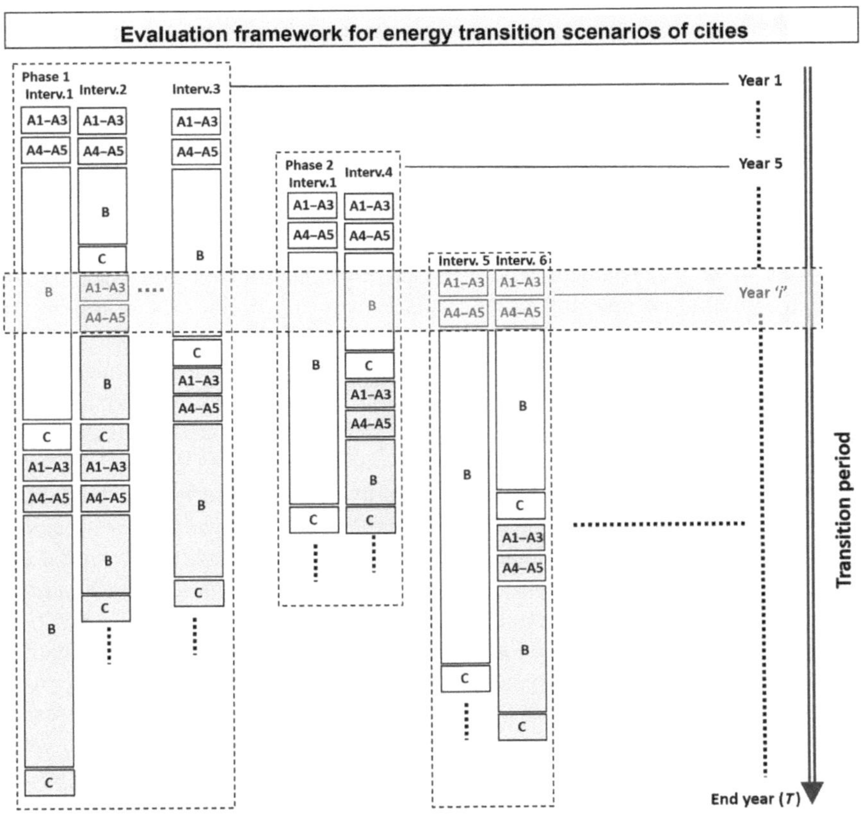

Figure 16.3 Framework for the consideration of the phases and information modules in the impact assessment of energy transition scenarios of cities.

measures will be deployed throughout the transition period. In addition, it allows the fact that a technology can be implemented in two or more temporary phases to be considered, as in the case of a district heating that can be launched initially for a certain area, and a few years later could be extended to cover a new area of the city.

Commencing from the left (see Fig. 16.3), it is observed the simultaneous implementation of the first group of interventions in the first year. Each of them with their corresponding phases of the life cycle and lifetime. It is clear that, depending on the length of time of the transition period under consideration, the complete renewal of some of the mentioned interventions may be required (phases marked in gray). Similarly, after a certain time period (in the fifth year in the example), the implementation of the next set of interventions should be contemplated. This procedure is successively replicated until all the interventions foreseen for the transition period are implemented.

Therefore the impact assessment should be calculated year by year as described here. For the year "*i*" the analysis should include phase B of intervention 1 (in its first implementation phase), phases A1–A5 of intervention 2, phase B of interventions 3, 4, and 1 (in its second implementation phase), and finally phases A1–A5 of interventions 5 and 6 that are implemented that year. This process will enable each of the phases of each intervention to be given specific characteristics for the year evaluated. The way of denominating the different phases of the scheme is $(\text{phase})_{int.x,I}$, where "*phase*" refers to the life cycle phase analyzed, "*int.x*" is the identifier of each intervention evaluated, and "*i*" is the year in which the phases corresponding to that intervention are considered. As an example, phase $(\text{A1}-\text{A3})_{int.1,1}$ refers to the phases A1, A2, and A3 of intervention 1 that is implemented in the first year.

Impact assessment indicators:

Once the framework for the definition of the phases and information modules has been reformulated, the next step focuses on the definition and reformulation of the impact assessment indicators according to it. In this regard, several impact assessment indicators can be defined for city scale and for the purpose of comparing alternative ET scenarios. The authors propose social and economic indicators based on the life cycle perspective such as the *cumulative net present value of the energy transition scenario ($CNPV_{TS}$), the dynamic payback period (DPP_{TS}), the cumulative net present cost-social ($CNPC\text{-}S_{TS}$), or the cumulative net present cost of public/private companies ($CNPC\text{-}PPC_{TS}$)*. For the energy and environmental dimension indicators such as the *cumulative GWP reduction of the energy transition scenario ($CGWPR_{TS}$), the cumulative nonrenewable primary energyPE use reduction ($CNR\text{-}PER_{TS}$) or the cumulative local renewable electricity generation increase ($CLREG_{TS}$)* are proposed. Detailed definitions and the adaptation process of all the indicators mentioned can be consulted in [20].

Eq. (16.1) is intended to serve as an example of the adaptation of the GWP indicator as can be used in conventional LCA for the analysis of energy technologies to the $CGWPR_{TS}$.

$$CGWPR_{TS} = -\sum_{i=1}^{T} \frac{\left[\sum_{int=1}^{X}\left(GWP_{int,i}^{A1-A5}\right) + \sum_{int=1}^{Y}\left(GWP_{int,i}^{B}\right) + \sum_{int=1}^{Z}\left(GWP_{int,i}^{C1-C4}\right)\right]}{FU} \tag{16.1}$$

Where Eq. (16.2) explains in detail the procedure for calculating phase B of the scenario.

$$\begin{aligned}
\sum_{int=1}^{Y}\left(GWP_{int,t}^{B}\right) &= \sum_{int=1}^{Y}\left(GWP_{int,i}^{B2} + GWP_{int,i}^{B4} + GWP_{int,i}^{B6,3} - GWP_{int,i}^{B6,4}\right)\\
GWP_{int,i}^{B6,3} &= \sum_{y=1}^{K}\left(EC_{int,i}^{y} \times GWPf^{y}\right)\\
GWP_{int,i}^{B6,4} &= \sum_{y=1}^{L}\left(ER_{int,i}^{y} \times GWPf^{y}\right)
\end{aligned} \tag{16.2}$$

And where $GWP_{int,i}^{A1-A5}$ refers to the GWP impact of phases A1−A5 of the corresponding intervention during time period i. "T" refers to the last year of the transition period, and "X," "Y," and "Z" refer respectively to the total number of interventions that have the phases A1−A5, B, and C1−C4 active for the year evaluated. $B_{6.3}$ refers to the operational energy use of the new intervention implemented and $B_{6.4}$ to the operational energy savings obtained with the new intervention respect to the operational energy use of the interventions replaced. $EC_{int,i}^{y}$ refers to energy consumption of the fuel and the corresponding intervention in time period i. $ER_{int,i}^{y}$ refers to the energy consumption of the fuel "y" equivalent to that which has been replaced or avoided by the intervention implemented in the period of time i (either by energy generation or by improving energy efficiency), "K" refers to the number of fuels consumed, "L" to the number of fuels replaced, and $GWPf^{y}$ to the conversion factor to GWP of the fuel y.

16.3.3 Environmental and economic impact assessment of low-carbon energy transition scenarios for the city of Donostia-San Sebastián

16.3.3.1 Case study and baseline assessment

This section presents the application of the methodology in a case study. The city evaluated is Donostia-San Sebastián, the capital of a province of the Basque Country in the north of Spain. The population of the city is 186,665 (2018), and it covers an area of 61 km^2. For the analysis, the city scale is understood as an aggregation of districts, and the area of study comprises six districts covering more than the 40% of the city's building stock (see Fig. 16.4). The districts evaluated are Amara (11,639 dwellings), Cortazar/Centro (8988 dwellings), Antiguo-Ondarreta

Figure 16.4 Categorization by age in Quantum Geographic Information System for the buildings within the area of study.

(3856 dwellings), Gros (9581 dwellings), Aldezaharra (3332 dwellings), and Txomin Enea (1300 new dwellings; new area under development). The sum of these districts encompasses the most relevant areas of the city in terms of building stock and population, and it provides a proper context for the simultaneous evaluation of different areas of the city with highly diverse characteristics. The focus of the analysis is on the energy supply and demand of the city's building sector (residential and office buildings).

Baseline energy characterization of the buildings has been carried out following a bottom-up approach, following methodologies described in [19,20].

The results obtained from the baseline energy analysis include the energy use of the six districts by building typology, by different energy end use, and by fuel type, all of them on an hourly basis. Table 16.1 shows the principal characteristics in terms of energy demand and use of fuels, including the new building developments at the Txomin Enea district.

The main consumption of the districts corresponds to natural gas and electricity use, with other types of fuels having lower usage values. The distribution of these energy consumptions by each of the districts is detailed in Fig. 16.5 Among other aspects, it is observed that the district of Cortazar, being the district with the highest number of old residential buildings, is responsible for 30% of heating consumption. Electricity use for cooling and other end uses is larger in Amara and Cortazar districts due to the higher number of office buildings in these districts.

The diagnosis highlights the relatively low degree of energy refurbishment of old buildings, the need to continue modernizing their energy systems, and the great potential for the integration of renewable energy technologies. Nevertheless, there are several aspects such as the heritage protection of buildings that should be

Table 16.1 Energy demand of the six districts by building typology, energy end use, and fuel type.

Energy demand (GWh/year)				
	Heating	**DHW**	**Cooling**	**Electricity**
Residential use	198.8	26.7	–	92.6
Office use	5.2	–	6.5	13.7

Use of fuel (GWh/year)					
	NG	**Electricity**	**Gas oil**	**Biomass**	**Other LF**
Residential and office use	222.6	152.8	33.3	2	2.1

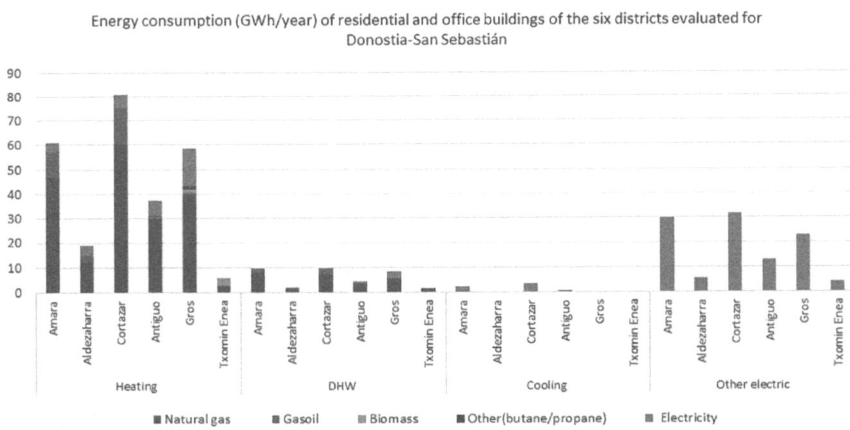

Figure 16.5 Distribution of energy consumptions by each of the districts evaluated for Donostia-San Sebastián.

considered when proposing scenarios for the city since those aspects can limit the deployment level of some measures.

16.3.3.2 Long-term energy scenario for the city of Donostia-San Sebastián

Two ET scenarios focused on reducing environmental emissions, decreasing the dependence on fossil fuels, and increasing the socioeconomic development of the city are compared in this section. Each scenario proposes the simultaneous implementation of several sustainable fuel technologies and measures and their staggered deployment across a transition period of 50 years beginning in 2017. The sustainable fuel technologies and measures included in the scenarios are listed as solar photovoltaic (SPV) systems, solar thermal (ST) systems, passive interventions for buildings (PIFBs), natural gas individual natural gas boilers (INGBs), central biomass boilers (CBBs), heat pumps (HPs), and district heating with natural gas and biomass boilers (DHNGB). In order to determine the degree of deployment of each technology, the maximum potential associated with each of them has been considered for the area of study as well as the main barriers that may limit their realistic implementation. The barriers include the limited options to replace existing gas oil-fired and natural gas-fired central heating systems with biomass central heating systems, the area available on building roofs for the installation of solar systems, and heritage conservation grade of buildings for refurbishment measures. Besides, a cut-off rule has been set for the scenario definition at a minimum of 65% of CO_2 equivalent emission savings at the end of the transition period with respect to the values from 2007. The most relevant characteristics of each scenario are summarized in Table 16.2.

The first scenario of Table 16.2 considers the implementation of a district heating network in Txomin Enea, whereas the second scenario proposes an expansion of it toward the district of Amara. In the latter case, the implementation of ST systems in the district of Amara is not contemplated. In addition, the first scenario evaluates the modernization of the energy systems of those buildings included in all districts except in Txomin Enea district. For this modernization, priority is given to replacing existing systems with HPs in the case of individual systems and with biomass boilers in the case of central systems. In contrast, the second scenario considers the replacement of the energy systems of all the districts except Txomin Enea and Amara, prioritizing a compromise solution that proposes a combination of condensing INGBs (to replace 45% of the individual systems), HPs (to replace 45% of the individual systems), and biomass central boilers (to replace all the central heating systems and 10% of individual systems). Regarding the SPV and building refurbishment, the two scenarios consider the same deployment level.

Other key parameters considered for defining the scenario as well as their evolution during the transition period are:

- *Annual implementation rate of the interventions*: A rate of 2% has been set for building refurbishment interventions [21]. The hypothesis that building energy system replacement follows the same rate is used in the study. Initial investments are considered to occur in the same year in which implementations take place. In the case of district heating, a connection rate to the district heating network of 9% is set for scenario 1 (as proposed in the feasibility study for the district heating of Txomin Enea), and a rate of 4% for scenario 2.
- *Energy price escalators*: The projections of energy price are adopted from the EU Energy, Transport and GHG Emissions Trends to 2050 [22].
- *Discount rate*: Based on EU energy trends to 2030 [23], for long-term planning, a discount rate of 3% has been considered.
- *Energy technology cost trends*: It is assumed that technology costs (in real terms) will decrease in the future. The evolution of the capital expenditure (CAPEX) and the operational expenditures (OPEXs) for the time frame of 2013 to 2050 have also been considered [24], [25]. In the case of central heating biomass boilers and building refurbishment-

Table 16.2 Technologies used for the different districts on the two energy transition scenarios.

Scenario 1	**Deployment in 50 years**	**Amara**	**Cortazar**	**Antiguo**	**Gros**	**Txomin Enea**	**Alde-Zaharra**
SPV	250 ($10^3 m^2$)	X	X	X	X	X	X
ST	34 ($10^3 m^2$)	X	X	X	X		X
PIFB	2.5 ($10^6 m^2$)	X	X	X	X	X	X
CBB, HP	216.5 (GWh)	X	X	X	X		X
DHNGB	5 (GWh)					X	
Scenario 2							
SPV	250 ($10^3 m^2$)	X	X	X	X	X	X
ST	25 ($10^3 m^2$)		X	X	X		X
PIFB	2.5 ($10^6 m^2$)	X	X	X	X	X	X
INGB, CBB, HP	166.3 (GWh)		X	X	X		X
DHNGB	55.6 (GWh)	X				X	

related interventions, no relevant cost reduction has been considered due to the current maturity level of the technology and the lack of reliable projections.

- *Shadow price of CO_2 emissions*: It has been considered that the price starts at 21.02 US\$/ton for the base year and rises to 63.62 US\$/ton in 2050 [26].

Regarding the economic aspects of both scenarios, energy prices (taxes included) defined by the Spanish Ministry of Development [26] have been considered for households of 0,236 US\$/kWh for electricity, 0.077 US\$/kWh for natural gas, 0.108 US\$/kWh for gas oil, and 0.052 US\$/kWh for biomass. From the district heating operator point of view, the cost of energy consumption and the selling price have been defined by direct interviews with operators. In this regard, energy costs of 0.06 US\$/kWh for natural gas, 0.034 US\$/kWh for biomass, and 0.15 US\$/kWh for electricity have been considered. Finally, a price of 0.012 US\$/kWh for the power generated for sale to the grid and a heat selling fixed price of 20.3 US\$ per dwelling and month and a variable price of 0.07 US\$/kWh have been defined.

On the other hand, the phases of the life cycle considered in this case study analysis follow the EN 15978 nomenclature and are A1–A3, A4–A5, B2, B4, and B6 for the economic analysis and A1–A3, B4, and B6 for the environmental analysis.

For the economic analysis, the CAPEX of the different technologies has provided the data for the sum of A1 to A5 phases, and the OPEX for the maintenance phase B2. The CAPEX of the technologies will also be a recurrent cost for the substitution phase B4. For the environmental analysis, specific LCA processes from the Ecoinvent database have been assigned for the different technologies assessed. The GWP of energy use during the operational phase (B6), the conversion factors recommended for the building sector in Spain, have been used [27].

Table 16.3 lists the most relevant features of the interventions included in the scenarios for the LCA and LCC.

16.3.3.3 Economic and environment life cycle impact assessment of scenarios

The environmental and economic impacts of the ET scenarios proposed for the city have been evaluated for the different stages of the life cycle. According to this, and based on the procedure described in Section 16.3, two indicators are evaluated. The CGWPR indicator for the environmental dimension shows the total reduction of the GWP indicator for each of the transition scenarios. The CNPV indicator is evaluated following the same logic used for the CGWPR indicator described in Section 16.3, and shows the present value of cash inflows and outflows for the whole transition scenarios. The main outcomes of the analysis are reported in this subsection.

The economic impact assessment shows that there are significant variations of up to 52% in the CNPV obtained at the end of the period for each of the scenarios evaluated. Scenario 1 with a cumulative cost of 235.9 MUS\$ (including investments, reinvestments, and operation and maintenance costs occurring throughout the transition period and discounted to the base year) reaches a CNPV value of 29.2 MUS\$.

Table 16.3 Main characteristics for the LCA and LCC of the scenarios evaluated.

Intervention	CAPEX	OPEX	Source	Lifetime	LCA process	LCA source
SPV	315.2 US\$/m^2	0.5%	[26,27]	30	3 kWp flat roof installation, single-Si, on roof	Ecoinvent database
ST	493.8 US\$/m^2	1.5%	[28]	30	Solar system, flat plate collector, multiple dwelling, hot water	Ecoinvent database
PIFB	0.51 US\$/m^{2}a envelope, 1 US\$/m^{2}a window	0.25 €/m^{2}a envelope, 0.47 (€/m^{2}a window	[29]	50	Internal energy refurbishment. Efficient	[29]
INGB	61 US\$/kW (INGB),	1.6%	[26]	20	Gas boiler 10 KW	Ecoinvent database
CBB	282.5 US\$/kW (CBB)	1.6%	[30]	20	Furnace, pellets, 50 kW	Ecoinvent database
HP	565 US\$/kWe (HP)	4%	[25,29]	20	Heat pump 30 kW	Ecoinvent database
DHNGB	67.8 US\$/KW (NGB), 565 US\$/kW (BB)	–	[31]	50	Supply side, trench-works, principal pipes, surface box, tap, pump, service pipes, components in buildings	[32]

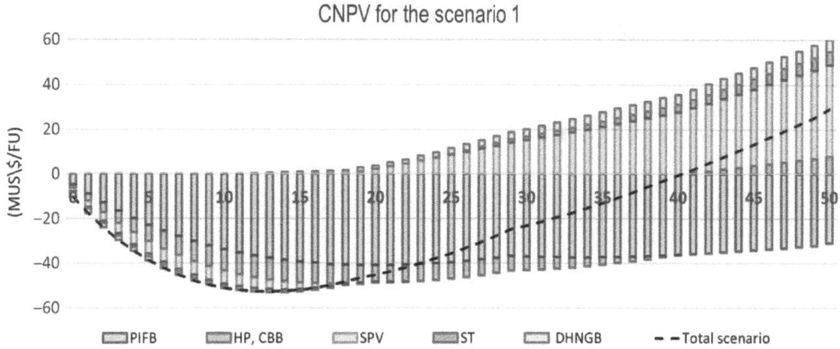

Figure 16.6 Evolution of the CNPV indicator for scenario 1 proposed for Donos tia-San Sebastián over the transition period.

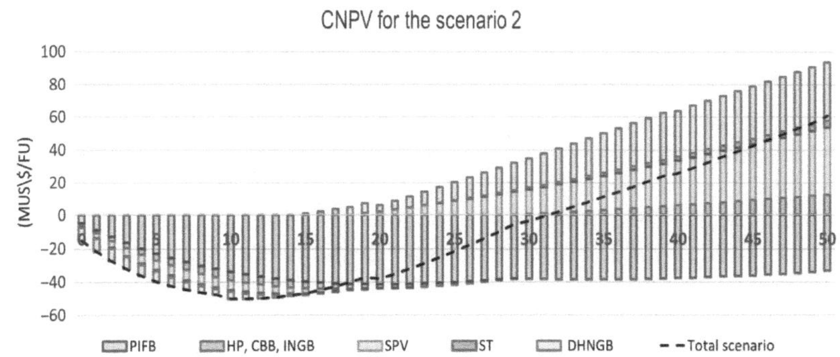

Figure 16.7 Evolution of the CNPV indicator for scenario 2 proposed for Do nostia-San Sebastián over the transition period.

Scenario 2, on the other hand, with a slightly higher cumulative cost of 241.8 MUS\$, obtains a significantly better result with a CNPV value of 60.5 MUS\$.

Figs. 16.6 and 16.7 show the evolution of the CNPV indicator for the two scenarios proposed for Donostia-San Sebastián over the transition period. The application of the methodology allows for greater evaluation of the annual evolution of the indicator for each scenario and the contribution of each intervention that compounds them over the transition period.

The first observation reveals that scenario 2 not only obtains a better economic performance at the end of the 50-year period, but also returns the investment earlier, in the year 32, as compared to scenario 1, which has a dynamic payback period (DPP) of 41 years. This is influenced by the typology and the degree of deployment of the interventions included in each scenario. More specifically, it is strongly conditioned in this case by the relationship between the deployment of building energy systems and the district heating-based solutions. Both interventions have large differences with respect to the business models that are most commonly adopted.

For solutions based on district heating, these indicators depend on the selling price for heating and domestic hot water (DHW). Those prices have been designed in this study in such a way that a compromise solution is achieved between the owner/operator and the citizen. However, as the business need to be inevitably profitable for the DHW owner and the operator, in the long-term, the economic benefits are evident.

In any case, it can be seen for the two scenarios that there are interventions such as building energy retrofitting (understood as the renewal of the entire aged building stock) for which the costs will not be recovered from the energy savings. However, as building refurbishment entails various other benefits besides energy related economic savings (e.g., health and safety, comfort, environmental performance), it is a key element in any ET scenario, and it has been considered that the profits obtained in the rest of the interventions contribute to achieving the building refurbishment goals. In this respect, photovoltaic solar systems (with a DPP of 19 years) contribute most to this compensation and represent 68% of the result of CNPV in scenario 1. The deployment of biomass boilers and HPs (with a DPP of 41 years) instead contribute 13% to the final CNPV result, while ST systems (with a DPP of 23 years), installed to a lesser extent in the city, and the relatively small district heating of Txomin Enea (with a DPP of 12 years) contribute 9% each. In the case of scenario 2, the contribution of each intervention is more balanced, with the photovoltaic solar systems (with a DPP of 19 years) being mainly responsible with 44%. These are followed by the district heating (with a DPP of 15 years) with 38%, which in this case covering a greater demand in the city. The effects of new investments associated with the extension of the district heating system can be seen for the entire scenario in year 10 as well as in years 20, 30, and 40 due to the renewal of a significant part of its components. The replacement of equipment with condensing gas boilers, HPs, and central biomass boilers (with a DPP of 30 years) represents 13% of the final CNPV, and finally, ST systems (with a DPP of 23 years) represent 4%.

The CGWPR indicator shows that CO_2 equivalent emission savings in the city vary from 76% in scenario 1 to 67% in scenario 2 with respect to the values of the year 2007, which is the reference year adopted by the city for defining the targets. This is a significative difference that, in this case, is mainly provoked by a better environmental performance of the HPs and biomass boilers deployed in scenario 1 in comparison to the district heating and the higher contribution of natural gas boilers from scenario 2.

Figs. 16.8 and 16.9 show the evolution of the CGWPR indicator in the city across the transition period, distinguishing the contribution of each of the interventions to the overall results of the scenarios.

In contrast to what occurs in the economic dimension, building renovation plays a crucial role in reducing environmental emissions, contributing, in this case, 47% to the total savings in both scenarios. SPV systems contribute 13% and 16% to this emission reduction in scenario 1 and scenario 2 respectively. The combination of HPs and biomass boilers contributes 32% in scenario 1, in contrast to scenario 2, in which the combination of HPs, biomass boilers, and natural gas boilers contributes

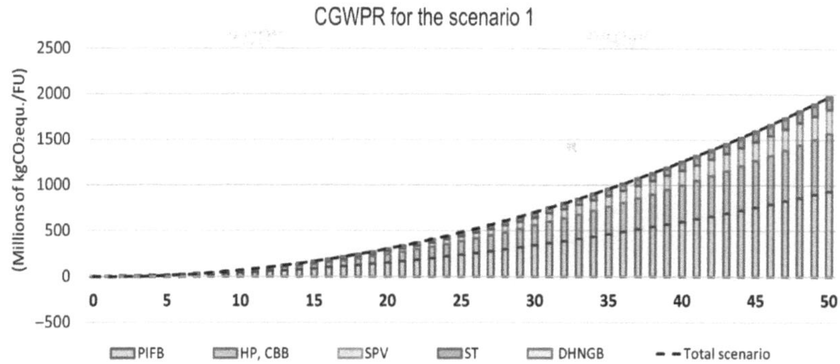

Figure 16.8 Evolution of the CGWPR indicator for scenario 1 proposed for Donostia-San Sebastián over the transition period.

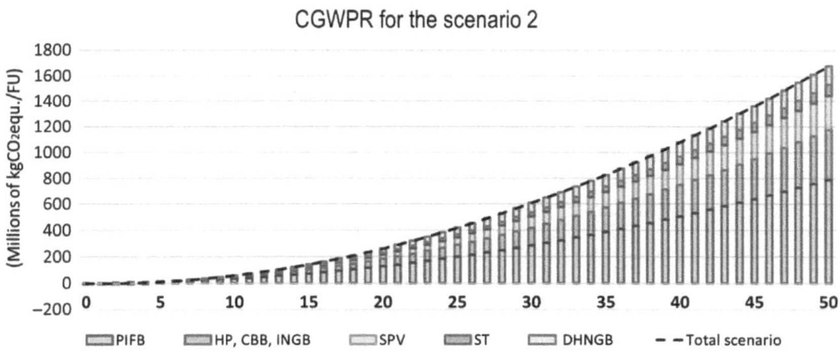

Figure 16.9 Evolution of the CGWPR indicator for scenario 2 proposed for Donostia-San Sebastián over the transition period.

23%. ST systems contribute 6% and 5% in scenario 1 and scenario 2 respectively. Finally, the district heating contributes 1% in scenario 1 and 9% in scenario 2.

At the beginning of the transition period, it can be observed that the CGWPR is negative. This means that the CO_2 equivalent emissions during these first years of the transition scenarios are increasing, which is due to the emissions related to stages A1–A5 of the life cycle of the interventions, so-called "embodied" emissions, which have not yet been compensated for from the emissions saving in the use stage. In fact, these negative emissions occur throughout the transition period, but in later years, they are rapidly compensated for by the cumulative emission savings gained in the use stage. Both ET scenarios start saving CO_2 equivalent emissions in the city after the fourth year. It is also observed that the life cycle environmental emissions of the rest of the interventions occur more gradually than in the case of district heating. The underlying reason is that although all interventions are implemented gradually in different areas of the city, district heating requires a significant amount of infrastructure and equipment to be installed from the first year of operation.

Fig. 16.10 shows the environmental emission reduction (at the end of the transition period) of each intervention included in the two scenarios evaluated, distinguishing the operational phase from the rest of the phases of the life cycle considered.

This analysis discloses that in scenario 1, building refurbishment interventions show the most favorable ratio between the CO_2 emissions reduction of the operational phase with respect to the emissions originating in the rest of the phases. This is why, from an environmental point of view, it is recommendable to start by improving the thermal envelope of buildings with passive measures before incorporating the necessary sustainable fuel technologies. The deployment of ST technology is in the second position followed by a combination of HPs and biomass boilers, the district heating solution, and SPV systems. In scenario 2, however, the solution based on district heating competes for the first position with building refurbishment. Those are followed by ST systems, the combination of HPs, biomass boilers, and natural gas boilers, and SPV systems. The improvement in the district heating intervention in scenario 2 is mainly due to a higher number of energy intensive buildings in the Amara district, which would be covered with a district heating source with lower associated emissions than actual systems.

The analysis shows that the prioritization of strategies and scenarios for the transformation of cities needs to be based on the evaluation of a number of criteria that cannot be optimized simultaneously. Thus trying to anticipate which of the scenarios assessed would be "optimal" to guide the city's transition to a more

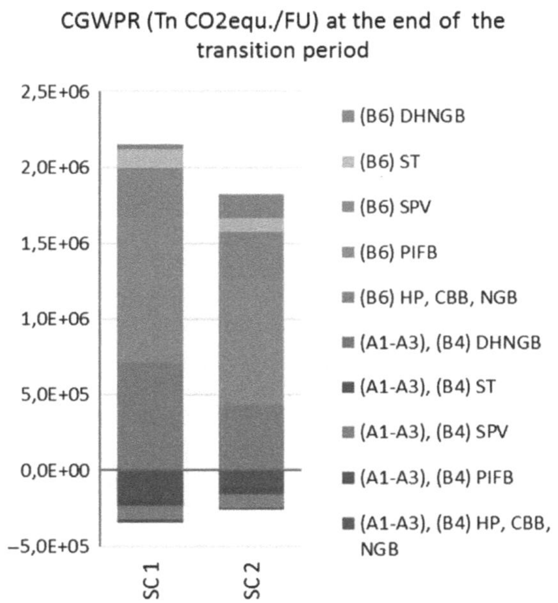

Figure 16.10 Environmental emission reduction at the end of the transition period of each intervention included in the two scenarios proposed for Donostia-San Sebastián.

sustainable future is a complex issue. The results obtained show that the configuration of interventions proposed in scenario 1 is more effective in reducing the city's environmental emissions to the detriment of the overall economic profitability as opposed to what occurs in scenario 2.

This presents the problem of trying to optimize scenarios based on criteria (impact indicators) that are potentially conflicting with each other. This problem will increase when considering a greater number of impact indicators, which is in principle recommended (always trying to keep the total number of indicators/criteria as low as possible) in order to adequately consider the effects on the three dimensions of sustainability. In this respect, once both scenarios ensure the economic viability and the accomplishment of emission reduction targets, the prioritization and final selection of the optimal scenario can only be obtained on the basis of these criteria by giving them a specific weight.

At this point, different methodologies are needed to support the decision-making process once the evaluation studies are completed. To this end, MCDA offers a wide range of methodologies and tools to assist decision makers, enabling them to combine their own preferences with the quantitative data available from the impact assessment, to arrive at their own findings and conclusions in a structured and coherent way.

16.4 Conclusion

The identification and evaluation of sustainable fuel technologies and their integration into ET scenarios that will shape the path toward a low-carbon future for our cities constitute a research line of growing interest. Although the awareness of municipalities about the need of transforming the way in which cities consume and generate energy is increasing, the planning process is still complex due to the lack of consensus on the way in which to assess and prioritize different alternatives under a holistic sustainability framework.

This chapter proposes a general framework for long-term city energy planning and a methodology capable of comparing the sustainability of different ET scenarios. The methodology is developed in detail for building sector energy planning, including energy generation, distribution, and building operation energy use. It seeks to identify and overcome the current difficulties linked to LCA and LCC methodologies for their application in ET scenarios. With this objective in mind, a life cycle assessment framework has been adapted to the new objective and scope of the study, defining the main procedures for establishing links between different methodologies, and developing and adapting indicators for impact assessment.

Connecting long-term energy modeling at district and city scales with the adapted methodologies for impact assessment is identified as one of the main challenges. In this regard, the authors also want to stress that ensuring the flexibility of new methods will be a critical aspect in order to facilitate their connection with the multiple tools and methodologies that are used in the planning processes.

The application of the methodology in the case study of Donostia-San Sebastián has illustrated its practical use and has demonstrated the importance of simultaneously considering various potentially conflicting criteria for the prioritization of scenarios. From the analysis of the results, it is possible to conclude that the adaptation of the method has proven effective for considering not only the temporal aspects, but also the distributional characteristics of each scenario. The method favors flexibility in the definition of scenarios and facilitates the disaggregation of results on a yearly basis. This latter aspect is potentially interesting for municipalities that can evaluate in detail the key moments to make investments. It also allows for relevant business models to be identified, and in general, it facilitates managing the cashflow with a long-term vision. Moreover, it provides a better understanding of the influence of some decisions on the pace at which the established emission reduction targets will be reached.

It is concluded that the impact assessment of low-carbon ET scenarios for cities requires great methodological development and harmonization efforts from the scientific community as well as a greater awareness and involvement of municipalities. In this regard, it is essential to foster collaborative works between municipalities and modelers to overcome the existing gaps between the long-term planning horizons and the short-term vision of many actions that are carried out in our cities.

References

[1] F. García-Guaita, S. González-García, P. Villanueva-Rey, M.T. Moreira, G. Feijoo, Integrating urban metabolism, material flow analysis and life cycle assessment in the environmental evaluation of Santiago de Compostela, Sustain. Cities Soc. 40 (April) (2018) 569–580.

[2] C. Lausselet, V. Borgnes, H. Brattebø, LCA modelling for zero emission neighbourhoods in early stage planning, Build Environ. 149 (2019) 379–389.

[3] E. Loiseau, L. Aissani, S. Le Féon, F. Laurent, J. Cerceau, S. Sala, et al., Territorial Life Cycle Assessment (LCA): what exactly is it about? A proposal towards using a common terminology and a research agenda, J. Clean Prod. 176 (2018) 474–485.

[4] J. Albertí, C. Brodhag, P. Fullana-i-Palmer, First steps in life cycle assessments of cities with a sustainability perspective: a proposal for goal, function, functional unit, and reference flow, Sci. Total Environ. 646 (2019) 1516–1527.

[5] J. Albertí, M. Roca, C. Brodhag, P. Fullana-i-Palmer, Allocation and system boundary in life cycle assessments of cities, Habitat Int. 83 (2019) 41–54.

[6] D. García-Gusano, M. Martín-Gamboa, D. Iribarren, J. Dufour, Prospective analysis of life-cycle indicators through endogenous integration into a national power generation model, Resources 5 (4) (2016).

[7] D. García-Gusano, D. Iribarren, J. Dufour, Is coal extension a sensible option for energy planning? A combined energy systems modelling and life cycle assessment approach, Energy Policy. 114 (2018).

[8] European Commission, Communication from the Commission to the Council—Integrated Product Policy—Building on Environmental Life-Cycle Thinking, 2003.

[9] California Energy Commission, Energy Research and Development Division—Final Report—Life-Cycle Assessment and Urban Sustainability, 2010, pp. 500–2013.
[10] CITYKEYS Project, European Union HORIZON 2020 programme. Available from: <http://citykeys-project.eu/>.
[11] The Anatomy Indicators were developed by the work team on Data Interoperability and City Indicators (DICI) of the City Protocol Task Force, City Protocol Agreement—Anatomy indicators, Developed by TAFT—DICI Indicators, July 2015.
[12] Siemens Green City Index. Available from: <http://www.siemens.com/entry/cc/en/greencityindex.htm>.
[13] Planning for Energy Efficient Cities (PLEEC) project. Available from: <http://www.pleecproject.eu/>.
[14] European Smart Cities v1.0, Available from: <www.smart-cities.eu>, 2007.
[15] ISO 37120—Sustainable development in communities—Indicators for city services and quality of life. Prepared by Technical Committee ISO/TC 268, Sustainable development in communities Working Group 2 Urban Indicators.
[16] UN General Assembly, Transforming our world: the 2030 Agenda for Sustainable Development, October 21, 2015, A/RES/70/1, Available at: <https://www.refworld.org/docid/57b6e3e44.html>. (accessed 17.02.20).
[17] J. Albertí, C. Brodhag, P. Fullana-I-Palmer, Science of the total environment first steps in life cycle assessments of cities with a sustainability perspective: a proposal for goal, function, functional unit, and reference flow, Sci. Total Environ. 646 (2019) 1516–1527. Available from:. Available from: https://doi.org/10.1016/j.scitotenv.2018.07.377.
[18] European Commission—Joint Research Centre—Institute for Environment and Sustainability, International Reference Life Cycle Data System (ILCD), Handbook—General Guide for Life Cycle Assessment—Detailed Guidance, first ed., Publications Office of the European Union, Luxembourg, 2010.
[19] D. Manjarres, L. Mabe, X. Oregi, I. Landa-torres and E. Arrizabalaga, A Multi-objective harmony search algorithm for optimal energy and environmental refurbishment at district level scale, In: J. Del Ser, (Ed), Harmony Search Algorithm. ICHSA 2017. Advances in Intelligent Systems and Computing, 514, 2017, Springer, Singapore.
[20] E. Arrizabalaga, Multi-Criteria Methodology for the Prioritisation of Alternative Energy Transition Scenarios of Cities, University of the Basque Country, 2017.
[21] Y. Saheb, Energy Transition of the EU Building Stock, 2016;
[22] European Commission, Directorate-General for Energy, Directorate-General forClimate Action, Directorate-General for Mobility and Transport. EU Energy, Transport and Ghg Emissions Trends to 2050 Reference Scenario 2013, 2013.
[23] EU Energy Trends to 2030, Update 2009, European Commission Directorate-General for Energy in Collaboration with Climate Action DG and Mobility and Transport DG, 2009.
[24] ETRI, Energy Technology Reference Indicator Projections for 2010–2050. Science and Policy Report by the Joint Research Centre of the European Commission, 2014.
[25] DEAc, Danish Energy Agency Technology Data [Internet]. 2017. Available from: <https://ens.dk/en>.
[26] International Energy Agency, Energy Technology Perspectives. Scenarios & Strategies to 2050, 2008.
[27] G.P. Smestad, The basic economics of photovoltaics, Sol. Energy New Mater. Nanostructur. Devices High Effic. 1, 2008, STuC8
[28] SDH, Solar District Heating Guidelines, 2012.

[29] X. Oregi, Techno-economic evaluation of building energy refurbishment processes from a life cycle perspective. Ph.D. Thesis, University of the Basque Country, Donostia, Spain, 2015.
[30] Fomento de San Sebastían, Estudio de Viabilidad Técnico Económico District Heating Txomin (San Sebastián), 2015.
[31] Danish Energy Agency and Energinet.dk, Generation of Electricity and District Heating, Energy Storage and Energy Carrier Generation and Conversion, Technology Data for Energy Plants, 2012.
[32] J. Oliver-Solà, X. Gabarrell, J. Rieradevall, Environmental impacts of the infrastructure for district heating in urban neighbourhoods, Energy Policy. 37 (11) (2009) 4711–4719.

Section XII

Green & Sustainable Future in Energy Sector (Zero Waste & Zero Emissions)

Oxy-fuel power cycles promoting the transition to green and sustainable future in the energy sector

17

N.D. Rogalev[1], *Yu. K. Petrenya*[2], *V.O. Kindra*[3] *and A.N. Rogalev*[3]
[1]Department of Thermal Power Plants, National Research University "Moscow Power Engineering Institute", Moscow, Russia, [2]PJSC "Power Machines", St. Petersburg, Russia, [3]Department of Innovative Technologies of High-Tech Industries, National Research University "Moscow Power Engineering Institute", Moscow, Russia

17.1 Introduction

During the past century, the Earth's population and primary power consumption have grown 4 and 16 times, respectively [1]. The continuous power consumption increase is followed by toxic agents and greenhouse gas emissions. According to numerous observations and investigations, the latter remarkably influences climate change [2].

The abnormal mean temperature increase of the Earth's climate system over the past century is generally accepted, but the amount and role of the anthropogenic factor in this process are being discussed for decades. The problem is the complicated consideration of numerous cross-influencing processes on the Earth, and out of it. Specifically, the possible sources of the global climate changes may be changes of sizes, reliefs, and relative locations of continents and oceans, the amount of heat accumulated in oceans, tectonic processes, change of the atmosphere chemical contents, parameters of the Earth orbit and axis, solar activity, and so forth.

For a very long time supporters of natural reasons for global warming determined solar activity as the main factor but supporters of anthropogenic influence emphasized the effect of the increasing greenhouse emission caused by the development of production activities. In order to prove this assumption scientists measure and calculate such quantities as the climate system mean temperature, the atmospheric carbon dioxide contents, and the solar activity level.

The Mauna-Loa (Hawaii) observatory has continuously measured the CO_2 atmospheric contents since the middle of the previous century. It is worth mentioning that the long-term observation results presented in the Keeling curve form show a subsequent CO_2 increase from 318 to 406 ppm during the 1958–2018 period [3]. In other words, over the past 60 years CO_2 concentration has increased by about 30%.

Sustainable Fuel Technologies Handbook. DOI: https://doi.org/10.1016/B978-0-12-822989-7.00019-6

Such a rapid increase is considered abnormal and its most probable source is the anthropogenic factor [4].

The world community's concern with climate changes caused a few international agreements that require advanced countries and transient-economy countries to stabilize or reduce greenhouse gas emissions. In 1997 the Kyoto Protocol and in 2015 the Paris Agreement were signed. However, to achieve the effect of CO_2 emission mitigation, the leading industrial countries must make proper agreements.

The four countries that lead CO_2 emissions are China, the USA, India, and Russia (Fig. 17.1) [5]. The European Union has a comparable power consumption level, but its contribution of about 14% is remarkably smaller than that of the USA due to the intensive involvement of renewable power sources.

According to the forecast of the International Energy agency by 2040, the world power consumption will increase by 30% against 2016 one. The amount of burned hydrocarbons will also grow, which will produce an increase of toxic and greenhouse gas emissions. The prospect of heavier anthropogenic pressure on nature in the 21st century aggravates the problem of sustainable development of single countries and the world.

If green technologies are not properly introduced, and the economic growth rate is kept stable, it will be impossible to slow the CO_2 emission growth. Climatology scientists give unfavorable forecasts for this case. If the existing growth rate of 2.2 ppm/year will be stable, the CO_2 concentration will reach 500 ppm by 2060. The climate models and the paleoclimate data show that reaching the 550 ppm level will produce serious irreversible effects on the whole planet [6]. Glaciers melting and ocean level increases, changes in the atmospheric precipitation amount and distribution followed by floods, droughts, hurricanes, and other cataclysms are only a few results already experienced by people [7].

An approximate forecast of the CO_2 concentration for the oncoming decades may be based on its past values in the regions with maximal emissions and the most significant influence factors. Since 1970 the CO_2 emission levels in the USA,

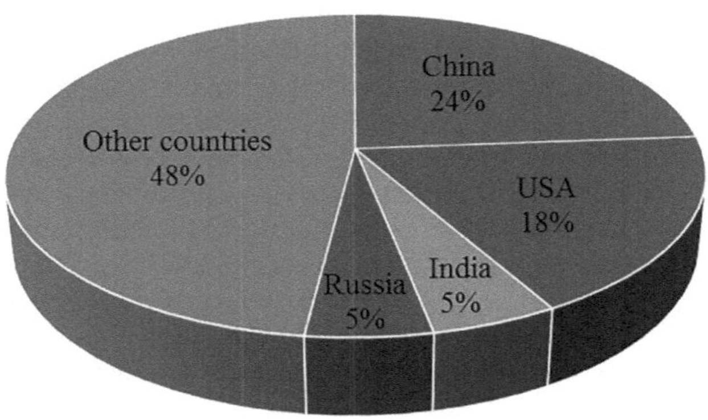

Figure 17.1 Distribution of carbon dioxide emissions among countries in 2010.

Russia, and the EU were nearly constant [8]. On the other hand, in China and India this factor increased a few times over, and the most significant growth may be seen at the beginning of 21 century due to these countries' high industrialization rate.

Regarding this swift CO_2 emission growth, its reduction or at least level stabilization will require remarkable time and investments into the industry modernization and innovation. The scenario for the CO_2 concentration stabilization described in Wigley et al. [9] shows the problems related with reaching the 450 ppm level during this period because now the tendency for CO_2 concentration increase is seen [10]. The 550 ppm scenario assumes that starting from the total anthropogenic CO_2 emission will be intensively reduced. In this case, stable economic growth is impossible without the development of environmentally clean and profitable technologies.

Effective emissions mitigation requires determining its major sources. About 75% of the anthropogenic CO_2 is produced by mining and combustion of oil, natural gas, and coal [11]. Power plants produce about 25% of world emissions [12]. In the USA the contribution of the power production industry in the total anthropogenic CO_2 emission is 35%, 6% in China, 31% in Europe, and 33% in Russia [13,14].

The reason for this exceptional power industry contribution in the CO_2 emission structure is the wide application of the traditional Rankine and Brighton-Rankine cycles where heat is supplied by the hydrocarbon fuel combustion in the presence of air. The problems of energy supply and environmental safety caused the active introduction of the technologies based on renewable sources, the global power network, the hydrogen power industry, and others [15–17].

17.2 Reducing CO_2 emissions in the power production industry

The main directions of the green technology development are investigations and design of power facilities that utilize the power of the sun, wind, geothermal energy, and that from sea waves. Large success in this field is confirmed by the continuous production of prospective types of energy from power-generation units.

In spite of this success and large effort spent on nonconventional power production development including renewable sources, the world power production industry is still based on fossil fuel power plants and their number is continuously increasing [18]. The main reasons are low-power density for renewable energy sources, complicated power control, and energy accumulation.

Another way is the mitigation of harmful agents produced by thermal power plants. The problem of reduction of toxic substance production including oxides of nitrogen and sulfur as well as ash was investigated in detail [19,20] and how the concerned technologies successfully operate in the existing power plants. The search for methods for the reduction of greenhouse gas emissions is topical. The goal may be achieved by two methods.

The first approach is an efficiency increase of existing fossil fuel power plants, which leads to an emission reduction per kilowatt of produced power. Development of cogeneration technology [21,22], or construction of combined cycle power plants and ultrasupercritical steam power plants [23,24] are prospective technical solutions that allow high parameters of power-production efficiency. This approach can slow down the increase of carbon dioxide concentration in the atmosphere, but it cannot stabilize the concentration because the combustion products that mostly consist of greenhouse gases will be exhausted into the atmosphere.

CO_2 emissions of traditional fossil fuel cycles may be reduced by the application of the sequester technology. It involves the subsequent capture of carbon dioxide, its compression up to the supercritical state, pipeline transportation to the storage site, and injection to vessels located deep underground [25]. The geologic formations for CO_2 storage usually are oil and gas reservoirs, deep saliferous formations, or coal layers that have no industrial application. The CO_2 may be stored below 800 m deep where the temperature and pressure transform the gas into its liquid phase.

If 100% of the CO_2 amount produced by power plants can be captured, the capacity of oil and gas fields will be sufficient for its storage from 350 to 470 years [26]. A large part of regions prospective for CO_2 storage is located in the counties that remarkably need new power production facilities, China, the USA, European countries, and Russia [27]. Therefore the sequester technology development devoted to the reduction of CO_2 emissions in fossil fuels power production is very prospective for the oncoming decades.

17.3 Carbon capture technologies for thermal power plants

Three main methods of CO_2 capture have reached the state of industrial demonstration. The first is CO_2 capture after combustion when a chemical solvent like amine or ammonia is used for the postcombustion CO_2 absorption from flue gas. The second method is precombustion absorption when a solid fuel passes the gasification process that produces synthetic gas fuels that mostly consist of the carbon monoxide (CO) and hydrogen (H_2). Water steam is used for the transformation of CO to CO_2 that is absorbed by the solvent application. The third technology is oxy-fuel combustion where the oxidizer is pure oxygen and the combustion product is the gas mixture with high CO_2 concentration that is ready for sequestering after the water steam is removed by condensing.

17.3.1 Postcombustion carbon capture

Postcombustion carbon capture is similar to the absorption of solid particles, and sulfur and nitrogen oxides. The method involves a system of CO_2 absorption from flue gas (Fig. 17.2) that is attractive in the modernization of existing power facilities and the construction of new facilities that operate in available technologies. CO_2 may be absorbed with special membranes or with the chemical absorption method that uses liquid dissolving agents such as monoethanolamine or ammonia.

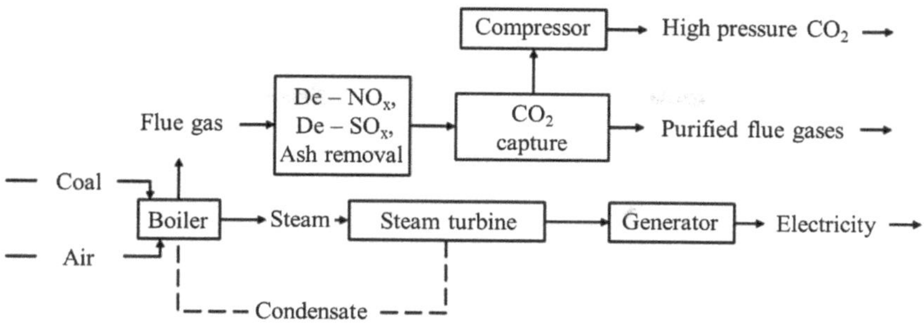

Figure 17.2 Flow cart of carbon dioxide capture from flue gas.

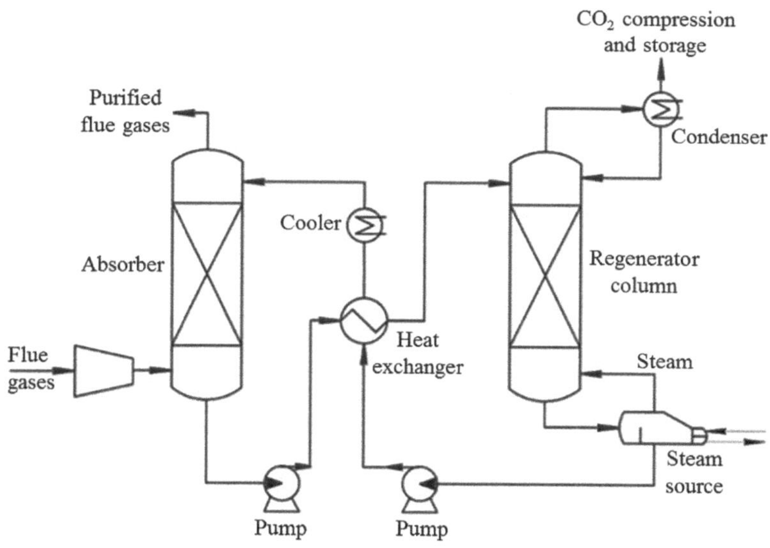

Figure 17.3 Flow chart of carbon dioxide capture by chemical absorption.

A typical chemical absorption process with the application of different dissolving agents is shown in Fig. 17.3. The combustion products are supplied to the lower part of the absorber and the solvent agent is supplied in the counter-flow direction. The solvent agent reacts with the flue gas and some CO_2 is absorbed by the amine solvent. The solution enriched with the absorbed CO_2 is pumped into the upper compartment of the second-regenerator column. In the second column, the enriched solution flows down and is heated with the counter-flow steam. The heat breaks the chemical bonds between CO_2 and the solving agent. The released CO_2 rises upward together with the steam flow, and then goes to the condenser where the steam is cooled and returned back to the column. The resulting 99% pure CO_2 is compressed for transportation to the storage site. After the lower compartment of the regeneration column, the absorbent is supplied to the absorber upper compartment. Before entering the absorber, the regenerated absorbent passes the heat exchanger where it exchanges a part of its heat to the absorbent enriched with CO_2, which returns the heat and reduces the consumption of the steam used for the regeneration process.

In the absorber and regenerator, the required contact surface is provided with the special filling. The CO_2 absorption degree is equal to the ratio of CO_2 compressor massflow and CO_2 content in the flue gas. An important process performance that describes the capability of the CO_2 capturing system. In most of the examples, this value is below 90% [28,29].

Chemical solvents are widely used for natural gas treatment and in oil refineries for CO_2 capturing. Combustion uses air oxidation so the main (i.e., 70%) part of flue gas consists of N_2 with a relatively low CO_2 concentration not higher than 15%, thus the precipitation process requires large-size equipment. One more problem is concerned with the large consumption of the power used for the solvent regeneration or for the flue-gas cooling. More than that, it is necessary to fulfill the SO_x, NO_x, and the flue ash limits to reduce amine consumption.

Fig. 17.4 presents a scheme of CO_2 capture from flue gas by the use of membrane separation. The wet scrubber that cools the flue gas down to the membrane operating temperature is installed upstream of the membrane module. Inside the module, some CO_2 passes through the membrane. The flue gas part that has not passed the membrane is called the retentant gas, or the concentrate. The CO_2 partial pressure is very low, so the increase of filtered CO_2 pressure requires the application of a compressor and vacuum pump.

Like the CO_2 chemical absorption process, the separation process efficiency is evaluated by the separation rate that is the part of CO_2 separated from the flue gas [30]. The membrane separation process is simpler than chemical absorption because it involves fewer components. The main power consumptions are in the compressor and vacuum pump.

In certain conditions the postcombustion carbon capture is financially reasonable. The main problem of this case is the rather low concentration of the captured substance.

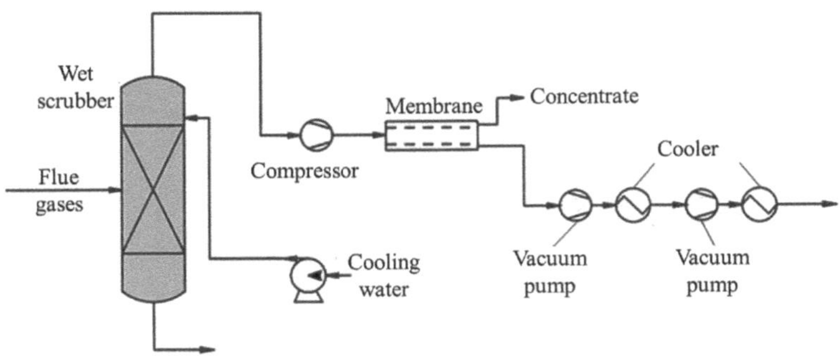

Figure 17.4 Flow chart of carbon dioxide capture from flue gas with a membrane separator.

17.3.2 Precombustion carbon capture

One more prospective method for reduction of CO_2 emissions is its precombustion removal from the coal, oil, or natural gas contents when the final fuel is hydrogen. The main commercial technologies for the hydrogen production from methane are the methane steam conversion or reforming, methane autothermal reforming (oxygen or air conversion), methane cracking, or the partial methane oxidation. and combinations of these methods [31–34].

Gas treatment for changing the initial gas mixture content is called gas conversion. Usually the conversion is used for the production of hydrogen or its mixture with carbon monoxide from gaseous hydrocarbons, methane, and its homologs. Such mixtures are used for the organic products synthesis, in metallurgy as the gaseous deoxidizers, or are further treated for H_2 production. The conversion involves different oxidation agents such as O_2, H_2O, CO_2, and their mixtures. In addition, it is possible to use metal oxides for this purpose. The most profitable raw product for the conversion is methane contained in natural gas.

Currently most industrial-scale H_2 is produced by methane steam conversion. The H_2 produced by this process is mostly used as an oil refinery agent, nitrogen fertilizer component, and the rocket industry.

In addition, the produced hydrogen may be used as the main fuel in steam turbine power plants (Fig. 17.5) or as additional fuel for the steam superheating in special oxy-hydrogen combustors. These combustors are key elements that need special development for this type of power facility [35].

Oxy-hydrogen combustion makes it possible to create a steam turbine unit with the turbine inlet temperature of 1400°C–1700°C. The scheme of the perspective hydrogen-fired power plant presented in Fig. 17.6 was elaborated in the framework of the World Energy Network (WE-NET) [36–38]. It has a single intermediate steam overheating pipeline and a regenerator that produce steam of supercritical parameters. The turbine inlet and intermediate temperatures are 900°C and 1600°C respectively, and the inlet and intermediate pressures are 25 and 2.4 MPa, respectively.

Steam is produced in the regenerator feeding by low- and high-pressure feedwater pumps. The high-pressure steam at the outlet of regenerator with a temperature of 400°C and pressure of 260 bar enters the high-pressure oxy-hydrogen combustion chamber together with oxygen and hydrogen in a stoichiometric ratio. The high-pressure steam at the outlet of the combustion chamber with a temperature of 900°C and a pressure of 250 bar enters the uncooled high-pressure turbine. The steam at the outlet of the high-pressure turbine with a temperature of 500°C and pressure of 25 bar mixes with steam with a temperature of 650°C, which used for medium-pressure turbine cooling. The steam at the inlet of the medium-pressure turbine has a temperature of 1600°C and a pressure of 24 bar, and the steam at the outlet of the medium-pressure turbine has a temperature of 800°C and a pressure of 1 bar. The steam at the outlet of the medium-pressure turbine enters the regenerator and is used for steam generation. After the regenerator, steam with a temperature of 110°C at a pressure of 1 bar enters the low-pressure turbine, where it expands to a

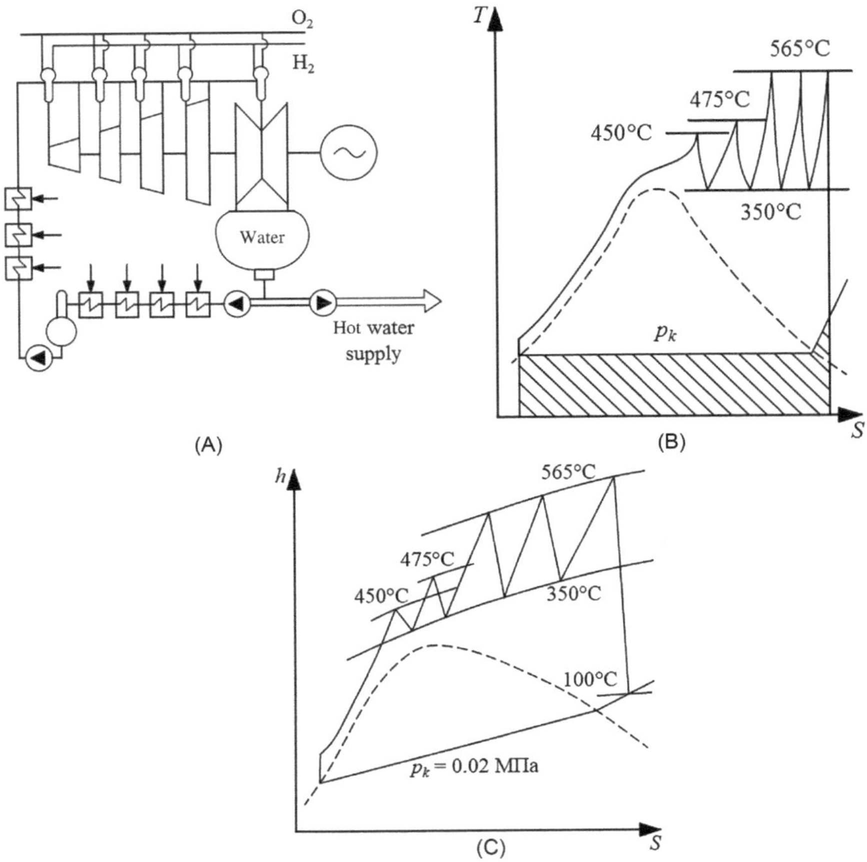

Figure 17.5 Flow chart and thermodynamic cycle diagrams for hydrogen-fired steam turbine power plant.

pressure of 0.05 bar and then enters the condenser. The calculated efficiency of a power unit with an initial steam temperature of 900°C is 73%. The layout of the equipment of the hydrogen-fired power plant is shown in Fig. 17.7.

17.4 Oxy-fuel combustion cycles with carbon dioxide recirculation: energy efficiency and emission levels

An effective method to reduce carbon dioxide emissions at a constant economic growth rate is the creation of the semiclosed thermodynamic cycles with oxy-fuel combustion. The specific feature and advantage of these new power production technologies against the traditional one is oxy-fuel combustion that produces water steam and carbon dioxide. This chemical composition of the combustion products

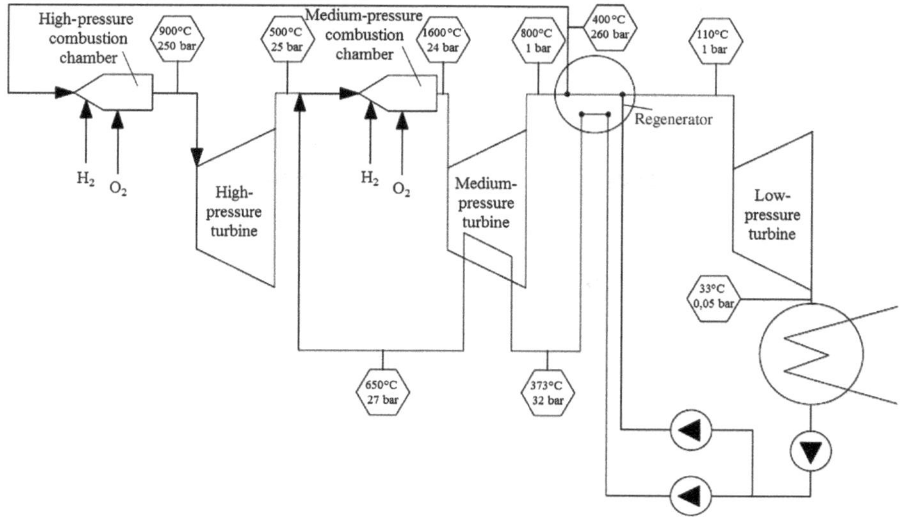

Figure 17.6 Flow chart of the perspective hydrogen-fired power plant.

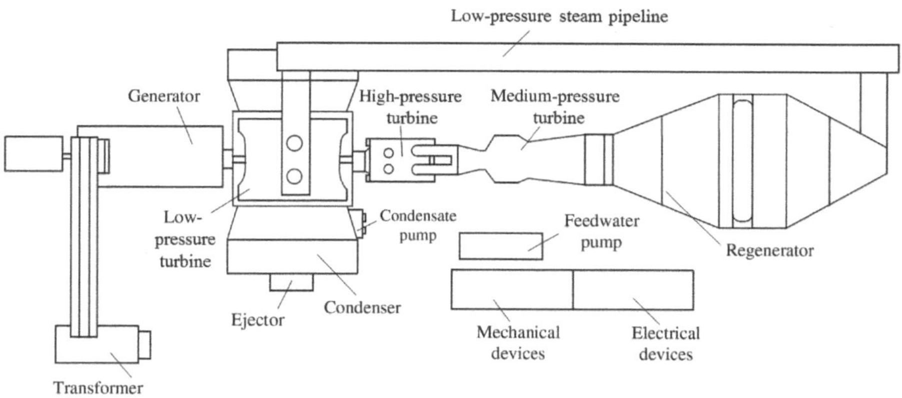

Figure 17.7 The layout of the equipment for the hydrogen-fired power plant.

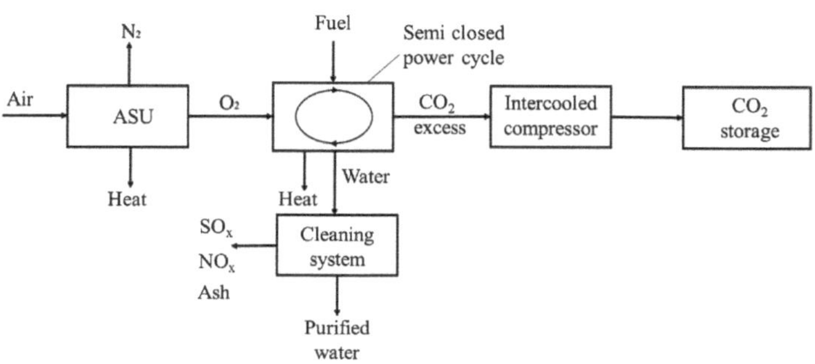

Figure 17.8 The oxy-fuel power cycle concept.

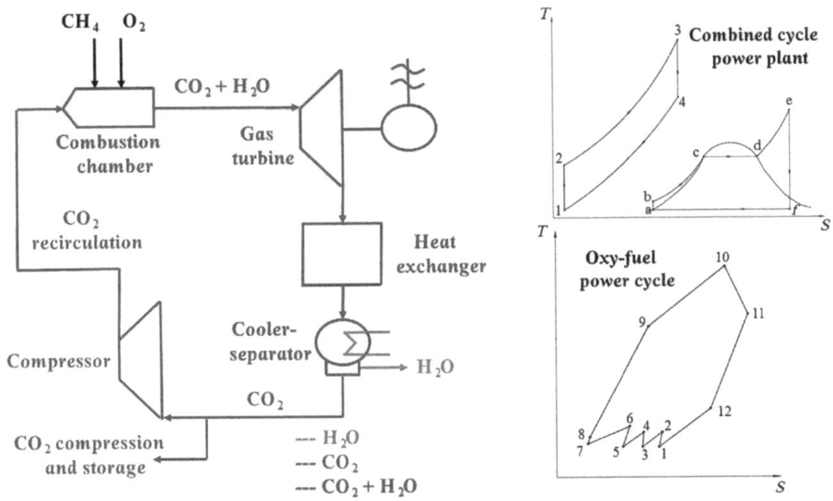

Figure 17.9 The oxy-fuel combustion cycle features.

allows the carbon dioxide separation from the water steam with minimal power consumption by the steam condensing and further sequester processing. More than that, in the combustion zone the absence of other air components and specifically of N_2 provides near-zero toxic emissions. A concept of an oxy-fuel power cycle and its thermodynamic cycle is shown in Figs. 17.8 and 17.9.

Ambient air enters the air separation unit (ASU) to produce high-purity O_2. The ASU auxiliary products are N_2 and heat. Oxygen is supplied to the closed-cycle combustor where it is used for gas-fuel oxidation. The combustion products mostly consist of CO_2 and H_2O. The products enter the turbine mounted on the power generator shaft, produce the work, and enter the heat exchanger that may be a steam generator or a regenerator. In the heat exchanger, the working fluid releases most of its heat and enters the cooling separator where the mixture is separated into the two components, CO_2 and H_2O. The water extracted in the heat exchanger passes the clarifier after which it may be reused. After the cooling separator, a part of the flow that mostly consists of CO_2 is evacuated from the cycle circuit for further storage and the main part of the flow enters the compressor where the fluid pressure is increased before entering the combustor as to limit the maximal temperature. The evacuated part of CO_2 is compressed in an intercooled compressor.

There are more than twenty oxy-fuel power cycles, for example, the semiclosed oxy-fuel combustion combined cycle (SCOC-CC), the MATIANT cycle, the Allam cycle, the Graz cycle, CES cycle, AZEP, and ZEITMOP cycles [39,40]. The cycles may be classified as the following. SCOC-CC, MATIANT, and Allam cycles employ carbon dioxide recirculation aimed at limiting the combustor temperature. In these cases, carbon dioxide is the main component of the working fluid. The main component of the working fluid for the Graz and CES cycles is water. The membrane cycles employ high-temperature membranes integrated into the scheme for oxygen production. From the technology availability point of view, the most developed are the oxy-fuel combustion cycles with carbon dioxide recirculation.

17.4.1 Semiclosed oxy-fuel combustion combined cycle

The semiclosed oxy-fuel combustion combined cycle has a simple configuration and may be considered as a combined Brighton-Rankine cycle with the oxy-fuel combustion and the carbon dioxide recirculation [41]. The main advantage of the SCOC-CC cycle among others is its simple concept (Fig. 17.10).

The gas fuel and the high-purity oxygen are supplied into the combustor at the stoichiometric rate. The flame temperature of this mixture may be up to 3500°C. The third flow is supplied into the combustor, which is a mixture with high CO_2 concentration to limit the maximal temperature. The produced gas that has about 80% CO_2 concentration at 1300°C–1400°C temperature is directed to the gas turbine, which drives the power generator. The working fluid expands in the gas turbine and enters the heat recovery steam generator (HRSG). The HRSG utilizes the gas turbine exhaust heat and produces the superheated steam that drives the steam turbine, which also produces power. The HRSG exhaust gas is directed into the condenser, where most of the water is separated by condensing at nearly atmospheric pressure. Then a part of the CO_2 flow is evacuated from the cycle for further sequestering. The residual flow that mostly consists of the carbon dioxide is recirculated and is supplied to the compressor inlet.

In traditional gas turbine cycles the main component of the working fluid is nitrogen. On the contrary, in the SCOC-CC, 80%–96% of the working fluid is carbon dioxide. The working fluid contents depend upon many factors such as type of fuel, oxidizer purity, thermodynamic parameters of the recirculation flow, and the air excess required for efficient combustion.

The initial working fluid parameters at the gas turbine inlet mostly determine the SCOC-CC cycle efficiency. According to the study results [42], the SCOC-CC has

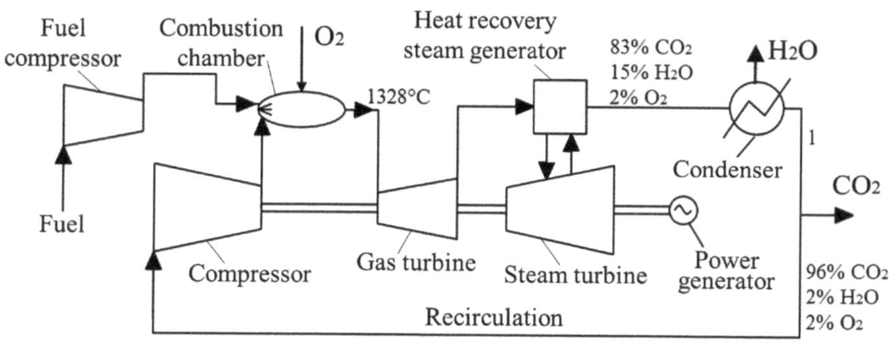

Figure 17.10 Flow chart of the SCOC-CC.

an efficiency of 45%–48% at a turbine inlet temperature of 1300°C–1400°C and pressure ratio of 30–45.

This cycle is semiclosed, so its minimal pressure could be not equal to the atmospheric one. An increase of the minimal pressure reduces the envelope dimensions of turbomachines and HRSG, but it increases the equipment wall thicknesses. The minimal pressure parametric studies show that the minimal pressure increase from 1 to 10 bars changes the efficiency for only 0.5% [43]. An important problem related to the pressure increase is the gas turbine blade cooling. The increased pressure reduces the volumetric flow and thus the cooled surface area. On the other hand, the heat transfer coefficients on both sides of the cooled channels become higher and the total heat flow through the blade walls also grows. This increases the temperature difference on the blade walls and limits the acceptable temperature increase of the coolant agent. As a result, the same heat flow release requires a larger flow of the cooling agent. These factors show that it is reasonable to keep the SCOC-CC outlet gas turbine pressure level equal to the atmospheric pressure.

Parametric studies of the recirculation flow temperature in the SCOC-CC show its specific influence on efficiency and structure features. Reduction of the recirculation flow temperature reduces the compressor work but, on the other hand, it increases losses in the cold source, first of all because of the increase of the condensed moisture part [44,45]. Also, the study shows that this parameter does not show a distinct optimum and it is located in the 40°C–70°C range for the gas turbine outlet temperature of 620°C. So cycle maximal efficiency may be reached at the recirculation flow temperature slightly higher than the cold source temperature.

17.4.2 Modified semiclosed oxy-fuel combustion combined cycle with turbine coolant precooling

The turbine coolant flow reaches the value of 36.3% for optimal SCOC-CC parameters [46]. One of the main reasons for such high-mass flow is the high-initial temperature of the cooling agent that is taken from the compressor bleeding.

The most effective method of the turbine coolant massflow reduction is the flow temperature decrease at the inlet of the cooling channels. This may be achieved by the preliminary cooling of the coolant in a surface heat exchanger or by the water injection into the coolant [47].

The first scheme is shown in Fig. 17.11. In terms of use of the heat taken from the coolant flow some condensate from the steam turbine exhaust may be heated. The main condensate flow is heated in the precooling surface heat exchanger and returned into the steam turbine flow.

The second scheme is shown in Fig. 17.12. Water injection reduces the coolant flow temperature by evaporation. The water may be taken from the cooler-separator so the expenses for its preparation are minimal.

The higher the working fluid turbine inlet temperature, the larger the coolant precooling effect. In the investigated temperature range the maximal net efficiency improvement of 3.8% is reached by the coolant temperature reduction down to

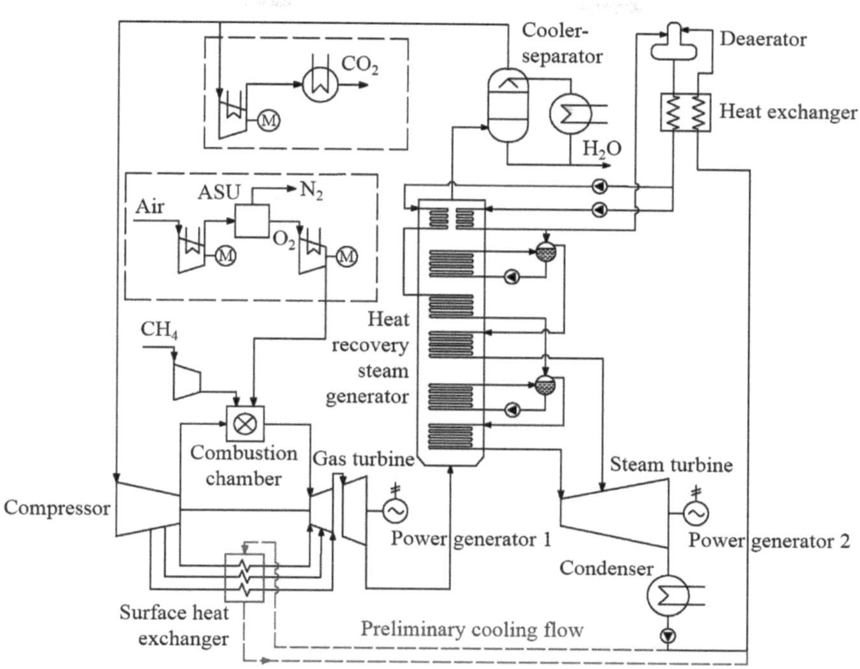

Figure 17.11 Preliminary cooling of coolant in a surface heat exchanger for the SCOC-CC.

100°C in the surface heat exchanger (Fig. 17.13). On the other hand, the coolant temperature reduction below 150°C is not recommended because of the risk of water condensate formation in the coolant flow. The maximal net efficiency increases at cooling to 150°C is 3.2%.

The maximal net efficiency improvement with water injection is 1.5%. The injection water massflow varies from 0.4% to 2.2% for the turbine inlet temperature from 1100°C to 1700°C, respectively (Fig. 17.13). The small water massflow values produce small changes in the coolant chemical composition. The cycle efficiency increase is mostly due to the coolant massflow reduction and the resulting reduction of cooling losses.

In the case of deep coolant cooling down to 150°C–250°C it is reasonable to use cooling in a surface heat exchanger with the steam turbine unit condensate. The net efficiency increase may be up to 3.2%.

In the case of precooling to a temperature above 250°C, injection of the water taken from the gas cycle cooler-separator may be recommended. The power facility net efficiency increase may be up to 1.5%.

The relative water injection flow, or the ratio of injection water massflow to the working fluid flow into the turbine header, is from 0.4% to 2.2% for the operating fluid temperature at the CO_2 turbine inlet from 1100°C to 1700°C, respectively (Fig. 17.14). The injection expenditures for the block of tens megawatt power are

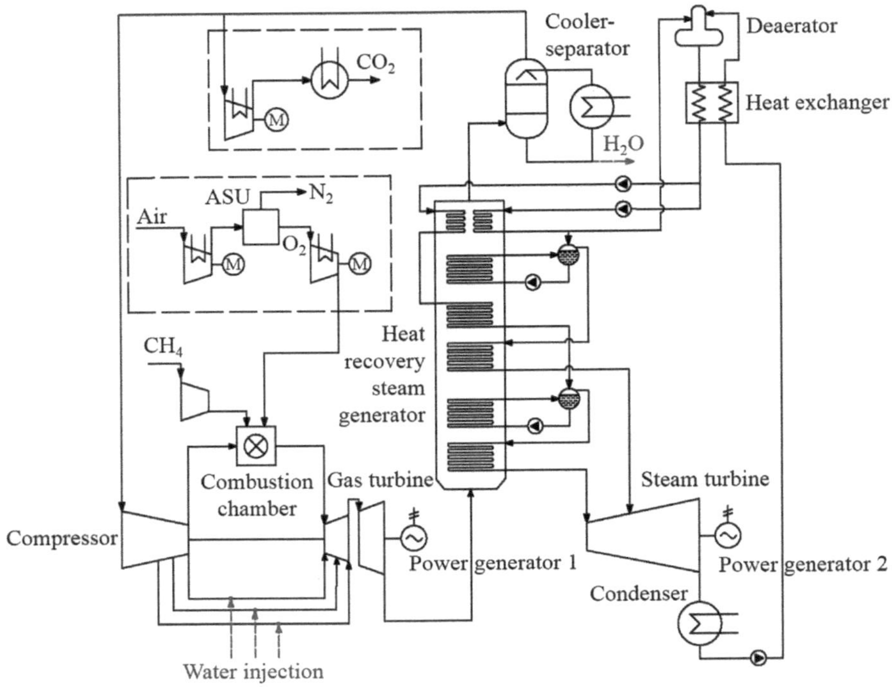

Figure 17.12 Preliminary cooling of the coolant by water injection for the SCOC-CC.

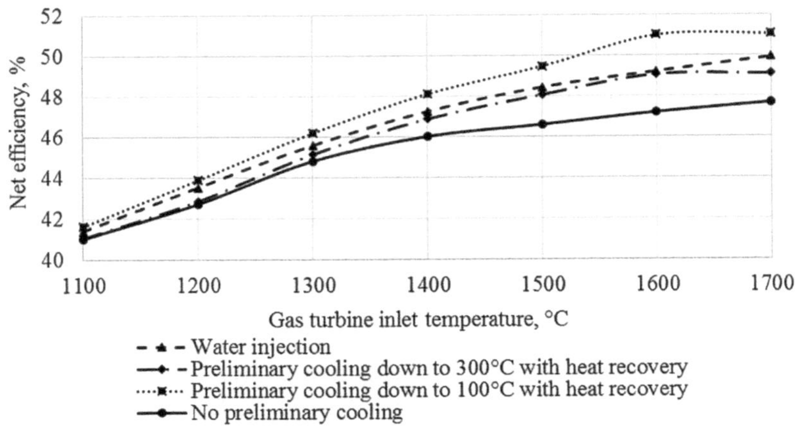

Figure 17.13 Influence of the precooling methods upon the cycle net efficiency.

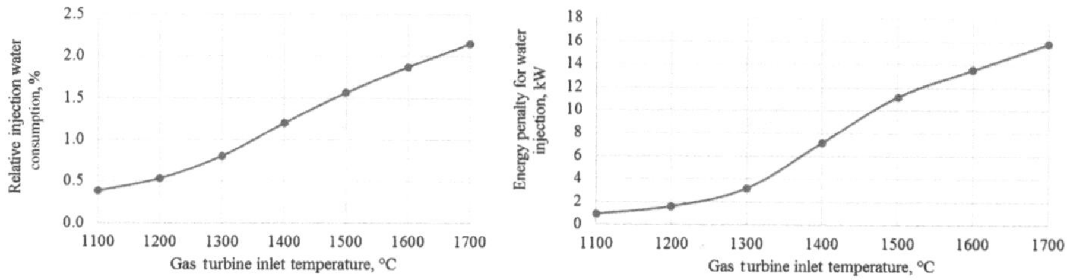

Figure 17.14 The SCOC-CC performance with water injection onto the coolant massflow.

relatively small and could be not higher than 16 kW. The increase of internal power consumption for the additional cooling water pumping in the cooler-separator is negligible.

The efficiency of the available and prospective power production technologies may be compared at equal power production levels and the working fluid initial temperatures. The analysis results show that the efficiency of the SCOC-CC with the preliminary cooling of the coolant is 2.9% higher than that of a combined cycle power plant with carbon dioxide capture. The SCOC-CC cycle optimal initial pressures are higher than those of the combined cycle power plant ones. The optimal carbon dioxide compressor pressure ratio at the initial working fluid temperatures of 1500°C–1700°C is about 70 bar, which is remarkably higher than that of the traditional power plants.

17.4.3 MATIANT cycles

The MATIANT cycle was developed by Philippe Mathieu and Evgeny Yantovsky [48]. The working fluid with high carbon dioxide content expands in high-pressure, intermediate-pressure, and low-pressure turbines (Figs. 17.15, 17.16).

The intercooled compressor increases the carbon dioxide pressure up to 300 bar and supplies it to the regenerator. The heated gas goes to the high-pressure turbine, expands to 40 bar, and returns to the regenerator for the repeat heating. After the regenerator, the flow is directed to the first cooled combustor where its temperature rises up to 1300°C. Then the hot-combustion products expand in the cooled intermediate-pressure turbine and enter the second cooled combustor. After the combustor, the working fluid is directed to the cooled low-pressure turbine where it expands down to the atmospheric pressure. After the low-pressure turbine, the flow goes to the regenerator and then to the condenser where the working fluid is cooled by the heat transfer to the ambient air and water is condensed with the later separation.

Thermodynamic studies show that the internal compressor efficiency remarkably influences the MATIANT cycle efficiency. An increase of the compressor efficiency from 75% to 90% increases the power cycle net efficiency by 5.5% [49]. Reduction of the maximal pressure or the HPT inlet pressure from 300 to 150 bar reduces the net efficiency for about 1%, but an increase of the IPT inlet pressure from 40 to 100 bar causes the efficiency growth for more than 3%.

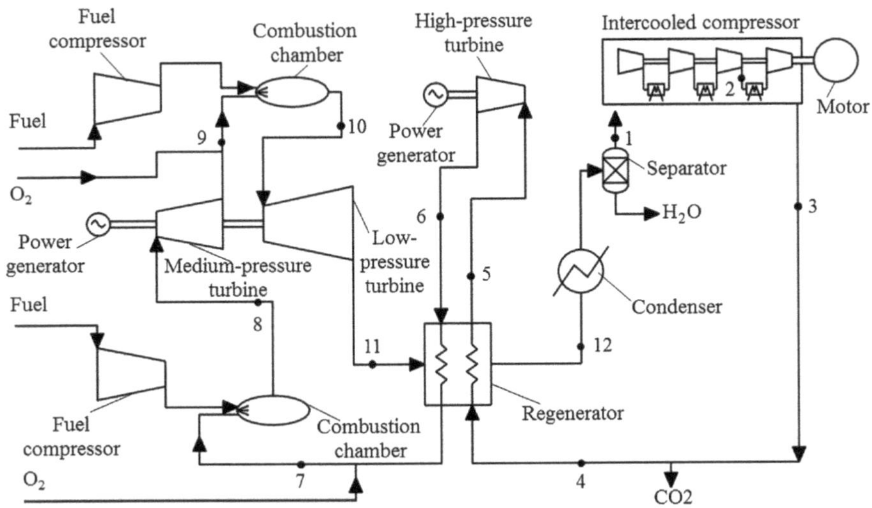

Figure 17.15 Flow chart of the MATIANT cycle.

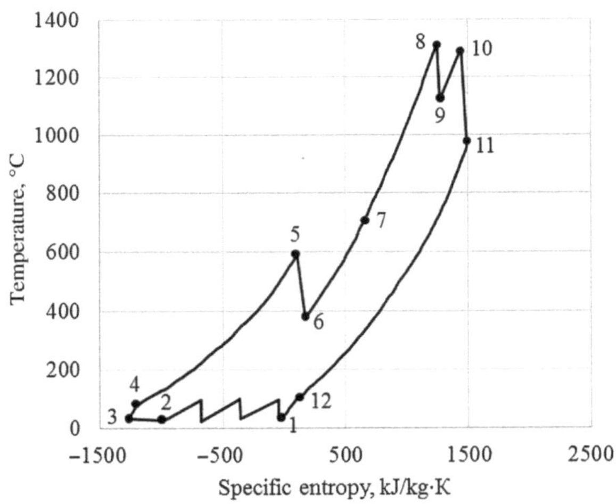

Figure 17.16 The diagram of the MATIANT cycle.

A study by Mathieu et al. [50] proposes a modification of this technology called the E-MATIANT cycle. This new version has the following advantages: a smaller number of elements, lower maximal pressure, and higher efficiency due to smaller losses for compression (Figs. 17.17, 17.18). The working fluid expands in two cooled turbines: the high-pressure and low-pressure turbines. In the inlets of both turbines, combustors are installed where in addition to the working fluid oxygen

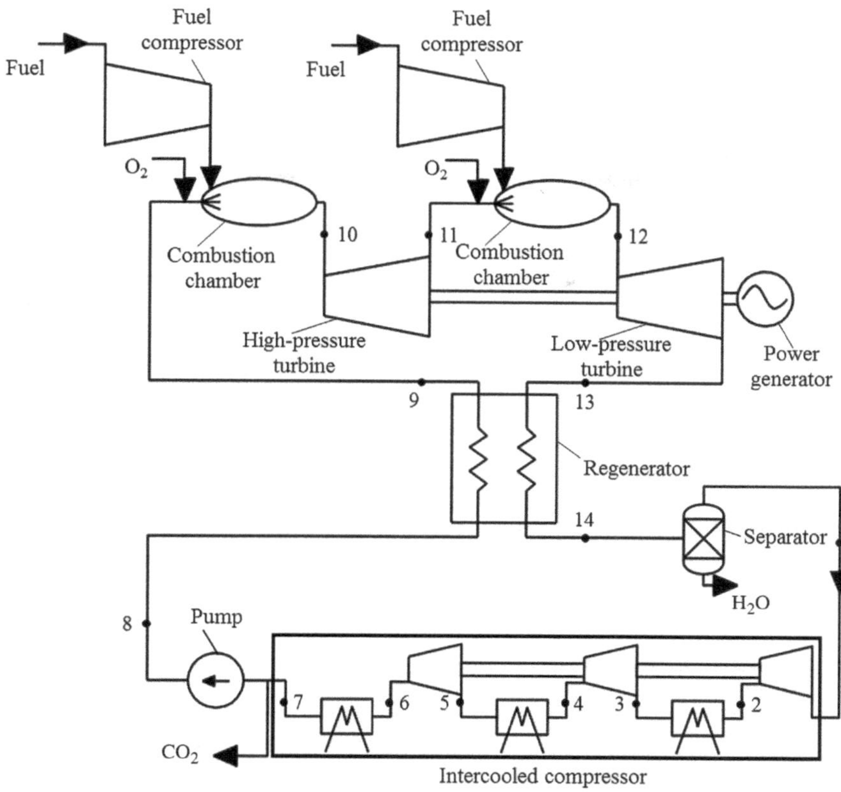

Figure 17.17 Flow chart of the E-MATIANT cycle.

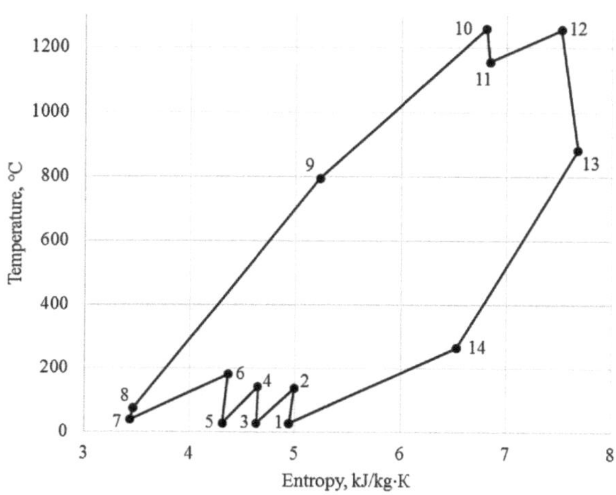

Figure 17.18 The diagram of the E-MATIANT cycle.

and fuel are supplied. The LPT exhaust flow is directed to the regenerator where it heats the compressed carbon dioxide after the pump exit.

The minimal E-MATIANT cycle pressure is almost atmospheric like in the prototype and the optimal HPT inlet pressure is 60 bar [50]. The optimization was done at constant HPT and LPT inlet temperatures of 1300°C and the regenerator underheating of 20°C. The compressor isentropic efficiency was assumed as 85% in the first three stages and 80% in the last stage. The maximal E-MATIANT cycle efficiency is reached for the HPT inlet pressure of 60 bar and equal to 47% without considering turbine cooling losses.

Like the SCOC-CC cycle, the key parameters of the E-MATIANT cycle that determine its efficiency are the initial temperature and pressure at the inlet of HPT and LPT. The efficiency losses caused by the turbines cooling were assumed equal to 2.5%. Consideration of the working fluid overheating at the IPT and LPT entrances shows a remarkable influence of the total cooling massflow upon the cycle efficiency at high initial parameters. This determines the remarkable error of the presented evaluation of technology efficiency.

17.4.4 Modified MATIANT cycles with turbine coolant precooling

The cooling flow source for the E-MATIANT cycle may be assumed as the flow heated in the regenerator (Fig. 17.19) [51]. Its pressure is sufficient to overcome hydraulic resistances of the blade internal channels and the turbine flowpath injection and the temperature is below the maximal acceptable blade-surface temperature.

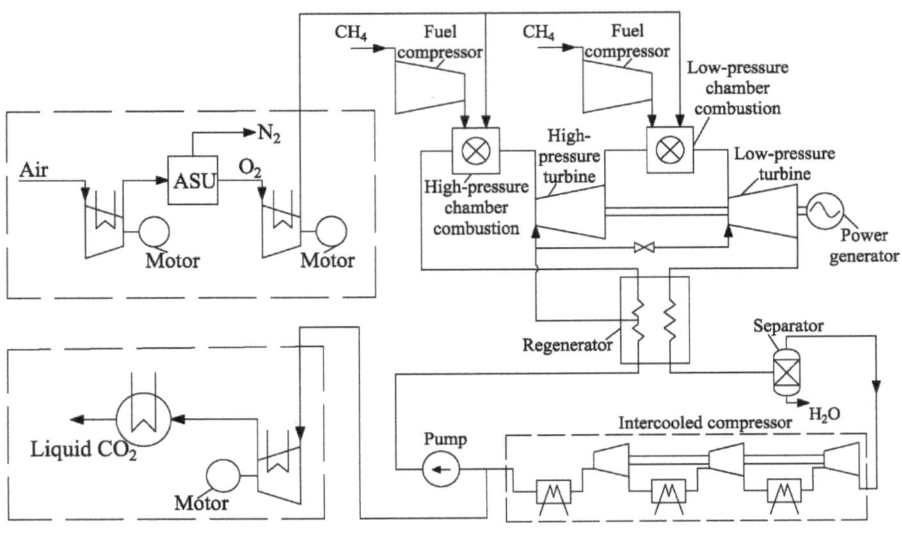

Figure 17.19 The E-MATIANT cycle with turbine cooling.

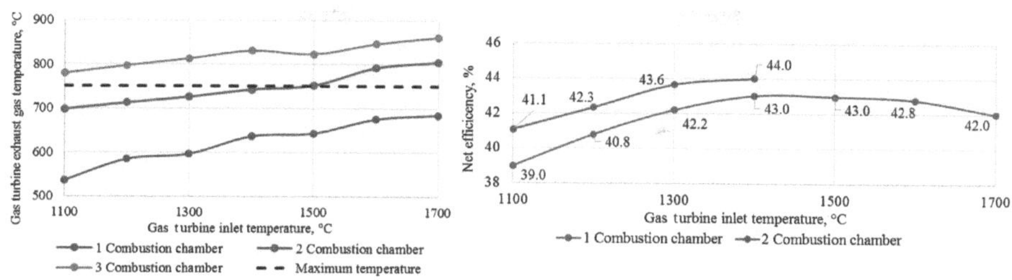

Figure 17.20 Optimization results of the E-MATIANT cycle calculated with consideration of the turbine cooling losses and the regenerator limitations.

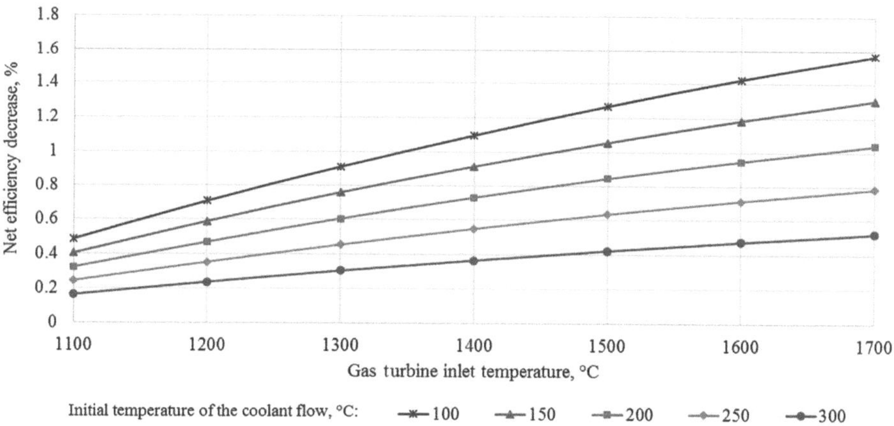

Figure 17.21 Influence of the turbine inlet temperature on the E-MATIANT cycle net efficiency for the different coolant flow initial temperature.

The E-MATIANT cycle net efficiency calculated with the turbine cooling losses and the limited regenerator inlet temperature of 750°C is shown in Fig. 17.20. The maximal net efficiency of 44% is reached at the initial parameters equal to 1400°C and 4 MPa. At these conditions, the coolant relative massflow and temperature are 40.3% and 400°C respectively.

So, in spite of the intermediate cooling and the superheating that improve the thermal efficiency, the E-MATIANT cycle has a lower efficiency than the SCOC-CC. This is caused by the larger losses for gas turbine cooling determined by the high working fluid mean temperature in the flowpath.

The investigation results of the coolant temperature on the efficiency of the E-MATIANT cycle with two combustion chambers are presented in Fig. 17.21. The decrease of the coolant temperature accompanied by the decrease of the cycle net efficiency despite the reduction of the coolant flow and turbine cooling losses. The decrease in the amount of heat transferred to the regenerator is the reason. That's why the thermal scheme with water injection in the coolant of the LPT was proposed (Fig. 17.22).

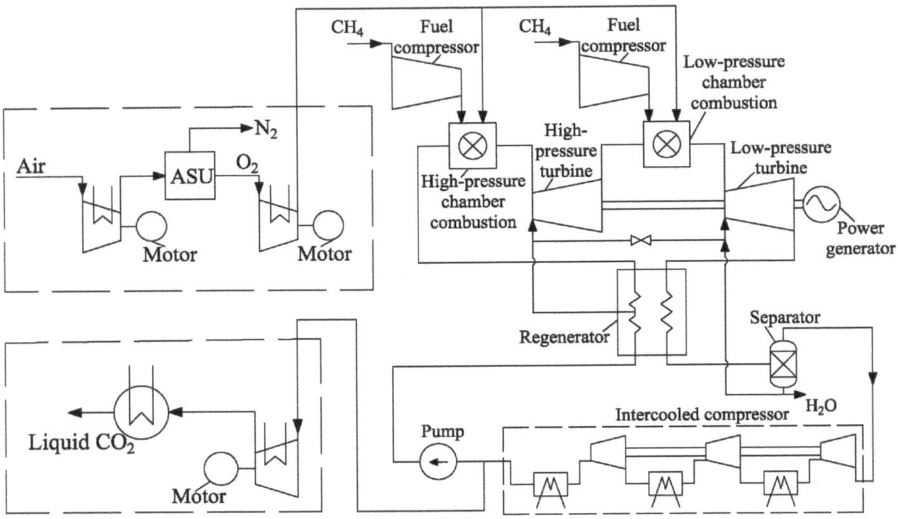

Figure 17.22 The E-MATJIAHT cycle with water infection in the coolant flow.

The E-MATIANT net efficiency increased with the injection of water taken from the cooler-separator in the coolant flows. When the cold water contacts hot streams of coolant, the evaporation occur and the coolant temperature decreased. In order to avoid the danger of moisture contact with the details of the hot-flow path, water injection is carried out only in the low-pressure turbine, where the partial pressure of water is relatively low.

The relative mass flow of the injection water should be in the range of 2.3%–2.7% (regarding the mass flow at the inlet of the LPT). The E-MATIANT cycle with two combustion chambers at turbine inlet temperatures from 1100°C to 1700°C has a possibility for 0.15%–0.80% net efficiency increase by means of water injection into the LPT coolant (Fig. 17.23) [52].

Analysis of the existing and prospective power production technologies shows that at equal power facilities output, the E-MATIANT cycle efficiency is 3.2% lower than that of a combined cycle power plant with the carbon dioxide capturing. In the case of the coolant flow preliminary cooling. In this analysis, the initial temperatures of the E-MATIANT cycle and the combined cycle facility were assumed as 1400°C and 1700°C, respectively.

17.4.5 The Allam cycle

The Allam cycle concept was patented in 2010 by its inventor Rodney John Allam [53]. A simplified concept of the natural gas-firing Allam cycle is shown in Fig. 17.24. This technology allows the record net efficiencies of 58.9% and 51.44% with natural gas and coal fuels, respectively, including the expenses for carbon dioxide storage [54]. This cycle may be compared against the combined cycle

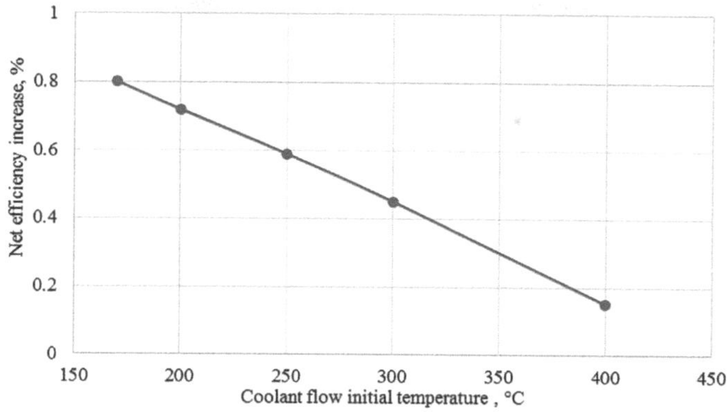

Figure 17.23 Influence of the coolant flow initial temperature on the net efficiency increase of the E-MATIANT cycle.

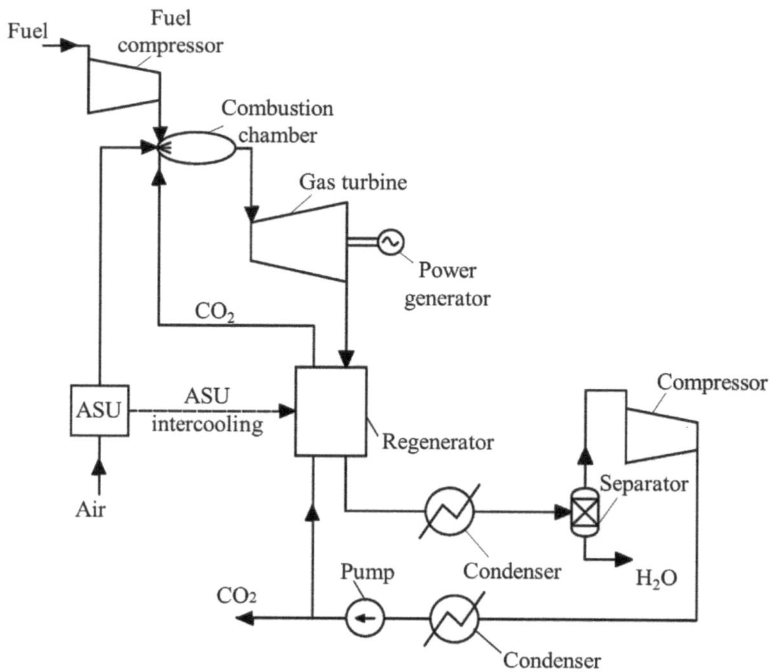

Figure 17.24 Flow chart of the Allam cycle.

power plant with sequestering technology net efficiency of 48% and the combined cycle power plant with gasification and sequestering one of 39% [55].

Carbon dioxide is compressed in an intercooled compressor up to 80 bar and then supplied to a pump where it reaches the cycle maximal pressure of

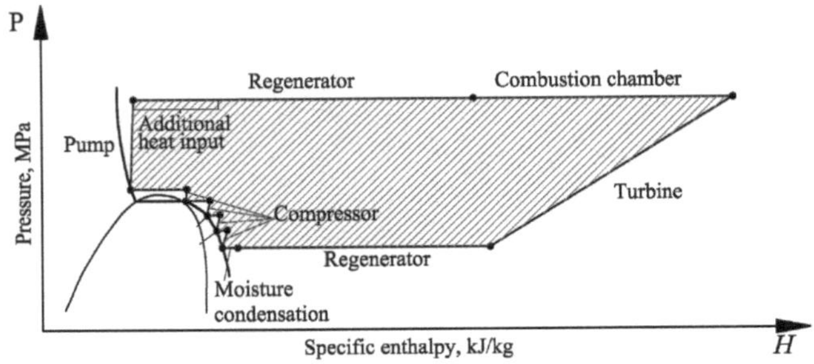

Figure 17.25 A diagram of the Allam cycle.

200–400 bar (Fig. 17.25). After the pump, the carbon dioxide is supplied to the regenerator where it is heated up to 750°C by the heat of the turbine exhaust and low-temperature heat from the ASU. After the regenerator, the larger part of carbon dioxide flow is supplied to the combustor to limit the maximal temperature. The smaller part is used for turbine cooling. The residual carbon dioxide flow is mixed with oxygen and supplied to the combustor. In the combustor, the oxy-fuel mixture burns, and its temperature grows up to 1150°C. The carbon dioxide recirculation limits the turbine inlet temperature, and 90% of the working fluid is carbon dioxide. In the turbine flowpath, the fluid expands down to 20–30 bar, which is below the carbon dioxide critical pressure. The optimal turbine inlet pressure is 200–400 bar, and the turbine pressure drop is from 6 to 12 [56].

The competitive specific price of the installed power is an important advantage of this technology. These parameters for the natural gas and coal gasification versions of the Allam cycle are 800–1000 US$/kW and 1500–1800 US$/kW respectively [51]. On the contrary, the current price of combined cycle power plants with gasification is at least 5000 US$/kW [56]. Such a low price of the Allam cycle is due to the equipment compactness caused by the high value of the minimal working fluid pressure equal to 20–40 bar and the absence of the combined cycle traditional elements, steam turbine, heat recovery boiler, steam pipelines, synthesis gas cooler, catalyzer, devices for emission mitigation of nitrogen oxides, carbon monoxide, sulfur, and so forth.

The results of thermodynamic investigations of the turbine inlet parameters influence on the Allam cycle net efficiency are presented in Fig. 17.26 [57]. The maximum value of cycle net efficiency of 56.5% (including air separation unit penalty and carbon capture and storage at 100 bar) is achieved for the turbine inlet temperature of 1083°C and pressure of 300 bar. The turbine outlet pressure value was fixed at 30 bar and turbine coolant temperature—at 200°C during the optimization of turbine inlet parameters.

The turbine inlet optimum temperature at a fixed pressure may be explained by specific features of the high-temperature regenerator thermodynamic process. Usually the turbine inlet temperature increase is followed by growth of the cycle mean integral heat intake temperature that increases the equivalent Carnot cycle

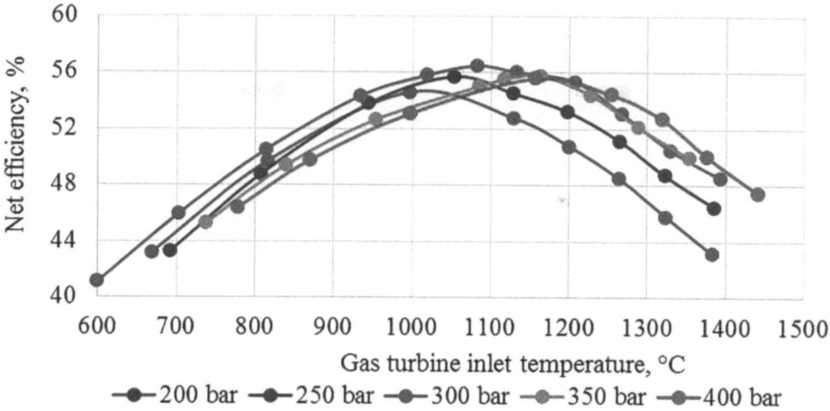

Figure 17.26 Net efficiency as a function of the turbine inlet temperature for the Allam cycle.

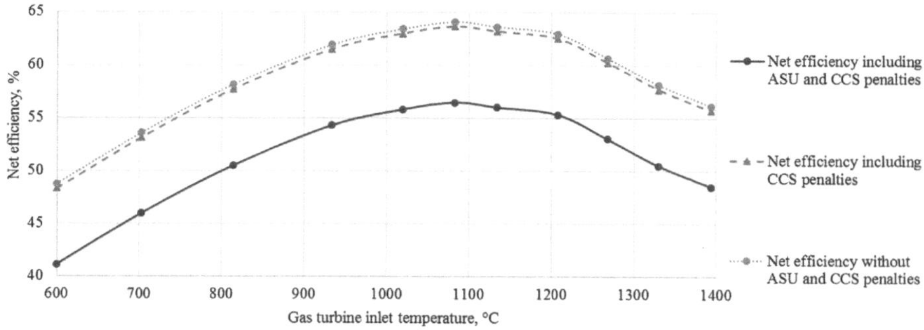

Figure 17.27 Influence of the oxygen production and compression and the carbon dioxide compression before storage upon the Allam cycle efficiency.

efficiency, and the cycle thermodynamic efficiency also grows. On the other hand, the regenerator analysis shows that the excessive increase of the turbine inlet temperature increases its exhaust temperature that changes the pinch point.

Fig. 17.27 illustrates the power consumption for oxygen production and compression and carbon dioxide storage at 100 bar pressure. The solid line is similar to the curve in Fig. 17.26 at the turbine inlet pressure of 300 bar. The oxygen purity was 91.25% and the ASU specific power consumption was 900 kW/(kg/s). Production and compression of the oxygen reduce cycle net efficiency by 7.2%. The compression of carbon dioxide before storage decreases cycle net efficiency by 0.4% [58]. The low energy consumption of CO_2 compressor is due to the high cycle minimal pressure, which is equal to 30 bar.

17.5 Conclusion

The development of the oxy-fuel power cycles in the power production industry allows the reduction of toxic and greenhouse gas emissions down to nearly zero level. Among the reviewed technologies, the most attractive is the Allam cycle that allows the best environment and efficiency performance. Its net efficiency is 8.5% higher than that of the CC facility including the carbon dioxide capturing degree of 98.9% in the Allam cycle and 89% in CC.

The large-scale introduction of the oxy-fuel power plants must be preceded by test development and construction of pilot facilities. The most advanced of these are the postcombustion carbon capture technology and the Allam cycle power plant. In particular, NET Power LLC, the Shaw Group, Toshiba Corporation, and Exelon Corporation have partnered to commercialize the NET Power system through the development of a 50 MW natural gas power plant [56]. This plant has a final operational goal of continuous operation with no performance degradation. Following initial startup, the plant will go through extensive testing to include a demonstration of all components; establishment of start-up and shut-down procedures, ramp rates, and safety and control systems; as well as measurement of flow stream chemistry and temperatures for future system optimization. Once converted to commercial operation, the plant will generate revenue through sales of electricity and CO2 for enhanced oil recovery.

The first full-scale commercial plant should enter operation in the near future [59]. The gasified solid fuel and coal-based system offer CO2 free power at high efficiencies and costs lower than comparable systems which do not employ carbon capture. This system is currently being advanced. The disrupting economics of the Allam cycle and its ability to integrate with a variety of processes presents several viable and unique commercialization opportunities under continual development.

Acknowledgment

This study conducted by Moscow Power Engineering Institute was supported by the Russian Science Foundation under Agreement No. 17–79-20371 dated July 28, 2017.

References

[1] J.R. McNeill, Something New Under the Sun: An Environmental History of the Twentieth-Century World (the Global Century Series), WW Norton & Company, New York, 2001.

[2] J.T. Houghton, et al., Climate Change 2001: The Scientific Basis, The Press Syndicate of the University of Cambridge, Cambridge, 2001.

[3] S. Rahmstorf, et al., Recent climate observations compared to projections, Science 316 (5825) (2007) 709.

[4] R. Lal, et al., Management to mitigate and adapt to climate change, J. Soil. Water Conserv. 66 (4) (2011) 276–282.

[5] P. Nejat, et al., A global review of energy consumption, CO2 emissions and policy in the residential sector (with an overview of the top ten CO2 emitting countries), Renew. Sustain. Energy Rev. 43 (2015) 843–862.

[6] M.I. Hoffert, C. Covey, Deriving global climate sensitivity from palaeoclimate reconstructions, Nature 360 (6404) (1992) 573–576.

[7] M.L. Weitzman, On modeling and interpreting the economics of catastrophic climate change, Rev. Econ. Stat. 91 (1) (2009) 1–19.

[8] V. Quaschning, Development of global carbon dioxide emissions and concentration in atmosphere. <https://www.volker-quaschning.de/datserv/CO2/index_e.php>. (accessed 03.03.18).

[9] T.M.L. Wigley, R. Richels, J.A. Edmonds, Economic and environmental choices in the stabilization of atmospheric CO2 concentrations, Nature 379 (6562) (1996) 240–243.

[10] T.J. Blasing, Recent Greenhouse Gas Concentrations, Carbon Dioxide Information Analysis Center, 2016.

[11] Global Environment Outlook 5: Environment for the Future We Want, United Nations Environment, 2012.

[12] O. Edenhofer, Climate Change 2014: Mitigation of Climate Change, Cambridge University Press, Cambridge, 2015.

[13] V. Marchal, et al., OECD environmental outlook to 2050, Organ. Econ. Co-op. Dev. 8 (2011) 397–413.

[14] K. Feng, et al., Drivers of the US CO2 emissions 1997–2013, Nat. Commun. 6 (2015) 7714.

[15] H. Lund, Renewable energy strategies for sustainable development, Energy 32 (6) (2007) 912–919.

[16] S. Dutta, A review on production, storage of hydrogen and its utilization as an energy resource, J. Ind. Eng. Chem. 20 (4) (2014) 1148–1156.

[17] S. Dutta, Hydrogen as sustainable and green energy resource, Kirk-Othmer Encycl. Chem. Technol. (2018) 1–23.

[18] R.S. Middleton, J.K. Eccles, The complex future of CO2 capture and storage: variable electricity generation and fossil fuel power, Appl. Energy 108 (2013) 66–73.

[19] B. de Foy, et al., Estimates of power plant NOx emissions and lifetimes from OMI NO2 satellite retrievals, Atmos. Environ. 116 (2015) 1–11.

[20] D. Quiroga, E. Sauma, D. Pozo, Power system expansion planning under global and local emission mitigation policies, Appl. Energy 239 (2019) 1250–1264.

[21] J.A.L. Asmar, et al., Power generation and cogeneration management algorithm with renewable energy integration, Energy Procedia 74 (2015) 1394–1401.

[22] E. Gholamian, S.M.S. Mahmoudi, V. Zare, Proposal, exergy analysis and optimization of a new biomass-based cogeneration system, Appl. Therm. Eng. 93 (2016) 223–235.

[23] M.K. Mohammed, et al., The optimum performance of the combined cycle power plant: a comprehensive review, Renew. Sustain. Energy Rev. 79 (2017) 459–474.

[24] Z. Zhao, Exergy analysis of the turbine system in a 1000 MW double reheat ultra-supercritical power plant, Energy 119 (2017) 540–548.

[25] M. Muratori, et al., Carbon capture and storage across fuels and sectors in energy system transformation pathways, Int. J. Greenh. Gas. Control. 57 (2017) 34–41.

[26] R. Wennersten, Q. Sun, H. Li, The future potential for carbon capture and storage in climate change mitigation–an overview from perspectives of technology, economy and risk, J. Clean. Prod. 103 (2015) 724–736.

[27] K.S. Lackner, A guide to CO_2 sequestration, Science 300 (5626) (2003) 1677–1678.

[28] I.P. Koronaki, L. Prentza, V. Papaefthimiou, Modeling of CO2 capture via chemical absorption processes – an extensive literature review, Renew. Sustain. Energy Rev. 50 (2015) 547–566.
[29] S. Vasudevan, et al., Energy penalty estimates for CO2 capture: comparison between fuel types and capture-combustion modes, Energy 103 (2016) 709–714.
[30] I. Pfaff, J. Oexmann, A. Kather, Optimised integration of post-combustion CO_2 capture process in greenfield power plants, Energy 35 (10) (2010) 4030–4041.
[31] J. Tong, Y. Matsumura, Pure hydrogen production by methane steam reforming with hydrogen-permeable membrane reactor, Catal. Today 111 (3–4) (2006) 147–152.
[32] E. García-Díez, Optimization of hydrogen production with CO2 capture by autothermal chemical-looping reforming using different bioethanol purities, Appl. Energy 169 (2016) 491–498.
[33] Z. Wang, Exergy analysis of methane cracking thermally coupled with chemical looping combustion for hydrogen production, Appl. Energy 168 (2016) 1–12.
[34] Y.K. Krisnandi, et al., Partial oxidation of methane to methanol over heterogeneous catalyst Co/ZSM-5, Procedia Chem. 14 (2015) 508–515.
[35] A.H. Рогалёв, Разработка научно-методологических основ создания перспективных высокотемпературных энергетических комплексов: докторская: May 14, 01, 2018.
[36] R.L. Bannister, R.A. Newby, W.C. Yang, Development of hydrogen-fueled combustion turbine cycle for power generation, J. Eng. Gas. Turbine. Power 120 (2) (1998) 276–283.
[37] A. Miller, J. Milewski, S. Kiryk, Remarks on hydrogen-fuelled combustion turbine cycle, in: Proceedings of the 2nd International Scientific Symposium COMPOWER 2000, Gdansk, Poland, 2000.
[38] A. Miller et al., New efficient hydrogen-fuelled combustion turbine cycle – a study of configuration and performance, in: Proceedings of the 14th World Hydrogen Energy Conference, Montreal, Canada, 2002.
[39] B.J. Buhre, et al., Oxy-fuel combustion technology for coal-fired power generation, Prog. Energy Combust. Sci. 31 (4) (2005) 283–307.
[40] F.C. Barba, et al., A technical evaluation, performance analysis and risk assessment of multiple novel oxy-turbine power cycles with complete CO_2 capture, J. Clean. Prod. 133 (2016) 971–985.
[41] O. Bolland, S. Saether, New concepts for natural gas fired power plants which simplify the recovery of carbon dioxide, Energy Convers. Manag. 33 (5–8) (1992) 467–475.
[42] O. Bolland, P. Mathieu, Comparison of two CO2 removal options in combined cycle power plants, Energy Convers. Manag. 39 (16–18) (1998) 1653–1663.
[43] P. Chiesa, G. Lozza, CO2 emission abatement in IGCC power plants by semiclosed cycles: part a-with oxygen-blown combustion, J. Eng. Gas. Turbine. Power 121 (1999) 635–641.
[44] A. Dahlquist et al., Optimization of an oxyfuel combined cycle regarding performance and complexity level, in: Proceedings of the ASME Turbo Expo 2013, San Antonio, TX, 2013.
[45] H.J. Yang, et al., Evaluation of design performance of the semi-closed oxy-fuel combustion combined cycle, J. Eng. Gas. Turbine. Power 134 (11) (2012) 111702.
[46] A. Rogalev, V. Kindra, S. Osipov, Modeling methods for oxy-fuel combustion cycles with multicomponent working fluid, in: AIP Conference Proceedings, 2018.
[47] A.N. Rogalev, et al., Methods for efficiency improvement of the semi-closed oxy-fuel combustion combined cycle, J. Phys. Conf. Ser. 1111 (1) (2018) 012003.

[48] P. Mathieu, R. Nihart, Sensitivity analysis of the MATIANT cycle, Energy Convers. Manag. 40 (15–16) (1999) 1687–1700.
[49] P. Mathieu, R. Nihart, Zero-emission MATIANT cycle, J. Eng. Gas. Turbine. Power 121 (1999) 116–120.
[50] P. Mathieu et al., New concept of CO_2 removal technologies in power generation, combined with fossil fuel recovery and long term CO_2 sequestration, in: Proceedings of the ASME Turbo Expo 2000, Munich, Germany, 2000.
[51] A. Rogalev, et al., Evaluation of bleed flow precooling influence on the efficiency of the E-MATIANT cycle, Mech. Mech. Eng. 22 (2) (2018) 593–602.
[52] V. Kindra, S. Osipov, A. Vegera, Structural and parametric investigations of oxy-fuel combustion cycles with the aim of increasing efficiency of electricity production, in: 2019 International Science and Technology Conference «EastConf», Vladivostok, Russia, 2019.
[53] Pat. US 8596075 B2, F02C 1/00, F02C 7/08, System and method for high efficiency power generation using a carbon dioxide circulating working fluid/R. Allam et al., Assignees Palmer Labs, LLC; filed August 31, 2010; publish July 28, 2011, 46 p.
[54] R.J. Allam, et al., High efficiency and low cost of electricity generation from fossil fuels while eliminating atmospheric emissions, including carbon dioxide, Energy Procedia 37 (2013) 1135–1149.
[55] A. Bhattacharya, A. Datta, Effects of supplementary biomass firing on the performance of combined cycle power generation: a comparison between NGCC and IGCC plants, Biomass Bioenergy 54 (2013) 239–249.
[56] J.M. Klara, J.E. Plunkett, The potential of advanced technologies to reduce carbon capture costs in future IGCC power plants, Int. J. Greenh. Gas. Control. 4 (2) (2010) 112–118.
[57] A. Rogalev, et al., Thermodynamic analysis of the net power oxy-combustion cycle, J. Clean. Prod. 236 (1) (2019) 117592.
[58] A. Rogalev, et al., Thermodynamic optimization and equipment development for a high efficient fossil fuel power plant with zero emissions, J. Clean. Prod. 236 (2019) 117592.
[59] R. Allam, et al., Demonstration of the Allam cycle: an update on the development status of a high efficiency supercritical carbon dioxide power process employing full carbon capture, Energy Procedia 114 (2017) 5948–5966.

Integrated photocatalytic hydrogen production and pollutants or wastes treatment: prospects and challenges

18

Rohini Singh[1] *and Suman Dutta*[2]
[1]Department of Chemical Engineering, Sitarambhai Naranji Patel Institute of Technology & Research Centre, Bardoli, India, [2]Department of Chemical Engineering, Indian Institute of Technology (ISM), Dhanbad, India

18.1 Introduction

Effective pollutant abatement and hydrogen generation are major concerns for the future environmental conservation and energy crisis that needs to be solved on the priority basis. Hydrogen is one of the essential sustainable, environment-friendly, and sustainable future fuels that possesses high calorific value. Various conventional methods have been adopted such as steam reforming and coal gasification for the production of hydrogen. However, hydrogen generation via photocatalytic water splitting can be considered as an economical and clean method as greenhouse gases are not being produced during the entire process. In addition, water and sunlight are the major sources for hydrogen evolution assisted by low-cost photocatalysts that makes the process economical. Photocatalysis, an advanced oxidation process, facilitates the environment-friendly solution simultaneously via oxidation and reduction reactions that occur on the surface of the selected photocatalysts. The overall mechanism of the dual functioned redox reactions can be explained as below [1].

Photoexcitation

$$TiO_2 + h\nu \rightarrow e^- + h^+ \tag{18.1}$$

Oxidation

$$H_2O_{(ads)} \xrightarrow{Photocatalyst} H^+_{(ads)} + OH^-_{(ads)} \tag{18.2}$$

$$OH^-_{(ads)} + h^+ \xrightarrow{Photocatalyst} OH^\bullet \tag{18.3}$$

Sustainable Fuel Technologies Handbook. DOI: https://doi.org/10.1016/B978-0-12-822989-7.00020-2

$$OH^{\bullet} \text{ or } h^{+}_{VB} + \text{pollutant} \rightarrow \text{degradation or oxidation of pollutant} \quad (18.4)$$

Reduction

$$2H^{+} + 2e^{-} \rightarrow H_2 \quad (18.5)$$

Several photocatalysts have been developed for efficient environmental remediation and water splitting under ultraviolet or visible light irradiation. Some of the photocatalysts that exhibit enhanced photo-response include semiconductors such as TiO_2, ZnO, and many other transition metal oxides. Biomass can be effectively utilized for hydrogen generation. Thus solid waste management and energy generation could be executed in single reaction time. Heterogeneous photocatalysis ensures sustainable water remediation and energy management. A few significant outcomes revealed by research worldwide are related to photocatalytic wastewater treatment and solar hydrogen generation utilizing ZnO as photocatalyst are listed in Tables 18.1 and 18.2.

Electrons (e^{-}) and holes (h^{+}) participate in the photocatalytic processes as charge carriers. The electron (e^{-}) resides in the valence band (VB). As the photocatalysts receive the energy higher than the band-gap, the electrons get excited in the VB and move to the conduction band. In this way electron–hole pairs (e^{-}/h^{+}) are formed. According to Eqs. (18.1–18.5), the holes participate in the oxidation reaction and degrade the toxic pollutants. On the other hand, the electrons participate in the reduction reaction and generate hydrogen. This chapter highlights the attempts executed by the researchers worldwide for photo-assisted hydrogen generation integrated with simultaneous waste treatment as a clean method with environmental sustainability. Advantages and disadvantages of the integrated systems are discussed, especially utilizing hazardous waste as sacrificial agents. In addition, the overall efficiency of the integrated systems would are discussed in terms of pollutant treatment and hydrogen evolution.

18.2 Utilization of pollutants or waste as sacrificial agents

Numerous dyes such as methylene blue, methyl red, methyl orange, rhodamine 6 G, rhodamine B and many more exhibit acute toxicity and possess an adverse effect on people's respiratory and digestive systems. In addition, the dyes are carcinogenic. However, visible light–active dyes and various other pollutants such as phenol, nitrates, and others can be utilized as sacrificial agents to enhance hydrogen evolution via photocatalytic water splitting. Sacrificial agents are the electron donors or hole scavengers that reduce the recombination tendency of electrons and holes and accelerates the rate of hydrogen generation. Dye-sensitized photocatalytic water splitting can be considered as an efficient technology for hydrogen generation along with the degradation of hazardous visible light–active dyes utilized as sensitizers. Application of

Table 18.1 Properties and photocatalytic pollutant degradation efficiency of ZnO.

Photocatalyst	Method of preparation	Method of modification	Morphology/size	Contaminant	Irradiation source	Degradation efficiency (%) (duration)	References
ZnO	Low temp. coprecipitation	–	Nanoparticle	Rhodamine B	Visible light	Up to 80 (120 min)	[2]
ZnO/Ag	Solution based approach	–	Flower-like aggregated nanosheet	Methylene blue	Visible light	20–90	[3]
Graphene/ZnO hybrids	Commercial	Ultrasonication and hydrothermal	Wrapped	Deoxynivalenol	UV light (365 nm)	20–100 (120)	[4]
ZnO	Solution based method	–	Lotus-like	Methylene blue	UV light and sunlight	98.99 (60) in UV light 98.25(120) in sunlight	[5]
ZnO/Ag	–	One-pot nonaqueous route	Nanorods	Rhodamine B	300 W Xe lamp ($\lambda > 420$ nm)	Up to 100 (3 h)	[6]
Au, Ag and Pd doped ZnO	Microwave-assisted hydrothermal process	Borohydride reduction method	Irregular nanoplate like	Congo red	UV irradiation	Up to 100 (160)	[7]
Ag/ZnO and Pt/ZnO	–	Solvothermal method	Nanowires, quasi-spherical	Methylene blue	UV irradiation	Up to 100 (120)	[8]

UV, Ultraviolet.

Table 18.2 Properties and photocatalytic hydrogen evolution capacity of ZnO.

Photocatalyst (dosage)	Method of preparation	Method of modification	Morphology	Sacrificial agent	Irradiation source	Hydrogen evolution	References
Pt-doped TiO_2-ZnO	Solution based approach	Solution based approach	Non uniform-shape	Methanol	400 W mercury arc lamp (visible light source)	2150 μmol/h/g	[9]
g-C_3N_4/ZnO (0.1 g/100 mL)	Commercial	Solution conversion method	Needle-like	Triethanolamine (TEOA)	300 W Xenon lamp (visible light source)	5–35 μmol/h	[10]
ZnO (20 mg/75 mL)	–	Green synthesis using Moringaoleifera natural extract	Spherical	Na_2S and Na_2SO_3	400 W Xenon lamp (UV-Vis light source)	Up to 350 μmol/g (120 min)	[11]
MoS_2-RGO/ZnO (50 mg/100 mL)	Modified solvothermal method	Hydrothermal	Irregular	NaS_2 and $NaSO_3$	300 W Xe lamp	Up to 300 μmol/g/h	[12]
Ce-doped ZnO/ZnS (0.1 g/100 mL)	–	Chemical precipitation	Almost spherical	Poly-sulfides (Na_2S and Na_2SO_3)	UV lamp	1200 μmol/g/h	[13]

UV, Ultraviolet.

several visible light–active dyes such as MB, Rh6G, and Eosin-Y among others as sensitizers enhances the rate of hydrogen generation, [14] and Erythrosin B dye is utilized to sensitize Zr-containing MOFs of UiO-66 octahedrons for enhanced photocatalytic hydrogen evolution. Table 18.3 lists a summary of recent research reports regarding simultaneous pollutant degradation and hydrogen generation.

18.3 Pros and cons of integrated systems

The advantages and disadvantages of the integrated system include:

1. Simultaneous hydrogen generation and degradation of hazardous pollutants by utilizing a single reactor set-up.
2. Solid waste could be reduced by the utilization as the sacrificial agent to enhance the rate of hydrogen evolution.
3. Specialized know-how is required to design the reactor for the efficient execution of the simultaneous process of pollutant abatement and hydrogen generation. Effective photoreactors must be intended to restrict the leakage of evolved gases entirely. In addition, a sample collection port must be provided in such a way that the process doesn't get disturbed during the entire operation tenure.
4. Risk of the generation of toxic intermediate products or by-products in some cases while utilizing waste materials for the significant increment in the hydrogen evolution after oxidation. Moreover, the release of greenhouse gases must be avoided. So the choice of waste materials to be used as sensitizers must be selected very carefully.

18.4 Reactors for an integrated system

Based on several research works executed globally, the dual functioned photoassisted reactions (i.e., hazardous contaminant remediation and hydrogen evolution) could be conducted in single and twin photoreactors. Researchers [17,18] have successfully designed a dual functioned novel photocatalytic twin-reactor (see Fig. 18.1) to execute simultaneous phenol degradation and hydrogen evolution. The photoreactor has been separated in two slots via Nafion membrane for the separate evolution of H_2 and O_2.

A low-cost silicone microreactor that exhibited stable hydrogen generation has been designed by Castedo et al. [21] with the microchannels of 500 μm width coated with Au/TiO_2 as the photocatalyst. Jaswel et al. [22] executed the photocatalytic water splitting reaction in a high temperature and pressure reactor. Ma et al. [23] designed a pyrex reactor system with photocatalyst membranes installed inside and irradiated with simulated visible light.

18.5 Prospects and challenges

Synthesis of visible light–active photocatalyst: To harness sunlight effectively, synthesis of visible light-responsive photocatalysts is required since a major portion of the solar spectrum consists of radiation in visible range.

Table 18.3 Summary of simultaneous pollutant degradation and hydrogen evolution.

Sr. no.	Photocatalysts	Amount of hydrogen produced	Pollutant degraded or sacrificial reagent utilized	Source of irradiation	References
1	Sn_3O_4 (100 mg)	350 μmol in 240 min	Phenol (upto100%, 6 ppm, 200 min)	Sunlight	[15]
2	Copper loaded titania (P-25) nanoparticles (45 mg)	Up to 14 μM in 240 min	Nitrate, nitrite, and ammonia (over 93%, 150 ppm, 120 min)	UV-A/Vis	[16]
3	Calcined Rh-doped SrTiO3	11.37 μmols/g	Phenol (200 μmol/L, 6 h)	Simulated sunlight	[17,18]
4	Oxidized graphitic carbon nitride doped with Cu, Ni and Cu	Up to 1050 μmols/g/h	Various industrial effluent	Simulated solar light	[19]
5	ZnO-ZnS/graphene	1070 μmol/h/g	Glycerol solution	Simulated solar light	[20]
6	P-25 Degussa TiO_2 nanoparticles	19.9 ml H_2/g pinewood	Pinewood	Simulated solar light	

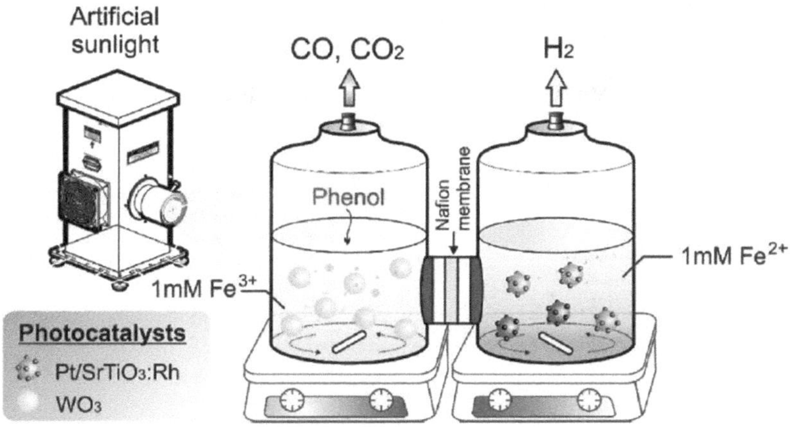

Figure 18.1 Illustration of dual function reaction conducted in a twin-reactor.
Source: Appl. Catal. B: Environ. 239 (2018) 268–279.

Design of reactor: Effective design of the photoreactors to execute the dual function of pollutant degradation and hydrogen generation can be considered as a great challenge for researchers. Leakage and reverse reaction of the produced gas should be avoided for acquiring enhanced efficiency.

Separation of the photocatalyst after treatment: Nanosized photocatalyst separation can be considered as an extremely challenging and expensive task. Thus low-cost membranes should be fabricated and successfully incorporated in the designed photoreactors so that no separate photocatalysts filtration set-up would be required.

Separation of oxygen and hydrogen after the reaction: During the overall process of photocatalytic water splitting, a constant risk of the back mixing exists. Therefore various efforts can be made by researchers such as (1) design of reactors for the separate evolution of H_2 and O_2, and (2) synthesis of photocatalysts for the selective generation of H_2 and O_2.

Storage of hydrogen: Due to its extreme flammability, the storage of hydrogen should be executed carefully. Leakage should be strictly avoided during storage and transportation of hydrogen cylinders.

Production of toxic chemicals in the water stream due to the utilization of certain sacrificial agents: Selection of an appropriate sacrificial agent is an essential factor that determines the overall performance of the integrated system. The secretion of harmful contaminants during utilization of certain sacrificial agents must be identified and restricted.

18.6 Conclusion

The utilization of waste materials as sacrificial agents during hydrogen generation via photocatalysis is an excellent method. In this way, we can produce hydrogen parallel with pollution control in a single reactor. However, the selection of catalyst and sacrificial agent are major challenges for this integrated method. Proper catalysts need to be synthesized for the utilization of solar irradiation. Many other factors such as reactor design, H2 purification, and H2 storage play a vital role in the success of this method. Practical application of this method needs special attention to the production of any toxic intermediate or by-product.

Acknowledgments

This work was partly supported by the Department of Chemical Engineering, Indian Institute of Technology (ISM), Dhanbad.

References

[1] R. Singh, S. Dutta, A review on H_2 production through photocatalytic reactions using TiO2/TiO_2-assisted catalysts, Fuel 220 (2018) 607–620.
[2] S. Akir, A. Barras, Y. Coffinier, M. Bououdina, R. Boukherroub, A.D. Omrani, Eco-friendly synthesis of ZnO nanoparticles with different morphologies and their visible light photocatalytic performance for the degradation of Rhodamine B, Ceram. Int. 42 (2016) 10259–10265.
[3] Y. Yang, H. Li, F. Hou, J. Hu, X. Zhang, Y. Wang, Facile synthesis of ZnO/Ag nanocomposites with enhanced photocatalytic properties under visible light, Mater. Lett. 180 (2016) 97–100.
[4] X. Bai, C. Sun, D. Liu, X. Luo, D. Li, J. Wang, et al., Photocatalytic degradation of deoxynivalenol using graphene/ZnO hybrids in aqueous suspension, Appl. Catal. B Environ. 204 (2017) 11–20.
[5] C. Hao, J. Wang, Q. Cheng, Y. Bai, X. Wang, Y. Yang, Anionic surfactants-assisted solution-phase synthesis of ZnO with improved photocatalytic performance, J. Photochem. Photobiol. A: Chem. 332 (2017) 384–390.
[6] X. Hou, ZnO/Ag heterostructured nanoassemblies: wet-chemical preparation and improved visible-light photocatalytic performance, Mater. Lett. 139 (2015) 201–204.
[7] N. Güy, M. Özacar, The influence of noble metals on photocatalytic activity of ZnO for Congo red degradation, Int. J. Hydrog. Energy 41 (2016) 20100–20112.
[8] L. Muñoz-Fernandez, A. Sierra-Fernandez, O. Milošević, M.E. Rabanal, Solvothermal synthesis of Ag/ZnO and Pt/ZnO nanocomposites and comparison of their photocatalytic behaviors on dyes degradation, Adv. Powder Technol. 27 (2016) 983–993.
[9] M.Y. Xie, K.Y. Su, X.Y. Peng, R.J. Wu, M. Chavali, W.C. Chang, Hydrogen production by photocatalytic water-splitting on Pt-doped TiO_2–ZnO under visible light, J. Taiwan Inst. Chem. Eng. 70 (2017) 161–167.
[10] J. Wang, Y. Xia, H. Zhao, G. Wang, L. Xiang, J. Xu, et al., Oxygen defects-mediated Z-scheme charge separation in g-C_3N_4/ZnO photocatalysts for enhanced visible-light degradation of 4-chlorophenol and hydrogen evolution, Appl. Catal. B: Environ. 206 (2017) 406–416.
[11] B. Archana, K. Manjunath, G. Nagaraju, K.B. Chandra Sekhar, N. Kottam, Enhanced photocatalytic hydrogen generation and photostability of ZnO nanoparticles obtained via green synthesis, Int. J. Hydrog. Energy 42 (2017) 5125–5131.

[12] Z. Guan, P. Wang, Q. Li, Y. Li, X. Fu, J. Yang, Remarkable enhancement in solar hydrogen generation from MoS_2-RGO/ZnO composite photocatalyst by constructing a robust electron transport pathway, Chem. Eng. J. 327 (2017) 397–405.
[13] C.J. Chang, K.L. Huang, J.K. Chen, K.W. Chu, M.H. Hsu, Improved photocatalytic hydrogen production of ZnO/ZnS based photocatalysts by Ce doping, J. Taiwan Inst. Chem. Eng. 55 (2015) 82–89.
[14] Y.P. Yuan, L.S. Yin, S.W. Cao, G.S. Xu, C.H. Li, C. Xue, Improving photocatalytic hydrogen production of metal–organic framework UiO-66 octahedrons by dye-sensitization, Appl. Catal. B: Environ. 168 (2015) 572–576.
[15] S. Balgude, Y. Sethi, B. Kale, D. Amalnerkar, P. Adhyapak, Sn_3O_4 microballs as highly efficient photocatalyst for hydrogen generation and degradation of phenol under solar light irradiation, Mater. Chem. Phys. 221 (2019) 493–500.
[16] R. Lucchetti, L. Onotri, L. Clarizia, F. Di Natale, I. Di Somma, R. Andreozzi, et al., Removal of nitrate and simultaneous hydrogen generation through photocatalytic reforming of glycerol over "in situ" prepared zero-valent nano copper/P25, Appl. Catal. B: Environ. 202 (2017) 539–549.
[17] D. Li, J.C.C. Yu, V.H. Nguyen, J.C. Wu, X. Wang, A dual-function photocatalytic system for simultaneous separating hydrogen from water splitting and photocatalytic degradation of phenol in a twin-reactor, Appl. Catal. B: Environ. 239 (2018) 268–279.
[18] M. Li, H. Liu, Y. Song, J. Gao, Design and constructing of mutually independent crystal facet exposed TiO_2 homojunction and improving synergistic effects for photoelectrochemical hydrogen generation and pollutant degradation, Int. J. Energy Res. 42 (15) (2018) 4625–4641.
[19] A. Speltini, F. Gualco, F. Maraschi, M. Sturini, D. Dondi, L. Malavasi, et al., Photocatalytic hydrogen evolution assisted by aqueous (waste) biomass under simulated solar light: oxidized g-C_3N_4 vs. P25 titanium dioxide, Int. J. Hydrog. Energy 44 (8) (2019) 4072–4078.
[20] C.J. Chang, Y.G. Lin, H.T. Weng, Y.H. Wei, Photocatalytic hydrogen production from glycerol solution at room temperature by ZnO-ZnS/graphene photocatalysts, Appl. Surf. Sci. 451 (2018) 198–206.
[21] A. Castedo, E. Mendoza, I. Angurell, J. Llorca, Silicone microreactors for the photocatalytic generation of hydrogen, Catal. Today 273 (2016) 106–111.
[22] R. Jaswal, R. Shende, W. Nan, A. Shende, Photocatalytic reforming of pinewood (*Pinus ponderosa*) acid hydrolysate for hydrogen generation, Int. J. Hydrog. Energy 42 (5) (2017) 2839–2848.
[23] C. Ma, Y. Li, H. Zhang, Y. Chen, C. Lu, J. Wang, Photocatalytic hydrogen evolution with simultaneous photocatalytic reforming of biomass by Er^{3+}: $YAlO_3$/Pt–TiO_2 membranes under visible light driving, Chem. Eng. J. 273 (2015) 277–285.

Conclusion

Concluding remarks by the editors

In Chapter 1, Overview of Sustainable Fuel and Energy Technologies, the authors address the need for sustainable and renewable fuel and energy resources to minimize environmental pollution. They also discuss the methods of production, storage, and utilization of various sustainable fuel and energy resources. Chapter 2, The Second- and Third-Generation Biofuel Technologies: Comparative Perspectives deliberates on a comparative study of 2nd- and 3rd-generation biofuel technologies. Various biorefineries and feedstock required for bioethanol production are summarized. Processes for bioethanol production such as thermo-chemical conversion and bio-chemical conversion along with the microorganisms required are incorporated in this chapter. After that, cost estimation, environmental aspects, and SWOT analysis are performed for these biofuels. In Chapter 3, Biomass, Biorefinery, and Biofuels, the authors give an overview of biomass, biorefinery, and biofuels. They discuss traditional biofuels like biodiesel, bioethanol, and biobutanol as well as various methods of hydrocarbon-biofuel production such as hydrodeoxygenation of vegetable oil, fast pyrolysis of lignocellulosic biomass, and Fischer-Tropsch synthesis. Fuel additives improve the quality of fuel, therefore various fuel additives, for example, ɣ-Valerolactone, furanic compounds, 5-Ethoxymethylfurfural, Alkyl levulinates, glycerol acetals, and dimethyl ether are discussed in this chapter.

Hydropower is the most efficient and abundantly available renewable energy source. Several parameters influence the activity of the hydropower plant, including the water head, dams, turbine, discharge, and generators. Chapter 4, Hydropower Technology, this chapter elucidates various aspects of hydropower like dam, reservoir, turbines. The benefits and shortcomings of micro-hydropower plants are discussed in detail. Wind power is one of the primordial forms of energy known to humanity. The historical background of wind power is presented in Chapter 5, Wind Power Technology. It also explains how to estimate the available power in the wind. This chapter discusses the construction and working principle of several modern turbines. Readers will get an idea about site selection, the effect of wind speed on power generation, and different types of generators. Subsequently, the environmental impact, wind power economics, and offshore wind power are discussed. Chapter 6, Enabling Solar Photovoltaics Penetration in Highly Dependent African Fossil Fuel Markets, addresses the role of solar photovoltaic cell power and its suitability to replace the fossil-fuel market in African countries. The authors address the role and responsibilities of the government and individuals to use renewable and clean-energy resources. It also considers how solar energy utilization helps to mitigate CO_2 emission. It is observed that proper pricing and participation from the private sector makes these projects successful.

In Chapter 7, An Up-to-Date Perspective of Geothermal Power Technology, the authors describe the perspective of geothermal power technology. A preliminary survey of the regional area of interest is essential for any geothermal power development project. Various aspects such as geothermal exploration methods, geothermal fluids, geothermal reservoir, and well-drilling are described in this chapter. It also includes a feasibility study and risk assessment of geothermal power plants. The ocean contains a large amount of energy within it. Ocean waves bare tremendous potential as a source of renewable energy; it is enormous, reliable, and considered to have the highest energy density compared to wind and solar power. Chapter 8, Marine Power Technology—Wave Energy, discusses how wave energy is harvested using a wave-energy converter. Various types of wave energy concepts are explained along with the power take-off systems for such converters. The chapter is concluded with the future of these WEC applications.

Chapter 9, Renewable Hydrogen Production by Water Electrolysis, discusses hydrogen production methods from various renewable and clean sources. Water electrolysis is one of the most environment-friendly ways to produce hydrogen from renewable sources. This chapter explains the theoretical bases, important characteristics, and the leading technologies of water electrolysis (e.g., alkaline, PEM, and solid oxide) to produce hydrogen. A review of commercial electrolyzers with cost analysis and the main projects on their integration with renewable energy sources are also discussed. Chapter 10, Microbial Production of Hydrogen, explains the microbial biohydrogen technology via photosynthetic microorganisms, dark fermentation, and microbial electrolysis cells. The fundamentals of each technology along with principles and the role of microorganisms are included in this chapter. Different bioreactors and each type's advantages, disadvantages, potential, and limitations toward scaling-up are also discussed. Finally, methods of purification and storage of microbiologic-produced biohydrogen are explained. Chapter 11, Hydrogen Energy, includes production methods of hydrogen from various feedstocks such as water, fossil fuels, and biomass. The combustion properties of premixed hydrogen are analyzed. The characteristics of the flame structure, mechanism of flame instabilities, nexus between flame stretch and flame speed, explosion characteristics of hydrogen under laminar, and turbulence are discussed. Finally, the formation and emission of NOx during hydrogen combustion are reported.

Chapter 12, Multifaceted usage of Miniaturized Energy Technologies for a Sustainable Energy Harvesting, discusses the development of miniaturized power harvesters with capabilities of drawing high-density power from their vicinity. Triboelectric, Piezoelectric, Electromagnetic, and Thermoelectric energy harvesters are described in this chapter. The authors have highlighted the working principles, salient features, and further scope of these micro- to nanoscale technologies. The Internet of Things is also discussed in this chapter. In Chapter 13, Energy Efficiency: Developing a Low-Cost, Real-Time, SMS Feedback Tool for Off-Grid Household Electricity Consumption, the authors introduce a low-cost Real-Time SMS Feedback Tool for off-grid household electricity consumption. Energy efficiency is enhanced by informing users of their electricity consumption through a short text message on their phones. This chapter elaborates how this information can be leveraged to improve solar systems' performance by influencing user behaviors. The developed algorithm collects data from three main parts of the solar system, namely power generation, storage, and consumption. Hydrogen can be utilized as fuel most efficiently via a fuel cell. Chapter 14, Recent Developments in Hydrogen Fuel Cells: Strengths and Weaknesses, explains various aspects of the hydrogen fuel cell. Various applications of fuel cells such as stationary and transport applications are discussed. The techno-economic barriers of the hydrogen fuel cell are also discussed in this chapter. The authors conclude with the future prospects of the H_2 fuel cell.

An energy-efficient building helps us to save energy in many ways. Chapter 15, Energy Efficiency in Buildings, discusses how to construct an energy-efficient building, such as the utilization of solar irradiation to maintain a comfortable temperature inside the building. This chapter shows that efficient landscape design can be integrated with building design to save energy. It is concluded that good design can maintain a comfortable temperature in winter and summer efficiently. Finally, the concept of green building is described. Chapter 16, Toward Sustainable Long-Term Energy Planning for Cities: An Economic and Environmental Assessment of Sustainable Fuel Technologies in the City of Donostia-San Sebastián, includes a case study on the city of Donostia, San Sebastián. The authors performed an economic and environmental assessment of sustainable fuel technologies via long-term energy planning. The Life Cycle Assessment (LCA) rationale and the evaluating metrics derived from it, extend the sustainability assessment when discussing energy-planning scenarios at the city level. This chapter explains the adaptation of proposed methodologies based on the life-cycle approach for the comparison of energy transition scenarios of cities.

Chapter 17, Oxy-Fuel Power Cycles Promoting the Transition to Green and Sustainable Future in the Energy Sector, discusses the reduction of CO_2 emission from thermal power plants using various carbon-capture technologies. The authors have paid special attention to the oxy-fuel combustion technology of power generation. The chapter includes well-known cycles like Semi-closed oxy-fuel combustion combined cycle (SCOC-CC), E-MATIANT cycle, and Allam cycle. The chapter shows that efficiency is increased via thermodynamic analysis of these cycles. Finally, the information about the recent development of a pilot oxy-fuel power cycle is presented. Integrated photocatalytic H_2 generation and pollutant removal is a challenging way to develop sustainable fuel technology. Chapter 18, Integrated Photocatalytic Hydrogen Production and Pollutants or Wastes Treatment: Prospects and Challenges, elucidates various aspects of simultaneous hydrogen generation and wastewater purification. It is found that the selection of catalysts and sacrificial agents is the major challenge for this integrated method. This chapter shows that transition metal oxides are mostly used as photocatalysts, and various pollutants can be utilized as sacrificial agents.

Overall, the book elucidates various sustainable methods of producing clean fuel and energy resources. It is observed that the minimization of energy utilization is necessary for the development of sustainable fuel and energy technologies. This can be achieved by developing energy-efficient systems, suitable energy harvesting, and storage devices.

Suman Dutta
Chaudhery Mustansar Hussain

Index

Note: Page numbers followed by "*f*" and "*t*" refer to figures and tables, respectively.

A

Absorption cooling systems, 468
Absorption heat transformer (AHT), 11
Acceptor (A), 316–317
Accessibility of site, 92–93
Acetalization of glycerol, 78, 78*f*
Acetone-butanol-ethanol fermentation (ABE fermentation), 59–60
Acetyl coenzyme A (acetyl-CoA), 326–327
Active cooling, 468
Active solar techniques, 466–468. *See also* Passive solar techniques
  active cooling, 468
  district heating and cooling systems, 468
  solar collectors, 467–468
Active stall control, 164
Actuators, 370
Aerobic respiration, 315
Aerodynamic forces, 152–153
  drag-type wind turbine, 152
  lift-type wind turbine, 152–153
Aesthetics, 164–165
Africa
  energy landscape, 174
  energy transition in, 176–185
  proven oil reserves for select African countries, 175*t*
Agricultural/agriculture, 181
  and hardwood waste, 37
Agro residue, 6
Air separation unit (ASU), 522
Aircraft architecture, 432
Alcohol, 55–56
Aldol-condensation, 70–71, 70*f*
Algae, 41
  conversion to biofuels, 42, 43*f*
Algal biomass, extraction of oil from, 42
Algal oil extraction technique, 41–43, 42*f*
  conversion of algae to biofuels, 42
  extraction of oil from algal biomass, 42
  steps in production of biofuels from algae, 42–43
Algeria, energy transition in Africa, 178–179
Alkaline electrolysis, 288, 302
  components of alkaline electrolysis cell
    electrodes and catalysts, 283–284
    electrolyte, 285–286
    separators, 284–285
Alkaline electrolyzers, 450
  main cell designs, 286–287
  stack configuration, 287–288
Alkaline exchange membrane electrolysis cell, 287*f*
Alkaline water electrolysis, 340
  components of alkaline electrolysis cell, 283–286
  configuration of alkaline electrolyzer, 286–288
  developments, limitations, and future perspectives, 288–289
  technology and operating principle, 280–282
Alkaline water electrolyzer (AWE), 450
5-Alkoxymethylfurfurals, 77, 77*f*
Alkyl levulinates (AL), 18, 77–78, 78*f*
Allam cycle, 522–523, 532–535, 533*f*, 534*f*
Alternating current machines (AC machines), 155–156
Ammonia fiber explosion (AFEX), 36–37
Ammonia recycle percolation (ARP), 36–37
Ammonia water, 37
Angola
  electricity access rates and renewable energy capacity, 182*f*
  energy transition in Africa, 182–183

Anionic exchange membrane (AEM), 280, 285
Anthropogenic CO_2, 515
Application Programming Interface (API), 416
Appropriateness, 470–473
Arabinosidases, 37–38
Arc gravity dam, 100
Arch dam, 99
Archaebacteria, 317
Archimedes screw turbine, 10
Archimedes wave swing (AWS), 245, 246*f*
Archimedes' principle, 354
Artificial storages, 96
Atmospheric nitrogen fixation, 320
Atomic force microscopy (AFM), 379
Attenuator(s), 243–244
  wave energy converter, 247–249, 248*f*
Auctions, 190
Axial thrust and torque on blades, 133–134

B
Backup generators (BUGs), 174–176
Bacteria, 317
Bagasse, 37
Balance of plant (BoP), 300–302, 301*f*, 442–443
Ballard Power Systems, 436–438
Bank-side reservoirs, 103
Barrage dam, 100
$BaTiO_3$ (BTO), 373
Bharat Stage VI norms in India, 4
Bidirectional impulse turbine, 254–257, 255*f*
  principle of operation of impulse turbine, 255–257
Binary cycle power plants, 231–232, 232*f*
Bio-butanol, 56–57
  ABE fermentation, 59–60
  economics, 61
  fermentation engineering, 60
  fuel properties, 59
  separation of ABE, 60–61
Bio-oil, 63–64
  properties of bio-oil and hydroprocessed bio-oil, 64*t*
Bio-olefins, 68
Biocatalysts, 17–18
Biochemical pathway for biofuel generation, 35–36, 36*f*
Biodiesel, 54–56
  catalysts, 55
  process and economics, 55–56
  process variables, 55
  specification, 56
Bioethanol, 30–31, 40–41, 56–59
  challenges and economics, 58–59
  CO_2 cycle for, 41*f*
  enzymatic saccharification and fermentation, 38–39
  fermentation technique of lignocellulosic hydrolysates, 38
  fuel properties, 56–57
  generation, 40–41
  inhibitors and impact on microorganisms, 37–38
  pretreatment, 36–37
  production, 31–32, 34–39, 58
    biochemical conversion, 35–36
    thermochemical conversion, 34–35
  raw feedstock for, 34
Biofuels, 32–33, 41, 51, 53–54, 54*f*
  production from algae, 42–43
  traditional
    bio-butanol, 56–57
    bio-ethanol, 56–59
    biodiesel, 54–56
Biological routes for fuel and chemicals production, 17–18
Biological water–gas shift, 323
Biomass, 12–13, 51–53, 54*f*, 63–64, 542
  biomass-derived fuels, 18
  biomass-to-liquid process, 68
  chemical structures of, 52*f*
  combustion and challenges, 14
  gasification, 14–15
  lignocellulose, 53
  pyrolysis, 15–16
  sugar and starch, 52
  triglyceride, 52
Biophotolysis, 321–322
  biophotolysis-based hydrogen production, 321
  direct, 321
  indirect, 322
Bioreactors
  fermentors, 330

MEC configurations, 330–331
photobioreactors, 329–330
Biorefinery technology, 30–32, 32*f*, 53
Biorenewable energy resources, 12–13
Bipolar
configuration, 288
plates, 292–293, 293*f*, 442–443
Bird collisions, 166
Blade's angle of attack, 162, 164*f*
Blades, axial thrust and torque on, 133–134, 135*f*
Blue-green algae. *See* Cyanobacteria
Bottom-up approach, 488
Braking jet, 108
Bucket and runner, 108
Bulb turbine, 111, 111*f*
Butanol, 59
Butyl butyrate (BB), 17–18
Butyl levulinate (BL), 18
Butyric acid fermentation, 324

C

Candida, 38
Capital expenditures (CAPEX), 305, 499–501
Capturing wind energy and challenges, 7–8
Carbon
capture after combustion, 516
capture technologies for thermal power plants, 516–520
postcombustion, 516–518
precombustion, 519–520
C–C bond formation reactions, 70–72
aldol-condensation, 70–71
HAA reaction, 71–72
emissions, 177–178, 458, 514
distribution, 514*f*
reducing CO_2 emissions in power production industry, 515–516
intensity, 184–185
pricing and monitoring, 188
Carbon dioxide (CO_2), 299, 533–534
emissions, 241
oxy-fuel combustion cycles with carbon dioxide recirculation, 520–535, 521*f*, 522*f*
Carbon monoxide (CO), 14, 516
Carbon nanotube (CNT), 383–384
Carnot cycle efficiency, 431
Casing, 108
Catalytic/catalysts, 55–56, 62–63, 67, 283–284, 290–291, 442
decomposition, 339–341
routes for fuels and chemicals production, 18
Category-based allocation, 485–486
CBH. *See* Exo-cellobiohydrolase (CBH)
Cellulase enzyme, 58
Cellulosic bioethanol generation, 39
Central biomass boilers (CBBs), 498–499
Cermet, 297
Chemical oxygen demand (COD), 328–329
Chemical pretreatment methods, 16
Chemical solvents, 518
Chloroplast-ferredoxin-hydrogenase system, 321
City baseline modeling, 488
City contextualization, 487
City energy planning, conventional approaches in, 485–486
City Protocol, 487
City-related LCA, 485
"Citycentric" approach, 483
CITYKEYS project, 487
Climate change, 124
mitigation, 187
Climate models, 514
Climate resilience, 187
"Closed-loop" frameworks, 96–97
Clostridial-type fermentation, 324, 325*f*
Coal, 3
Coastal reservoirs, 103
Cogeneration technology, 515–516
Combined heat and power (CHP), 436
Combustion, 14
properties of hydrogen, 346–359
flame stretch and speed, 356–359
flame structure and instabilities, 347–356
Commercial electrolyzers, 302–304, 303*t*
Communication mode, 414
Community engagement, 412
Compact fluorescent lamps (CFL), 192
Composite climate, 470
Compound annual growth rate (CAGR), 451
Compressed natural gas (CNG), 446
Concentrating PV, 5–6
Conceptual model, geothermal system, 216

Concrete face rockfill dam (CFR dam), 101
Constant speed operation, 157–158, 158*f*
  strategies for, 158–161
    fixed speed system, 159
Constant-angle dams, 99
Constant-radius dam, 99
Construction-based development, 104
Consumer-based allocation method, 485–486
Continuous stirred tank reactor (CSTR), 330
Conventional approaches in city energy planning, 485–486
Cool dry climate, 470
Cost
  estimation, 39–40
  perspective, 305–306, 305*f*
  prediction, 39–40
Coulombic efficiency (CE), 328–329
Counterparty risk, 190–191
Cross-flow turbines, 109–110, 110*f*
  design structure, 109–110
  operational activity, 110
Crude oil resources, 29–30
Crystalline silicon, 5–6
Cumulative GWP reduction of energy transition scenario ($CGWPR_{TS}$), 495
Cumulative local renewable electricity generation increase ($CLREG_{TS}$), 495
Cumulative net present cost of public/private companies ($CNPC\text{-}PPC_{TS}$), 495
Cumulative net present cost-social ($CNPC\text{-}S_{TS}$), 495
Cumulative net present value of energy transition scenario ($CNPV_{TS}$), 495
Cumulative nonrenewable primary energyPE use reduction ($CNR\text{-}PER_{TS}$), 495
Cup anemometer, 137–138, 138*f*
Cut-in wind speed, 129
Cut-out or furling wind speed, 129
Cyanobacteria, 41, 316

D

Dam(s), 99–102
  arc gravity, 100
  arch, 99
  barrage, 100
  CFR, 101
  earthfill, 101–102
  embankment, 101
  fixed-crest, 102
  gravity-based, 100
  rockfill, 101
Dammed valley, 102
"Danish" concept, 159
Dark fermentation, 324–327, 331
Darrieus type wind turbine, 148–149, 149*f*
Darrieus-Landau instability (D-L instability). *See* Hydrodynamic instability
DC generator, 155–156, 156*f*
Decarbonization, 306–307
Decentralized frameworks, 98
Design-Builder, 460–461
Diaphragms, 284–285
Diesel, 51
1,1-Diethoxyethane, 77
Dimethyl ether (DME), 53–54, 79
2,5-Dimethylfuran (DMF), 18, 53–54, 76
2,5-Dimethyltetrahydrofuran (2,5-DMTHF), 53–54, 76
Direct biophotolysis, 321
Direct electrical drive systems, 258–259, 259*f*
Direct gain system, 465
Direct mechanical drive systems, 257–258
Direct methanol fuel cell (DMFC), 440
District heating with natural gas and biomass boilers (DHNGB), 498–499
DOE. *See* US Department of Energy (DOE)
Domestic hot water (DHW), 503–504
Donor (D), 316–317
Donostia-San Sebastián city
  case study and baseline assessment, 496–498
  economic and environment life cycle impact assessment of scenarios, 501–507
  long-term energy scenario for, 498–501
Double flash power plants, 230, 230*f*
Doubly-fed induction generator (DFIG), 153, 260
  for variable speed operation, 160–161
  wind turbine system, 161*f*
Downwind wind turbine, 149–151
Draft tube, 115
Drag-type wind turbine, 152
Drilling production wells, 202
Drop-in fuels, 435
Dry steam technology, 228–229, 228*f*

Drying pond fly ash, 6
Dutch type wind turbine, 142–144, 145*f*
Dye(s), 542–545
Dye-sensitized photocatalytic water splitting, 542–545
Dynamic payback period (DPP), 495, 503–504

E
E-MATIANT cycle, 528–530, 529*f*
influence of turbine inlet temperature on, 531*f*
with turbine cooling, 530*f*
with water infection in coolant flow, 532*f*
e-QUEST, 460–461
Earth's climate system, 513
Earth's fossil fuel resources, 124
Earthen dam. *See* Earthfill dam
Earthfill dam, 101–102
Economic(s), 61
and commercial processes, 63
efficiency, 96–97
impact assessment of low-carbon energy transition scenarios, 496–507
life cycle impact assessment of scenarios, 501–507
Efficiency of power unit, 519–520
Electrically excited synchronous generator, 156–157
Electricity generation process, 89–90
Electrification, 173–174
Electrocardiography (ECG), 394
Electrodes, 283–284, 290–291, 296–297
Electrolysis, 339–341
electrolysis system technology
manufacturers and review of commercial electrolyzers, 302–304
perspective of costs, 305–306
system lay-out and BoP, 300–302
reaction, 273
of water, 448
hydrogen generation by, 448
international scenario, 448–450
types of water electrolyzer technology, 449*t*
Electrolyte, 285–286, 291–292, 296
Electrolytic splitting of water, 448
Electrolyzer(s), 448
models, 276
Electromagnetic energy harvesters, 386–391. *See also* Piezoelectric energy harvesters; Thermoelectric energy harvesters
applications, 387–391
future perspectives, 391
Electrons (e^-), 542
acceptors, 315
donor, 315–316
Embankment dam, 101
"Embodied" emissions, 505
Endoglucanase (EG), 37–38
Energy, 3, 458
audits, 458–462
awareness, 420–423
conversion, 228–234
binary cycle power plants, 231–232
development of geothermal power plant, 233–234
dry steam technology, 228–229
flash steam power plants, 229–231
flash/binary combined cycle, 232
crops, 51–52
energy-conserving landscape strategies, 469
energy-Plus, 460–461
harvesters
electromagnetic, 386–391
piezoelectric, 378–386
thermoelectric, 392–398
triboelectric, 371–378
management program, 461
model, 460–461
modeling, 458–462
and power from wind, 125–126
axial thrust and torque on blades, 133–134
extractable limits of wind power, 129–133
modern wind turbine, 126–127
power output and efficiency considerations, 127–128
power *vs.* wind speed characteristics, 128–129
solidity, 134
TSR, 135–136
scenario, 29–30, 30*f*
sector, 174
sources, 241

Energy (*Continued*)
storage, 400
use in buildings, 458–462
Energy Act (2006), 179–180
Energy Conservation Building Code (ECBC), 478
Energy consumption, 458
consumption information access in off-grid households, 412–414
energy efficiency, 411*f*
frequency and time of notification, 425
general process setup, 418–426
input section, 418
output section, 419–423
priority 1 message, 420
priority 2 message, 420
priority 3 message, 420–422
priority 4 message, 423
processing section, 418–419
global population share without grid access, 410*f*
potential benefits, 426
solar power system, 414–417
variation of least-cost option for energy access, 411*f*
Energy efficiency (EE), 409–410, 520–535
Energy research and development association (ERDA), 450
Energy Sector Management Assistance Program (ESMAP), 174
Energy system modeling (ESM), 484
Energy transition process (ET process), 483
in Africa, 176–185
Algeria, 178–179
Angola, 182–183
carbon intensity, 178*f*
carbon intensity and Paris agreement, 184–185
country insights, 177–183
electricity supply by type, source, and scenario in sub-Saharan Africa, 177*f*
GDP per capita growth for selected African countries, 179*f*
Kenya, 179–180
Nigeria, 180–182
subsidies as enablers of fossil fuel dependency, 183–184
scenarios, 488
Energy-conserving opportunities (ECOs), 461
Energy-efficient buildings, 457
active solar techniques, 466–468
building in wake of energy crisis, 457–458
concept of green buildings, 476–480
energy use in buildings, 458–462
energy audit, 461–462
energy model, 460–461
energy-efficient landscaping, 468–476
passive solar techniques, 462–466
Energy-efficient landscaping, 468–476
climate and appropriate landscaping, 469–470
for comfort in summer and winter, 473–474
achieving winter sun penetration, 474
blocking cold winter winds, 474
capturing summer breezes, 474
points to achieve summer shading, 473
reducing glare and ground temperature, 473
methods and appropriateness, 470–473
growth, 471–473
selecting and planting trees and shrubs, 470–471
shape characteristics, 471
reducing mechanical heating and cooling costs, 475
water and role in energy-efficient landscaping, 475–476
Engineered geothermal systems (EGS), 201
Enterobacter, 319
Enterobacterial-type fermentation, 324, 325*f*
Entrained-flow gasification technology, 342
Environment(al)
environment-friendly approach, 43–44
impact, 43–44
assessment of low-carbon energy transition scenarios, 496–507
and public perception, 164–166
life cycle impact assessment of scenarios, 501–507
metrics, 424–425
remediations, 376
sustainability, 177–178
Enzymatic hydrolysis, 37–38

Enzymatic saccharification and fermentation, 38–39
Equilibrium conditions, 273, 273*f*
Erythrosin B dye, 542–545
Esterases, 37–38
Ethanol, 17–18, 326–327
5-Ethoxymethylfurfural (EMF), 18, 77
Ethyl levulinate (EL), 18, 77
Eukaryotic phototrophic microorganism, 324
Euro VI in Europe, 4
European Commission (EC), 486
European Union (EU), 33
Exo-cellobiohydrolase (CBH), 37–38
Exploratory drilling exploratory drilling, 218–219
Explosion characteristics of hydrogen gas, 359–361
Extractable limits of wind power, 129–133
  velocity and pressure profiles, 130*f*
Extraction of oil from algal biomass, 42
ExxonMobil. *See* Mobil Oil

F

Facultative aerobic hydrogen producing microorganisms, 319
Faraday efficiency, 278
Faraday's law, 388–389
  of electrolysis, 277
  of electromagnetic induction, 386–387
Fast pyrolysis, 16
Fee-for-service models, 193
Feedback
  messages, 414
  systems, 426
Feedstocks, 58–59
  for biofuel production, 30–31, 31*f*
Fermentation, 346
  engineering, 60
  technique of lignocellulosic hydrolysates, 38
Fermentative hydrogen production, 324, 330–331
Fermentors, 330
Ferredoxin, 315–316
  ferredoxin-linked redox reductants, 319
Finite element analysis, 388–390
First-generation biorefineries, 32–33
  process of biofuel production in first-generation biorefinery, 33*f*
Fischer–Tropsch reactions, 33
Fischer–Tropsch synthesis (FTS), 53–54, 67–68
  catalysts, 67
  commercial processes and economics, 68
  process conditions, 67–68
  reaction, 13
Fixed speed system, 159, 159*f*
Fixed tariffs for small producers, 189–190
  Kenya's FiT structure for projects, 190*t*
Fixed-crest dam, 102
Fixed-speed generators, 260–261
Flame
  flame-front, 354
  stretch and speed, 356–359
  structure and instabilities, 347–356
Flaming combustion, 14
Flash
  flash/binary combined cycle, 232, 233*f*
  steam power plants, 229–231
Flash pyrolysis. *See* Fast pyrolysis
Fluidized-bed gasification technology, 342
Fly ash, 6
Food processing, construction, and technology, 181
Forest-based woody by-products, 34
Fossil fuel(s), 51, 174, 369
  dependency, 183–184
  power plants, 515–516
Fossil-derived crude oil, 3
Francis turbine, 10, 114–116, 115*f*
  applications, 115–116
  draft tube, 115
  guide and stay vanes, 114
  operation activity, 115
  runner blades, 114
  spiral packaging, 114
Free-stream turbine. *See* Kinetic turbine
Frequency of notification, 425
Friction coefficient, 137
  for terrain types, 138*t*
Frozen center dam, 101–102
Fuel cell driven electric vehicles (FCEV), 444
Fuel cell vehicles (FCV), 441, 444–446
Fuels, 3, 431
  additives, 72–79
    5-EMF, 77
    alkyl levulinates, 77–78

Fuels (*Continued*)
DME, 79
furanic compounds, 72–76
glycerol acetals, 78
GVL, 72
properties of, 73*t*
cells
based on hydrogen, 431–451
for stationary applications, 436–438, 437*t*
properties, 56–57
bio-butanol, 59
of traditional biofuels, 57*t*
Functional unit (FU), 485–486, 493
Furanic compounds, 72–76, 75*f*
2-MF, 74
2-MTHF, 74–76
2,5-DMF, 76
2,5-DMTHF, 76
Furfural, 53–54

G

Gap cell, 286
Gas conversion, 519
Gasification, 14–16, 341–344
Gasoline, 51
Generators
wind power plants on basis of, 161
working principle of, 153, 155*f*
Geochemical data, 208, 209*t*
Geothermal
energy, 10–11, 201
exploration, 204
fluids, 206
power plant development, 233–234
reservoir, 206
Geothermal power technology. *See also* Marine power technology
classification of geothermal systems, 202*t*
energy conversion, 228–234
exploration stage, 204–217
conceptual model, 216
geochemistry, 207–208
geology, 206–207
geophysics, 208–210
lab measurements on rock samples, 210
mechanical stress field, 210–216
numerical model, 216–217
exploratory drilling, 218–219
field development, 223–226, 225*f*
prefeasibility stage, 217–218
preliminary survey stage, 202–203
project planning, 223
reservoir monitoring at start-up and operation, 226–228
well drilling, 219–223
Gibbs free energy (ΔG), 274, 278–280, 431
Glass fiber–reinforced polyphenylene sulfide, 284
Global warming, 40–41, 124, 458, 513
Global warming potential (GWP), 486, 488
"Globalization of architecture", 457
Glowing combustion, 14
β-Glucosidase (BGL), 37–38
Glycerol, 53–54
acetals, 78
Government of Kenya (GoK), 180
Government-backed bulk electricity trader (NBET), 189–190
Grafana software, 415–416
Gravitation water vortex (GWV), 93–94
Gravitational vortex, 93–94
type of water turbine, 94
Gravity-based dam, 100
Green buildings, 476–480
green rating systems prevailing in India, 478–479
green roofs, 479–480
Green house gas emissions, 432
Green rating systems prevailing in India, 478–479
Green roofs, 479–480, 479*f*
Green technologies, 514
Greenhouse gas emissions (GHG emissions), 40–41, 409, 438–439, 513–514
GRIHA, 478
Gross domestic product (GDP), 177–178, 492
Guide vanes (GVs), 114, 251–252

H

Heat pumps (HPs), 498–499
Heat recovery steam generator (HRSG), 523
Heating, ventilation, and air-conditioning controls (HVAC controls), 460, 477
Heavy freight trucks (HFTs), 444
Hemicellulase enzyme, 58
Hemicellulose, 53

Heterocystes, 316
Heterogeneous catalysts, 70–71
Heterogeneous photocatalysis, 542
Heteropolyacids, 77–78
Hexose sugar, 58
High temperature alkaline electrolyzers, 289
High-pressure separator, 230
High-temperature
 biomass conversion technologies, 13–16
  biomass combustion and challenges, 14
  biomass gasification, 14–15
  biomass pyrolysis, 15–16
 electrolyzers, 302
High-temperature FT (HTFT), 67–68
High-temperature fuel cell (HTFC), 435
Holes (h^+), 542
Homogeneous catalysts, 78
Hoover dam, 90–91, 91*f*
Horizontal axis wind turbine (HAWT), 142, 143*f*, 150*f*, 151–152
Hot-dry climate, 470
Household energy, 409
Human–machine interface systems, 375–376
Hydraulic PTOs, 257, 258*f*
Hydraulic retention time (HRT), 327
Hydrocarbons, 343–344
 biofuels, 53–54
  C–C bond formation reactions, 70–72
  fast pyrolysis of lignocellulosic biomass, 63–64
  Fischer–Tropsch synthesis, 67–68
  HDO of vegetable oil, 58
  oligomerization of olefins, 68–70
 fuels, 435
 hydrocarbon-analogous biofuels, 53–54
Hydrochloric acid (HCl), 341
Hydrodeoxygenation (HDO), 53–54
 of vegetable oil, 58
  catalysts, 62–63
  economics and commercial processes, 63
  fuel properties of green diesel, 62*t*
  reaction mechanism, 58–59
  supports, 63
Hydrodynamic instability, 350–351
Hydroelectricity, 458
Hydrogen (H_2), 14, 271, 444, 516, 541
 combustion properties of hydrogen, 346–359
 generation, 541
 hydrogen-producing microorganisms, 317–321
  facultative aerobic hydrogen producing microorganisms, 319
  obligatory anaerobic hydrogen producing microorganisms, 317–319
  photosynthetic hydrogen producing microorganisms, 320–321
 neutrality, 450–451
 problems relating to using, 359–362
 production, 271–272, 272*f*
  electrolysis and catalytic decomposition, 339–341
  gasification and reforming, 341–344
  technologies for hydrogen gas, 339–346
  thermochemical and biological process, 344–346
Hydrogen evolution reaction (HER), 272–273, 279–280, 279*f*
Hydrogen fuel cells, 431–451
 classification of fuel cells, 433*t*
 cost reduction and hydrogen neutrality, 450–451
 history and utilization, 431–432
 hydrogen fuel, 432–435
 infrastructure and commercialization, 446–450
 paramount importance, 435–436
  for stationary applications, 436–438, 437*t*
  for transport applications, 438–442
 road ahead, 444–446
 strategic energy plan, 452
 techno commercial barriers, 442–444
 technology barriers, 452–453
 world view of fuel cell, 444
Hydrogenases, 316–317
 biochemical classification, 318*t*
Hydrolysis, 12–13
Hydropower energy capturing and challenges, 8–10
Hydropower plant on power scale, 103–104
Hydropower technology, 4–5, 90–99. *See also* Wind power technology

Hydropower technology (*Continued*)
benefits and shortcomings of micro hydropower plants, 104–106
classification of hydropower plant on power scale, 103–104
construction-based development, 104
dams, 99–102
gravitational vortex, 93–94
impoundment hydropower, 90–91
pumped storage, 94–99
reservoirs, 102–103
ROR hydroelectricity plant, 91–93
small hydro, 106–107
turbines, 107–116
Hydropower turbines, 10
Hydroprocessing of bio-oil, 65–67
Hydroxyalkylation–alkylation reaction (HAA reaction), 70–72, 71*f*
5-Hydroxymethylfurfural (HMF), 53–54
Hyperthermophilic archaebacterium *Pyrococcus furiosus*, 319

I

Ice accumulation, 166
IES Virtual Environment, 460–461
Impact assessment
indicators, 495
methodology, 491
Impoundment
agency, 90–91
hydropower, 90–91
conventional impoundment dam, 90*f*
Impulse turbines, 10, 107, 254–255
principle of operation of, 255–257
Independent power producers (IPPs), 190–191
India, green rating systems prevailing in, 478–479
Indian wave energy plant, 244*f*
Indirect biophotolysis, 322
Indirect gain system, 465–466
Individual natural gas boilers (INGBs), 498–499
Indoor space planning, 464
Induction generator, 157, 158*f*
Inedible oil, 52
Inequality, 483–484
Influx database, 416
Influx DB, 415–416
Integrated photocatalytic hydrogen production
pros and cons of integrated systems, 545
prospects and challenges, 545–547
dual function reaction conducted in twin-reactor, 547*f*
simultaneous pollutant degradation and hydrogen evolution, 546*t*
reactors for integrated system, 545
utilization of pollutants or waste as sacrificial agents, 542–545
Integrated-design approach, 460
Interconnectors, 297–298
Internal combustion engines (IC engines), 431
International Energy Agency, 458, 514
Internet of things (IoT), 370, 398–400
comparison of METs for powering IoT devices, 399*t*
Investment growth in off-grid industry, 412–413
Irradiance, 414

J

Jet fuel, 51

K

Kaplan turbine, 106, 111–114, 113*f*
applications, 113–114
operational activity, 112–113
Karana
device, 416
framework, 415–416
Kenya, energy transition in Africa, 179–180
Kinetic battery model (KiBaM), 415
Kinetic energy (KE), 124
Kinetic losses and overpotentials, 275–277
Kinetic turbine, 10, 116
KNN-based alkaline niobate particles, 385
Kyoto Protocol, 514

L

Lactate, 326–327
Land installed marine energy transformer (LIMPET), 249

Landscaping methods, 469–473
Lanthanum manganite ($LaMnO_3$), 297
Leadership in Energy and Environmental Design rating system (LEED rating system), 460, 478
Least-developed countries (LDC), 182
Levulinic acid (LA), 53–54
Lewis number, 353–354
Life cycle assessment (LCA), 40–41, 484–485
  innovative LCA adaptation for city planning, 492–496
Life cycle costing (LCC), 484
Life cycle sustainability assessment (LCSA), 484
Lift-type wind turbine, 152–153, 154*t*
Light-emitting diodes (LED), 370
Lignin, 53
Lignocellulose, 53
Lignocellulosic biomass, 12–13, 34
  pyrolysis of, 63–64
    hydroprocessing of bio-oil, 65–67
    process conditions, 65
    reactors and commercial processes, 65
Lignocellulosic feedstock, 30, 37
Lignocellulosic hydrolysates, 38
Lignocellulosic raw feedstock, 36
Lipases, 17–18
Liquid biofuel generation, 35–36
Liquid dissolving agents, 516
Liquid hot water (LHW), 36–37
Liquid–gas diffusion layers (LGDL), 292
Lithium-ion batteries (Li–ion batteries), 369
Lithium-polymer batteries (Li–polymer batteries), 369
Long-chain hydrocarbons, 18
Low-carbon electricity, 452
Low-embodied energy, 457
Low-grade heat production, 341
Low-temperature biomass conversion processes, 17–18
  biological routes for fuel and chemicals production, 17–18
  catalytic routes for fuels and chemicals production, 18
Low-temperature FT (LTFT), 67–68
Low-temperature fuel cells (LTFC), 435

M
Marine energy, 11–12
Marine power technology. *See also* Geothermal power technology
  classification of ocean energy, 242*f*
  future of wave energy, 262
  PTO system for WECs, 250–262
  types of WECs, 243–250
  wave energy, 242–243
Material flow analysis, 485
MATIANT cycle, 522–523, 527–530, 528*f*
Mechanical stress field, 210–216
Membrane-electrode assembly (MEA), 290–292, 291*f*, 442–443
  electrodes and catalysts, 290–291
  electrolyte, 291–292
Membranes, 284–285
Mesophilic species, 322–323
Mesoporous catalyst, 63
Message interpretation, 414
Metalloenzymes, 315
Methane, 343
Methanol, 79
Methyl levulinate (ML), 18
Methyl orange, 542–545
Methyl red, 542–545
Methylene blue, 542–545
2-Methylfuran (MF), 18, 53–54, 74
2-Methyltetrahydrofuran (2-MTHF), 53–54, 74–76
Methyltrophic bacteria, 319
Micro hydropower plants, 104–106
  conditions for, 105–106
  turbines utilized in micro hydro plants, 106
Micro-hydropower plants, 104
Microbial electrolysis cell (MEC), 327–329
  configurations, 330–331
  typical dual-chamber MEC, 328*f*
Microbial fuel cell (MFC), 327
Microbial hydrogen production, 315–317
  biophotolysis, 321–322
  bioreactors, 329–331
  dark fermentation, 324–327
  hydrogen-producing microorganisms, 317–321
  MEC, 327–329
  microbial water–gas shift, 322–323

Microbial hydrogen production (*Continued*)
molecular basis of microbial hydrogen production, 315–317
hydrogenases, 316–317
nitrogenases, 315–316
photofermentative hydrogen production, 323–324
processes of microbial hydrogen production, 321–329
storage and purification, 329–330
Microbial metabolism, 315
Microbial water–gas shift, 322–323
Microclimate, 470
Microorganisms
enzymatic hydrolysis, 37–38
inhibitors and impact on, 37–38
Microwave-assisted reaction, 55
Miniaturized energy technologies (MET), 369
Internet of Things, 398–400
types, 371–398
electromagnetic energy harvesters, 386–391
piezoelectric energy harvesters, 378–386
thermoelectric energy harvesters, 392–398
triboelectric energy harvesters, 371–378
MIRAI, 440–441
Miscanthus giganteus, 34
Mitigation, climate change, 187
Mixed-acid fermentation, 324
Mobil Oil, 69–70
Mobile phone, 413–414
Modified MATIANT cycles with turbine coolant precooling, 530–532
Modified semiclosed oxy-fuel combustion combined cycle, 524–527
preliminary cooling of coolant by water injection, 526*f*
preliminary cooling of coolant in surface heat exchanger, 525*f*
SCOC-CC performance with water injection, 527*f*
Molecular hydrogen production, 315
Monetary-based allocation method, 485–486
Monopolar configuration, 288
Monosaccharides, 33, 37
Multibladed type wind turbine, 145, 146*f*
Multicriteria decision analysis (MCDA), 484, 507
Multicriteria impact assessment of energy transition scenarios, 488
Multilayered TENG device, 375–376

N
National Institute of Ocean Technology (NIOT), 242–243
Natural gas, 342, 458
Neodymium iron boron (NdFeB), 390–391
Newton's second law, 131
Nickel foams (Ni foams), 284, 284*f*
Nigeria, energy transition in Africa, 180–182
Nigeria Electrification Project (NEP), 181–182
Nitrogenases, 315–316, 345
Noise pollution, 164
Noncondensable gas (NCG), 231
Nonedible biomass, 30
Nonlignocellulosic biomass, 12–13
Nonrenewable energy consumption, 40–41
NOx emissions of hydrogen gas, 361–362
Nuclear fuels, 4–5
Numerical model, 216–217
with geological layers, 217*f*

O
Ocean energy, 241, 242*f*
Ocean Power Technologies (OPT), 247
Off-grid
consumption information access in off-grid households, 412–414
feasibility of SMS-based feedback tool in rural Africa, 413–414
investment growth in off-grid industry, 412–413
mode of communication and message interpretation, 414
household energy, 411–412
PV for improved electricity access, 191–192, 192*f*
systems, 410
Offshore system, 11–12

Offshore wind
energy, 166
farm, 166*f*
Ohmic resistance, 284
Oil
extraction from algal biomass, 42
producing African countries, 186–187
Olefins, oligomerization of, 68–70
Oligomerization of olefins, 68–70
commercial processes, 69–70
process parameters, 68–69
zeolite catalyst, 69
One-bladed wind turbine, 146–147, 147*f*
One-dimensional nanomaterials (1D nanomaterials), 380–381
"One-size-fits-all" approach, 414
Operational expenditures (OPEX), 305, 499–501
Organic acids, 323
Organic PV, 5–6
Organic Rankine cycle (ORC), 11
Oscillating water column (OWC), 242–243, 252*f*
Oscillating water column–wave energy converter (OWC–WEC), 242–243, 249–250, 250*f*. *See also* Wave energy converter (WEC)
PTO system for, 251–257
bidirectional impulse turbine, 254–257
wells turbine, 252–254
Overall efficiency, 278
Overpotentials, 275–277
51/49 Ownership policy of Algeria, 193
Oxidation, 541–542
agents, 519
Oxy-fuel combustion, 516, 519
Oxy-fuel power cycles
carbon capture technologies for thermal power plants, 516–520
oxy-fuel combustion cycles with carbon dioxide recirculation, 520–535, 521*f*, 522*f*
Allam cycle, 532–535
MATIANT cycles, 527–530
modified MATIANT cycles with turbine coolant precooling, 530–532
modified semiclosed oxy-fuel combustion combined cycle, 524–527
SCOC-CC, 523–524
reducing CO_2 emissions in power production industry, 515–516
Oxygen evolution reaction (OER), 272–273

P
Paleoclimate, 514
Palladium, 331
Paris Agreement, 184–185, 483–484, 514
breakdown of emissions and climate commitments, 185*t*
Passive interventions for buildings (PIFBs), 498–499
Passive solar techniques, 462–466. *See also* Active solar techniques
direct gain, 465
indirect gain, 465–466
passive solar heating design considerations
building shape and orientation, 463–464
indoor space planning, 464
openings, 464–465
site consideration, 462–463
sunspaces, 466
Pay-as-you-go model (PAYG model), 193, 412–413, 426
Pelton turbines, 10, 105*f*, 108–109
braking jet, 108
bucket and runner, 108
casing, 108
nozzle and flow regulation, 108
working of, 109
Pentose sugar, 58
Permanent magnet DC (PMDC), 155–156
Permanent magnet synchronous generators (PMSGs), 153, 260
Permeability, 206
Petroleum refinery, 31, 53
Phlogiston theory, 339
Phosphotungstic acid, 77–78
Photobioreactors, 329–330
Photocatalytic/photocatalysts, 541–542
processes, 272
twin-reactor, 545, 547
water splitting, 541–545
Photochemical efficiency (PE), 324
Photoexcitation, 541
Photofermentation process, 323–324, 331

Photoreactors, 545
Photosynthetic hydrogen producing microorganisms, 320–321
Photosynthetic organisms, 345
Photovoltaics (PV), 179–180
  application, 5
  PV-wind hybrid configuration, 167
  system, 414
Pichia, 38
Piezoelectric energy harvesters, 378–386. *See also* Electromagnetic energy harvesters; Thermoelectric energy harvesters; Triboelectric energy harvesters
  applications, 379–385
  future perspectives, 385–386
Piezoelectricity, 378–379
Piston, 257
Pitch-controlled turbines, 162
Platinum (Pt), 291
PLEEC project, 487
Point absorbers, 247, 248*f*
Polarization curve, 276–277, 276*f*
Policy making, 486
Pollutants
  missions, 271
  utilization as sacrificial agents, 542–545
  ZnO
    properties and photocatalytic hydrogen evolution capacity of, 544*t*
    properties and photocatalytic pollutant degradation efficiency of, 543*t*
Poly (3-hydroxyalkanoate acid) (PHA/PHB), 32
Poly (vinylidenefluoride-*co*-polyethylene) {P(VDF-TrFE)}, 373
Polymer electrolyte membrane fuel cells (PEMFC), 432
  commercialization of, 441
  systems deployment, 436
Postcombustion carbon capture, 516–518
Potassium hydroxide (KOH), 280, 285*f*, 340–341
Poverty, 483–484
Power
  banks, 369
  coefficient, 127
  generation, 414
  output and efficiency considerations, 127–128
  power-to-gas systems, 307
  power-to-X processes, 271–272
  speed characteristics, 128–129
  storage, 415
Power law index (PLI), 137
Power take-off system (PTO system), 242–243
  for WECs, 250–262
    direct electrical drive systems, 258–259
    direct mechanical drive systems, 257–258
    hydraulic PTOs, 257
    for OWC–WEC, 251–257
    rotary electrical generators, 259–262
Precombustion
  absorption, 516
  carbon capture, 519–520
    hydrogen-fired power plant, 520*f*
Primary energy (PE), 488
Priority 1 message (P_1M), 420, 421*f*
Priority 2 message (P_2M), 420, 422*f*
Priority 3 message (P_3M), 420–422, 423*f*
Priority 4 message (P_4M), 423, 424*f*
Private sector participation, 192–193
Producer-based allocation method, 485–486
Production costs, 39–40
  prediction, 39–40
Prokaryotic microorganisms, 317–319, 324
Propeller
  anemometer, 138–139, 139*f*
  turbine, 110–111
  type wind turbine, 145–148
    one-bladed wind turbine, 146–147
    three-bladed wind turbine, 148
    two-bladed wind turbine, 147–148
Property tax incentives, 189
Propyl levulinate (PL), 18
Proton exchange membrane (PEM), 279–280, 289, 291–292
  electrolysis cell, 294
    bipolar plates, 292–293
    LGDL, 292
    MEA, 290–292
  electrolyzer configuration, 293–294

configuration of PEM electrolysis stack, 294*f*
developments, limitations, and future perspectives, 294–295
fuel cells, 436
water electrolysis
components of PEM electrolysis cell, 290–293
configuration of PEM electrolyzer, 293–294
technology and operating principle, 289–290, 290*f*
Public–private partnerships (PPPs), 183
Pump backed hydroelectric dams, 97
Pump-as-turbine systems (PAT systems), 96
Pumped hydropower energy storing (PHES), 94–95
Pumped storage, 94–99
decentralized frameworks, 98
economic efficiency, 96–97
facilities at small scale, 97
plant, 95*f*
potential advancements, 97
pump backed hydroelectric dams, 97
regular or artificial storages, 96
underground stores, 97–98
underwater reservoirs, 98–99
Pumped storage hydroelectricity (PSH), 94–95
Pumping water, 124
Purification, 329–330
Purple sulfur bacteria, and nonsulfur bacteria (PNSB), 324
Pyrex reactor system, 545
Pyrolysis, 13, 16
of lignocellulosic biomass, 63–64, 64*t*
hydroprocessing of bio-oil, 65–67
process conditions, 65
reactors and commercial processes, 65
Python, 416
PZT fibers, 380, 381*f*

R

Rainwater harvesting, 476
Rankine cycle, 11
Rated wind speed, 129
Raw feedstock for bioethanol production, 34, 35*f*
Reaction turbine, 110
Reactors
and commercial processes, 65, 66*t*
for integrated system, 545
Rebound effect, 184
Redox reactions, 541
Reforming, 341–344
Regulatory Indicators for Sustainable Energy (RISE), 176, 187–188, 188*f*
Renewable energy (RE), 173–174, 271
integration with, 306–307, 306*f*
policy network, 29–30
sources, 241
Renewable energy technologies (RET), 176–177
Renewable hydrogen production by water electrolysis
alkaline water electrolysis, 280–289
electrolysis system technology, 300–306
integration with renewable energies, 306–307
PEM water electrolysis, 289–295
principles of water electrolysis, 272–280
solid oxide water electrolysis, 295–300
Renewable resources, 29–30
Renewable sources, 515
of energy, 458
Research, development and demonstration (RD&D), 446
Reservoirs, 97–98, 102–103
bank-side, 103
coastal, 103
dammed valley, 102
monitoring at start-up and operation, 226–228
stages of business plan for geothermal project, 227*t*
service repositories, 103
Resources, 476
Reverse-biased Schottky diode, 380–381
Reversible voltage, 274
Rhodamine 6 G, 542–545
Rhodamine B, 542–545
Ring oxygenates, 18
Rockfill dam, 101
Rolled-earth dam. *See* Earthfill dam
Roof-based heating system, 466
Rotary electrical generators, 259–262
fixed-speed generators, 260–261
variable-speed generators, 261–262

Run-of-river (ROR), 91–92
advantages, 92
inclusive advantages, 92
less flooding/store, 92
disadvantages, 92–93
accessibility of site, 92–93
ecological effects, 93
hydroelectricity plant, 91–93
Runner blades, 114
Rural electrification, 187

S
Saccharification and fermentation (SHF), 38
Saccharomyces, 38
S. cereviseae, 37, 58
Sacrificial agents, 542–545
utilization of pollutants or waste as, 542–545
Salinity-gradient energy, 12
Savonius type wind turbine, 148, 149*f*
Scenario analysis (SA), 484
Second-generation biorefineries, 33–41
advantages and disadvantages, 40
cost estimation, 39–40
environmental aspect, 40–41
process of bioethanol production, 34–39
raw feedstock for bioethanol production, 34
Seebeck coefficient, 392–393
Self-rectifying air turbine, 249–250
Self-sufficient energy harvesting technologies, 370
Semiclosed oxy-fuel combustion combined cycle (SCOC-CC), 522–524, 523*f*
Separators, 284–285
AEM, 285
diaphragms, 284
Service repositories, 103
Settling tank, 10
Shading, 475
Shoreline system, 11–12
Siemens Green City Index, 487
Signal emitters, 370
Simultaneously hydrolysis and fermentation (SSF), 38
Single flash steam power plant, 229–230, 229*f*
Single-bladed wind turbines, 146–147
Slow pyrolysis, 16
Small hydro, 106–107
Small hydropower ventures (SHP ventures), 106–107
Small-scale hydropower plant, 103
Smoldering combustion, 14
SMS-based feedback tool feasibility in rural Africa, 413–414
Social LCA (SLCA), 484
Sodium hydroxide (NaOH), 280, 285*f*
Sodium-ion batteries (Na–ion batteries), 369
Solar cells, 5–6
Solar collectors, 467–468
Solar drying of wet or pond fly ash, 6
Solar energy, 4–5
capturing and usage, 5–7
cooling devices, 6–7
Solar flux tracking devices, 5–6
Solar heating, 6
Solar home systems (SHS), 409–410, 415
Solar photovoltaic systems (SPV systems), 498–499
Solar power system, 414–417
consumption, 415–416
message sending and integration, 416–417
power generation, 414
power storage, 415
Solar PV deployment in oil producing African countries, 186–187
boost to businesses and rural electrification, 187
climate resilience, 187
rationale for solar-powered Africa, 186–187
Solar PV penetration in highly dependent African fossil fuel markets
complexity of Africa's energy landscape, 174
energy transition in Africa, 176–185
financing solar projects and private sector participation, 192–193
institutional and economic barriers to solar deployment, 187–191
attributes of financial and regulatory incentives, 189–190
auctions, 190
carbon pricing and monitoring, 188
counterparty risk, 190–191
fixed tariffs for small producers, 189–190

network connection and use, 189
market creation, 193
off-grid PV for improved electricity access, 191–192
research contribution, 174–176
solar PV deployment in oil producing African countries, 186–187
Solar radiation, 464–466
Solar thermal (ST)
solar thermal-mechanical refrigerators, 6–7
systems, 498–499
Solar-powered Africa, 186–187
Solarium, 466
Solid oxide electrolysis cell
electrodes, 296–297
electrolyte, 296
electrolyzers, 280
interconnectors, 297–298
and stack configuration, 298
Solid oxide electrolyzer cell (SOEC), 297, 298*f*
Solid oxide fuel cells (SOFC), 295, 432
Solid oxide water electrolysis
components of solid oxide electrolysis cell, 296–298
developments, limitations, and future perspectives, 299–300
solid oxide electrolysis cell and stack configuration, 298
technology and operating principle, 295
Solid polymer electrolyte (SPE), 279–280
Solidity, 134
Solids retention time (SRT), 330
Sonic anemometer, 139
Sorption refrigerators, 6–7
Spaces, 460–461
Speed control strategies for wind turbines, 161–164, 165*t*
Spiral packaging, 114
Splitter, 108
Squirrel cage induction generator (SCIG), 260
Stack configuration, 287–288, 288*f*
Stall-controlled machines, 162–163
Starch, 52
starch-based raw material, 30–31
State of charge (SOC), 415, 419
Stationary applications, fuel cells for, 436–438, 437*t*
Stationary fuel cell development, 438
Stator, 153
Stay vanes, 114
Steam explosion, 36–37
Steam methane reforming (SMR), 271
Stored Energy at Sea (StEnSea), 98–99
Straflo turbine, 111, 112*f*
Strategic energy plan, 452
Straw, 37
Stretch rate, 356
SubSaharan Africa (SSA), 411–412
Subsidy, 184
as enablers of fossil fuel dependency, 183–184
Sucrose feedstock, 58
Sugar, 52
Sunlight, 465
flicker, 166
Sunspaces, 466
Supply chain analysis, 492
Sustainability assessment methods, 485
Sustainable Development Goals (SDGs), 484
Sustainable Development Goal 7, 409
Sustainable energy, 5
harvesting technologies, 370
supply systems, 468
Sustainable Energy for All (SEforAll), 409
Sustainable long-term energy planning for cities, 483–485
conventional approaches in city energy planning, 485–486
new conceptual framework for long-term energy scenarios for cities, 486–507
environmental and economic impact assessment of low-carbon energy transition scenarios, 496–507
focus on sustainability assessment, 492–496
sustainability dimension, 489–492
Sustainable technologies
for energy, fuel and chemicals production, 12–18
high-temperature biomass conversion technologies, 13–16
low-temperature biomass conversion processes, 17–18
for energy production, 4–12
capturing wind energy and challenges, 7–8

Sustainable technologies (*Continued*)
clean and sustainable energy sources and production technologies, 10–12
hydropower energy capturing and challenges, 8–10
solar energy capturing and usage, 5–7
SWOT analysis, 44–45
Synchronous generators, 156–157, 156*f*, 261
Synthetic fuels, 435
Synthetic gas (syngas), 13, 15, 33

T

Tangential flow impulse turbine, 108
Tarbela Dam, 101–102
Techno commercial barriers, 442–444
Techno–economic assessment (TEA), 484
Thermal comfort conditions, 464
Thermal power plants, carbon capture technologies for, 516–520
Thermo-catalytic reforming, 16
Thermo-siphon systems, 466–467
Thermochemical
conversion, 34–35, 35*f*
cycles, 272
Thermodynamics of water electrolysis
equilibrium conditions, 273
minimum cell voltage, 274
theoretical operating ranges in water electrolysis, 275
Thermoelectric energy harvesters, 392–398. *See also* Electromagnetic energy harvesters; Piezoelectric energy harvesters
applications, 393–397
future perspectives, 398
Thermoneutral voltage, 274
Thermophilic species, 322–323
Thin-film, 5–6
Third-generation bioethanol (TGB), 41
Third-generation biorefineries, 41–44
environmental impact, 43–44
technique of algal oil extraction, 41–43
Third-generation PV, 5–6
Thomson coefficient, 392–393
Thomson effect, 392–393
Three-bladed wind turbine, 148, 148*f*
Three-phase boundary, 296–297
Time of notification, 425
Time-series data, 418
Tip speed ratio (TSR), 135–136, 136*f*
Top-down approach, 488
Torque on blades, 133–134
Total final energy consumption (TFEC), 181–182
"Tower shadow effect", 150
Toxic gases, 40–41
Traditional Rankine and Brighton-Rankine cycles, 515
Trane TRACE, 460–461
Transesterification
process, 55–56
reaction, 55
Transient pressure test, 222
Transition process, 174–176
Transport applications, fuel cells for, 438–442
Triacylglycerol, 42, 42*f*
Triboelectric energy harvesters, 371–378. *See also* Electromagnetic energy harvesters; Piezoelectric energy harvesters
applications, 371–376
future perspectives, 376–378
Triboelectric nanogenerators (TENG), 371, 372*f*
Triboelectricity, 376
Trichoderma reesei, 37–38
Triglyceride, 52
Triple flash system power plants, 230–231, 231*f*
Trombe Wall, 465
Tube turbines, 111, 113*f*
Turbine(s), 107–116
arrangement in wind farm, 141–142
bulb, 111
coolant precooling
modified MATIANT cycles with turbine coolant precooling, 530–532
modified semiclosed oxy-fuel combustion combined cycle, 524–527
cross-flow, 109–110
Francis, 114–116
impulse, 107
kaplan, 111–114
kinetic, 116
pelton, 108–109
propeller, 110–111
reaction, 110

straflo, 111
tube, 111
utilized in micro hydro plants, 106
Turgo turbine, 10
Twilio, 416
Two-bladed wind turbine, 147–148, 147*f*

U

Ulsan Techno Park project (UTP project), 436–438
Ultrasonic anemometer, 139–140, 140*f*
Underground stores, 97–98
Underwater reservoirs, 98–99
Unmanned aerial vehicles (UAV), 441
Upwind wind turbine, 149–151
Urban LCA, 485–486
Urban metabolism, 485
Urbanization, 483
US Department of Agriculture (USDA), 30
US Department of Energy (DOE), 30, 442, 450–451

V

Vadodara, 450
Valence band (VB), 542
γ-Valerolactone (GVL), 18, 53–54, 72, 74*f*
Valley-based repository, 102
Value added (VA), 492
Vapor absorption chiller (VAC), 11
Variable renewable energy (VRE), 436
Variable speed operation, 157–158, 158*f*
strategies for, 158–161
DFIG for variable speed operation, 160–161
variable speed (adjustable speed) generators, 159–160, 160*f*
Variable-speed generators, 261–262
Vegetable oil, 55–56
HDO of, 58
Vertical axis wind turbine (VAWT), 142, 143*f*, 150*f*, 151–152
Visible light–active dyes, 542–545
Visible light–active photocatalyst synthesis, 545
Voltage efficiency, 277, 278*f*

W

Warm-humid climate, 470
Waste as sacrificial agents, 542–545
Water
molecules, 272
and role in energy-efficient landscaping, 475–476
fountains, 475
rainwater harvesting, 476
water gardens, 475–476
steam, 516
turbine
gravitational vortex type of, 94, 94*f*
sectional view of runner, 95*f*
water-storage systems, 10
water–gas shift reaction, 322
Water electrolysis, 272, 306–307
principles of, 272–280
characterization of water electrolysis process, 275–278
efficiency and electrochemical performance, 277–278
Faraday law of electrolysis, 277
kinetic losses and overpotentials, 275–277
thermodynamics of, 273–275
types of, 278–280
technologies for, 281*t*
Water gas, 341
Water Wall, 465
Water–gas shift reaction (WGS reaction), 343
Wave climate at Indian coast, 242–243
Wave dragon, 245–246, 247*f*
Wave energy, 11–12, 242–243
future of, 262
wave climate at Indian coast, 242–243
Wave energy converter (WEC), 242–243
conversion stages of, 251*f*
PTO system for, 250–262
types of, 243–250
attenuator type wave energy converter, 247–249
AWS, 245
on location, 244–245
OWC–WEC, 249–250
point absorbers, 247
wave dragon, 245–246
on working principle, 245–250
Weather data, 418
Well drilling, 219–223, 220*f*
correlation of well information, 221*f*
graph showing behavior of transient pressure test, 222*f*

Wells turbine, 252–254, 253*f*, 254*f*
principle of operation of, 252–254
Wet oxidation, 37
Wickerhamomyces, 38
Wind electrical generators
classification, 153–154, 155*f*
wind power plants on basis of generators, 161
constant speed and variable speed operations of wind turbines, 157–158
DC generator, 155–156
induction generator, 157
strategies for constant speed and variable speed operations, 158–161
synchronous generator, 156–157
working principle of generators, 153
Wind energy, 4–5, 8
applications, 167
types of wind energy conversion devices, 142
aerodynamic forces, 152–153
Darrieus type wind turbine, 148–149
Dutch type wind turbine, 142–144
HAWT *vs*. VAWT, 151–152
multibladed type wind turbine, 145
propeller type wind turbine, 145–148
savonius type wind turbine, 148
upwind and downwind wind turbine, 149–151
Wind farm, turbines arrangement in, 141–142
Wind measurement, 137–140
compass points, 140*f*
time of flight calculation of ultra sound, 141*f*
Wind power density (WPD), 137
Wind power plants on basis of generators, 161, 162*f*, 163*t*
Wind power technology. *See also* Hydropower technology
energy and power from wind, 125–136
environmental impact and public perception, 164–166
historical background, 123–124
offshore wind energy, 166
speed control strategies for wind turbines, 161–164
wind electrical generators, 153–161
wind energy applications, 167
wind energy conversion devices, 142–153
wind resource feasibility, 136–142
wind turbine
of Charles F. Brush, 125*f*
economics, 167–168
of James Blyth, 124*f*
Wind resource feasibility, 136–142
arrangement of turbines in wind farm, 141–142
evaluation of suitable sites, 141
variation of wind speed with height, 137
wind measurement, 137–140
WPD, 137
Wind speed
characteristics, 128–129
variation of wind speed with height, 137, 138*f*
Wind turbine(s), 124*f*, 125*f*, 126–127
classification of, 144*f*
constant speed and variable speed operations of, 157–158
Darrieus type, 148–149
Dutch type, 142–144
economics, 167–168
modern, 126–127, 126*f*
multibladed type, 145
power flow of, 127*f*
power–speed characteristics of, 128*f*
propeller type, 145–148
Savonius type, 148
spacing, 142*f*
speed control strategies for, 161–164
upwind and downwind, 149–151
Windmills, 123–124
World Energy Network (WE-NET), 519
Wound rotor synchronous generator (WRSG), 156–157

Y

Yaw and tilt control, 164
Yttria-stabilized zirconia (YSZ), 296

Z

Zeolite catalyst, 69
Zero-gap cell configuration, 287
Zirconium oxide stabilized with yttrium oxide (ZrO2/Y2O3), 296
Zirfon PERL, 284
ZnO nanowires, 381–383

FSC
www.fsc.org
MIX
Papier | Fördert
gute Waldnutzung
FSC® C083411

Kolibri 360 GmbH
Ferdinand-Jühlke-Straße 7
99095 Erfurt, Deutschland
produktsicherheit@kolibri360.de